Himalayan Weather and Climate and their Impact on the Environment

A. P. Dimri • B. Bookhagen • M. Stoffel
T. Yasunari
Editors

Himalayan Weather and Climate and their Impact on the Environment

 Springer

Editors
A. P. Dimri
School of Environmental Sciences
Jawaharlal Nehru University
New Delhi, India

B. Bookhagen
Institute of Geosciences
University of Potsdam
Potsdam, Germany

M. Stoffel
Institute of Environmental Sciences
University of Geneva
Carouge, Switzerland

T. Yasunari
Research Institute for Humanity & Nature
Kyoto, Japan

ISBN 978-3-030-29683-4   ISBN 978-3-030-29684-1   (eBook)
https://doi.org/10.1007/978-3-030-29684-1

This Springer imprint is published by the registered company Springer Nature Switzerland AG.
The registered company address is: Gewerbestrasse 11, 6330 Cham, Switzerland

Preface

In Indian mythology, the Himalayas is abode of Indian Gods. The Himalayas is revered in philosophical and historical literatures and is having spiritual influences. Many religious Indian shrines and places of pilgrimages are situated here. Himalayan environment has fascinated mankind all through the ages. It attracted and fascinated many wonderers, adventurers, and travelers in hue of its majestic nature. Geographically, the majestic Himalayas is the highest mountainous region and is characterized by variable topography and heterogeneous landuse/landcover. It has a huge and significant influence of weather and climate patterns across the world and in particular over the South Asian region. The Himalayan massif protruding out of the globe influences weather and climate at different spatial and temporal scale. Its positioning steers weather and climate with its manifestations on regional hydrological, glaciological, ecological regime, etc. It is water source of important rivers of the nations' surrounding it. Himalayas being a fragile, delicate, and highly susceptible ecosystem is prone to even minor changes in its complex environmental parameters with high degree of dependence on changes in global and local climate factors. It is surrounded by nations having largest human populations thus affected by these changes and impacts.

Thus, an effort is made to synthesize available knowledge in an integrated edited book volume *Himalayan Weather, Climate and Its Manifestations*. Various researchers have tried understanding associated complexities and manifestations from various angles to answer respective scientific questions. It will provide a synthesis document for researchers and students interested in understanding complexities, thus so far understood, over the Himalayas. It discusses the science, causes, and impacts in view of latest understanding of weather and climate change with specific focus on associated impacts on glaciers, ecology, forest, habitat, disasters, extremes, floods, etc. The present volume provides extensive and excellent details on its climatic history; Himalayan weather systems in past, present, and future; and corresponding changes over the region. The volume begins with an overview of Himalayan weather and climate analysis based on data trends and international initiatives. Weather and climate systems over the Himalayas are analyzed and detailed through climate models, seasonal observations, and overviews of various climate scenarios. Then the volume segues into paleoclimate changes over the Himalayas based on proxy records through tree ring, sediment analysis, isotopes analysis, etc. Further signature of these changes seen over the snow, glaciers, and associated Himalayan hydrology is discussed extensively. The book then finally discusses climate change impacts and signatures specific to the issues of sustainability, challenges, and risks.

The book is particularly useful for environmental and disaster managers, researchers, and graduate students, as well as policymakers. This book will appeal to researchers studying climate science, climatology, environmental scientists, and policymakers.

Many scientific studies and particularly the IPCC reports on Himalayas have highlighted the irreversible nature of the impact of climate change on the mountain regions, especially the melting of snow and an accelerated pace of retreating glaciers. The global and regional climate changes over the last century in the form of increase in air and sea surface temperatures, glaciers recession/advancement, changing/shifting of climate regimes, increased extreme events, and sea level changes are reported. These changes have posed threat to resilience, lives, and livelihoods at global, regional, and local levels. Recently, major ecosystems of the world have experienced several climate-induced disasters. This volume provides latest insights on relevant aspects of weather and climate, paleoclimate, snow, glacier and hydrology, and ecology/forestry among other topics associated with the Himalayas. It includes studies on rainfall and temperature trends, floods and drought disasters, weather- and climatic-related disasters in mountains, changes in plant activities, risk assessment, and responses in different ecosystems of the world. The modern scientific pursuit to understand its natural processes has led to gather knowledge to understand its evolution and future changes. Though the Himalayas has paucity of observations, with possible monitoring and generation of global data sets, enhancements in the basic theoretical understanding of complex orographic processes have provided important inputs in the role of human interference in climate change. In addition, anthropogenic influences in the last few decades due to unprecedented activity led to understand issues of climate change impacts, mitigation, adaptability, and remedial measures at the forefront with a focus on conserving the Himalayan environment and the sustainability of natural resources. The Belmont Forum of Collaborative Research Action (2015) quote "Mountains are the sentinels of change" indicates crucial role of Himalayas as sensitive recorders of climate-driven and tectonic impacts on environment. As a result, permafrost thawing, snow cover changes, and glacial retreat/glacial advance need scrutiny. Earthquakes and associated effects (e.g., mass movements, natural dam breaks, and floods) laying over extreme events are difficult to forewarn. These changes will seriously impact downstream hydrological regime and sediment delivery and thus forcing of landuse/landcover changes. In view of this, communities living in downstream region thus will be increasingly impacted and challenged by a rapidly evolving Earth surface.

Himalayan environment is of paramount importance for physical, chemical, and biological processes despite of its low areal coverage. Sediments eroded and delivered to forelands and oceans provide nutrients for downstream farming, generate high weathering fluxes, collect the majority of precipitation, and support an exceptionally high biodiversity. The Himalayas, the tectonically active mountain, is characterized by pronounced gradients in climate and surface processes. Minor changes in this sensitive system (e.g., changes in the amount and frequency of rainfall, recurrent earthquakes, or a combination thereof) will likely lead to long-lasting perturbations that cause major changes of natural environments, where recovery will be difficult. These aspects may be exacerbated, and the vulnerability for populations in these areas may increase through multiple socioeconomic trends: the effects of population increase and expansion of settlements into high-risk areas, the extraction of natural resources and hydropower generation, or the construction of critical infrastructure (power lines, pipelines) and recreational facilities.

We sincerely wish that the researchers interested on the Himalayas will find this book useful and informative. It is just a humble beginning as far as the Himalayas are concerned. The next generation of researchers will enlarge and add on the aspects and information provided here with higher resolution of data, methods, and models. And their commands, efforts, and contributions will continue.

New Delhi, India   A. P. Dimri
Potsdam, Germany   B. Bookhagen
Carouge, Switzerland   Markus Stoffel
Kyoto, Japan   T. Yasunari
November 2019

Contents

About the Editors

Dr. A. P. Dimri is a professor at the School of Environmental Sciences, Jawaharlal Nehru University, New Delhi, India (www.crsl.jnu.ac.in). His research interests include Indian weather and climate using observations and modeling tools, regional climate dynamics and its variability, statistical and dynamical downscaling of numerical model outputs, and extreme events and their physical understanding, particularly over the Himalayan massif. He has contributed important dimension on Western disturbances and Indian winter monsoon to explain the precipitation mechanism occurring during winter. His emphasis on studies of the snowfall during winters leading to glacier replenishment provides an insight for recharging of the snow-fed rivers and overall ecology of the region. He is member of many committees and published more than 100 peer-reviewed papers in international and national journals and is editor of *Meteorology and Atmospheric Physics*, *Theoretical and Applied Climatology*, and *PLOS One*.

Dr. B. Bookhagen is a professor at the Institute of Geosciences, University of Potsdam, Germany (http://www.geo.uni-potsdam.de/member-details/show/524/Bodo_Bookhagen.html). His research interests are understanding of quaternary climate change, geomorphic processes, landscape evolution, and tectonic processes through integrated studies involving cosmogenic radionuclide dating (He, Ne, Be, Al, Cl), recent and past climatic records, remote sensing, numerical modeling, and field observations. Spatial scales range from hillslopes (10^0 km^2) to entire mountain ranges (~10^3 km). For smaller-scale analysis, he uses a terrestrial laser scanner (RIEGL) to create high-resolution, cm-scale digital elevation models for erosion process deciphering. He has number of publications to his credit.

Dr. Markus Stoffel is a full professor at the Institute for Environmental Sciences (ISE) and is director of the Swiss Tree-Ring Lab (www.dendrolab.ch) at the Department F.-A. Forel for Environmental and Aquatic Sciences and the Department of Earth Sciences, University of Geneva. His research interests are in hydrogeomorphic and Earth surface processes, climate change impacts, and dendroecology. He has authored more than 200 peer-reviewed papers on geomorphic, hydrologic, cryospheric, and geologic processes in mountain and hillslope environments, with a focus on time series of frequency and magnitude and process dynamics, as well as dendroecology and wood anatomy of trees and shrubs, and integrated water resource management. He has collaborated with other important coeditors in chief of *Geomorphology* (Elsevier) and guest editors of scientific journals. He as well coedited books and encyclopedia.

Prof. T. Yasunari is director general of the Research Institute for Humanity and Nature, Kyoto, Japan (http://www.chikyu.ac.jp/yasunari/yasunari.bak/index-e. html). He holds many positions in international scientific programs and contributed key theories to the Indian summer monsoon. He has many publications to his credit.

Part I
Weather and Climate

Chapter 1
Inter-Comparison of High Resolution Satellite Estimates for Cloudburst Events in the Northwest Himalaya

Garima Dahiya, Pravat Jena, Sourabh Garg, and Sarita Azad

Abstract The Tropical Rainfall Measuring Mission (TRMM) 3B42 version 7 precipitation data has been extensively used for inter-comparison with observations and model validation. The rain distribution over the Northwest Himalaya (NWH) were found to be accurate with a strong positive correlation of 0.88 between TRMM and India Meteorological Department (IMD) station data, supporting the use of 3B42 V7 for the study of extreme rainfall events (ERE's) over the region. However, many high resolution satellite data sets were made available in the recent past and their potential have not been evaluated for ERE's like cloudbursts in the NWH. The present endeavor aims to provide guidance to the choice of global precipitation data sets (GPDs). In particular, this study is conducted to evaluate three recent satellite-based rainfall products, i.e. Global Precipitation Measurements (GPM), Indian National Satellite System (INSAT 3D), and CPC Morphing Technique (CMORPH), against the highly used TRMM-RT 3B42 V7 precipitation data for the estimation of rainfall episodes in the recent years (2014–2016). Our results reveal that the magnitude of precipitation and location of peak rainfall are biased in INSAT 3D, whereas CMORPH and the high resolution GPM product capture it with relatively higher values of the employed statistical metrics. Also, the rainfall estimates from GPM and CMORPH are in good agreement with TRMM for cloudbursts events. Particularly, high resolution GPM is useful for monitoring the extreme rainfall event in the region.

G. Dahiya · P. Jena · S. Garg · S. Azad (✉)
School of Basic Sciences, Indian Institute of Technology Mandi,
Kamand, Himachal Pradesh, India
e-mail: sarita@iitmandi.ac.in

© Springer Nature Switzerland AG 2020   3
A. P. Dimri et al. (eds.), *Himalayan Weather and Climate and their Impact on the Environment*, https://doi.org/10.1007/978-3-030-29684-1_1

1.1 Introduction

Heavy rainfall and associated flash floods cause tremendous damage to life and property across most of the mountainous regions of the world, including the Himalaya. The Himalayan ranges are prone to heavy and prolonged rainfall events and associated flooding, particularly during the summer rainy months of June to September (monsoon season). The NWH is a mountainous region highly prone to ERE's and cloudbursts due to its extremely intricate topography and altitude-dependent climate (Bharti 2015). There are sharp weather fluctuations in different sectors of mountains which can be both unpredictable and harsh leading to sudden occurrences of heavy rainfall events of short to long duration (Nandargi and Dhar 2011). Cloudbursts and associated flash floods are one of the most potent disasters in NWH (Thayyen et al. 2013). Colloquially defined as a sudden copious downpour with a vehement force usually for a very short duration over a restricted region, the NWH has witnessed many colossal disasters initiated by cloudbursts in the recent times causing immense human and economic losses.

The cloudbursts at Dharampur (2015), Srinagar (2014), Pithoragarh (2016), Banjar, Kullu (2015), Marchula, Pauri (2016), Mendhar Belt, Poonch (2015) and Purala (2014) are only a few of the catastrophic disasters occurred in the recent times. Several past studies attempting to study ERE's over Indian subcontinent largely excluded Himalayan region due to non-availability of data or have been done on a very coarser resolution (Bharti 2015). The remoteness of the area and a very sparse coverage of rain gauges and automatic weather stations across the mountainous terrain are the major factors responsible for making the prediction and observation of precipitation incredibly difficult in the region.

Remote sensing has emerged as an attractive approach to study precipitation, offering high spatial and temporal sampling density unattainable through any other means over complex terrains (McCabe and Wood 2006). The latest advancements in meteorological satellites and improved precipitation estimation algorithms have facilitated the research on extreme weather events.

The TRMM, a joint mission between the National Aeronautics and Space Administration (NASA) and the Japan Aerospace Exploration Agency (JAXA), was the first satellite primarily dedicated to study rain structure and monitor precipitation distribution over tropics and sub-tropics. The TRMM has been the only satellite providing inter-calibrated precipitation data routinely since December 1997 at such fine spatial and temporal resolution compared to any other space-borne precipitation products (Bharti et al. 2016). Moreover, the latest released version 7 data offer improved precipitation estimations with significantly lower bias over complex terrain and outperform previous versions (Huffman et al. 2010; Qiao et al. 2014; Zulkafli et al. 2014; Prakash et al. 2015). Various past studies were conducted to assess the capability of TRMM 3B42 V7 in precipitation studies. According to

Zulkafli et al. (2014) 3B42 V7 correlates better with rain gauge data than version 6. Similar results were also reported in Mantas et al. (2014). Rahman et al. (2009) studied the spatial distribution of rainfall over India including Himalaya using TRMM 3B42V6 and IMD gridded data and reported that TRMM well captures the orographic rainfall over Himalayan region. However, no study has assessed the recently available high resolution satellite data over this region. Therefore, this study is an attempt to examine the potential of high resolution GPM and INSAT 3D against TRMM-RT 3B42 V7 which is highly correlated with IMD station data (Bharti et al. 2016).

1.2 Study Area

The Himalayan mountain range is divided into three major fold axes: the outer Himalaya, the Lesser Himalaya, and the Greater Himalaya (Pant and Kumar 1997). The study area includes North-western part of the Himalayan mountain range extending from 28°–37°N and 72°–82°E encompassing the three states of India – Uttarakhand (UK), Himachal Pradesh (H.P.), and Jammu and Kashmir (J&K). Various reports on the cloudbursts in NWH lead to the inclusion of seven events in this study as listed in Table 1.1.

1.3 Data

The following satellite products are used in the present study which is listed in Table 1.2.

Table 1.1 List of cloudburst events occurred in the NWH region during 2014–2016 taken for the analysis

S.no.	Year	Days	Location	State	Latitude	Longitude	Altitude (m)
1.	2014	1–7 Sept	Srinagar	J&K	34.084	74.797	1585
2.	2014	14–15 Aug	Purala Bairagarh Village, Pauri	UK	30.147	78.775	1524
3.	2015	7–9 Aug	Dharampur, Mandi	H.P.	31.803	76.760	1189
4.	2015	25 July	Jibhi village, Banjar, Kullu	H.P.	31.63	77.34	1356
5.	2015	23 July	Mendhar Belt, Poonch	J&K	33.60	74.11	1580
6.	2016	30 June–2 July	Pithoragarh	UK	30.081	80.365	1514
7.	2016	20 Aug	Marchula Village, Pauri	UK	29.606	79.092	1814

Table 1.2 Summary of satellite rainfall products used for the study

S.No	Satellite Product	Spatial resolution	Temporal resolution	Availability	Range
1.	TRMM TMPA, 3B42-RT	0.25° × 0.25°	3 hourly	February 2000-present	(60°, −180°, −60°, 180°)
2.	CMORPH	0.25° × 0.25°	3 hourly	December 2002-Present	(60°, −180°, −60°, 180°)
3.	GPM, IMERG	0.10° × 0.10°	Half hourly	March 2014-Present	(90°, −180°, −90°, 180°)
4.	INSAT 3D, HEM	0.10° × 0.10°	Half hourly	October 2013-Present	(81°, 0.8°, −81°, 163°)

1.3.1 TRMM TMPA-RT

The TRMM Multi-Satellite Precipitation Analysis (TMPA) (Huffman et al. 2007) is daily derived precipitation and 3 hourly rain rate data. The TRMM multi-satellite precipitation analyses have a spatial resolution of 0.25° × 0.25° available from 1998 onwards. TRMM data sets focus on tropical region precipitation and thus have a horizontal coverage of 180°W–180°E and 50°S–50°N and are downloaded from ftp://disc2.nascom.nasa.gov/data/TRMM/Gridded/.

1.3.2 CMORPH

CMORPH (Joyce et al. 2004) is generated precipitation dataset. This provides global precipitation estimated from passive microwave and infrared satellite data available from 2002 onwards. CMORPH data is giving daily precipitation estimates at 0.25° × 0.25° spatial resolution with a temporal resolution of 3 h. This is the reprocessed data denoted CMORPH version 1.0 with gauge and satellite blended precipitation estimates and is downloaded from ftp://ftp.cpc.ncep.noaa.gov/precip/CMORPH_V1.0/CRT/.

1.3.3 GPM

The GPM provides global precipitation data at a spatial resolution of 0.10° × 0.10° available from 2014 onwards. It uses gauge analysis from distinct sources for bias correction in their respective research version products. The capability of the Integrated Multi-satellite Retrievals for GPM (IMERG) product in heavy rainfall detection was recently assessed against gauge-based observations over India (Prakash et al. 2016b). A preliminary analysis for the southwest monsoon season of 2014 showed a notable improvement in IMERG over TMPA-3B42 for heavy rainfall detection. The missed and false precipitation biases were also noticeably reduced in the GPM-based satellite precipitation products (SPPs) (Prakash et al. 2016a). Even though the errors in the GPM-based SPPs were reduced overall, the SPPs still have rather larger uncertainty over the orographic regions. The dataset is downloadable from NASA website https://earthdata.nasa.gov/.

1.3.4 INSAT 3D

INSAT is a series of multipurpose geo-stationary satellites launched by ISRO to satisfy the telecommunications, broadcasting, meteorology, and search and rescue needs of India. Commissioned in 1983, INSAT is the largest domestic communication system in the Asia Pacific region. INSAT-3D is an advanced meteorological geostationary satellite which was launched on 26 July 2013 and located at 82°E degrees east. From this satellite 3 types of rainfall estimation, namely, Quantitative Precipitation Estimation (QPE), INSAT Multi-Spectral Rainfall Algorithm (IMSRA) & Hydro Estimator (HE) are obtained. Hydro Estimator (HE) rain rate at pixel level are estimated at half hourly basis using different algorithms. It is a high spatial and temporal scale rain estimation technique which was initially developed by NOAA/STAR for GOES (Kumar and Varma 2017). The dataset is downloadable from Meteorological and Oceanographic Satellite Data Archival Centre, Space Application Centre, ISRO websitehttp://www.mosdac.gov.in/.

1.4   Methodology

The main purpose of the present work is to evaluate the performance of high resolution satellite products, namely, GPM, INSAT 3D, and CMORPH against TRMM. The precipitation data for GPM, INSAT 3D, CMORPH and TRMM, from HDF5 files and NetCDF files is extracted at particular latitude and longitude of a station. Thereafter, the rainfall value is obtained by finding the sum at nearby grid points in which the desired latitude and longitude lies. A validation study of GPM, INSAT3D, and CMORPH against TRMM data has been performed to analyze the potentiality of satellite data sets in capturing the precipitation events. The data validation methodology uses the following statistical measures (Tawde and Singh 2015).

1.4.1 Correlation Coefficient (CC)

The correlation analysis has been performed to determine the bias in satellite datasets with respect to TRMM data using Karl Pearson's coefficient of correlation and scatter plot. The Pearson's coefficient of correlation denoted as r, can be calculated using the following mathematical expression:

$$r = \frac{\sum xy}{N \sigma_X \sigma_Y} \tag{1.1}$$

Where $x = X - \bar{X}; y = Y - \bar{Y}$

σ_X=Standard deviation of series X
σ_Y=Standard deviation of series Y
N=Number of pairs of observations
Where X represents the time series of the satellites and Y is taken as observed data. Here, TRMM is considered as the reference dataset.

1.4.2 Root Mean Square Error (RMSE)

It is routinely used as the measure of the distances between model values (satellite data) and actually observed values (TRMM data), and it generally called as residual or error of the model. The RMSE serves to aggregate the magnitudes of the errors in model for various times into a single measure of estimating power. The RMSE computed as:

$$RMSE = \sqrt{\frac{\sum_{i=1}^{i=n}\left(X_i - O_i\right)^2}{n}} \tag{1.2}$$

Where X_i is the ith time estimated value of (satellite data) and O_i is the ith time value of the observed data (TRMM).

1.4.3 Mean Bias (MB)

Mean Bias indicates that the method is overestimated or underestimated or exactly matched with the actual observed climatology. It is computed as:

$$MB = \bar{X} - \bar{O} \tag{1.3}$$

Where $\bar{X}$ indicates the climatologic mean value of the method (satellite data) and $\bar{O}$ indicates the climatologic mean value of reference method (TRMM data).

1.4.4 Index of Agreement (IOA)

It is a standardized measure of the degree of model (satellite data) error with observations (TRMM data). Value "1" indicates a perfect agreement between estimated values of model and actual observation. The IOA represents the ratio between the mean square error and potential error. The potential error is the sum of the squared absolute values of the distances from the estimated values to the mean of the actual observed values and distances from actual observed values to the mean actual observed values. The IOA detects additive and proportional differences in the observed and estimated means and variances. It is computed as:

$$IOA = 1 - \left(\frac{\sum_{i=1}^{i=n} \left(X_i - O_i \right)^2}{\sum_{i=1}^{i=n} \left(\left| X_i - \bar{O} \right| + \left| O_i - \bar{O} \right| \right)} \right) \qquad (1.4)$$

Where X_i and O_i indicate the ith time value of the satellite and the observed data respectively. And also $\bar{O}$ represents climatologically mean of the observed data for the given time period.

1.4.5 Nash-Sutcliffe Efficiency Coefficient (NSEC)

It is generally used to assess the estimation power of the model (Nash and Sutcliffe 1970). In addition, it can be used to describe quantitatively the accuracy of the model. It is defined as:

$$E = 1 - \left(\frac{\sum_{i=1}^{i=n} \left(O_i - X_i \right)^2}{\sum_{i=1}^{i=n} \left(O_i - \bar{O} \right)^2} \right) \qquad (1.5)$$

The skill score ranges from $-\infty$ to 1, where value 1 indicates a perfect match between estimated values of model and actual observations. The NSEC less than zero $(-\infty < NSEC < 0)$ occurs when the actual observed mean is a better predictor than the models.

Where X_i and O_i indicate the ith time value of the satellite and the observed data respectively. And also $\bar{O}$ represents climatologically mean of the observed data for the given time period.

1.5   Results and Discussions

The seven reported cloudburst events are shown in Fig. 1.1. It is seen that all the events belong to the Southern belt of lower Himalaya and they have occurred within the range of 1382.50–2765.0 meters.

Rainfall estimates from all the satellite products under study is obtained. Out of seven, the event occurred on 30th June to 2nd July, 2016 at Pithoragarh is spatially depicted in Fig. 1.2. It is seen from the figure that the satellite TRMM estimates around 120 mm rainfall at the location of the event. In comparison to this, GPM is closest, whereas, CMORPH underestimates the rainfall amount and INSAT 3D overestimates.

Further, the amount of rainfall recorded during cloudbursts event occurred at Dharampur, Pithoragarh, Srinagar, Banjar, Purala (Pauri District), Mendhar Belt (Poonch District), and Marchula (Pauri District) is plotted in Fig. 1.3. The Fig. 1.3a shows daily accumulated rainfall of each event. The maximum rainfall during the seven events of all satellites is shown in the Fig. 1.3b. For all the seven cloudburst events, the accumulated rainfall for total duration of the events recorded by GPM is in close match with TRMM as compared to CMORPH and INSAT 3D. It is seen in Fig. 1.3a that the lowest accumulated rainfall is captured by the satellite INSAT 3D in comparison with others. Moreover, for all the events, INSAT 3D has more bias, whereas GPM has comparatively less bias in capturing the accumulated rainfall.

Figure 1.4a–g shows the time series of hourly rainfall for all the seven cloudburst events. It is shown that all the satellites except INSAT 3D are able to capture the timings of the events and amount of accumulated rainfall with some errors.

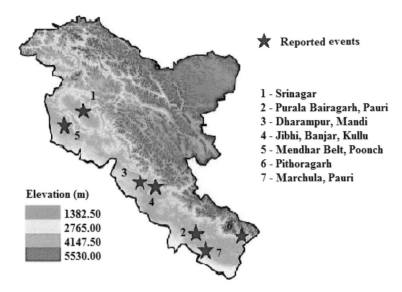

Fig. 1.1 Location of seven cloudburst events occurred in the NWH region during 2014–2016 taken for the analysis

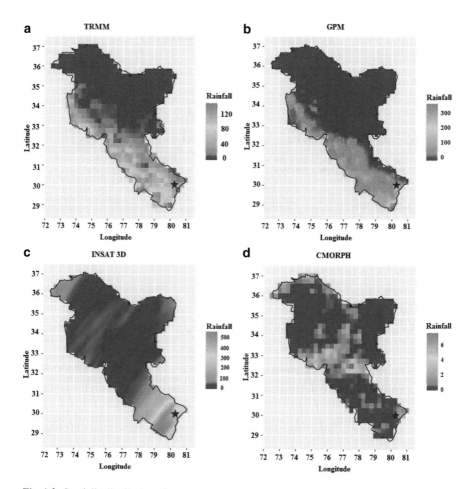

Fig. 1.2   Spatially distribution of rainfall obtained from: (**a**) TRMM; (**b**) GPM; (**c**) INSAT 3D; and (**d**) CMORPH, for the event at Pithoragarh marked with ∗)

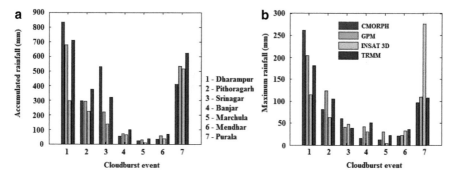

Fig. 1.3   (**a**) Accumulated precipitation captured during the seven cloudburst event by satellites; (**b**) the maximum rainfall occurred during the seven events

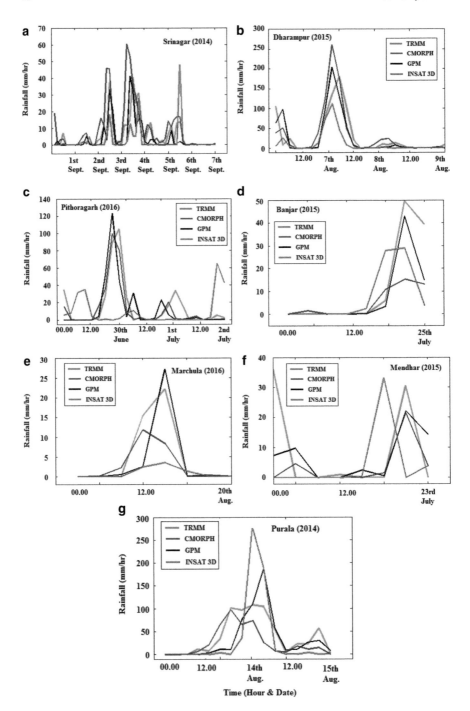

Fig. 1.4 Time series plot of seven cloudburst events

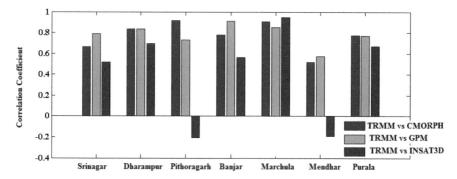

Fig. 1.5 Correlation Coefficient between the three satellite precipitation data and TRMM

Moreover, the pattern of the graph of GPM and CMORPH match with TRMM, but unable to capture the exact amount of rainfall at the peaks.

In particular, GPM overestimated the amount of rainfall at the peak at Pithoragarh (30th June, 2016), Purala (15th August, 2014) and Marchula (12:00 on 20th August, 2016) whereas, CMORPH at Dharampur (7th August, 2015) and Srinagar (3rd and 4th Sept. 2014).

In order to quantify the behavior of satellites in capturing the ERE's, five statistical tests have been employed. The CC of the GPM, CMORPH and INSAT 3D with TRMM for the seven extreme events is listed in the Fig. 1.5. The correlation coefficient of the GPM is observed to be relatively higher than the other satellites at Srinagar, Dharampur, Banjar and Mendhar. Relatively higher value of the statistical score is seen for CMORPH at the event Pithoragarh and Purala cloudburst.

The Nash-Sutcliffe efficiency coefficient of GPM, CMORPH, and INSAT 3D is shown in the Fig. 1.6a which shows that the GPM performs better at each of the event except for Srinagar. In the case of the cloudburst at Srinagar, the satellite CMORPH has relatively higher NSEC value against the other satellites. Similarly, Fig. 1.6b shows the degree of closeness of the results obtained by TRMM and with the other satellites. It is observed that GPM has highest IOA value implying the highest agreement with TRMM.

Further, the mean bias and root mean square error is least for GPM as compared to CMORPH and INSAT 3D, against TRMM, as shown in Fig. 1.6c and d, respectively. Using the above statistical measure the satellite GPM positioned as rank 1 followed by CMORPH.

The summary of the statistical measures are shown in Table 1.3 for each individual event. The statistics show that the GPM has a high CC of 0.90 and low RMSE (2.9%) for Banjar, Kullu event. In fact CC of GPM with TRMM is higher for all the events, except for Pithoragarh cloudburst, where CC of CMORPH with TRMM is high as 0.91 and low RMSE (3.7%), while CMORPH had a slightly lower CC (0.90) and RMSE (1.6%) for Marchula, Pauri cloudburst. Further, MB is low in case of GPM and TRMM for most of the events.

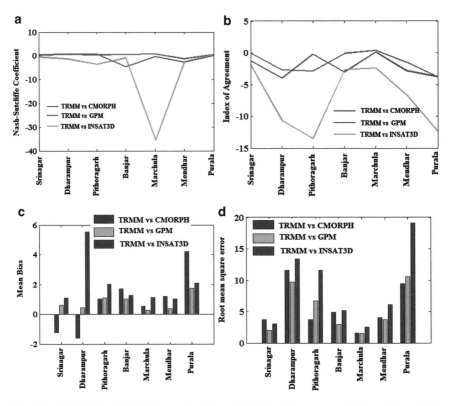

Fig. 1.6 (**a**) Nash-Sutcliffe Efficiency Coefficient, (**b**) Index of Agreement, (**c**) Mean Bias and (**d**) Root Mean Square Error between the three satellite precipitation data and TRMM

Table 1.3 Comparison of satellite products against TRMM using various statistical tests

S.N	Metric	TRMM Vs. CMORPH	TRMM Vs. GPM	TRMM Vs. INSAT 3D
Srinagar cloudburst				
1.	CC	0.66618	0.79265	0.5149
2.	RMSE	3.7221	2.0209	3.0142
3.	MB	−1.2427	0.5996	1.0805
4.	NSEC	0.3738	0.5173	−0.3634
5.	IOA	−1.2333	−0.0640	−1.8700
Dharampur cloudburst				
1.	CC	0.8334	0.8368	0.6937
2.	RMSE	11.5561	9.6728	13.3917
3.	MB	−1.6236	0.4382	5.5419
4.	NSEC	0.6873	0.6840	−1.2598
5.	IOA	−3.9617	−2.7127	−10.6244
Pithoragarh cloudburst				
1.	CC	0.9166	0.7279	−0.2035
2.	RMSE	3.7253	6.6781	11.5657

(continued)

Table. 1.3 (continued)

S.N	Metric	TRMM Vs. CMORPH	TRMM Vs. GPM	TRMM Vs. INSAT 3D
3.	MB	1.0374	1.0826	2.0329
4.	NSEC	0.7962	0.4396	−3.4941
5.	IOA	−0.2496	−2.8731	−13.4814
Banjar, Kullu cloudburst				
1.	CC	0.7785	0.9097	0.5606
2.	RMSE	4.9371	2.9186	5.2046
3.	MB	1.6948	1.0127	1.2714
4.	NSEC	−4.5872	0.5720	−0.8544
5.	IOA	−3.0085	−0.1118	−2.6465
Marchula, Pauri cloudburst				
1.	CC	0.9049	0.8506	0.9493
2.	RMSE	1.6075	1.5282	2.5169
3.	MB	0.5432	0.2669	1.1179
4.	NSEC	−0.3465	0.7070	−35.4418
5.	IOA	0.1287	0.3811	−2.3526
Mendhar, Poonch cloudburst				
1.	CC	0.5176	0.5716	−0.1918
2.	RMSE	4.0229	3.6951	6.1299
3.	MB	1.2116	0.3762	1.0220
4.	NSEC	−2.6184	−1.3156	−2.4683
5.	IOA	−2.8448	−1.5271	−6.6410
Purala, Pauri cloudburst				
1.	CC	0.7734	0.7717	0.6694
2.	RMSE	9.4814	10.5679	19.1140
3.	MB	4.1995	1.7361	2.0797
4.	NSEC	0.0939	0.5819	0.4226
5.	IOA	−3.7673	3.8353	−12.4150

Taking GPM, of higher resolution of 10 km, into consideration the drawn-out conclusion reflects the usability of GPM for various climatologically studies in NWH where weather station network isn't dense and high resolution observational data is needed.

1.6 Conclusions

Reliable estimate of the precipitation is crucial for several applications ranging from hydrometeorology to climatology. After the launch of the TRMM satellite, precipitation estimation techniques got rapid boost and several high resolution satellite products were developed to study the tropical and subtropical precipitation characteristics. Recently, several studies were performed to characterize the errors in the different TRMM-era high-resolution global or quasi global SPPs over the Indian subcontinent. In this study, evaluations of high resolution SPPs over NWH region are performed. In general, the TMPA-3B42 product is proven to be superior to other TRMM-era SPPs. Two finer resolution SPPs, GPM IMERG and INSAT 3D are analyzed in this study. Both SPPs were compared to TMPA-3B42 observations for seven ERE's across NWH.

The results showed that the GPM-based estimates of higher resolution $(0.10° \times 0.10°)$ are closely related to TRMM over the INSAT 3D of higher resolution, and CMORPH of the same resolution as TRMM. The correlation between TRMM and GPM seems to be reducing with altitude. A more comprehensive evaluation of GPM-based multi-satellite precipitation estimates for longer period is further required for its wide usage and applications in various sectors. Hence, continuous evaluation of the updated SPPs is essential.

References

Bharti VIDHI (2015) Investigation of extreme rainfall events over the northwest Himalaya region using satellite data. University of Twente Faculty of Geo-Information and Earth Observation (ITC), Enschede

Bharti V, Singh C, Ettema J, Turkington TAR (2016) Spatiotemporal characteristics of extreme rainfall events over the northwest Himalaya using satellite data. Int J Climatol 36(12):3949–3962

Huffman GJ, Bolvin DT, Nelkin EJ, Wolff DB, Adler RF, Gu G, Stocker EF (2007) The TRMM multi-satellite precipitation analysis (TMPA): quasi-global, multiyear, combined-sensor precipitation estimates at fine scales. J Hydrometeorol 8(1):38–55

Huffman GJ, Bolvin DT, Nelkin EJ, Adler RF (2010) Highlights of version 7 TRMM multi-satellite precipitation analysis (TMPA). In: 5th international precipitation working group workshop, workshop program and proceedings, pp 11–15

Joyce RJ, Janowiak JE, Arkin PA, Xie P (2004) CMORPH: a method that produces global precipitation estimates from passive microwave and infrared data at high spatial and temporal resolution. J Hydrometeorol 5(3):487–503

Kumar P, Varma AK (2017) Assimilation of INSAT-3D hydro-estimator method retrieved rainfall for short-range weather prediction. Q J R Meteorol Soc 143(702):384–394

Mantas VM, Liu Z, Caro C, Pereira AJSC (2014) Validation of TRMM multi-satellite precipitation analysis (TMPA) products in the Peruvian Andes. Atmos Res 163:132–145. https://doi.org/10.1016/j.atmosres.2014.11.012

McCabe MF, Wood EF (2006) Scale influences on the remote estimation of evapotranspiration using multiple satellite sensors. Remote Sens Environ 105(4):271–285

Nandargi S, Dhar ON (2011) Extreme rainfall events over the Himalayas between 1871 and 2007. Hydrol Sci J 56(6):930–945

Nash JE, Sutcliffe JV (1970) River flow forecasting through conceptual models part I—a discussion of principles. J Hydrol 10(3):282–290

Pant GB, Kumar KR (1997) Climates of South Asia. Wiley, Chichester

Prakash S, Mitra AK, Momin IM, Pai DS, Rajagopal EN, Basu S (2015) Comparison of TMPA-3B42 versions 6 and 7 precipitation products with gauge-based data over India for the southwest monsoon period. J Hydrometerol 16(1):346–362

Prakash S, Mitra AK, AghaKouchak A, Liu Z, Norouzi H, Pai DS (2016a) A preliminary assessment of GPM-based multi-satellite precipitation estimates over a monsoon dominated region. J Hydrol. https://doi.org/10.1016/j.jhydrol.2016.01.029

Prakash S, Mitra AK, Pai DS, AghaKouchak A (2016b) From TRMM to GPM: how well can heavy rainfall be detected from space? Adv Water Resour 88:1–7

Qiao L, Hong Y, Chen S, Zou CB, Gourley JJ, Yong B (2014) Performance assessment of the successive version 6 and version 7 TMPA products over the climate-transitional zone in the southern Great Plains, USA. J Hydrol 513:446–456

Rahman SH, Sengupta D, Ravichandran M (2009) Variability of Indian summer monsoon rainfall in daily data from gauge and satellite. J Geophys Res 114(D17)

Tawde SA, Singh C (2015) Investigation of orographic features influencing spatial distribution of rainfall over the Western Ghats of India using satellite data. Int J Climatol 35(9):2280–2293

Thayyen RJ, Dimri AP, Kumar P, Agnihotri G (2013) Study of cloudburst and flash floods around Leh, India, during August 4–6, 2010. Nat Hazards 65(3):2175–2204

Zulkafli Z, Buytaert W, Onof C, Manz B, Tarnavsky E, Lavado W, Guyot JL (2014) A comparative performance analysis of TRMM 3B42 (TMPA) versions 6 and 7 for hydrological applications over Andean–Amazon river basins. J Hydrometeorol 15(2):581–592

Chapter 2
Impact of Nino Phases on the Summer Monsoon of Northwestern and Eastern Himalaya

Sandipan Mukherjee, Vaibhav Gosavi, Ranjan Joshi, and Kireet Kumar

Abstract Summer monsoon rainfall (SMR) and it's ENSO linkages over Indian Himalayan Region (IHR) is investigated assuming that local effects of complex Himalayan terrain can substantially attenuate the coupling effect of SMR and ENSO. Hence, relationships between the SMR of northwestern (NWH) and eastern Himalayan (EH) region and nino indices (Nino 3.0 and Nino 3.4 indicating tropical Pacific Ocean sea surface temperature) are investigated using six cases of El-Nino (EN), La-Nina (LN) and normal (NN) events during 1981–2005. Particular objectives of this study are: (i) to compare impact of three nino phases (i.e. EN, LN and NN) on the monthly average rainfall of NWH and EH region and (ii) to assess relationships between dominant modes of two nino phases (i.e. EN and LN) and rainfall of NWH and EH region. The relationships are further investigated with respect to latitudinal transacts representing changes in the terrain characteristics. Results of this study indicate existence of an inverse relationship between monthly rainfall and nino indices for NWH and EH region. Over the NWH region, the area averaged monthly rainfall index is found to have statistically significant (p-value <0.05) correlation coefficients of −0.48 and −0.49 for EN cases of Nino 3.0 and Nino 3.4; similarly, over the EH region, statistically significant (p-value <0.05) correlation coefficients of 0.46 and 0.57 for LN cases of Nino 3.0 and Nino 3.4 are observed. It is noted that the rainfall modes of NWH, obtained from the Empirical Orthogonal Function analysis, associated to LN events have higher spread of heavy rainfall towards higher latitudes in Uttarakhand than the rainfall modes associated to EN events. Generically, the dominant modes of EN events are found to negatively impact rainfall distribution of mountainous regions of Himachal Pradesh and Uttarakhand states of India, whereas, marginally positive impact on the rainfall distribution of Khasi-Garo hills and Brahmaputra river basin area of EH region is

S. Mukherjee (✉) · V. Gosavi · R. Joshi · K. Kumar
G.B. Pant National Institute of Himalayan Environment and Sustainable Development, Almora, Uttarakhand, India
e-mail: sandipan@gbpihed.nic.in

© Springer Nature Switzerland AG 2020   19
A. P. Dimri et al. (eds.), *Himalayan Weather and Climate and their Impact on the Environment*, https://doi.org/10.1007/978-3-030-29684-1_2

observed. The transect-wise analysis of correlation coefficients indicated that the EN events having negative impacts on rainfall distribution of NWH, whereas, the LN events had positive impacts on rainfall distribution of EH region.

2.1 Introduction

The early studies of Walker (1923, 1924) indicated a relationship between tropical Pacific Ocean sea surface temperature (SST) and Asian monsoon. Subsequently, correlation between the summer monsoon rainfall (SMR) of Indian subcontinent and El-Nino-Southern Oscillation (ENSO) events, associated to irregular SST and wind variation over Pacific Ocean on an inter annual time scale, was well elaborated by Walker and Bliss (1932). They had noted a weak SMR under the influence of low southern oscillation index, i.e. increase in the drought events over Indian subcontinent during warm phases of ENSO (i.e. El-nino event) and excess rainfall during the opposite phase (i.e. La-nina event). This coupling between atmospheric Walker circulation and SST of tropical Pacific Ocean was further studied by Bjerknes (1969). Henceforth, the connections and dynamics of ENSO-monsoon, particularly ENSO-SMR, were studied by many researchers using observed and model simulated data (Sikka 1980; Rasmusson and Carpenter 1983; Shukla and Paolina 1983; Mooley and Parthasarathy 1984; Webster and Yang 1992; Kirtman and Shukla 1997; Krishna Kumar et al. 1999; Yadav 2009a, b; Yadav et al. 2010). The generic outcome of these studies indicate that SMR is weaker before the peak of an El-Nino winter and, vice-versa for La-nina, and summer monsoon circulations over south Asia is weaker than normal periods during El-nino summers. However, the coupling of SMR and ENSO is not always found to be consistent as it weakened during the second decade of nineteenth century, strengthening during 1960s and, for consecutive 14 years since 1988 no droughts were noted in India despite of El-nino events (Gadgil et al. 2004; Maraun and Kurths 2005). Moreover, during the strongest El-nino event of the last century in 1997–1998, no significant depreciation in total SMR was noted over India (Krishna Kumar et al. 1999). Subsequently, a detail Indian SMR-ENSO feedback analysis by Kumar et al. (2002) revealed that there were only 11(8) cases during 1871–2001 resulting significant deficit/excess of monsoon rain-fall concurrent with ENSO events.

Although the scientific understanding of SMR-ENSO interaction has improved significantly during last couple of decades, studies related to summer monsoon rainfall over 'Indian Himalayan region (IHR)' and its association to ENSO events are substantially low. Here, particular emphasis to 'Indian Himalayan region' is provided to accentuate that summer monsoon rainfall within IHR is significantly modulated by local effects. The complex terrain of IHR significantly modulates rainfall process by locally modulating altitudinal variation of vapour flux, cloud water content and wind distribution, subsequently, influence the actual amount of precipitation (Barry 2008). Moreover, Himalaya also acts as a northern most barrier to the low level monsoon circulation of Indian sub- continent, and during peak of the summer monsoon season, monsoon trough resides at the Himalayan foothill. Large horizontal extent of the Himalayan region also impacts the rainfall climatology and dynamics resulting total rainfall of monsoon period over east (EH) and northwestern Himalaya (NWH) to be around 1115 and 274 mm, respectively (Mukherjee et al. 2015), and prevalence of 30–60 and 10–20 day monsoon intra-seasonal oscillation modes over NWH and EH region, respectively (Mukherjee et al. 2016). In view of dominance of the local terrain induced effect on SMR over IHR which can attenuate an existing SMR-ENSO coupling, role of regional scale circulations on monsoon rainfall over IHR, particularly the coupling effect of SMR and ENSO, needs to be quantified. This study, therefore, is an attempt to quantify role of tropical Pacific Ocean sea surface temperature variation on SMR over IHR. Here variation in the tropical Pacific Ocean SST through usages of Nino 3.0 and 3.4 indices are considered as the signal of ENSO.

Hence, particular objectives of this study are (i) to compare impact of three nino phases (i.e. El-Nino (EN), La Nina (LN) and neutral (NN)) on the monthly average rainfall of northwestern and eastern Himalaya and (ii) to assess relationships between dominant modes of two nino phases (i.e. EN and LN) and rainfall of NWH and EH using empirical orthogonal function (EOF) analysis. The relation-ships are further investigated with respect to latitudinal transacts roughly representing changes in the elevation.

2.2 Data Description

Six monsoon season months of June to September (JJAS) of EN, LN and NN were selected during 1981–2005. The El-nino years were 1983, 87, 92, 95, 98, and 2003; similarly, La-Nina years were 1984, 86, 91, 94, 97 and 2002; and neutral years were 1985, 88, 89, 90, 99 and 2000. Daily JJAS sea surface temperature (SST) standardised anomaly, estimated from NOAA-OI SST-V2 high resolution data, for the selected years of EN, LN and NN, area averaged over 150°W–90°W and 5°N–5°S and 170°W–120°W and 5°N–5°S, were used to produce Nino 3.0 and 3.4 indices, respectively, and these areas were further termed as 'Nino 3.0 and Nino 3.4 region'. Daily SSTs were also used for EOF analysis. The rainfall data was the daily gridded product of the Asian Precipitation Highly Resolved Observational Data Integration Towards the Evaluation of Water Resources (APHRODITE) project (Yatagai et al. 2009, 2012). The APHRO MA V1101R2 data were used for the selected seasons of 1981–2005, each season having 122 days of 1st June to 30th September of individual year. The area averaged data were produced using gridded data over the region (i) 72.0°E–82.325°E and 27.525°N–37.125°N, and (ii) 88.0°E–98°E and 22.0°N–30°N to represent Northwestern Himalayan region (NWH) and Eastern Himalayan region (EH), respectively (Fig. 2.1).

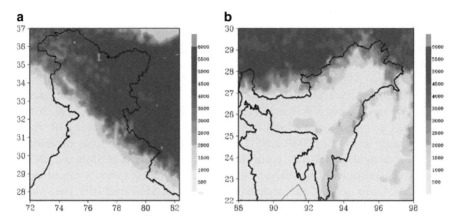

Fig. 2.1 Subplots (**a**) and (**b**) represent elevation of the study area of northwestern and eastern Himalayan region, respectively

2.3   Methods

Before addressing objective one of this study, impact of different nino phases on the seasonal rainfall of NWH and EH region was assessed using area averaged seasonal rainfall anomalies. The long-term seasonal mean was computed using seasonal rainfall data of 1981–2005. In order to assess significant change in the area averaged seasonal rainfall anomalies associated to EN, LN and NN years, two sample t-test was carried out between EN-LN, EN-NN and LN-NN cases. The null hypothesis for the analysis was: both samples come from independent random samples from normal distributions with equal but unknown variance and tested against a p-value of 0.05. The test result having any p-value equal to or less than 0.05 is considered statistically significant and subsequent changes in seasonal rainfall anomalies were highlighted in the result section. To address objective one of this study, monthly area averaged standardized indices of Nino 3.0 and 3.4 were produced for EN, LN and NN phases. Similarly, monthly area averaged standardized indices of rainfall were produced for NWH and EH regions and correlation coefficients were estimated. Before addressing the second objective, dominant modes of daily rainfall of each season associated to EN and LN years were computed using empirical orthogonal functions (EOF) for identifying spatial differences in rainfall pattern. Subsequently, to address the second objective of this study, dominant modes of tropical Pacific Ocean SST over Nino 3.0 and 3.4 region and EN and LN years were computed using EOF analysis. Next, the first principal components (PC) of the daily SST of Nino 3.0 and 3.4 regions for six selected years of EN and LN were used to compute spatial correlation. The spatial correlations were further investigated with respect to 7 (74°N–80°N with 1°interval) and 8 (89°N–96°N with 1°interval) latitudinal transects of NWH and EH region, respectively. These transects were assumed to be the proxies for change in terrain characteristics of the IHR.

2.4 Results and Discussion

In order to assess impact of EN, LN and NN events on the total seasonal rainfall of NWH and EH, area averaged seasonal rainfall anomalies were produced (Fig. 2.2). The average total seasonal rainfall, produced as averages of gridded total rainfall of monsoon seasons, during 1981–2005 for the NWH and EH regions were found to be 308.1 (51.2) mm and 1059.6 (111.1) mm, whereas the same were found to be 274.2 (56.6) mm and 1115.8 (110.6) mm, respectively, for the long-term period of 1951–2005 by Mukherjee et al. (2015). One can see from Fig. 2.2 that out of six EN events, only 1987 and 1992 experienced negative seasonal anomaly over NWH, and 1983, 1992 and 2003 experienced negative seasonal anomaly over EH. Similarly, three (1984, 1993, 1996) and four (1984, 1991, 1997, 2002) positive seasonal anomalies were observed over NWH and EH, respectively, associated to LN. The average total seasonal rainfall for NWH region associated to six EN, LN and NN events were found to be 298.5 (41.4), 318.6 (72.0) and 330.8 (47.6) mm. Although the EN years were found to receive ~ 20.1 and ~ 32.3 mm less rainfall than the LN and NN years, in terms of total seasonal rainfall variation between LN and NN years, no significant impact of forced Walker circulation on SMR was noted for NWH region. Similarly, the average total seasonal rainfall for EH region associated to six EN, LN and NN events were found to be 1067.3 (156.8), 1092.4 (116.7) and 1090.8 (42.6) mm. Similar to the NWH region, the EN years were found to receive ~ 25.1 and ~ 23.5 mm less rainfall than the LN and NN years, however, marginally higher average total seasonal rainfall was observed over EH for LN years than NN years. When the total seasonal rainfall for the EN, LN and NN years of NWH and EH regions were used for two sample t-test, it was found that the null hypothesis, i.e. both samples came from independent random samples from normal distributions with equal but unknown variance, was accepted for both regions but not at the 5% significance level.

To compare impact of EN, LN and NN events on the monthly average rainfall of NWH and EH region, monthly area averaged standardized indices of Nino 3.0 and

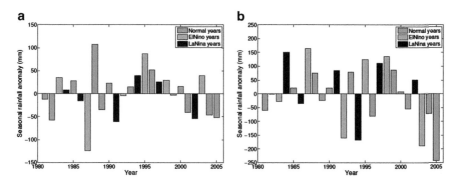

Fig. 2.2 Area averaged seasonal rainfall anomaly is represented for (**a**) northwestern and (**b**) eastern Himalaya

3.4 were compared with the monthly area averaged standardized indices of rainfall of NWH and EH regions, and correlation coefficients were estimated (Fig. 2.3). It can be noted from Fig. 2.3 that irrespective of nino indices, area averaged monthly rainfall indices of NWH and EH region have inverse relationship with tropical Pacific Ocean SST anomaly for the selected cases of EN, LN and NN events as area

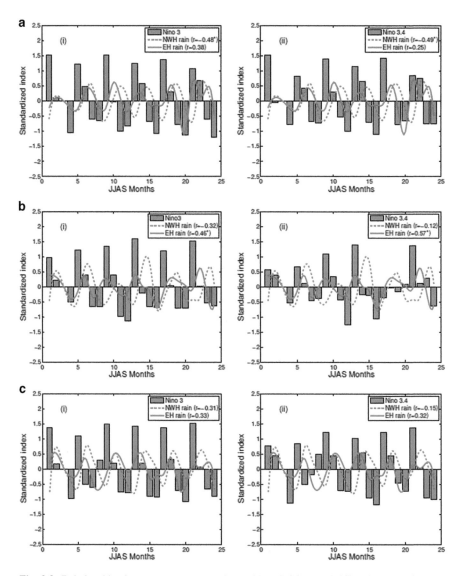

Fig. 2.3 Relationships between area averaged monthly rainfall and (i) Nino 3 & (ii) Nino 3.4 indices of selected (**a**) El-nino (1983, 1987, 1992, 1995, 1998, 2003) (**b**) La-nina (1984, 1986, 1991, 1994, 1997, 2002) and (**c**) normal years (1985, 1988, 1989, 1990, 1999, 2000). Values marked with asterisks are statistically significant at p-value <0.05

averaged monthly rainfall index of NWH was always negatively correlated to Nino 3.0 and 3.4 indices, whereas, the same of EH was always positively correlated to Nino 3.0 and 3.4 indices. Such inverse relationship between monthly rainfall and nino indices for NWH and EH region indicates dominance of two different nonlinear forcings controlling monsoon rainfall. Over the NWH region, the area averaged monthly rainfall index was found to have statistically significant (p-value <0.05) correlation coefficients of −0.48 and −0.49 for EN cases of Nino 3.0 and Nino 3.4; similarly, over the EH region, the area averaged monthly rainfall index was found to have statistically significant (p-value <0.05) correlation coefficients of 0.46 and 0.57 for LN cases of Nino 3.0 and Nino 3.4. However, no such statistically significant relationship between the area averaged monthly rainfall index and nino indices were observed for the NN cases.

To identify any spatial differences in rainfall pattern, composite results of the EOF − 1 of NWH region daily rainfall associated to six cases of EN and LN events are presented in Fig. 2.4a and b. One can note from Fig. 2.4a and b, representing dominant modes of rainfall associated to EN and LN events over NWH region,

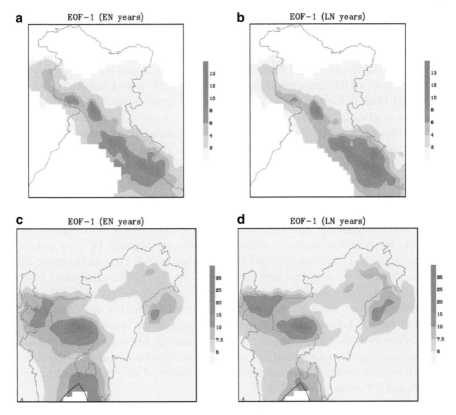

Fig. 2.4 EOF-1 of daily rainfall for (**a, c**) El Nino and (**b, d**) La Nina years of (**a, b**) NWH and (**c, d**) EH are represented

irrespective of occurrence of EN and LN events, two heavy rainfall regions are concentrated near the foothills of Uttarakhand and surrounding Dharmashala, Chamba and Dalhousie region of Himachal Pradesh. Only significant difference between the rainfall modes associated to EN and LN events over NWH was the higher spread of heavy rainfall zone towards higher latitudes in Uttarakhand during LN years. Similar to the NWH region, composite results of the EOF – 1 of EH region daily rainfall associated to six cases of EN and LN events are presented in Fig. 2.4 c and d. Irrespective of occurrence of EN and LN events, three dominant rainfall regions were found to be present near the foothills of Sikkim and Bhutan, region surrounding the Garo-Khasi hills of Meghalaya and a region east of Mizoram. Only significant difference between the rainfall modes associated to EN and LN events over EH was the spread of heavy rainfall zone to upper Brahmaputra river valley in Assam and to the Dooars region of West Bengal during LN years.

To assess relationships between dominant modes of nino phases and rainfall of NWH and EH region, PC-1s of tropical Pacific Ocean SST over Nino 3.0 and 3.4 regions associated to EN and LN years were produced. Subsequently, these PC-1s were correlated to spatial rainfall distribution of NWH and EH region for EN and LN years (Fig. 2.5) to assess significant relationships between dominant modes of nino phases and rainfall of NWH and EH region. It can be noted from Fig. 2.5a, b, e and f that irrespective of EN and LN years rainfall over the complex terrains of Himachal Pradesh and Uttarakhand was negatively correlated with PC-1 of both tropical Pacific Ocean SST of Nino 3.0 and 3.4 regions having statistically significant (p-value <0.05) correlation coefficient > −0.25 near Chamba region of Himachal Pradesh for EN years. Surprisingly, the same region was found have a positive correlation with PC-1 for LN years. Such a localized disparity in correlation coefficient cannot be directly linked to regional scale circulations rather heterogeneity in the surface characteristics, such as terrain orientation, moisture convergence etc., might play a role. Irrespective of EN and LN years, statistically insignificant correlation coefficients between PC-1 of tropical Pacific Ocean SST of Nino 3.0 and 3.4 region and rainfall were found to change from negative to positive in the higher latitudes of Jammu and Kashmir state of India. Similar to NWH region, it can be noted from Fig. 2.5c, d, g and h that irrespective of EN and LN years, statistically insignificant positive correlation coefficients between PC-1 of tropical Pacific Ocean SST over Nino 3.0 and 3.4 region and rainfall exist for the entire EH region. However, no dominant spatial characteristic was identified. Overall, albeit any statistical significance, the EN events were found to negatively impact the rainfall distribution of mountainous regions of Himachal Pradesh and Uttarakhand states of India, whereas, marginally positive impact on the rainfall distribution of Khasi-Garo hills and Brahmaputra river basin area of EH region was observed. Moreover, except for minor localized alteration, no significant change in the correlation coefficient patterns between the dominant modes of nino indices and rainfall for EN and LN events were noted.

To further assess the relationships between dominant modes of nino phases and rainfall of NWH and EH region, latitudinal transact-wise variation in the correlation coefficients between PC-1s of tropical Pacific Ocean SST over Nino 3.0 and 3.4

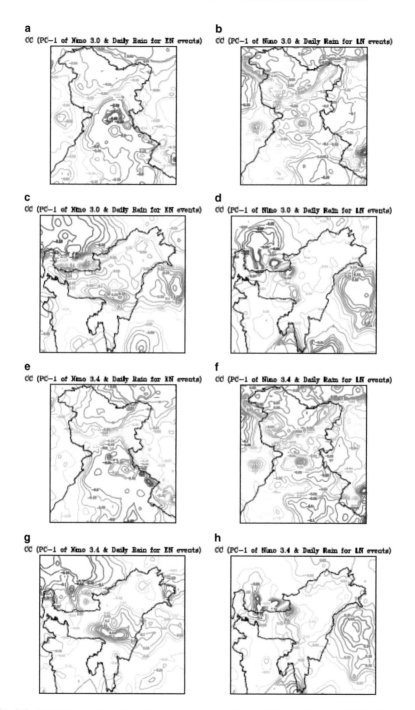

Fig. 2.5 Subplots (**a–d**) and (**e–h**) represent correlation coefficients between PC-1 of tropical Pacific Ocean SST over Nino 3.0 and Nino 3.4 region and rainfall, respectively. Subplots (**a, c, e, g**) and (**b, d, f, h**) are for EN and LN years, respectively

regions and rainfall on NWH and EH regions were estimated. These transects, as indicated in the method section, represent changes in the terrain characteristics. The transect-wise distribution of correlation coefficients, irrespective of the nino regions, were found to indicate that the entire NWH region had received lower rainfall during EN years (Fig. 2.6a, b) whereas, the EH region had mixed responses as median correlation coefficients were found to change from negative to positive for west to east progression of transects (Fig. 2.6c, d), however, the median correlation coefficients for EH were small and median values varied between −0.05 and 0.05. Similarly, irrespective of the nino regions, correlation coefficients were found to indicate that the entire NWH region had not received higher rainfall during LN years (Fig. 2.6e, f) whereas, the EH region had positive impacts of LN events on rainfall distribution as correlation coefficients were greater than zero (Fig. 2.6g, h), however, the median correlation coefficients for EH were small and varied between 0.0 and 0.09. Generically, it was noted that the EN events having negative impacts on rainfall distribution of NWH, whereas, the LN events had positive impacts on rainfall distribution of EH region.

2.5   Conclusion

In spite a plethora of studies on the linkages between SMR and ENSO, focused research on the monsoon rainfall variability over IHR and its linkages to regional scale circulation, such as variation in the tropical Pacific Ocean SST generally represented by Nino indices, remained largely unaddressed except for the study by Dimri (2013) and Yadav et al. (2013) who have quantified role of ENSO on controlling winter rainfall of NWH region. SMR over IHR and its linkages to ENSO is specially emphasized here as significant modulation of monsoon rainfall process within IHR by local terrain induced effects can attenuate an existing SMR-ENSO coupling mechanism. Hence, role of regional scale circulations on monsoon rainfall over IHR, particularly the coupling effect of SMR and ENSO, are quantified. This study, therefore, is an attempt to assess relationship between the SMR and ENSO over IHR using six respective cases of El-nino (EN), La-nina (LN) and normal (NN) years during 1981–2005. Standardized tropical Pacific Ocean SST anomalies over Nino 3.0 and 3.4 regions were considered as representative of ENSO events.

Analysis of six cases of EN, LN and NN events with the rainfall distribution of NWH and EH region indicated existence of an inverse relationship between monthly rainfall and nino indices for NWH and EH region. Irrespective of NWH and EH, the EN years were found to receive approx. < 21.8 and 28.7 mm rainfall than the LN and NN years, in terms of total seasonal rainfall variation between LN and NN years, having no significant impact of thermally forced Walker circulation on SMR over NWH region. Moreover, it was noted that the rainfall modes of NWH, obtained from the EOF analysis, associated to LN events had higher spread of heavy rainfall towards higher latitudes in Uttarakhand than the rainfall modes associated to EN events; similarly for EH region, larger spatial extent of heavy rainfall to upper

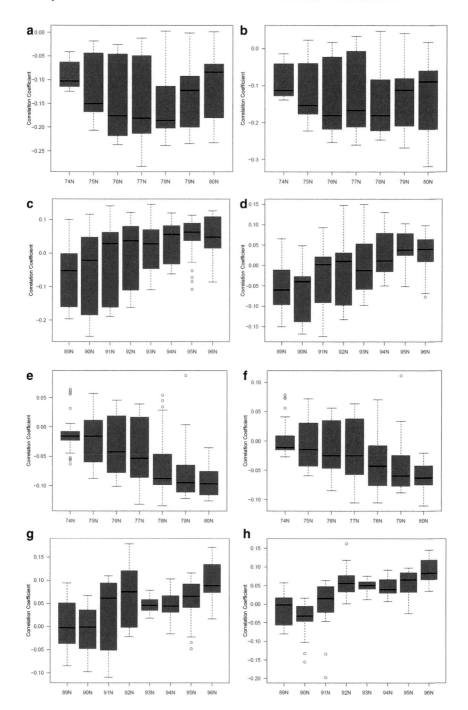

Fig. 2.6 Transact-wise distribution of correlation coefficients between PC-1 of tropical Pacific Ocean SST over Nino 3.0 and Nino 3.4 region and rainfall are presented. Subplots (**a–d**) and (**e–h**) are respectively for EN and LN years. Subplots (**a–b**), (**e–f**) are for NWH region; (**c–d**), (**g–h**) are for EH region. Subplots (**a, c, e, g**) and (**b, d, f, h**) are for Nino 3.0 and 3.4 regions, respectively

Brahmaputra river valley in Assam and to the Dooars region of West Bengal was noted during LN years. Overall, the dominant modes of EN events were found to negatively impact the rainfall distribution of mountainous regions of Himachal Pradesh and Uttarakhand states of India, whereas, marginally positive impact on the rainfall distribution of Khasi-Garo hills and Brahmaputra river basin area of EH region was observed. The dynamics of contradictory signatures of EN and LN events on the SMR over IHR is not discussed in this article. However, this is to be noted that one of the shortcomings of this study is the small number of ENSO cases considered for analysis. Results of this study can be improved in a further study including comprehensive ENSO cases from last six to seven decades.

Acknowledgements The Director, GBPNIHESD Kosi-Katarmal, Almora, India is gratefully acknowledged for providing the computational facility.

References

Barry R (2008) Mountain weather and climate. Cambridge University Press, Cambridge

Bjerknes J (1969) Atmospheric tele-connections from the equatorial Pacific. Mon Weather Rev 97:163–172

Dimri A (2013) Relationship between ENSO phases with Northwest India winter precipitation. Int J Climatol 33(8):1917–1923. https://doi.org/10.1002/joc.3559

Gadgil S, Vinaychandran P, Francis P, Gadgil S (2004) Extremes of the Indian summer monsoon rainfall, ENSO and equatorial Indian Ocean oscillation. Geophys Res Lett 31(12):L12, 213. https://doi.org/10.1029/2004GL019733

Kirtman B, Shukla J (1997) Influence of the Indian summer monsoon on the ENSO. Q J R Meteorol Soc 126:213–239

Krishna Kumar K, Rajagopalan B, Cane M (1999) On the weakening relationship between the Indian summer monsoon and ENSO. Science 284:2156–2159

Kumar R, Krishna Kumar K, Asrit R, Patwardhan S, Pant G (2002) Climate change in India (Shukla J et al (ed)). Tata McGraw Hill, New Delhi

Maraun D, Kurths J (2005) Epochs of phase coherence between El-nino/Southern Oscillation and Indian monsoon. Geophys Res Lett 32(L15):709. https://doi.org/10.1029/2005GL023225

Mooley D, Parthasarathy B (1984) Indian summer monsoon and El-Nino. Pure Appl Geophys 121:339–352

Mukherjee S, Joshi R, Prasad R, Vishvakarma S, Kumar K (2015) Summer monsoon rainfall trends in the Indian Himalayan region. Theor Appl Climatol 121(3–4):789–802. https://doi.org/10.1007/s00704-014-1273-1

Mukherjee S, Ballav S, Soni S, Kumar K, De UK (2016) Investigation of dominant modes of monsoon ISO in the northwest and eastern Himalayan region. Theor Appl Climatol 125(3–4):489–498. https://doi.org/10.1007/s00704-015-1512-0

Rasmusson E, Carpenter T (1983) The relationship between eastern equatorial Pacific Sea surface temperature and rainfall over India and Sri Lanka. Mon Weather Rev 111:517–528

Shukla J, Paolina D (1983) The Southern Oscillation and long range forecasting of the summer monsoon rainfall over India. Mon Weather Rev 111:1830–1837

Sikka D (1980) Some aspect of the large scale fluctuations of summer monsoon rainfall over India in relation to fluctuations in the planetary and regional scale circulation parameters. Proc Indian Acad Sci 89:179–195

Walker G (1923) Correlation in seasonal variations of weather, VIII: a preliminary study of world weather. Mem India Meteorol Dep 24:75–131

Walker G (1924) Correlation in seasonal variations of weather, IX: a further study of world weather. Mem India Meteorol Dep 24:275–333

Walker G, Bliss E (1932) World weather v. Mem R Meteorol Soc 4:53–84

Webster P, Yang S (1992) Monsoon and ENSO: selective interactive systems. Q J R Meteorol Soc 118:877–926

Yadav R (2009a) Changes in the large-scale features associated with the Indian summer monsoon in the recent decades. Int J Climatol 29(1):117–133. https://doi.org/10.1002/joc.1698

Yadav R (2009b) Role of equatorial central Pacific and northwest of North Atlantic 2-meter surface temperature in modulating Indian summer monsoon variability. Clim Dyn 32(4):549–563. https://doi.org/10.1007/s00382-008-0410-x

Yadav R, Yoo J, Kucharski F, Abid M (2010) Why is ENSO influencing northwest India winter precipitation in recent decades? J Clim 23:1979–1993. https://doi.org/10.1175/2009JCLI3202.1

Yadav R, Ramu D, Dimri A (2013) On the relationship between ENSO patterns and winter precipitation over north and central India. Glob Planet Chang 107:50–58. https://doi.org/10.1016/j.gloplacha.2013.04.006

Yatagai A, Arakawa O, Kamiguchi K, Kawamoto H, Nodzu M, Hamada A (2009) A 44-year daily gridded precipitation dataset for Asia based on a dense network of rain gauges. Sci Online Lett Atmos 5:137–140

Yatagai A, Kamiguchi K, Arakawa O, Hamada A, Yasutomi N, Kitoh A (2012) APHRODITE: constructing a long-term daily gridded precipitation dataset for Asia based on a dense network of rain gauges. Bull Am Meteorol Soc 93:1401–1415

Chapter 3
Changes in the Large-Scale Circulations Over North-West India

Ramesh Kumar Yadav

Abstract The northwestern part of India occupies a vast landmass which roughly lies in the area bounded by 70.5°E–80.5°E longitudes and 27°N–37°N latitudes of South Asia. This is an important region of food-grain production in the country. The summer season (June to September) contributes about 75% of annual precipitation and the winter season from December to March 15–20%. These precipitations are very important for the crops and maintaining the western Himalayas Glaciers. The interannual variability of summer and winter precipitation are examined using observed and reanalysis datasets for the period of 1948–2015. The analysis shows changes in teleconnection pattern around the late-1970s, when the major 'climate shift' was observed in the Indo-Pacific Oceans. The summer precipitation teleconnection change is related to the change in the shape and position of the equatorial Pacific warming. And, the winter precipitation is mostly influenced by the two major weather phenomenon Arctic Oscillation/North Atlantic Oscillation (AO/NAO) and El-Niño–Southern Oscillation (ENSO), which exert strong control on the weather/climate of the Northern Hemisphere particularly in the boreal winter. The AO/NAO phenomenon were more influencing in the earlier decades, while the ENSO in the recent decades.

R. K. Yadav (✉)
Indian Institute of Tropical Meteorology, Pune, Maharashtra, India
e-mail: yadav@tropmet.res.in

© Springer Nature Switzerland AG 2020   33
A. P. Dimri et al. (eds.), *Himalayan Weather and Climate and their Impact on the Environment*, https://doi.org/10.1007/978-3-030-29684-1_3

3.1 Introduction

India gets 80% of rainfall during the summer season. The large landmasses of Asia and the Indian subcontinent heat up, generating a low-pressure area elongated along North India, called as Monsoon Trough. The wind blows from south-west across the India Ocean, accumulating considerable moisture which is deposited as heavy precipitation during the season. In contrast, during winter season, the temperatures are at their lowest, the pressure decreases from north and are generally higher than summer over central India. Airflow reverses and wind blows northeasterly across the Indian Ocean. Clear skies, fine weather, light winds, low humidity, and large daytime variations of temperature are the normal features of the winter season in India. This is the driest season for the country as a whole, except in the extreme south-east of Peninsular India experience large rainfall, called the Indian north-east monsoon. Another area of experiencing large precipitation during winter is the north-west India from the mid-latitude cyclones, known in Indian meteorological parlance as 'western disturbances' (WDs).

The rainfall distribution over different regions of India during summer is, however, inhomogeneous due to influence of several local and remote factors. For example, the central Indian plain is influenced by the north south movement of the monsoon trough, the northwestern part is influenced by the extreme west-northwest ward movement of the lows and depressions and mid-tropospheric cyclones (MTCs) over west India (Miller and Keshavamurthy 1968; Mak 1975). MTC has a unique vertical structure as it is hardly detectable at the lower and upper troposphere. Its largest intensity is observed at 600-hPa level. It is one of the most important rain-producing systems in the west India. While the north India comprising, western Himalayas are influenced by the interaction between mid-latitude troughs and monsoon circulations, and also the local and external factors play a dominant role in the rainfall variability. While, during winter the north-western part of India and the extreme south-eastern part of Peninsular India experience precipitation due to WDs and Indian north-east monsoon, respectively. The rest of the country remains dry during this season.

The seasonal and intra-seasonal variability of north-west India is influenced by large-scale mid-latitude circulation. The intrusion of an anomalous high-amplitude mid-latitude westerly trough into the north-west India, with the interaction with the monsoonal flow, produces very heavy precipitation over north-west India during summer season (Dairaku and Emori 2006). The pioneering works by Ramaswamy (1956, 1962) on the dynamical aspects of monsoon breaks highlighted the importance of anomalous southward intruding large-amplitude westerly troughs, from the mid-latitudes into the Indo-Pakistan region, in causing the break situations over India. Raman and Rao (1981) noted that prolonged monsoon-break situations are typically associated with upper tropospheric blocking ridges over West and East Asia. Apart from the remote forcing from the El-Niño/Southern Oscillation (ENSO) phenomenon (Yadav 2009a, b), Atlantic Niño has shown its influence on north-west India rainfall (Yadav 2017a; Yadav et al. 2018). Also, the temperature and pressure of Middle-East region have shown their association with north-west India rainfall (Yadav 2016, 2017b).

During winter season, the precipitation is mainly associated with the sequence of mid-latitude synoptic systems, called 'western disturbances' (WDs). The WDs which move across India appear to have a life-history similar to that of the extra-tropical cyclones (Petterssen 1956). They originate over the Mediterranean and sometimes over the Atlantic, get modified over the Persian Gulf and Caspian Sea (Agnihotri and Singh 1982), and arrive over the Indian longitudes more or less in an occluded stage, cause inflow of moist and warm air from Arabian Sea ahead of the depressions. They converge at lower levels and contribute to convection and precipitation, while cold and dry air from northern latitudes sweeps in the rear of the disturbances resulting in cold waves affecting north of 20°N over northwest India (Pisharoty and Desai 1956; Mooley 1957).

3.2 Results and Discussion

3.2.1 North-West India Time Series Analyses

The northwestern parts of India occupy a vast landmass which lies in the area bounded by 70.5°E–80.5°E longitudes and 27°N–37°N latitudes of South Asia as shown in Fig. 3.1a. The area compromises the western Himalayas to the north and Thar Desert to the south-west, Indo-Gangetic plain in the middle, and the Aravalli Range to the south-east. The mean (averaged over the period 1948–2015) annual cycle of precipitation (India Meteorological Department 0.25° × 0.25° data) averaged over the box (70.5°E–80.5°E, 27°N–37°N) for north-west India shows two peaks: (1) primarily during the summer season, June through September, and (2) a secondary peak during winter season, December through March (Fig. 3.2a). These are the two rainy seasons, summer and winter, when it receives considerable amount of precipitation, separated by the two dry transitional seasons of spring (April and May) and autumn (October and November). Both summer and winter seasons span 4 months from June to September (JJAS) and December of current year to January, February and March of next year (DJFM), respectively. The precipitation climatology for summer and winter are shown in Fig. 3.1b and c, respectively. The north-west India summer rainfall (NWISR) and north-west India winter precipitation (NWIWP) time series, averaged over the box (70.5°E–80.5°E, 27°N–37°N) for the summer (JJAS) and winter (DJFM) seasons, have been plotted in the Fig. 3.2b and c, respectively. The NWISR has a long-term average of 119.3 mm/day, with a standard deviation (SD) of 24.6 mm and coefficient of variation (CV) of 20.6%, while NWIWP has long-term average of 45.1 mm/day, with SD of 15.7 mm and CV of 34.9%. The NWIWP mean is less, while variability is more when compared to the NWISR.

The NWISR and NWIWP display considerable inter-annual variability (Fig. 3.2b and c, respectively) and affect the summer and winter crops production over northwest India and water resources. Since, the major 'climate shift' in the distribu-

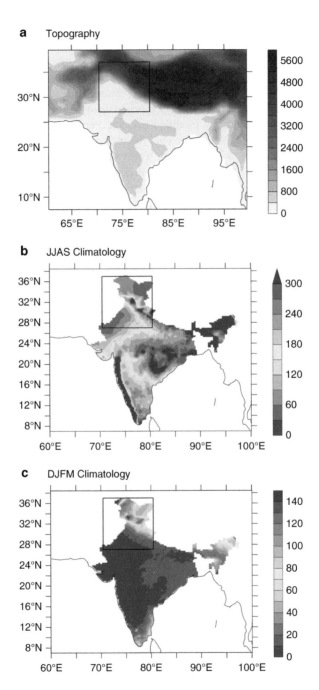

Fig. 3.1 (**a**) Topography of India, (**b**) JJAS and (**c**) DJFM precipitation climatology for the period 1978–2015

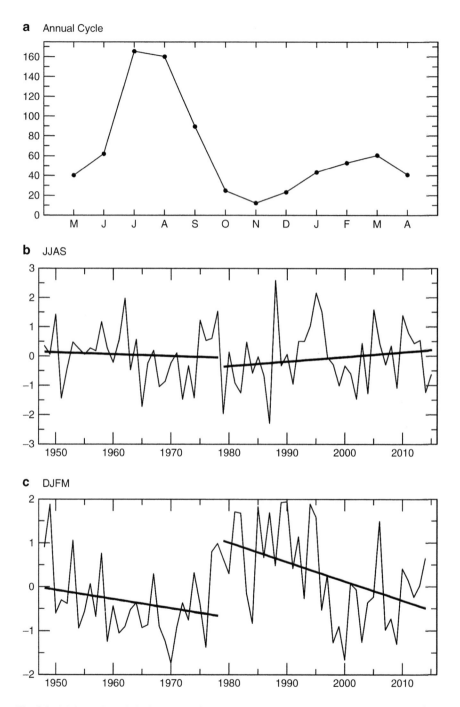

Fig. 3.2 (**a**) Annual precipitation cycle of north-west India starting from the month May to April averaged for the period 1948–2015. Time-series of (**b**) NWISR and (**c**) NWIWP. The straight thick lines are the linear trend line

Table 3.1 Trend, Mean and Standard Deviation (S.D.) for S1, S2, W1 and W2. The bold value represents the significant at 95% level. The values in bracket is the for the full data length (i.e. 1948–2015 for JJAS and 1948–2014 for DJFM). Significant value at 95% level is represented in **bold**

Period/Season	Trend (mm/year)	Mean (mm/day)	S.D. (mm)
S1 (1948–1978) JJAS	−0.006	121 (119.3)	21.9 (24.6)
S2 (1979–2015) JJAS	+0.016	117.9 (119.3)	26.55 (24.6)
W1 (1948–1978) DJFM	−0.02	40.05 (45)	13 (15.72)
W2 (1948–1978) DJFM	**−0.044**	49.48 (45)	16.45 (15.72)

tion of tropical Indo-Pacific sea surface temperature (SST) was observed in the late-seventies (Graham 1994; Trenberth and Hurrell 1994). Therefore, to study the influence of this 'climate shift' on the NWISR and NWIWP time-series, the linear trend analysis has been done for the period 1948–1978 (31-year), named hereafter as S1 and for 1979–2015 (37-year), named hereafter S2 for NWISR (Fig. 3.2b). Similarly, for NWIWP the trend analysis has been done for the period 1948–1978 (31-year), named hereafter W1 and for 1979–2014 (36-year), named hereafter W2 for the NWIWP. The trend analysis shows decreasing trend (0.006 mm/year) in S1 and increasing trend (0.016 mm/year) in S2, while decreasing trend in W1 (0.02 mm/ year) and W2 (0.044 mm/year; significant at 95% level). The mean was more and SD was less than normal for S1, while for S2 the mean was less and S.D. was more than normal. On the contrary, the mean and S.D. were less than normal for W1, while for W2 the mean and S.D. were more than normal (Table 3.1). This suggests that after the major 'climate shift' of the late-seventies the variability in the summer and winter had increased. The trend in precipitation had increased during summer, while the winter precipitation had decreased.

3.2.2   NWISR Teleconnections During 1948–1978: S1

The simultaneous spatial correlation pattern between NWISR for the period 1948–1978 (S1) and surface temperature (SST over the Ocean basins (ERSST v4 data) and 2-metre surface temperature (2mST) over land regions (NCEP/NCAR reanalysis data); Fig. 3.3c) shows significant negative correlation over central to eastern equatorial Pacific, Somalia coast and north equatorial Atlantic, and significant positive correlation over south-west tropical Pacific, west of Australia and south-east of South Africa. Significant negative correlation over north-west India is obvious as the rainfall decreases the surface temperature. The negative anomaly over equatorial central and eastern Pacific and positive SST anomaly over south-west tropical Pacific depicts the La-Niña pattern. The mean sea level pressure (MSLP; NCEP/ NCAR reanalysis data) correlation (Fig. 3.3d) shows dipole structure with significant negative correlation over tropical western Pacific and positive correlation over eastern Pacific, suggesting the positive phase of southern Oscillation. Significant negative correlation is observed over Indian landmass and tropical Indian Ocean. The negative correlation over Indian landmass infers the low-pressure areas and

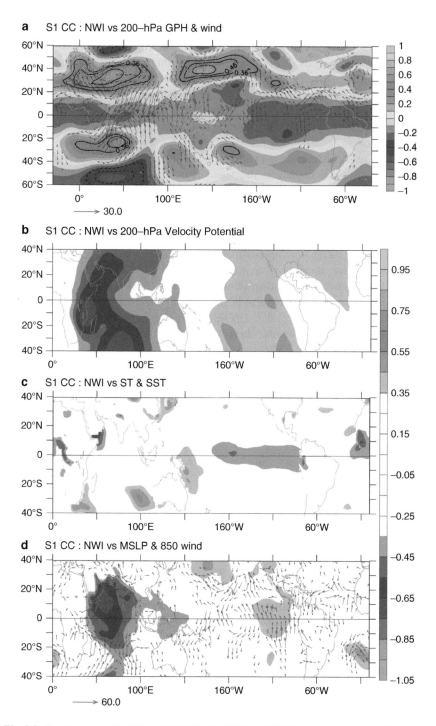

Fig. 3.3 Simultaneous CC of S1 with (**a**) 200-hPa GPH, (**b**) 200-hPa velocity potential, (**c**) surface temperature and (**d**) MSLP (color shade), regression of (**a**) 200-hPa and (**d**) 850-hPa wind (blue arrows) onto S1. Contour interval is 0.1 and greater than and equal to 0.33 and less than and equal to −0.33

over north Indian Ocean low pressure systems. The regression of 850-hPa wind (NCEP/NCAR reanalysis data) onto S1 shows strong cross-equatorial flow along the East African coast originating from the Mascarene High in the south Indian Ocean. A study by Krishnamurti and Bhalme (1976) highlighted the role of this cross-equatorial flow as an important feature in the monsoon circulation. The negative SST correlation over Somali coast is due to upwelling caused by strong cross-equatorial flow.

Figure 3.3a shows correlation pattern for the 200-hPa level geopotential height (GPH; NCEP/NCAR reanalysis data) and regression of 200-hPa wind (NCEP/NCAR reanalysis data) onto S1. The 200-hPa GPH patterns shows significant positive correlation over north-west of India, east Asia, North Pacific, North Atlantic and western Europe, suggesting a circum-global teleconnection (CGT) pattern as first noted by Hoskins and Ambrizzi (1993); Ambrizzi et al. (1995); and later defined by Ding and Wang (2005), with approximately structure of zonal wave number 5. The significant positive anomaly over north-west of India represents the anomalous Tibetan High due to extreme Indian summer monsoon (ISM). The relationships with west Europe and east Asia have also been shown by Yadav (2009a, b, 2016, 2017a, b), Kripalani and Kulkarni (1997, 1999), Kripalani et al. (1997). Ding and Wang (2005) have shown that when the interaction between the Indian summer monsoon (ISM) and ENSO is active, ENSO may influence northern China via the ISM and the CGT. Also, positive correlations are observed over southeast of Madagascar, south-west and east Australia, subtropical south Pacific, and south-east Atlantic, suggesting another CGT pattern in the Southern Hemisphere also observed by Ambrizzi et al. (1995), Yadav (2009b). 200-hPa wind anomalies shows associated anti-cyclonic circulation anomalies related to positive GPH anomalies of CGT. The anomalous north-easterlies over Peninsular India and west Indian Ocean suggests the cross equatorial flow from the anticyclone centered over north-west of India to the anticyclone centered over south-east of Madagascar. The significant cross equatorial flow across the west Indian Ocean excite the anti-cyclonic anomaly south-east of Madagascar. Also, the Southern Hemisphere subtropical westerly jet stream intensified during ENSO years (Yadav 2009b), acts as Rossby waveguide, have enhanced the geographical amplitude of low-frequency fluctuations downstream to the south Pacific and south Atlantic (Hoskins and Ambrizzi 1993; Ambrizzi et al. 1995).

The 200-hPa velocity potential (Fig. 3.3b; NCEP/NCAR reanalysis data) shows significant negative correlation core over south-west tropical Indian Ocean and south Indian Ocean and positive correlation core over equatorial central-east Pacific and south Pacific. This suggests the stronger southern oscillation associated with southward shift in the stronger Walker circulation from its normal position. The warming of the tropical south-western Pacific must have shifted and intensified the Walker circulation southwards. In summary, during La-Niña years, the SST over tropical south-west Pacific rises and the Southern Hemisphere subtropical westerly jet stream over south Indian Ocean, south-west Australia, and south-east Atlantic intensifies due to the consequence of thermal wind balance. The intensification of jet stream forms anti-cyclonic circulation anomaly which intensifies the Mascarene High at the lower level as the atmospheric response over the region is equivalent barotropic in nature. The Mascarene High intensifies the cross-equatorial flow and hence ISM circulation. This intensifies the Tibetan High, which excite downstream Rossby wave train extending to the North Pacific. The interaction between the ISM and La-Niña influence northern China via the ISM and the strong CGT is formed in the Northern Hemisphere. The intense Tibetan High also intensifies the anticyclone south-east of Madagascar through strong Hadley Cell. The intensified Southern Hemisphere jet, acts as Rossby waveguide and enhances the CGT in the Southern Hemisphere and vice-versa for El-Niño years (Yadav 2009b).

3.2.3 NWISR Teleconnections During 1979–2015: S2

Similar correlation and regression analysis have been done for NWISR for the period 1979–2015 (S2). Figure 3.4c shows the spatial correlation pattern between S2 and surface temperature. The surface temperature shows significant negative correlation over central equatorial Pacific and significant positive correlation over north-west equatorial Pacific and Iran. The significant negative correlation over north-west India is obvious as the rainfall decreases the surface temperature. The MSLP correlation (Fig. 3.4d) shows dipole structure with significant negative correlation over tropical Indian Ocean, western India, Middle-East and north Africa and positive correlation over eastern and south Pacific. The regression of 850-hPa wind onto S2 shows cyclonic circulation anomaly over Arabian Sea with anomalous south-easterly over north-west India. In S2, the significant relationship with the cross-equatorial flow was missing. Instead, a cyclonic circulation is observed over north Arabian Sea with convergence towards north-west India with abundant of moisture supply, favoring deep convection. This cyclonic circulation anomaly with northerly over Arab, westerly over north Arabian Sea and southerly over north-west India and Pakistan. As these winds are coming from the warm desert land of Arabian Peninsula, they can hold more moisture after blowing through the Persian Gulf and the Arabian Sea (Yadav 2009a, 2016, 2017a). The anomalous southerly observed over north Pakistan hinders the intrusion of mid-latitude cold and dry air towards Indian subcontinent favoring excess precipitation over north-west India. This anomalous cyclonic circulation also increases the frequency and intensity of mid-tropospheric cyclones (MTCs) over west India. These MTCs are one of the main systems for producing heavy rainfall over north-west India.

The 200-hPa GPH (Fig. 3.4a) shows significant positive correlation over north-west India, East Asia, south-west USA and North Africa. The pattern shows shorter wavelength with approximate structure of zonal wave number 7. The associated 200-hPa level wind flow shows anti-cyclonic circulation over significant positive GPH anomalies. The positive surface temperature over Iran modulates the Indian summer monsoon circulation during June–September (Yadav 2016, 2017a, 2009a). The surface temperature variability over Iran is perturbed by the passage of the mid-latitude wave train that propagates from north-west Europe and traverses through the Eurasian region to Iran. In some years, these waves split into two branches: one propagates eastwards towards the western Siberian plains troposphere (Ding and

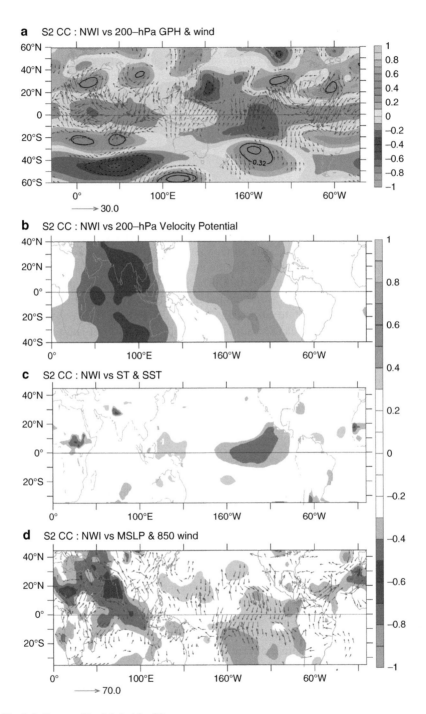

Fig. 3.4 Same as Fig. 3.3, but for S2

Wang 2005, 2007, 2009) and another propagates equator wards towards the Mediterranean and follows the Asian subtropical westerly jet stream, known as the 'Silk Road' pattern (Enomoto et al. 2003; Yadav 2017a). Both the branches converge over Iran and increase the GPH anomaly in the upper to middle. This increases the tropospheric temperature along the vertical cross-section of atmospheric column over Iran with maximum warming in the lower troposphere. The anomalous cyclonic circulation associated with the lower-level warming helps the convergence of winds toward north-west India (Yadav 2016, 2017a). Also, the strong upper-tropospheric GPH anomalies over Iran intensify the Tibetan High westward, which dynamically weakens the Tibetan High in the middle. This allows an anomalous high amplitude mid-latitude trough intrusion into the northern parts of the Indo-Pakistan region favoring very heavy rainfall events over north-west India (Yadav 2016).

The 200-hPa velocity potential (Fig. 3.4b) shows significant negative correlation core over south Asia and positive correlation core over the equatorial central Pacific. This suggests stronger Walker circulation with rising motion over Indian subcontinent and subsidence over the central Pacific. The shift in the La-Niña cooling towards the equatorial central Pacific are associated with warm SST anomaly over the tropical north-western Pacific owing to stronger rising motion over Cambodia and subsidence over the equatorial central Pacific. The warming of the tropical north-western Pacific during S2 must have shifted and intensified the inter-tropical convergence zone to the Cambodian latitude, which in turn intensifies the Walker circulation during the La-Niña years and vice versa during the El-Niño years. The upper-level divergence (velocity potential minimum) anomaly over south Asia reinforces the convective activity over Indian subcontinent including north-west India and vice versa for the El-Nino years (Yadav and Singh 2017).

3.2.4 Secular Variations of Simultaneous Relationships

Since, ENSO and Arctic Oscillation/North Atlantic Oscillation (AO/NAO) exert strong control on the climate of the Northern Hemisphere particularly in the boreal winter. The correlation of Niño-3.4 (representing the ENSO index averaged over SST over the equatorial central Pacific, commonly referred to as Niño-3.4 region obtained from ERSST.v4 data) and NAO (index taken as the difference in normalized sea level pressure between southwest Iceland and Gibraltar, obtained from the Climatic Research Unit, University of East Anglia) with NWIWP are 0.25 and 0.35 for the period 1948–2014, which are statistically significant at 95% and 99% level, respectively. The correlation between ENSO and NAO indices for the period 1902–2008 is very poor (0.03) and insignificant (Yadav et al. 2009a). This shows that both these indices are independent of each other, but they significantly exert their influence on NWIWP variability. Further, the secular variation of the relationship of these indices with NWIWP is shown by sliding correlations on a 21-year moving window (Fig. 3.5). The Niño-3.4 index shows significant correlation in the recent

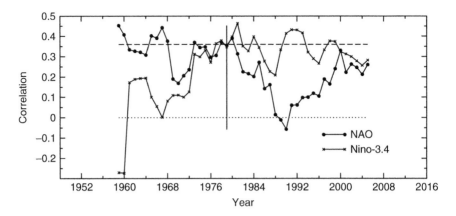

Fig. 3.5 Sliding correlations on a 21-year moving window among NWIWP and NAO and Nino-3.4 region SST indices for the period 1948–2014 for the simultaneous season of DJFM. Values are plotted at the center of 21-year period. The dashed line indicates 95% significance level, and zero correlation is indicated by dotted line

decades after 1979. Prior to 1978 the correlations are insignificant. On the other hand, the NAO index shows significant correlation prior to 1979. It is interesting to see that when the relationship with NAO was significant i.e. prior to 1979, the relationship with Niño-3.4 SST was insignificant. In the recent decades, when the relationship with Niño3.4 SST became significant, NAO relationship has dropped to insignificant level. While in the last decade both Niño-3.4 and NAO correlations are almost equal. The relationship suggests that ENSO and NAO are the two major modes of variability of NWIWP. The major climate shifts observed in Indo-Pacific Ocean in the late-1970s have connection with these observed changes.

3.2.5 NWIWP Teleconnections During 1948–1978: W1

Similar correlation and regression analyses, as done for NWISR (S1 & S2) have been done for NWIWP (W1 & W2) for the season DJFM. The correlation with MSLP for the period 1948–1978 i.e. W1 (Fig. 3.6d) shows significant strong positive correlation over the subtropical North Atlantic and west Africa and significant negative correlation over polar North Atlantic, Arctic, Greenland and north Asia. The negative and positive MSLP anomalies over polar and subtropical North Atlantic, respectively, show the relationship with positive phase of AO/NAO phenomena. The regression of 850-hPa wind onto W1 shows trough over north-west India. The correlation with surface temperature (Fig. 3.6c) shows significant positive correlation over east-central Asia and negative correlation over east Mediterranean. The 200-hPa zonal wind correlation patterns for W1 (Fig. 3.6a) shows patches of significant positive correlations north of 45°N over east Canada, north Atlantic, north-east Asia and north of Alaska. These positive zonal wind anomalies represent

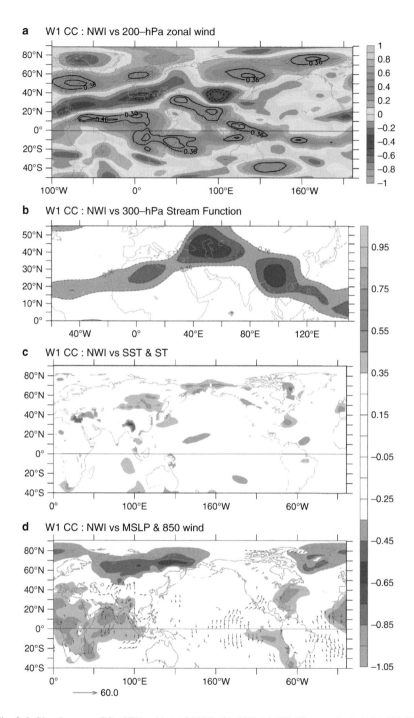

Fig. 3.6 Simultaneous CC of W1 with (**a**) MSLP, (**b**) SST, (**a**) 200-hPa zonal wind, (**b**) 300-hPa stream function, (**c**) surface temperature and (**d**) MSLP (color shade), regression of (**d**) 850-hPa wind (blue arrows) onto W1. Contour interval is 0.1 and greater than and equal to 0.33 and less than and equal to −0.33

the strong circumpolar zonal westerly arising due to a steep pressure gradient anomaly between extra-tropic and pole (Raman and Maliekal 1985). The strong circumpolar zonal westerlies are the indication of positive phase of AO/NAO phenomena. The strong circumpolar westerly does not allow the frigid winter air to spill into northern Eurasia keeping the surface temperature warmer than normal. Also, significant positive correlation is observed extending from north equatorial Atlantic to north Africa and from Middle-East to Myanmar. Significant negative correlation is observed over west Mediterranean and north-east of India. The upper-troposphere high-pressure anomaly over subtropical north Atlantic and low-pressure anomaly over north Africa during positive NAO phase, produces easterly anomaly over west Mediterranean and westerly anomaly over north Africa to Middle East, respectively. This indicates that the secondary maxima of the Asian subtropical jet stream centered over north Africa to Middle East strengthen (Yadav et al. 2009a).

Stream-function enhances signal in the low latitudes, where the subtropical westerly jet stream largely resides. The upper troposphere is the most important region for Rossby-wave propagation in the tropical and subtropical regions but, equivalent barotropic mid-latitude waves are probably best represented using the flow from a slightly lower level (Held et al. 1985). Therefore, 300-hPa stream-function have been analyzed (Fig. 3.6b; NCEP/NCAR reanalysis data) for searching the western disturbances (WDs) pattern. The negative (positive) anomalies over Northern (Southern) Hemisphere reveals cyclonic circulation and vice versa. W1 (Fig. 3.6d) shows negative correlation extending from tropical North Atlantic to north India along North Africa and Caspian Sea (Yadav et al. 2009a). This suggest that during the W1 when AO/NAO was more influencing NWIWP the WDs were originating over tropical North Atlantic, at the origin of the Asian jet, and traveling through Mediterranean Sea and Caspian Sea, picking moisture from there, to north-west India as documented by the earlier studies before the 1980s (Pisharoty and Desai 1956; Mooley 1957; Bhaskara Rao and Morey 1971; Dutta and Gupta 1967; Singh 1963, 1979; Singh and Kumar 1977).

3.2.6   NWIWP Teleconnections During 1979–2014: W2

During W2 i.e. for the period 1979–2014, the simultaneous MSLP correlation (Fig. 3.7d) shows strong significant positive correlation over eastern tropical hemisphere and east Asia (Yadav et al. 2009b) and significant negative correlation over tropical north central Pacific and Arctic. The large seesaw of MSLP anomaly between the Indo-Pacific Ocean shows the negative phase of Southern Oscillation relationship. The negative anomaly over Arctic suggests positive phase of AO. Also, the significant negative correlation north-west of India suggests frequency of deeper low-pressure systems of WDs during excess years of NWIWP during recent decades. The correlation with surface temperature (Fig. 3.7c) shows significant correlation over central equatorial Pacific and north Asia, and significant negative correlation over north-west and south Pacific. The negative correlation pattern over

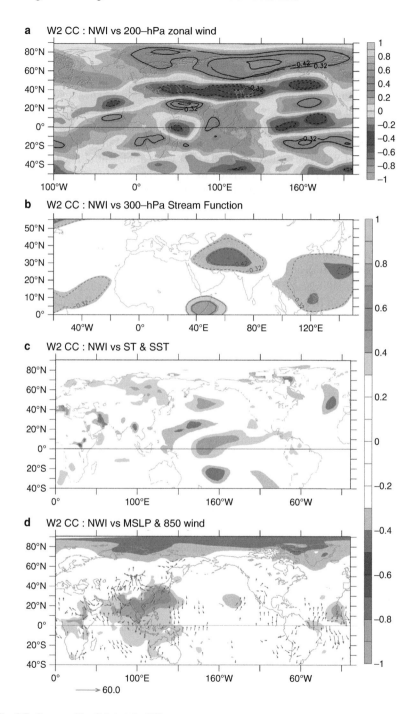

Fig. 3.7 Same as Fig. 3.6, but for W2

tropical Pacific resembles the "Horse Shoe" pattern indicating relationship with positive phase of ENSO.

For 200-hPa zonal wind correlation pattern (Fig. 3.7a) shows strong significant positive correlations over Arabian Sea, Arabian landmass, eastern Indian Ocean and along 80°N latitude over north Asia and Europe, and significant negative correlation over equatorial central Pacific and Black Sea extending zonally up to south Japan. The negative and positive correlations over east Asia indicates that the westerly jet stream strengthens and penetrates to lower latitudes from its normal position during excess years of W2. The easterly anomaly over equatorial central Pacific is indication of weak Walker Circulation related to warm phase of ENSO. The big zonal patch of positive zonal wind anomalies at 80°N represent the strong circumpolar zonal westerly arising of positive phase of AO phenomena. The strong circumpolar westerly does not allow the frigid winter air to spill into northern Eurasia keeping the surface temperature warmer than normal over north Asia.

300-hPa stream-function correlation pattern (Fig. 3.7b) shows strong significant negative correlation over northwest of India and north-west Pacific. In W2, when ENSO is more influencing NWIWP, WDs are originating over Iran, picking moisture from Caspian Sea and north Arabian Sea, moving to NW Pacific. A stronger westerly jet associated with the upper tropospheric cyclonic anomaly is increasing WDs (Hoskins and Ambrizzi 1993; Branstator 2002; Yadav et al. 2009a, b).

References

Agnihotri CL, Singh MS (1982) Satellite study of western disturbances. Mausam 33:249–254
Ambrizzi T, Hoskins BJ, Hsu H-H (1995) Rossby wave propagation and teleconnection patterns in the austral winter. J Atmos Sci 52:3661–3672
Bhaskara Rao NS, Morey PE (1971) Cloud systems associated with western disturbances—a preliminary study. Indian J Meteorol Geophys 22:413–420
Branstator G (2002) Circumglobal teleconnections, the jet stream waveguide, and the North Atlantic Oscillation. J Clim 15:1893–1910. https://doi.org/10.1175/1520-442(2002)015<1893:CTTJS W>2.0.CO;2
Dairaku K, Emori S (2006) Dynamic and thermodynamic influences on intensified daily rainfall during Asian summer monsoon under doubled atmospheric CO2 conditions. Geophys Res Lett 33:L010704. https://doi.org/10.1029/2005GL024754
Ding Q-H, Wang B (2005) Circumglobal teleconnection in the northern hemisphere summer. J Clim 18:3483–3505
Ding Q, Wang B (2007) Intraseasonal teleconnection between the summer Eurasian wave train and the Indian monsoon. J Clim 20:3751–3767
Ding Q, Wang B (2009) Predicting extreme phases of the Indian summer monsoon. J Clim 22:346–363
Dutta RK, Gupta MG (1967) Synoptic study of the formation and movement of western depressions, Indian J. Meteorol Geophys 18:45–50
Enomoto T, Hoskins BJ, Matsuda Y (2003) The formation mechanism of the Bonin high in August. Q J R Meteorol Soc 129:157–178. https://doi.org/10.1256/qj.01.211
Graham NE (1994) Decadal-scale climate variability in the tropical and North Pacific during the 1970s and 1980s: observations and model results. Clim Dyn 10:135–162

Held IM, Panetta RL, Pierrehumbert RT (1985) Stationary external Rossby waves in vertical shear. J Atmos Sci 42: 865–883. https://doi.org/10.1175/1520-0469(1985)042<0865:SERWIV>2.0.CO;2

Hoskins BJ, Ambrizzi T (1993) Rossby wave propagation on a realistic longitudinally varying flow. J Atmos Sci 50:1661–1671

Kripalani RH, Kulkarni A (1997) Rainfall variability over Southeast Asia-connections with Indian monsoon and ENSO extremes: new perspectives. Int J Climatol 17:1155–1168

Kripalani RH, Kulkarni A (1999) Climatology and variability of historical Soviet snow depth data: some new perspectives in snow-Indian monsoon teleconnections. Clim Dyn 15:475–489

Kripalani RH, Kulkarni A, Singh SV (1997) Association of the Indian summer monsoon with northern hemisphere mid-latitude circulation. Int J Climatol 17:1055–1067

Krishnamurti TN, Bhalme HN (1976) Oscillations of monsoon system. Part I: observational aspects. J Atmos Sci 45:1937–1954

Mak M-K (1975) The monsoonal mid-tropospheric cyclogenesis. J Atmos Sci 32:2246–2253

Miller FR, Keshavamurthy RN (1968) Structure of an Arabian Sea summer monsoon system, International Indian Ocean expedition meteorological monographs no. 1. East–West Center Press, Honolulu, 94 p

Mooley DA (1957) The role of western disturbances in the production of weather over India during different seasons. Indian J Meteorol Geophys 8:253–260

Petterssen S (1956) Weather analysis and forecasting, 2nd edn. McGraw-Hill, New York, p 422

Pisharoty PR, Desai BN (1956) Western disturbances and Indian weather. Indian J Meteorol Geophys 8:333–338

Raman CRV, Maliekal JA (1985) A 'northern oscillation' relating northern hemispheric pressure anomalies and the Indian summer monsoon? Nature 314:430–432. https://doi.org/10.1038/314430a0

Raman CRV, Rao YP (1981) Blocking highs over Asia and monsoon droughts over India. Nature 289:271–273

Ramaswamy C (1956) On the sub-tropical jet stream and its role in the development of largescale convection. Tellus 8:26–60

Ramaswamy C (1962) Breaks in the Indian summer monsoon as a phenomenon of interaction between the easterly and the subtropical westerly jet streams. Tellus 14:337–349

Singh MS (1963) Upper air circulation associated with western disturbance. Indian J Meteorol Geophys 1:156–172

Singh MS (1979) Westerly upper air troughs and development of western depression over India. Mausam 30(4):405–414

Singh MS, Kumar S (1977) Study of western disturbances. Indian J Meteorol Hydrol Geophys 28(2):233–242

Trenberth KE, Hurrell JW (1994) Decadal atmosphere–ocean variations in the Pacific. Clim Dyn 9:303–319

Yadav RK (2009a) Changes in the large-scale features associated with the Indian summer monsoon in the recent decades. Int J Climatol 29:117–133. https://doi.org/10.1002/joc.1698

Yadav RK (2009b) Role of equatorial central Pacific and northwest of North Atlantic 2-metre surface temperatures in modulating Indian summer monsoon variability. Clim Dyn 32:549–563. https://doi.org/10.1007/s00382-008-0410-x

Yadav RK (2016) On the relationship between Iran surface temperature and north-west India summer monsoon rainfall. Int J Climatol 36:4425–4438. https://doi.org/10.1002/joc.4648

Yadav RK (2017a) Mid-latitude Rossby wave modulation of the Indian summer monsoon. Q J R Meteorol Soc 143:2260–2271. https://doi.org/10.1002/qj.3083

Yadav RK (2017b) On the relationship between east equatorial Atlantic SST and ISM through Eurasian wave. Clim Dyn 48:281–295. https://doi.org/10.1007/s00382-016-3074-y

Yadav RK, Singh BB (2017) North equatorial Indian Ocean convection and Indian summer monsoon June progression: a case study of 2013 and 2014. Pure Appl Geophys 174(2):477–489. https://doi.org/10.1007/s00024-016-1341-9

Yadav RK, Rupa Kumar K, Rajeevan M (2009a) Increasing influence of ENSO and decreasing influence of AO/NAO in the recent decades over northwest India winter precipitation. J Geophys Res 114:D12112. https://doi.org/10.1029/2008JD011318

Yadav RK, Rupa Kumar K, Rajeevan M (2009b) Out-of-phase relationships between convection over north–west India and warm-pool region during winter season. Int J Climatol 29:1330–1338. https://doi.org/10.1002/joc.1783

Yadav RK, Srinivas G, Chowdary JS (2018) Atlantic Niño modulation of the Indian summer monsoon through Asian jet. npj Clim Atmos Sci 1:23. https://doi.org/10.1038/s41612-018-0029-5

Chapter 4
Projected Climate Change in the Himalayas during the Twenty-First Century

Imtiaz Rangwala, Elisa Palazzi, and James R. Miller

Abstract In this chapter we synthesize our current understanding of projected climate change in the greater Himalayan region during the twenty-first century. This understanding has been constrained by the sparsity of climate observations and relatively greater limitations of our current modeling framework to represent the complex topographical influence of this vast high elevation region. Here, we examine studies that have analyzed global and regional climate model experiments for the greater Himalayan region to assess and quantify (a) future increases in temperature and how this warming trend varies with elevation, (b) climate feedbacks that amplify the warming in these high mountain regions, (c) changes in large-scale circulation that transport moisture and energy into the region, and (d) the implications from all of the above on the nature of precipitation, i.e., phase, amount and extremes, and the fate of its cryosphere. Wherever plausible, we compare these model projections with observations from recent decades to better constrain, as well as further improve, our understanding of the perceived hydroclimatic changes in this region during the twenty-first century.

4.1 Introduction

The greater Himalayan region encompasses the Indian Himalayas, the Hindu Kush, the Karakoram, the Pamir and the Tibetan Plateau (see Fig. 4.1) – also referred to as the Hindu Kush-Himalaya (HKH) region by the International Centre for Integrated

I. Rangwala (✉)
Cooperative Institute for Research in Environmental Sciences, University of Colorado, Boulder, CO, USA

Physical Sciences Division, NOAA, Boulder, CO, USA
e-mail: imtiaz.rangwala@colorado.edu

E. Palazzi
Institute of Atmospheric Sciences and Climate (ISAC-CNR), Torino, Italy

J. R. Miller
Department of Marine and Coastal Sciences, Rutgers University, New Brunswick, NJ, USA

© Springer Nature Switzerland AG 2020
A. P. Dimri et al. (eds.), *Himalayan Weather and Climate and their Impact on the Environment*, https://doi.org/10.1007/978-3-030-29684-1_4

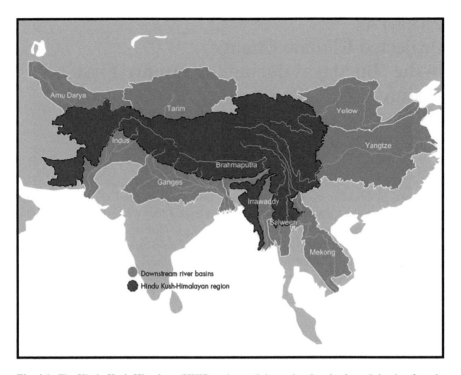

Fig. 4.1 The Hindu Kush-Himalaya (HKH) region and the major river basins originating from it. Reproduced from Schild 2008. © Copyright 2008 International Mountain Society and United Nations University

Mountain Development (ICIMOD). This region is a vast expanse of high elevation terrain and permanent ice cover with a varying degree of topographic complexity, orientation and exposure to the large-scale atmospheric flows of energy and moisture. The HKH region is the source of several large Asian rivers, identified in Fig. 4.1, that provide water, ecosystem services and basis for the livelihood of more than 200 million people directly. When including their downstream basins, these rivers support about a fifth of the world's population (Schild 2008).

Global and regional climate model (GCM and RCM) simulations have been analyzed for the HKH region to reproduce specific aspects of the current and past climate, perform process-oriented evaluations, and make projections of future climate evolution. GCMs provide useful insights on physical processes driving changes in temperature and other hydroclimatic variables including precipitation, snow, soil moisture and atmospheric water vapor from the anthropogenic climate forcings. The use of GCMs allows different processes occurring in the HKH region to be examined in relation to large-scale patterns of atmospheric variability and teleconnections, such as El Niño-Southern Oscillation (ENSO) and North Atlantic Oscillation (NAO), occurring elsewhere in the world. The large extent of high elevation terrain of the HKH region also helps in evaluating these changes and the mechanisms driving them as a function of elevation, even in coarse resolution climate models (Rangwala et al. 2016).

Because of the coarse vertical and horizontal model resolution, GCMs do not adequately represent the regional-scale forcing from steep topography (Palazzi et al. 2015) that influences processes such as orographic lifting and moisture entrainment. Such processes ultimately shape the observed precipitation climatology and its spatial variability in the HKH region. One study performed with GCMs from the latest Climate Model Intercomparison Project phase 5 (CMIP5), for example, showed that the GCMs have different performances in simulating the precipitation climatology (e.g., the climatological annual cycle) in different sub-regions of the HKH (Palazzi et al. 2015). In particular, the models more coherently simulate the precipitation annual cycle in the monsoon-dominated eastern Himalayan region. However, the model spread is considerably greater in the western areas where the precipitation cycle is influenced by both the westerly winds and the Indian Monsoon. Sperber and Annamalai (2014) and Sperber et al. (2013) investigated how well GCMs were able to represent monsoon behavior in the HKH region. They found that it is difficult to determine, within a given ensemble, the "best performing" GCM, i.e., the ability of one single model to represent best all different aspects of the monsoon, such as the annual cycle climatology, the interannual and intraseasonal variability, time of onset and other characteristics. Since no individual model can be labelled as the best, results from the whole ensemble are usually taken into account and summarized through the multi-model ensemble mean (or median) and the inter-model standard deviation. Even so, the multi-model mean should be regarded with caution when the ensemble is characterized by large spread and noticeable differences are found in the individual model outputs (Tebaldi and Knutti 2007; Palazzi et al. 2015). More recently, there are also several studies that have used RCMs to examine climate change in the HKH region; typically employing resolutions in the range of 10–50 km for hydrostatic RCMs (see also Chap. 6 in Part I). In most cases, these regional models are still too coarse to capture finer-scale topographic forcings. Mostly owing to the large computation effort required, the use of non-hydrostatic RCMs with resolutions up to 1 km is still in its infancy for climatic applications, while being common for shorter timescales, from days to years, or to simulate particular meteorological extreme events (e.g. Norris et al. 2015).

In the following sections, we discuss results from studies that have examined projections from both global and regional models for the HKH region, or sub-regions within it. We synthesize the emerging understanding on this topic over the last two decades, and specifically bring in insights gained from the analyses of the latest generation of GCMs (CMIP5 experiments) and regional modeling experiments focused on the HKH region. Section 4.2 discusses projections for temperature change, Sect. 4.3 examines selective feedbacks that amplify the greenhouse forced warming response in the HKH region, Sect. 4.4 discusses projected changes in the large-scale circulations relevant to the HKH region and Sect. 4.5 reviews implications of projected climate change on precipitation, snow and streamflow.

4.2 Projected Warming

Since the latter half of the twentieth century, high rates of warming have been reported for different parts of the HKH region (Shrestha et al. 1999; Liu and Chen 2000; Kothawale et al. 2010; Ohmura 2012; Yan and Liu 2014) which are significantly greater than the global average warming rates, particularly in the colder seasons, suggesting increased sensitivities for this region to the anthropogenic greenhouse warming. Climate model experiments have also corroborated such relatively higher warming rates in the HKH region (Liu et al. 2009; Rangwala et al. 2013; Palazzi et al. 2017).

Based on various studies examining global (Shrestha et al. 1999; Liu et al. 2009; Kang et al. 2010; Rangwala et al. 2010) and regional (Kulkarni et al. 2013; Sanjay et al. 2017) climate model projections for the HKH region, mean annual temperature is projected to increase in the range of 1–4 °C by mid-twenty-first century and 2–6 °C by late-twenty-first century relative to the late-twentieth century (see Fig. 4.2). These projections are based on modeling experiments that consider increasing atmospheric greenhouse gas concentration during the twenty-first century from moderate or high-end emission scenarios. Although the trajectory of the anthropogenic greenhouse gas emissions scenario during the twenty-first century will largely control the warming response by the end of the twenty-first century,

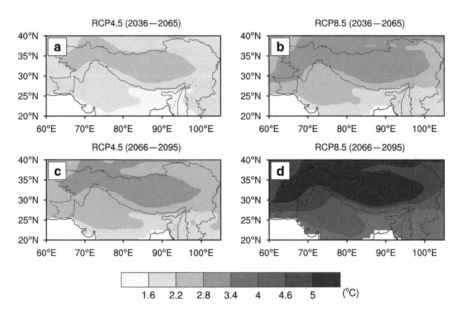

Fig. 4.2 Projected increases in surface air temperature in the HKH region by mid (2036–2065) and late (2066–2095) twenty-first century relative to the 1976–2005 period for the RCP4.5 (left) and RCP8.5 (right) emissions scenarios. The lighter boundary on the map demarcates the HKH region. Results are based on the ensemble mean of 21 CMIP5 models. Reproduced from Wu et al. 2017. Copyright © 2017, National Climate Center (China Meteorological Administration)

large uncertainties in the warming trends during the twenty-first century also arise from the differences in the climate models, i.e., their climate sensitivities and biases in representing the region's climate (Hawkins and Sutton 2009). Overall for the HKH region, the projected increases in the daily minimum temperatures are greater than in the daily maximum temperatures on an annual basis (Rangwala et al. 2013; Panday et al. 2015; Wu et al. 2017). Seasonally, the projected increases in the daily minimum temperatures are greater during winter and spring, while the opposite is true during summer and autumn (Rangwala et al. 2013).

There has also been significant research into examining the phenomenon of elevation dependent warming (EDW) in the HKH region. In general, observations (Liu and Chen 2000; Qin et al. 2009; Rangwala et al. 2009; Ohmura 2012; Wang et al. 2014) and models (Liu et al. 2009; Rangwala et al. 2013, 2016; Palazzi et al. 2017) suggest an amplified warming response at higher elevations (Fig. 4.3). Models also suggest that such an amplification of warming at higher elevations will further increase during the twenty-first century and is proportional to the amplitude of anthropogenic greenhouse radiative forcing (Rangwala et al. 2013). Chapter 9 in Part I examines projections of EDW in the Indian Himalayan region from a high-resolution model.

The relationship between projected warming rates and elevation is not necessarily linear. Enhanced warming rates can occur at intermediate elevations that are in the vicinity of the 0 °C isotherm, therefore influenced by strong positive feedbacks from snow and ice cover changes (Pepin and Lundquist 2008; Ceppi et al. 2012; Palazzi et al. 2017). When examining seasonally, the increasing trends in the daily minimum and maximum temperatures with elevation also vary significantly owing

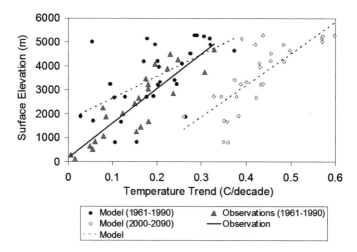

Fig. 4.3 Observed (1961–1990; from Liu and Chen 2000) and modeled (1961–1990 and 2000–2090; from the GISS AOM model analyzed in Rangwala et al. 2010) trends in surface temperature (°C/decade) in the Tibetan Plateau as related to the elevation of the observing station and the model grid, respectively. Reproduced from Rangwala and Miller 2012. © Springer Science+Business Media B.V. 2012. Part of Springer Nature

to their relationship to the components of the surface energy balance. This has motivated several studies to examine them separately and relate their response to specific physical processes and feedbacks as discussed in the next section (e.g., Rangwala et al. 2016; Palazzi et al. 2017).

4.3   Important Feedbacks Influencing Projected Warming

The high elevation regions such at the HKH are expected to have higher sensitivities to the anthropogenic greenhouse gas forcing, i.e. experience greater warming relative to other land areas particularly at comparable latitudes (e.g., Rangwala et al. 2016). These higher sensitivities occur largely because of various climate feedbacks that exist in high elevation regions and/or are relatively stronger there. Pepin et al. (2015) comprehensively explores these various feedbacks. In this section, we discuss two of them – (i) Snow/Ice Albedo Feedback and (ii) Downward Longwave Radiation-Water Vapor Feedback – which are discussed more extensively in the literature, and arguably contribute more strongly to the high sensitivities of high elevation regions to global climate change, particularly during the colder seasons.

4.3.1   Snow/Ice Albedo Feedback

The snow/ice albedo feedback loop is one of the strongest feedbacks in the climate system and works as follows: increasing temperatures melt snow, which reduces the surface albedo, thus leading to the absorption of more solar radiation which enhances the initial temperature increase. This positive feedback is strongest during the transition seasons in regions where snow does not persist year-round but that are cold enough to experience subzero temperatures for a significant period of time. Reviewing our current state of understanding into surface albedo feedback (SAF), Thackeray and Fletcher (2016) point out that although the SAF appears to be small on a global scale (amplitude ~ $0.1\ Wm^{-2}\ K^{-1}$), it is regionally important in the northern hemisphere over land where observation-based estimates suggest a peak feedback of about 1% decrease in surface albedo per degree of warming in the spring. They also suggest that the current generation of climate models do SAF well although some still use outdated parameterizations, and new regional models and large ensembles open up new insights into SAF.

Analysis of high elevation station data indicates that maximum warming rates tend to occur near the annual 0 °C isotherm. This is where snow cover decreases most rapidly, and studies suggest that snow/ice albedo feedbacks have a role in these enhanced warming rates often found there (Pepin and Lundquist 2008; Ceppi et al. 2012; Ohmura 2012). The specific temperature response from these feedbacks (i.e. increases in the daily minimum vs. maximum temperature) will depend on soil moisture conditions and how the excess sunlight absorption is balanced by the surface energy fluxes. Although minimum temperatures have generally increased and are projected to increase at a faster rate than maximum temperatures over most of the Tibetan Plateau (e.g., You et al. 2008; Liu et al. 2009; Wu et al. 2017), maximum temperatures appear to be increasing faster at some of the highest elevation sites on the edges of the Plateau. Kothawale et al. (2010) found that maximum temperatures were increasing at a faster rate than minimum temperatures in the western Himalayas between 1971 and 2007. Shrestha et al. (1999) suggested that the faster increase in maximum temperatures than minimum temperatures in the Nepal Himalayas was related to the combined effects of the snow/albedo effect and changes in the monsoonal circulation. Kattel and Yao (2013) also found that maximum temperatures have been increasing faster than minimum temperatures on the southern slopes of the central Himalayas during the last three decades. A recent model study by Yan et al. (2016) quadrupled atmospheric carbon dioxide concentrations and showed that the snow/albedo effect was, in part, responsible for EDW at higher elevations in the Tibetan Plateau. These studies are consistent with others that have found that the snow cover season has become shorter and that more precipitation is now falling as rain (Rikiishi and Nakasato 2006; Archer and Fowler 2004; Bhutiyani et al. 2010). Multidecadal observations of temperature and snow cover from satellite retrievals show that the warming rates tend to increase with elevation between 3000 and 4800 m, but tend to stabilize above that (Qin et al. 2009) suggesting an influence of snow/ice albedo feedbacks.

Climate models consistently elucidate the significant role played by snow/ice albedo feedbacks in driving warming trends in mountain regions (Giorgi et al. 1997; Rangwala et al. 2013; Minder et al. 2016). However, one concern related to GCMs is that they have used significantly different parameterizations of the albedo of snow-covered surfaces. The strength of snow/ice albedo feedbacks depends on the difference between the albedo of snow and that of the land surface beneath the snow. Models with larger differences have stronger feedbacks. Qu and Hall (2007) examined a suite of global models and found that this feedback strength could vary by a factor of three among the different models. Thackeray and Fletcher (2016) suggest that recent models do include improved parameterizations of snow albedo.

Ghatak et al. (2014) examined two CMIP5 GCM projections for the Himalayas and found the greatest increases in temperature during the twenty-first century occur at the highest elevations in the southwestern Tibetan Plateau associated with decreases in snow cover and corresponding increases in the absorption of incoming radiation. They also found the largest warming near the freezing line as it moves upward during the twenty-first century. A more recent study by Palazzi et al. (2017) which examined 27 global climate models for both historical and future climate trends showed enhanced rates of warming of both maximum and minimum temperatures at higher elevations in the Tibetan Plateau/Himalayan region. They also investigated several potential feedbacks and concluded that the snow/ice albedo feedback was the dominant mechanism causing enhanced warming at higher elevations with secondary effects from water vapor and cloud feedbacks.

Another factor that affects the snow/ice albedo effect in these mountains is the presence of aerosols. During the boreal spring, dust from deserts and regionally-emitted black carbon can be found in the atmosphere against the foothills of the Himalayas and Tibetan Plateau. Ramanathan and Carmichael (2008) suggested that this black carbon could be responsible for half the total warming there during the last several decades. Lau et al. (2010) investigated this effect with a global climate model and found that the black carbon layer absorbs solar radiation and warms the mid-troposphere which in turn causes snow to melt faster and enhances surface warming. Xu et al. (2009) suggested that atmospherically deposited black carbon can increase the absorption of visible radiation by 10–100% in the Tibetan glaciers. This is consistent with Ming et al. (2015) who used MODIS satellite retrievals and found that glaciers in the Hindu Kush, Karakoram, and Himalayas have been darkening since 2000, with the most rapid darkening above 6000 m. Black carbon can also cause decreases in cloud cover and affect the radiation budget at the surface (Hansen et al. 1997).

4.3.2   *Downward Longwave Radiation-Water Vapor Feedback*

Of all the atmospheric greenhouse gases that contribute to surface warming, water vapor is the most significant. The water vapor feedback works as follows: an initial increase in surface temperature leads to increases in atmospheric water vapor which increases the absorption of longwave radiation emitted from the Earth's surface and, thereby, increases the downward longwave radiation (DLR) back to the surface thus enhancing the initial temperature perturbation. Although increases in DLR occur in conjunction with globally increasing specific humidity, the sensitivity of DLR to changes in water vapor is non-linear with the sensitivity being relatively higher when the initial water vapor concentration is very low as found in the polar regions and at high elevations during the cold season.

There is evidence that increasing water vapor is partially responsible for higher winter warming rates in upland areas globally in both observations (e.g., Ruckstuhl et al. 2007, 2009) and climate models (e.g., Rangwala et al. 2010, 2013, 2016). Research suggests that increases in surface specific humidity (q) are partly responsible for an increase in surface warming across central Europe (Philipona et al. 2005) and the Tibetan Plateau (Rangwala et al. 2010; Ghatak et al. 2014; Palazzi et al. 2017) in the late twentieth century. These studies indicate that increases in specific humidity are linked with increases in DLR and lead to surface warming. Using observations from four different elevations in the Swiss Alps, Ruckstuhl et al. (2007) found that the sensitivity of DLR to changes in q is particularly high when q is below 5 g/kg, conditions which occurred more often at the higher elevations in winter. Rangwala et al. (2010) did a similar analysis using a global climate model simulation for the Tibetan Plateau and found a relationship similar to that of Ruckstuhl et al. (2007), namely greater sensitivities at higher elevations, particularly during the cold season. Similar relationships between DLR and specific humidity, under both clear and cloudy sky conditions, have also been found when examining high-elevation observations from the Colorado Rocky Mountains (Naud et al. 2013).

4.4   Changes in Large-Scale Circulation

The central and south-eastern sections of the HKH region receive their moisture primarily from the Indian Monsoon between May and September, the eastern section from the East Asian monsoon between June and August, while the northwestern and northern sections get a large proportion of their moisture from the mid-latitude westerly extratropical storms during winter and spring (Yao et al. 2012; Maussion et al. 2014; Rangwala et al. 2015). These different circulation patterns and associated moisture flows influence many aspects of the hydrological cycle in the area, such as the amount and distribution of precipitation, snow cover, the dynamics of glaciers and, as a consequence, the hydrological regimes, thus having an impact downstream. For example, glaciers in the Karakoram receive their input mainly in winter and early spring as snow, carried by moisture-laden westerly winds and melt in summer, while the dynamics of eastern Himalayan glaciers is controlled by the summer monsoon and there is almost no water input in winter.

4.4.1   Asian Monsoons

The Asian summer monsoons are the primary source of moisture and precipitation to much of the central, southern and eastern part of the HKH region. The Indian summer monsoon, part of the broader Asian-Australian monsoon system, also contributes to precipitation in the western part of the HKH even though about two-thirds of snowfall accumulation in this area occurs in winter and spring when mid-latitude westerlies dominate (e.g., Hewitt 2007).

There is evidence that GCMs have become better at simulating the Asian-Australian monsoon system, however they are still poor at modeling effects of large-scale teleconnections (e.g., El Niño-Southern Oscillation or ENSO) on the monsoons (Kitoh et al. 2013; Christensen et al. 2013; see also Chap. 2 in Part I that discusses impacts of ENSO phases on the summer monsoon precipitation in the Himalayas). They also have significant difficulty in reproducing the observed precipitation climatology, including the HKH region and in particular its westernmost part where the interaction of the monsoon with westerly winds occurs (Palazzi et al. 2015), and there is a large scatter across the models for projected change (Kitoh et al. 2013). Simulating the mean summer monsoon precipitation in the western HKH, such as in the upper Indus Basin and its vicinity, with state-of-the-art global climate models, still represents a major challenge (Palazzi et al. 2015; Hasson et al. 2016).

CMIP5 projections show that the strength of monsoon circulation weakens during the twenty-first century because of an increase in the condensational heating of the upper troposphere over the tropics such that the upper troposphere over the Indian Ocean warms more than that over the Tibetan Plateau thereby reducing the meridional thermal gradient between the land and ocean (e.g., Christensen et al.

2013). However, the overall precipitation from the monsoon increases because of the enhancement of moisture convergence from increased column water vapor and surface evaporation which offsets the weakening of the monsoon circulation (Kitoh et al. 2013; Christensen et al. 2013). As a consequence, CMIP5 models show increases in summer precipitation in the Himalayas during the twenty-first century (e.g., Palazzi et al. 2015). CMIP5 models project an increase in mean precipitation and in its interannual variability and extremes. Further analyses performed with one state-of-the-art GCM (EC-Earth, Hazeleger et al. 2012) showed that the positive trend in summer precipitation projected for the Himalayas through the twenty-first century is associated with a negative trend in the number of wet days and a positive trend in precipitation intensity, indicating a likely transition toward more episodic and intense monsoonal precipitation in the future (Palazzi et al. 2013). Overall, the monsoon season is expected to become longer than in the current climate, and with larger interannual variability in monsoon intensity (e.g. Sharmila et al. 2015). Regional model experiments, forced with boundary conditions from a selection of CMIP5 GCMs, show similar results for increases in the summer precipitation over much of the Himalayas and the Tibetan Plateau (Kulkarni et al. 2013; Sanjay et al. 2017). One recent study based on 13 regional model simulations found, on average, a 22% increase in the summer monsoon precipitation over the southeastern Himalayas and much of the Tibetan Plateau by the end of the twenty-first century under RCP8.5 scenario, with a large consensus among the models on this direction of change (Sanjay et al. 2017; see also Fig. 4.4).

4.4.2   Mid-Latitude Westerlies

The mid-latitude extratropical storms during winter and spring are a primary source of moisture to the northwestern and northern parts of the HKH region. The strength of these westerlies, which is correlated to the phase of the North Atlantic Oscillation (NAO) (e.g., Thompson and Wallace 2001; Filippi et al. 2014), determines the amount of winter precipitation and snowfall accumulation in the Karakoram glaciers. In particular, several observational studies have shown that above normal precipitation in winter in the western HKH is typically recorded during the positive NAO phase (Archer and Fowler 2004; Yadav et al. 2009; Syed et al. 2006). During this phase westerly winds are intensified in the region of the Middle East jet stream, from North Africa to southeastern Asia, in particular at longitudes between 40° and 70°E where most of the moisture transport to the western HKH takes place. In addition to this, during positive NAO phases, enhanced evaporation occurs from the Persian Gulf, the northern Arabian Sea, and the Red Sea (the main moisture sources for precipitation in the western HKH) owing to higher surface wind speed. Enhanced moisture and stronger westerlies result in enhanced moisture transport to the western HKH region (Filippi et al. 2014).

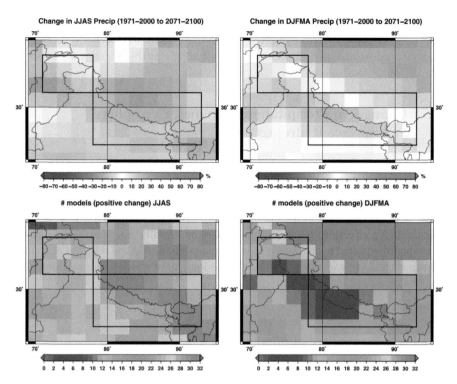

Fig. 4.4 (**Top Row**) Projected change in summer (left) and winter (right) precipitation (%) by 2071–2100 (RCP 8.5 scenario) relative to 1971–2100 for the multi-model mean of the CMIP5 ensemble. (**Bottom Row**) For each 2° × 2° pixel, number of models (out of 32) showing increases in future precipitation. Reproduced from Palazzi et al. 2015. © Springer-Verlag Berlin Heidelberg 2014. Part of Springer Nature. **Disclaimer:** The boundaries shown in the figure do not necessarily correspond to the actual political boundaries

As compared to the summer monsoons, GCMs show less agreement on the sign of the projected change for winter precipitation (Palazzi et al. 2015; see Fig. 4.4). Moreover, simulating the changes in precipitation over the Karakoram and north-western Himalayas, where precipitation is associated with both summer monsoon and winter-early spring westerly storms, is more challenging than for the eastern HKH, both using GCMs and RCMs.

During the latter half of the twentieth century, there is some evidence that winter precipitation has a small upward trend in the Upper Indus Basin (Archer and Fowler 2004), but no trend in the HKH region southeast of it (Bhutiyani et al. 2010). This could be part of a precipitation trend more broadly observed in the central Asian region, where the northern parts are experiencing an increase in precipitation while the southern regions are seeing a decrease as shown by Chen et al. (2011), who suggest this pattern of change is likely caused by a northward shift of the subtropical high and westerly jet. Huang et al. (2014) also found increases in precipitation in the northern parts of the HKH region (i.e., northern Central Asia, Tian Shan Mountains, and northern Tibet) from a selective analysis of CMIP5 projections caused by strengthened water vapor transport and enhanced precipitable water.

4.5   Changes in Precipitation Extremes, Cryosphere and Surface Hydrology

Precipitation is affected by strong intermittency at all scales and has strong orographic dependence, which makes it hard to model and to measure. The HKH region poses additional challenges because different sub-regions are exposed to different large-scale circulation patterns and associated moisture convergence which can trigger, through convection, the occurrence of extreme rainfall events. Further, increased warming is expected to affect the phase of precipitation, size and duration of annual snowpack and snow quality, and glacial mass loss, all of which will impact the timing of streamflow and ecosystem response.

4.5.1   Extreme Precipitation

Driven by heavy rainfall during the summer monsoon, a number of catastrophic flooding events have occurred in recent years (i.e., since 2010) in different parts of the Himalayan region – in particular over northern Pakistan, the Indus basin, and along the western flank of the South Asian Monsoon. Among them, the devastating flood in Pakistan in 2010 and the Uttarakhand flood in India in 2013 have become exemplary cases for the research on extreme rainfall (rainfall exceeding 100 mm day^{-1}) in the Himalayas and its possible causes. The different studies that have analyzed storms that have occurred in the western HKH in recent decades, for example, point toward a common primary cause of such events: the formation of anomalous moisture supply over the area (e.g., Rasmussen et al. 2015; Houze et al. 2011; Krishnamurti et al. 2017; Kumar et al. 2014). The study by Priya et al. (2017), who analyzed extreme precipitation events in historical rainfall records since the 1950s in the western Himalayas, suggests that the changes in the background circulations have likely facilitated increased activity of cyclonic troughs over the western Himalayas and enhanced input of moisture from the Arabian sea (one of the main moisture sources for the western HKH, see e.g., Filippi et al. 2014). In another study analyzing precipitation data measured during the period 1961–2012, Zhan et al. (2017) found that the amount and frequency of intense precipitation (precipitation above the 90th percentile) significantly increased over the Tibetan Plateau, while a more heterogeneous signal of change is found over the remainder of the HKH. The maximum 1-day, 3-day, and 5-day precipitation amounts averaged over the entire HKH, as well as consecutive wet-days in most parts of the region, all showed significant positive trends since the 1960s.

Global climate models indicate that precipitation patterns are likely to change in the HKH in the future, with changes of different magnitude across the whole area. Several studies indicate that the frequency and intensity of extreme rainfall events are generally projected to increase across the region. For example, Wu et al. (2017) found a future increase of extreme precipitation in the HKH, particularly in the central and northern Tibetan Plateau, with greater intensification in the RCP8.5 scenario than in the RCP4.5 scenario, corroborating other studies (e.g., Panday et al. 2015; Yang et al. 2012; Li et al. 2015; see Fig. 4.5). By analyzing extreme precipitation indices (5-day precipitation, precipitation greater than the 95th percentile, precipitation intensity) in CMIP3 and CMIP5 GCMs, Panday et al. (2015) suggest that more frequent extreme rainfall is expected during the summer monsoon throughout the twenty-first century, which confirms the behavior already observed in the recent decades in some portions of the area (e.g., Zhan et al. 2017). Sanjay et al. (2017) analyzed climate change projections for the HKH region using regional climate models from the Coordinated Regional Climate Downscaling Experiment (CORDEX) and also found an intensification of about 22% of the precipitation associated with the summer monsoon in the southeastern Himalayas and in the Tibetan Plateau.

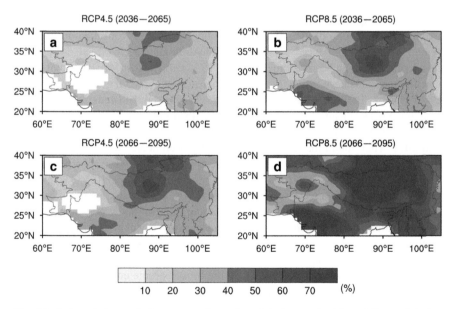

Fig. 4.5 Projected changes in the percent of annual precipitation received from daily precipitation exceeding the historical (1976–2005) 95th percentile value for daily precipitation (R95p) in the HKH region by mid (2036–2065) and late (2066–2095) twenty-first century for the RCP4.5 (left) and RCP8.5 (right) emissions scenarios. The lighter boundary on the map demarcates the HKH region. Results are based on the ensemble mean of 21 CMIP5 models, and only the changes significant at 95% level are shown. Reproduced from Wu et al. 2017. Copyright © 2017, National Climate Center (China Meteorological Administration)

4.5.2   Changes to Snow Processes and Glacial Mass

The ongoing and projected warming will influence the depth, quality and timing of mountain snowpack, although this response will be complex across the HKH region because of the presence of large elevation gradients. The atmospheric warming will increase the freezing level height which means that, all else being equal, more precipitation will fall as rain rather than snow. Lower elevations are particularly vulnerable to this phenomenon where we are likely to see a shrinking of the snowfall season. Bradley et al. (2009) find that freezing level height in the tropics has been increasing at the rate of ~2 m/year between 1948 and 2007. Bhutiyani et al. (2010) suggest that in recent decades a large proportion of winter precipitation has been falling as rain instead of snow on the windward side of the northwestern Indian Himalayas. Other studies have also noted this shift from snow to rain in the HKH region (Rikiishi and Nakasato 2006; Archer and Fowler 2004). More precipitation occurring as rain will reduce the snowpack thickness in wide sections of the HKH region, except those that are situated above the atmospheric freezing level and experience an increase in precipitation. Shallow snowpack will tend to melt earlier, and this melting will be further augmented by warming temperatures and atmospheric deposition of absorbing aerosols. Chapter 16 in Part III discusses observed changes in the cryosphere for northwest and central Himalayan regions.

Another important implication of climate change is the reduced snow cover extent and shortening of the season with ground snow cover for much of the HKH region which will influence both the regional hydrological cycle (see also Chaps. 19 and 20 in Part III) and ecosystem (see Chap. 19 in Part III) response. In recent years, satellite retrievals have begun to provide a multi-decadal record of albedo and snow cover changes in high elevations. Ming et al. (2015) used the Moderate Resolution Imaging Spectrometer (MODIS) to examine albedo changes over glaciers in the greater Himalayan region (Himalayas, Karakoram and Hindu Kush) between 2000 and 2011. They found that albedo has generally been decreasing (largest decrease above 6000 m) with a corresponding increase in surface shortwave absorption. They also found a mass loss for Himalayan glaciers estimated to be 10.4 GT/year that may contribute to about 1% of recent sea level rise. Other studies using MODIS over the Tibetan Plateau have found that the duration for snow persistence varies with elevation and tends to be longer at high elevations (Pu et al. 2007). Future projections performed using CMIP5 GCMs with spatial resolution in the range of 0.75°–1.25° indicate a significant decrease in winter snow depth in the HKH region by the end of the twenty-first century. In the most extreme emission scenario, the western part of the area is expected to undergo a reduction of about 40% with respect to current conditions, while toward the east, the Himalayas will face an even stronger decrease of up to 50% and a shift in the snow depth maximum from March to February, as reported by Terzago et al. (2014) (see Fig. 4.6).

Climate warming is also generally expected to decrease glacial mass across the HKH region during the twenty-first century (Barnett et al. 2005). Recent observations suggest that most HKH glaciers are losing mass at rates similar to those found

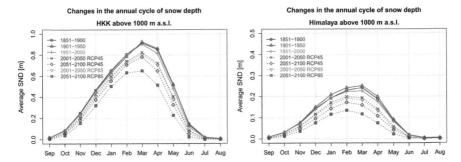

Fig. 4.6 Monthly mean values of snow depth, SND (m), in the Hindu Kush & Karakoram (left) and Himalayas (right) region above 1000 m mean sea level, for 50-year time slices in the historical and future periods. These results are based on the average of five high resolution (0.75°–1.25° zonal resolution) CMIP5 GCMs (CMCC-CM, EC-EARTH, BCC_CSM1.1(m), MRI-CGCM3, and CESM1-BGC), which include both RCP4.5 and RCP8.5 emission scenarios for the twenty-first century projections. Reproduced from Terzago et al. 2014. © Copyright 2014 American Meteorological Society

elsewhere globally, with the exception of the Karakoram where there is evidence for mass gain and expansion (Bolch et al. 2012; Fowler and Archer 2006). Given the projected increase in temperatures, continued and possibly accelerated mass loss from glaciers in the HKH is likely. For example, the study by Kraaijenbrink et al. (2017) found projected HKH's ice mass losses of 49 ± 7%, 51 ± 6% and 64 ± 5% by the end of the twenty-first century for the RCP4.5, RCP6.0 and RCP8.5 emission scenarios, respectively. Chapters 17 and 18 in Part III further examine impacts of climate change on Himalayan glaciers.

4.6 Conclusions

The greater Himalayan region is expected to have relatively higher sensitivities to a changing climate during the twenty-first century. Because of its vast extent, complex topography and elevation gradients, and varied interactions with different large-scale atmospheric circulations, the projected changes in temperature and precipitation, and their response on the cryosphere and surface hydrology will vary spatially. The region is projected to warm between 2 and 6 °C by the end of this century. The monsoonal precipitation in the region is generally expected to increase, although the penetration of monsoonal flow into the region may become somewhat attenuated. There is relatively less agreement on how the westerly storm-tracks will change, although there is some indication from both observations and models that they could shift northward, and create a north-south gradient in precipitation such that the northern part of the HKH region will see excess moisture while the southern part will experience a deficit. The warming trend will generally enhance precipitation extremes and the ratio of rainfall-to-snowfall, reduce glacial mass, snowpack depth and extent and duration of ground snow cover, and consequently influence stream and subsurface flow.

Simulating future hydroclimatic conditions in orographically complex regions such as the HKH has proven to be a challenging task, with uncertainties arising from multiple sources. One source is the use of empirical parameterizations through which the processes occurring at scales below the model grid size – such as radiation, heat transfer, cloud microphysics, convection, land surface processes – are incorporated in the models. Parameterizations approximate the reality and inevitably lead to uncertainty. Therefore, while enhancing the spatial model resolution of both GCMs and RCMs would allow to capture finer-scale details, improving model parameterizations, particularly those involving convection, surface processes and feedbacks in high-altitude regions will improve simulations of hydroclimate conditions in HKH. Another important source of uncertainty is the representation of land-use and aerosol loading at regional to local scales. Uncertainty in how and to what degree important large-scale circulations affecting the region (i.e., the monsoons and the westerly storm tracks) will change is significant enough to better constrain the projections of changes in the region's hydrologic cycle, and its spatial variation. Lastly but crucially, improved observational networks, both ground based and remotely sensed (e.g., see Chaps. 1 and 8 in Part I), are needed to better constrain our models and enhance our process level understanding of the region's changing hydroclimate.

HKH vs. Polar Regions: Sensitivities to Global Warming

The cryosphere is among the components of the climate system most affected by global warming. One reason for this is that there are positive feedbacks in the cryosphere, such as the ice-albedo feedback, that amplify existing warming trends. The HKH region is often called "The Third Pole" because it has the largest extent of permanent ice outside the two poles (Yao et al. 2012). In addition, it has cold and dry conditions similar to those found in the Arctic and the Antarctic. As such, scientists expect the sensitivities of the HKH region to anthropogenic climate change to have similarities with those in polar regions. A recent study by Wang et al. (2016) examines about 3000 global station observations to investigate whether high-elevation temperature amplification has similarities to Arctic amplification. Between 1961 and 2000, they found that warming rates in regions below 60°N increase with elevation and are significantly higher at higher elevations than the global average. For elevations below 500 m, they found that warming rates increase with increasing latitude and are significantly higher at higher latitudes than the global average, particularly in the Arctic. They suggest that the Arctic amplification is an integrated part of this latitudinal amplification trend. These similarities between altitudinal and latitudinal warming rates likely arise, in part, because some of the same positive feedbacks are found at high elevations and high latitudes, including those related to reduced snow and ice cover, to increases

(continued)

in atmospheric humidity, and to the Stefan-Boltzman law that intensifies temperature increase in colder climates (Pepin et al. 2015). There is some uncertainty in temperature trends in both the Arctic and at high-elevations because of the relatively sparse network of surface stations there, although satellite retrievals have helped fill in some of the gap. There is arguably even greater uncertainty regarding future projections of climate change in the HKH region because it is not adequately represented in most state-of-the-art climate models, and there is limited climate monitoring to provide high confidence in the magnitude of the recent historical warming, although there are indications for fast-paced changes (Qiu 2008). There is significant interest today on the implications of Arctic amplification on both Arctic circulation as well as the rest of the northern hemisphere, and we posit that rapid climate change in the HKH region, and potentially other high-elevation regions, will have consequences both locally and for global scale circulation, with possibilities for unforeseen impacts.

Acknowledgements IR acknowledges support from the Western Water Assessment and Cooperative Institute for Research in Environmental Sciences at the University of Colorado, Boulder. JRM acknowledges support from the New Jersey Agricultural Experiment Station and the USDA-National Institute for Food and Agriculture, Hatch project number NJ32103.

References

Archer DR, Fowler HJ (2004) Spatial and temporal variations in precipitation in the Upper Indus Basin, global teleconnections and hydrological implications. Hydrol Earth Syst Sci Discuss 8(1):47–61

Barnett TP, Adam JC, Lettenmaier DP (2005) Potential impacts of a warming climate on water availability in snow-dominated regions. Nature 438(7066):303–309

Bhutiyani MR, Kale VS, Pawar NJ (2010) Climate change and the precipitation variations in the northwestern Himalaya: 1866–2006. Int J Climatol 30(4):535–548

Bradley RS, Keimig FT, Diaz HF, Hardy DR (2009) Recent changes in freezing level heights in the tropics with implications for the deglacierization of high mountain regions. Geophys Res Lett 36(17)

Bolch T, Kulkarni A, Kääb A, Huggel C, Paul F, Cogley JG, Frey H, Kargel JS, Fujita K, Scheel M, Bajracharya S (2012) The state and fate of Himalayan glaciers. Science 336(6079):310–314

Ceppi P, Scherrer SC, Fischer AM, Appenzeller C (2012) Revisiting Swiss temperature trends 1959–2008. Int J Climatol 32(2):203–213

Chen F, Huang W, Jin L, Chen J, Wang J (2011) Spatiotemporal precipitation variations in the arid Central Asia in the context of global warming. Sci China Earth Sci 54(12):1812–1821

Christensen JH, Kanikicharla KK, Marshall G, Turner J (2013) Climate phenomena and their relevance for future regional climate change. In: Stocker TF et al (eds) Climate change 2013: the physical science basis. Working Group I contribution to the Fifth Assessment Report of the Intergovernmental Panel on Climate Change. Cambridge University Press, Cambridge

Filippi L, Palazzi E, von Hardenberg J, Provenzale A (2014) Multidecadal variations in the relationship between the NAO and winter precipitation in the Hindu Kush–Karakoram. J Clim 27(20):7890–7902

Fowler HJ, Archer DR (2006) Conflicting signals of climatic change in the Upper Indus Basin. J Clim 19(17):4276–4293

Ghatak D, Sinsky E, Miller J (2014) Role of snow-albedo feedback in higher elevation warming over the Himalayas, Tibetan Plateau and Central Asia. Environ Res Lett 9(11):114008

Giorgi F, Hurrell JW, Marinucci MR, Beniston M (1997) Elevation dependency of the surface climate change signal: a model study. J Clim 10(2):288–296

Hansen J, Sato M, Ruedy R (1997) Radiative forcing and climate response. J Geophys Res Atmos 102(D6):6831–6864

Hasson S, Pascale S, Lucarini V, Böhner J (2016) Seasonal cycle of precipitation over major river basins in South and Southeast Asia: a review of the CMIP5 climate models data for present climate and future climate projections. Atmos Res 180:42–63

Hawkins E, Sutton R (2009) The potential to narrow uncertainty in regional climate predictions. Bull Am Meteorol Soc 90(8):1095–1107

Hazeleger W, Wang X, Severijns C, Ştefănescu S, Bintanja R, Sterl A, Wyser K, Semmler T, Yang S, Van den Hurk B, Van Noije T (2012) EC-Earth V2. 2: description and validation of a new seamless earth system prediction model. Clim Dyn 39(11):2611–2629

Hewitt K (2007) Tributary glacier surges: an exceptional concentration at Panmah glacier, Karakoram Himalaya. J Glaciol 53(181):181–188

Houze RA Jr, Rasmussen KL, Medina S, Brodzik SR, Romatschke U (2011) Anomalous atmospheric events leading to the summer 2010 floods in Pakistan. Bull Am Meteorol Soc 92(3):291–298

Huang A, Zhou Y, Zhang Y, Huang D, Zhao Y, Wu H (2014) Changes of the annual precipitation over central Asia in the twenty-first century projected by multimodels of CMIP5. J Clim 27(17):6627–6646

Kang S, Xu Y, You Q, Flügel WA, Pepin N, Yao T (2010) Review of climate and cryospheric change in the Tibetan Plateau. Environ Res Lett 5(1):015101

Kattel DB, Yao T (2013) Recent temperature trends at mountain stations on the southern slope of the Central Himalayas. J Earth Syst Sci 122(1):215–227

Kitoh A, Endo H, Krishna Kumar K, Cavalcanti IF, Goswami P, Zhou T (2013) Monsoons in a changing world: a regional perspective in a global context. J Geophys Res Atmos 118(8):3053–3065

Kothawale DR, Munot AA, Kumar KK (2010) Surface air temperature variability over India during 1901–2007, and its association with ENSO. Clim Res 42(2):89–104

Kraaijenbrink PDA, Bierkens MFP, Lutz AF, Immerzeel WW (2017) Impact of a global temperature rise of 1.5 degrees Celsius on Asia's glaciers. Nature 549:257–260

Krishnamurti TN, Kumar V, Simon A, Thomas A, Bhardwaj A, Das S, Senroy S, Bhowmik SR (2017) March of buoyancy elements during extreme rainfall over India. Clim Dyn 48(5–6):1931–1951

Kulkarni A, Patwardhan S, Kumar KK, Ashok K, Krishnan R (2013) Projected climate change in the Hindu Kush–Himalayan region by using the high-resolution regional climate model PRECIS. Mt Res Dev 33(2):142–151

Kumar A, Houze RA Jr, Rasmussen KL, Peters-Lidard C (2014) Simulation of a flash flooding storm at the steep edge of the Himalayas. J Hydrometeorol 15(1):212–228

Lau WK, Kim MK, Kim KM, Lee WS (2010) Enhanced surface warming and accelerated snow-melt in the Himalayas and Tibetan Plateau induced by absorbing aerosols. Environ Res Lett 5(2):025204

Li S, Lü S, Gao Y, Ao Y (2015) The change of climate and terrestrial carbon cycle over Tibetan Plateau in CMIP5 models. Int J Climatol 35(14):4359–4369

Liu X, Chen B (2000) Climatic warming in the Tibetan Plateau during recent decades. Int J Climatol 20(14):1729–1742

Liu X, Cheng Z, Yan L, Yin ZY (2009) Elevation dependency of recent and future minimum surface air temperature trends in the Tibetan Plateau and its surroundings. Glob Planet Chang 68(3):164–174

Maussion F, Scherer D, Mölg T, Collier E, Curio J, Finkelnburg R (2014) Precipitation seasonality and variability over the Tibetan Plateau as resolved by the high Asia reanalysis. J Clim 27(5):1910–1927

Minder JR, Letcher TW, Skiles SM (2016) An evaluation of high-resolution regional climate model simulations of snow cover and albedo over the Rocky Mountains, with implications for the simulated snow-albedo feedback. J Geophys Res Atmos 121(15):9069–9088

Ming J, Wang Y, Du Z, Zhang T, Guo W, Xiao C, Xu X, Ding M, Zhang D, Yang W (2015) Widespread albedo decreasing and induced melting of Himalayan snow and ice in the early 21st century. PloS one 10(6):e0126235

Naud CM, Chen Y, Rangwala I, Miller JR (2013) Sensitivity of downward longwave surface radiation to moisture and cloud changes in a high-elevation region. J Geophys Res Atmos 118(17):10072–10081

Norris J, Carvalho LMV, Jones C, Cannon F (2015) WRF simulations of two extreme snowfall events associated with contrasting extratropical cyclones over the western and central Himalaya. J Geophys Res Atmos 120:3114–3138

Ohmura A (2012) Enhanced temperature variability in high-altitude climate change. Theor Appl Climatol 110(4):499–508

Palazzi E, Hardenberg JV, Provenzale A (2013) Precipitation in the Hindu-Kush Karakoram Himalaya: observations and future scenarios. J Geophys Res Atmos 118(1):85–100

Palazzi E, von Hardenberg J, Terzago S, Provenzale A (2015) Precipitation in the Karakoram-Himalaya: a CMIP5 view. Clim Dyn 45(1–2):21–45

Palazzi E, Filippi L, von Hardenberg J (2017) Insights into elevation-dependent warming in the Tibetan plateau-Himalayas from CMIP5 model simulations. Clim Dyn 48(11–12):3991–4008

Panday PK, Thibeault J, Frey KE (2015) Changing temperature and precipitation extremes in the Hindu Kush-Himalayan region: an analysis of CMIP3 and CMIP5 simulations and projections. Int J Climatol 35(10):3058–3077

Pepin NC, Lundquist JD (2008) Temperature trends at high elevations: patterns across the globe. Geophys Res Lett 35(14)

Pepin N, Bradley RS, Diaz HF, Baraer M, Caceres EB, Forsythe N, Fowler H, Greenwood G, Hashmi MZ, Liu XD, Miller JR, Ning L, Ohmura A, Palazzi E, Rangwala I, Schöner W, Severskiy I, Shahgedanova M, Wang MB, Williamson SN, Yang DQ (2015) Elevation-dependent warming in mountain regions of the world. Nat Clim Chang 5(5):424–430

Philipona R, Dürr B, Ohmura A, Ruckstuhl C (2005) Anthropogenic greenhouse forcing and strong water vapor feedback increase temperature in Europe. Geophys Res Lett 32(19)

Priya P, Krishnan R, Mujumdar M, Houze RA (2017) Changing monsoon and midlatitude circulation interactions over the Western Himalayas and possible links to occurrences of extreme precipitation. Clim Dyn 49(7–8):2351–2364

Pu Z, Xu L, Salomonson VV (2007) MODIS/Terra observed seasonal variations of snow cover over the Tibetan Plateau. Geophys Res Lett 34(6)

Qin J, Yang K, Liang S, Guo X (2009) The altitudinal dependence of recent rapid warming over the Tibetan Plateau. Clim Chang 97(1):321–327

Qiu J (2008) The third pole climate change is coming fast and furious to the Tibetan plateau. Nature 454:393–396

Qu X, Hall A (2007) What controls the strength of snow-albedo feedback? J Clim 20(15):3971–3981

Ramanathan V, Carmichael G (2008) Global and regional climate changes due to black carbon. Nat Geosci 1(4):221–227

Rangwala I, Miller JR (2012) Climate change in mountains: a review of elevation-dependent warming and its possible causes. Clim Chang 114(3–4):527–547

Rangwala I, Miller JR, Xu M (2009) Warming in the Tibetan Plateau: possible influences of the changes in surface water vapor. Geophys Res Lett 36(6)

Rangwala I, Miller JR, Russell GL, Xu M (2010) Using a global climate model to evaluate the influences of water vapor, snow cover and atmospheric aerosol on warming in the Tibetan Plateau during the twenty-first century. Clim Dyn 34(6):859–872

Rangwala I, Sinsky E, Miller JR (2013) Amplified warming projections for high altitude regions of the northern hemisphere mid-latitudes from CMIP5 models. Environ Res Lett 8(2):024040

Rangwala I, Pepin NC, Vuille M, Miller J (2015) Influence of climate variability and large-scale circulation on the mountain cryosphere. In: The high-mountain cryosphere: environmental changes and human risks. Cambridge University Press, Cambridge

Rangwala I, Sinsky E, Miller JR (2016) Variability in projected elevation dependent warming in boreal midlatitude winter in CMIP5 climate models and its potential drivers. Clim Dyn 46(7–8):2115–2122

Rasmussen KL, Hill AJ, Toma VE, Zuluaga MD, Webster PJ, Houze RA (2015) Multiscale analysis of three consecutive years of anomalous flooding in Pakistan. Q J R Meteorol Soc 141(689):1259–1276

Rikiishi K, Nakasato H (2006) Height dependence of the tendency for reduction in seasonal snow cover in the Himalaya and the Tibetan Plateau region, 1966–2001. Ann Glaciol 43(1):369–377

Ruckstuhl C, Philipona R, Morland J, Ohmura A (2007) Observed relationship between surface specific humidity, integrated water vapor, and longwave downward radiation at different altitudes. J Geophys Res Atmos 112(D3)

Sanjay J, Krishnan R, Shrestha AB, Rajbhandari R, Ren GY (2017) Downscaled climate change projections for the Hindu Kush Himalayan region using CORDEX South Asia regional climate models. Adv Clim Chang Res 8(3):185–198

Schild A (2008) ICIMOD's position on climate change and mountain systems: the case of the Hindu Kush–Himalayas. Mt Res Dev 28(3):328–331

Sharmila S, Joseph S, Sahai AK, Abhilash S, Chattopadhyay R (2015) Future projection of Indian summer monsoon variability under climate change scenario: an assessment from CMIP5 climate models. Glob Planet Chang 124:62–78

Shrestha AB, Wake CP, Mayewski PA, Dibb JE (1999) Maximum temperature trends in the Himalaya and its vicinity: an analysis based on temperature records from Nepal for the period 1971–94. J Clim 12(9):2775–2786

Sperber KR, Annamalai H (2014) The use of fractional accumulated precipitation for the evaluation of the annual cycle of monsoons. Clim Dyn 43(12):3219–3244

Sperber KR, Annamalai H, Kang IS, Kitoh A, Moise A, Turner A, Wang B, Zhou T (2013) The Asian summer monsoon: an intercomparison of CMIP5 vs. CMIP3 simulations of the late 20th century. Clim Dyn 41(9–10):2711–2744

Syed FS, Giorgi F, Pal JS, King MP (2006) Effect of remote forcings on the winter precipitation of central south- west Asia. Part 1: observations. Theor Appl Climatol 86:147–160

Tebaldi C, Knutti R (2007) The use of the multi-model ensemble in probabilistic climate projections. Philos Trans R Soc Lond A Math Phys Eng Sci 365(1857):2053–2075

Terzago S, von Hardenberg J, Palazzi E, Provenzale A (2014) Snowpack changes in the Hindu Kush–Karakoram–Himalaya from CMIP5 global climate models. J Hydrometeorol 15(6):2293–2313

Thackeray CW, Fletcher CG (2016) Snow albedo feedback: current knowledge, importance, outstanding issues and future directions. Prog Phys Geogr 40(3):392–408

Thompson DW, Wallace JM (2001) Regional climate impacts of the Northern Hemisphere annular mode. Science 293(5527):85–89

Wang Q, Fan X, Wang M (2014) Recent warming amplification over high elevation regions across the globe. Clim Dyn 43(1–2):87–101

Wang Q, Fan X, Wang M (2016) Evidence of high-elevation amplification versus Arctic amplification. Sci Rep 6:19219

Wu J, Xu Y, Gao XJ (2017) Projected changes in mean and extreme climates over Hindu Kush Himalayan region by 21 CMIP5 models. Adv Clim Chang Res 8:176–184

Xu BQ, Wang M, Joswiak DR, Cao JJ, Yao TD, Wu GJ, Yang W, Zhao HB (2009) Deposition of anthropogenic aerosols in a southeastern Tibetan glacier. J Geophys Res Atmos 114(D17):D17209

Yadav RK, Rupa Kumar K, Rajeevan M (2009) Increasing influence of ENSO and decreasing influ-
    ence of AO/NAO in the recent decades over northwest India winter precipitation. J Geophys
    Res 114:D12112. https://doi.org/10.1029/2008JD011318
Yan L, Liu X (2014) Has climatic warming over the Tibetan Plateau paused or continued in recent
    years. J Earth Ocean Atmos Sci 1:13–28
Yan L, Liu Z, Chen G, Kutzbach JE, Liu X (2016) Mechanisms of elevation-dependent warming
    over the Tibetan plateau in quadrupled CO2 experiments. Clim Chang 135(3–4):509–519
Yang T, Hao X, Shao Q, Xu CY, Zhao C, Chen X, Wang W (2012) Multi-model ensemble projec-
    tions in temperature and precipitation extremes of the Tibetan Plateau in the 21st century. Glob
    Planet Chang 80:1–13
Yao T, Thompson L, Yang W, Yu W, Gao Y, Guo X, Yang X, Duan K, Zhao H, Xu B, Pu J (2012)
    Different glacier status with atmospheric circulations in Tibetan Plateau and surroundings. Nat
    Clim Chang 2(9):663–667
You Q, Kang S, Aguilar E, Yan Y (2008) Changes in daily climate extremes in the eastern and
    central Tibetan Plateau during 1961–2005. J Geophys Res Atmos 113(D7)
Zhan YJ, Ren GY, Shrestha AB, Rajbhandari R, Ren YY, Sanjay J, Xu Y, Sun XB, You QL, Wang S
    (2017) Changes in extreme precipitation events over the Hindu Kush Himalayan region during
    1961–2012. Adv Clim Chang Res 8(3):166–175

Chapter 5
Diurnal Cycle of Precipitation in the Himalayan Foothills – Observations and Model Results

Bodo Ahrens, Thomas Meier, and Erwan Brisson

Abstract The interplay of the Indian Monsoon and the Himalayas is vital to many climatological aspects of the Himalayan foothill and foreland regions. A unique climate feature in the Himalayan foothill and foreland regions is a bi-modal diurnal cycle of precipitation with high rainfall amounts in the afternoon and around midnight. The reason for this night-time precipitation maximum is not yet fully understood, and current climate models do not well represent the regions' diurnal cycle of precipitation. Nevertheless, estimation of realistic spatiotemporal precipitation patterns is crucial for the climate community (e.g., for impact modeling). This study reviews discussions in literature, available observational findings, and simulation results with the regional climate model (RCM) COSMO-CLM. Our COSMO-CLM simulations indicate that the model is not able to recover the nighttime's precipitation behavior with currently typical horizontal RCM grid-spacings (e.g., 20 or 50 km), but it can do so with convection-permitting grid-spacing (~3 km) which sufficiently resolves the relevant orographic thermal wind together with the moist monsoonal flow characteristics in the area.

5.1 Introduction

The diurnal cycle of precipitation is a primary climatological and society-relevant parameter. For example, biting activities of mosquitoes including disease vectors are different for nights with or without rainfall. Diurnal variation of precipitation varies in space and seasonally in dependence on the varying importance of the different precipitation triggering processes which are still not understood in full

B. Ahrens (✉) · T. Meier · E. Brisson
Institute for Atmospheric and Environmental Sciences, Goethe University Frankfurt,
Frankfurt am Main, Germany
e-mail: Bodo.Ahrens@iau.uni-frankfurt.de

© Springer Nature Switzerland AG 2020   73
A. P. Dimri et al. (eds.), *Himalayan Weather and Climate and their Impact on the Environment*, https://doi.org/10.1007/978-3-030-29684-1_5

today. Therefore, representing realistically the diurnal cycle of precipitation is still challenging in weather forecasting and climate projection.

Kikuchi and Wang (2008) analyzed tropical rainfall observed by TRMM satellite observations, and they showed that precipitation with a day-time maximum (which occurs around 15:00 LT) explains the largest portion of the variance over most land areas. Thus, the maximum variation of precipitation often links with the peak in daytime heating which leads to sufficient and necessary buoyancy for convection to occur. However, and this is of interest in this study, their analysis also points to early morning precipitation maxima close to or downstream of major mountain ranges.

Such variations in precipitation's diurnal cycle are investigated in more detail by Sahany et al. (2010) for monsoon rainfall over the Indian region. Their analysis suggests that the timing of maximum rainfall is highly location dependent with peaks at 1430 IST over the Gangetic plains, at 1430–1730 IST over the Burmese mountains and the Western Ghats (west coast of India), and at about 0230 IST over the Himalayan foothills. The existence of the nocturnal/early morning maximum of rainfall at the foot of the Himalayas during the summer monsoon has been noted in a variety of studies (Barros and Lang 2003; Barros et al. 2000, 2004; Romatschke and Houze 2011; Romatschke et al. 2010), as well as during winter season (Sen Roy 2008). Also, Rüthrich et al. (2013) detected an increase of cloud frequencies until the early morning, and Albrecht et al. (2016) reported a late night/early morning maximum in lightning activity above the southern foothills of the Himalayas.

A bi-modal behavior of precipitation with a late afternoon and a nocturnal maximum over the southern Himalayan slopes based on TRMM remote sensing data is discussed in Bhatt and Nakamura (2006) and Sahany et al. (2010). Climate simulations done in earlier studies with horizontal grid-spacings of 50 km (Dobler and Ahrens 2008) or experimentally with 25 km showed the afternoon precipitation maximum, but a lack of precipitation along the southern Himalayan slopes. Missing nocturnal precipitation in the coarse-grid climate simulations explains the lack of precipitation in the simulations.

The literature discusses a variety of explanations for nocturnal precipitation over land. These studies show the broad range of processes which may be possible sources for nocturnal precipitation. For example, the summer U.S. precipitation is characterized by late afternoon maximum precipitation over the Southwest and the Rocky Mountains but by midnight maximum precipitation in the region east of the Rockies. Different studies explain this nocturnal maximum precipitation east of the Rockies by different mechanisms: (a) eastward propagation of mesoscale convective complexes (MCCs) and mesoscale convective systems (MCSs) starting in the afternoon over the Rockies and lasting until the early morning (Maddox 1980), (b) a diurnal cycle of low-level large-scale convergence enhanced by eastward propagation of late afternoon thunderstorms generated over the Rockies (Dai et al. 1999, Geerts et al. 2017), and/or (c) advection of evaporation-cooled layers from afternoon showers. The nocturnal precipitation maximum in the northeastern foothills of the central Andes is explained by Romatschke and Houze (2010) with moisture transport and wind shear by the nocturnal South American low-level jet. Trachte and Bendix (2012) discuss the hypothesis of the formation of convective clouds at the eastern Andes mountains because of a katabatic cold front. Their idealized modeling experiments supported this hypothesis. Pfeifroth et al. (2015) discuss a bi-modal diurnal cycle in central Benin, West Africa, with a locally initiated convective rainfall evening peak and a peak after midnight caused by westward-propagating, organized convective systems that originate from the Jos Plateau, Nigeria, and reach central Benin toward the ends of their lifetimes. Ethiopia's nocturnal convection is explained by cold air drainage flow which builds from 0200 to 0800 CAT over the Ethiopian mountains (Jury 2016).

Similar processes were hypothesized to trigger the nocturnal precipitation along the southern Himalayan slopes during monsoon season. This paragraph describes two hypotheses in more detail. More, less likely hypotheses were summarized in Houze (2012). The first hypothesis is that during the night, the prevailing low-level monsoon southwesterlies and the northern downslope katabatic flow converge at the base of the mountains and trigger convection (e.g., Sahany et al. 2010; Rüthrich et al. 2013). While downslope/drainage flow during the night observed in field experiments conducted at the Himalayan foothills (e.g., Egger et al. 2000; Barros and Lang 2003) support this hypothesis, the occurrence of nighttime maximum precipitation during winter (Sen Roy 2008) suggests that the southwesterly winds may not be an essential condition for nocturnal precipitation to occur. The second hypothesis assumes propagation of isolated convective cells generated over the Tibetan Plateau to the south-west, which develop into MCSs energized by moist air advected from the Arabian Sea and the Bay of Bengal rising from the south-western foothills of the Himalayas (Rasmussen and Houze 2012). While the cell propagation has been found in other orographic environments (see above) and may be considered as a straightforward process, Barros et al. (2004) reported strong nighttime activity over the foothills characterized by the development of short-lived convection (lifetimes of 1–3 h).

The question arises which of these hypotheses can be confirmed? Is the propagation of large convective systems responsible for nighttime precipitation for the full Himalayan arc or is the nocturnal downslope flow an equally valid trigger?

This paper will investigate remote satellite datasets with the goal to investigate hot-spots of nocturnal precipitation in the Himalayan foothill area. In addition, the diurnal precipitation cycle in coarse-grid and convection-permitting simulations with RCM COSMO-CLM will be investigated (with COSMO-CLM in convection permitting set-up well representing the diurnal cycle in Europe and Africa, Brisson et al. 2016; Prein et al. 2015; Thiery et al. 2016). With former coarse-grid simulations underestimating nocturnal rainfall, the simulations of the convection-permitting model will be evaluated and used for studying the diurnal precipitation processes.

5.2 Data and Methods

5.2.1 CMORPH Observational Data

In a former study (Pfeifroth et al. 2015), we compared and evaluated several satellite-based precipitation datasets and found the NOAA Climate Prediction Centre's satellite-based data-set CMORPH (Joyce et al. 2004) useful in investigating the diurnal cycle of precipitation. The data-set is available with a very high resolution of 30 min in time and 8 km in space online (http://rda.ucar.edu). The very high temporal resolution is an advantage over often used TRMM satellite data (Houze et al. 2015). In the following, we use data in the monsoonal period (June to September, JJAS) for the years 2011–2015. The five investigated years proved to be sufficient to detect the diurnal cycle of precipitation robustly.

5.2.2 COSMO-CLM Modelling

For this study, we applied the non-hydrostatic regional climate model COSMO-CLM as in previous studies in the area (e.g., Dobler and Ahrens 2008, 2011; Asharaf and Ahrens 2015), but here in an updated version 5.00clm7 (see http://www.clm-community.eu). Here, we discuss results of two set-ups. The first set-up used is a large-domain set-up with horizontal grid-spacing of 0.22° (~25 km) driven by the ERA-Interim reanalysis (Simmons et al. 2007). In this coarse-grid set-up the convection parameterization, a modified Tiedtke (1989)-scheme is switched on. The second fine-grid set-up is a convection-permitting set-up with 0.025° (~2.8 km) grid-spacing with deep convection parameterization switched off and improved description of orography compared to the coarse-grid set-up. Details of the setting can be found in Brisson et al. (2018) and Meier (2017). Meier (2017) showed that the details of the microphysics scheme and of switching on or off the shallow convection scheme do not change the results as discussed below significantly. The fine-grid convection-permitting simulations are nested without feedback into the coarse-grid simulations. The quite large resolution jump between coarse- and fine-grid nest was tested not to impact the results to be shown below.

Figure 5.1 shows the simulation domain and the orography as represented in the 0.22° coarse-grid COSMO-CLM set-up. The coarse-grid set-up does not define well even the major valleys in the Himalayas. The representation of the orography, esp. the deep valleys cutting into the Himalayas in about north-south direction, is improved in the fine-grid domains (Fig. 5.2). Observed interesting features in the CMORPH data-set (see the next section) motivated the locations of the fine-grid domains (called domains BT – Bhutan and UT-Uttarakhand).

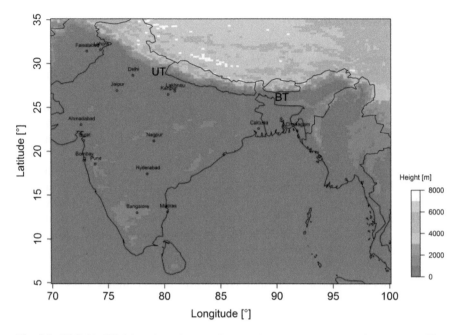

Fig. 5.1 COSMO-CLM domain and orography used in the coarse-grid (0.22°) simulation. The locations of the fine-grid domains are indicated with UT (Uttarakhand) and BT (Bhutan)

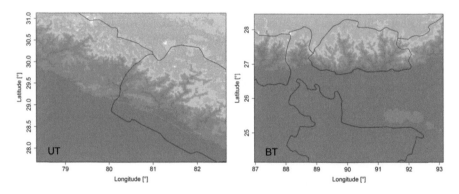

Fig. 5.2 Orographies used in the fine-grid simulations in the computing domain UT (Uttarakhand, left panel) and in BT (Bhutan, right panel). The color coding is the same as in Fig. 5.1

5.3 Results and Discussion

First, findings on the diurnal cycle of precipitation based on the satellite-based CMORPH observations will be shown and discussed in this section. Afterward, the limitations and added information by the COSMO-CLM simulations will be investigated.

5.3.1 Observations

Figure 5.3 shows the mean CMORPH precipitation in the area of interest for August. It illustrates clearly that the Himalayas build up a natural rainfall barrier. The area of maximum rainfall follows the Himalayan southern foothills nicely. Two regions with precipitation hotspots (BT-Bhutan, UT-Uttarakhand) are indicated and will be discussed in more detail below. The CMORPH precipitation pattern was similar in July, more heterogeneous in June, and precipitation was already vanishing in September along the foothills (not shown).

The time of daily maximum precipitation intensity is shown in Fig. 5.4. In large parts of the Indian subcontinent, the daily maximum was in the late afternoon, which is explained by delayed convective precipitation following the maximum of daily heating. However, in large parts of the Himalayan foothills, the maximum was around or after midnight. There were also large areas with no distinct diurnal cycle in the satellite-based CMORPH data. In the other monsoonal months, the patterns were similar.

The width of the area along the foothills with around midnight precipitation maximum was small with roughly 100 km. This band of midnight precipitation was broadened, for example, in the area of Nepal (cf. Fig. 5.4). Figure 5.5 shows a Hovmöller diagram for precipitation intensity at 90° longitude through the focus region BT. It shows a pronounced midnight maximum at about 26.5° latitude. Interestingly the diagram shows a weak secondary maximum at 1400 IST, which was somewhat earlier than the late afternoon maximum in Central India but is consistent with observations discussed in Sahany et al. (2010), who found an early afternoon maximum (around 1430 IST) over the Gangetic plains with TRMM satellite data.

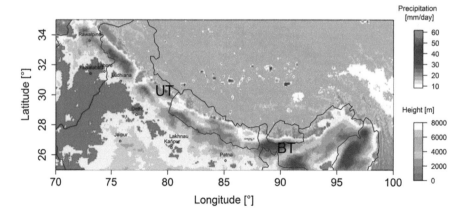

Fig. 5.3 Mean August precipitation as derived from CMORPH data. Only mean daily precipitation with values above 10 mm/d is shown. The regions marked UT and BT will be discussed in more detail later on

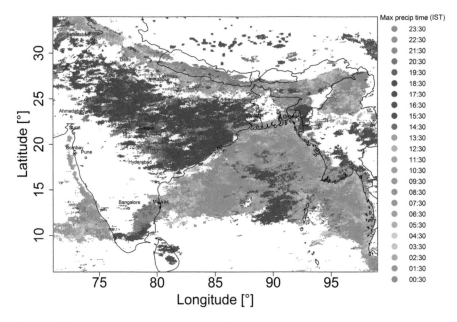

Fig. 5.4 Timing of maximum of diurnal precipitation intensity. Areas with no distinct maximum are in white shading

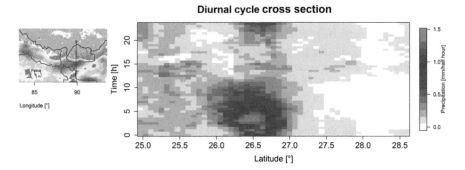

Fig. 5.5 Diurnal cycle of rainfall intensity along a cross-section at 90° longitude (i.e., cutting through the BT region)

5.3.2   Simulations

With all years showing a very similar pattern of diurnal precipitation behavior, we focus on the presentation of results in the year 2011. All discussed simulations were initialized on April the 1st and continuously ran until the end of September. The precipitation patterns and amounts of the coarse-grid simulation (grid-spacing ~25 km) were similar to the results of coarse-grid simulations with COSMO-CLM

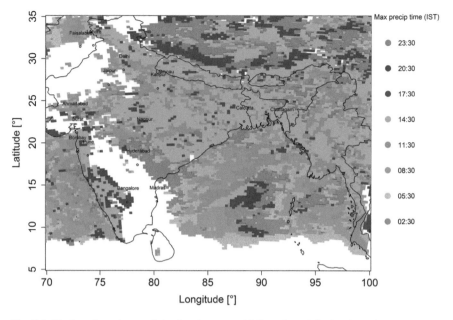

Fig. 5.6 Timing of maximum of simulated coarse-grid diurnal precipitation intensity

discussed in Lucas-Picher et al. (2011) or Kumar et al. (2013). Figure 5.6 shows the simulated timing of daily maximum coarse-grid intensity in July and August 2011. The pattern over the Bay of Bengal was reasonable (and as described in the literature, Kraus 1963) and the pattern over Central India shows the often detected (see Bechthold et al. 2004) too early precipitation maximum in simulations with parameterized deep convection. However, the early afternoon maximum of simulated precipitation in major parts of the Himalayan foothills is in contradiction to the observed maximum at about midnight (Fig. 5.4).

In the following, we discuss the ability of the fine-grid convection-permitting (i.e., without parameterization of deep convection) simulations to better represent the mean precipitation pattern and the diurnal cycle in the two sub-regions BT and UT. As an example, Fig. 5.7 compares the mean precipitation fields of the coarse-grid and fine-grid simulations in the BT area. The fine-grid pattern is closer to the CMORPH pattern (Fig. 5.3) than the coarse-grid pattern, although there exists substantial difference too. Since the CMORPH remote-sensed pattern also has some uncertainty, esp. in the mountainous areas, the deviation of the fine-grid pattern to the CMORPH pattern should not be over-interpreted here.

The timing of the maximum precipitation intensity for coarse- and fine-grid simulations in the area BT is compared in Fig. 5.8. The coarse-grid simulations showed a few pixels with early morning maximum along the edge of the foothills. However, the maximum pattern of the fine-grid simulation fits much better to the observed pattern in the region (Fig. 5.4). Figure 5.9 shows that in the UT area the diurnal cycle as simulated with coarse-grid did not fit at all. The fine-grid simulation showed

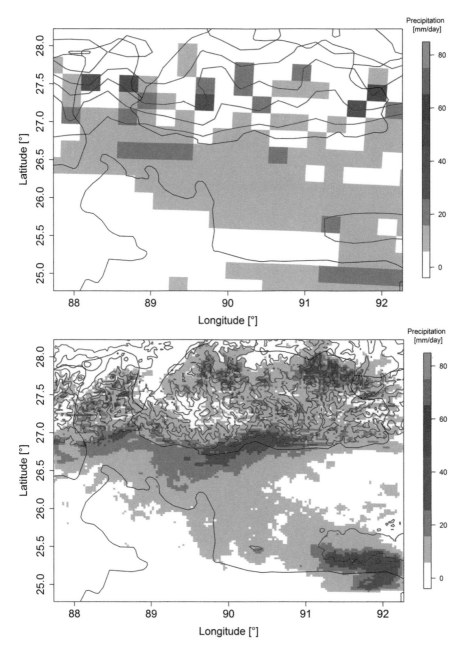

Fig. 5.7 Mean coarse- (top) and fine- (bottom) grid precipitation (in mm/day) in the domain BT. Green lines indicate orography in the simulations

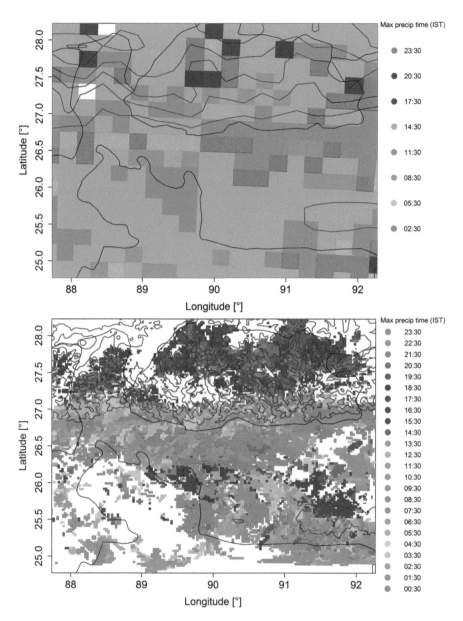

Fig. 5.8 Timing of maximum of simulated coarse- (top) and the fine-grid (bottom) simulated diurnal precipitation intensity in the area BT

a late evening maximum deep in the mountain range which moved southwards during the night. This general pattern fits well with the observed maximum pattern.

In the following, we discuss in more detail the diurnal cycle as simulated with the fine-grid set-up along two cross-sections (one in region BT and one in region UT,

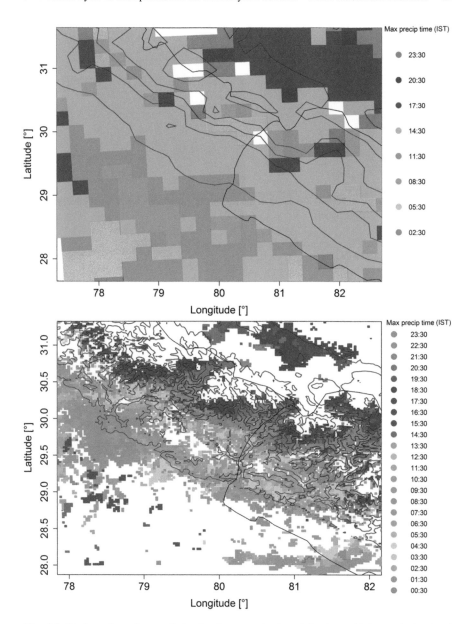

Fig. 5.9 Timing of maximum of simulated coarse- (top) and the fine-grid (bottom) simulated diurnal precipitation intensity in the area UT

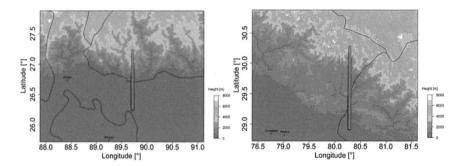

Fig. 5.10 Cross-sections (red) discussed in the region BT (longitude: 89.72°, left panel) and UT (longitude: 80.30°, right panel)

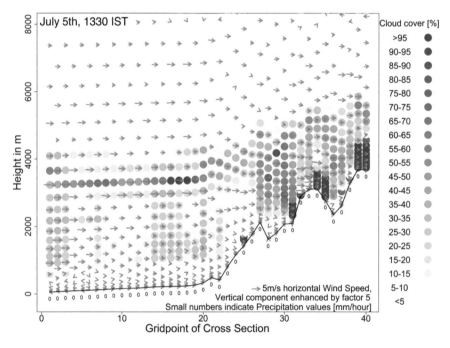

Fig. 5.11 Cloud cover, wind speed, and precipitation diurnal cycle for July 5th–6th along the cross-section in region BT at 1330, 2130, and 0130 IST

see Fig. 5.10) for one typical rainfall event in each region. The length of the cross-section in BT is 40 simulation grid-lengths or 112 km and covers an altitudinal range of almost 4000 m. In region BT the event of 5th to 6th July 2011 with maximum precipitation intensity after midnight was chosen as a typical representative event.

In Fig. 5.11, the cloud cover, wind field, and precipitation in BT are shown for three different times for these days. At midday, there were only a few shallow clouds in the Himalayan foreland. Over higher ridges, the clouds produced no precipitation but were a hint to moist air advected from the south and lifted by an anabatic flow. The CAPE values were highest (up to 1400 J/kg) at 1830 IST during this event. At 2130 IST the clouds over the higher mountains disappeared, and the wind vectors show a low-level katabatic flow (with 2–3 K lower potential temperature over the slopes compared to flatland temperature at same height levels) into the foreland. In the area of convergence of the northerly katabatic and the southerly monsoonal flow, convection was initiated yielding precipitation (about 10 mm/h). Later in the night, a stronger monsoonal flow pushed the convergence area to the north yielding precipitation at the foot of the mountains (at ~0130 IST) and further uphill later on. This diurnal cycle as simulated with the fine-grid model fits well, although not entirely, to the diurnal cycle observed in the area (Fig. 5.5), while it cannot be represented at all in the coarse-grid model. Indeed, the coarse-grid does not represent well enough the mountain-valley wind systems and thus cannot represent well enough the nighttime katabatic flows and their interaction with the monsoonal flow.

The triggering of convection by a nocturnal low-level jet is often-discussed in literature. But here, the simulated low-level zonal wind component of 2–4 m/s from the east showed only small changes of 1 m/s or less just before the onset of precipitation in the convergence area. The onset of a nocturnal low-level jet would imply more substantial wind changes. Following the northward-moving area of deep convection in the simulation, the easterly zonal wind flipped to a westerly zonal component. Overall, there is no indication that nocturnal low-level jets were triggering night-time maximum precipitation in the simulations.

It is interesting to note that during rainless nights a katabatic outflow also established but was diverted from the monsoonal flow by a robust zonal wind component (~10 m/s) from the east over the flatlands. This strong zonal wind also blocked the moisture transport from the south. Together this prevented nighttime rain in the flatland and the foothills.

Next, we discuss the precipitation along the cross-section in region UT. The length of cross-section UT is 55 simulation grid-lengths or 150 km and covers an altitudinal range of almost 4000 m with several small ridges along the section (Fig. 5.12). We discuss the rainy event of the 2nd to the 3rd of August as a typical example.

Moisture was transported along the cross-section and hindered the establishment of a stronger katabatic flow after sunset. Convection was triggered already in the late afternoon near the highest ridge in the section with a maximum intensity of above 10 mm/h. Therefore, precipitation built up earlier than in the BT region. The simulated evening precipitation over the central ridge and the Tibetan plateau was not observed in the CMORPH data-set but documented in the literature (e.g., Rasmussen and Houze 2012). Later in the night northerly wind was dominant in all height levels in the northern parts of the sections. Relatively dry air is flowing down from the Tibetan plateau. This together with katabatic and downdraft density currents pushed the precipitation maximum to the South during the night and longer-living self-organized systems developed leading to a precipitation maximum early morning in the foothills (as hypothesized by Rasmussen and Houze 2012).

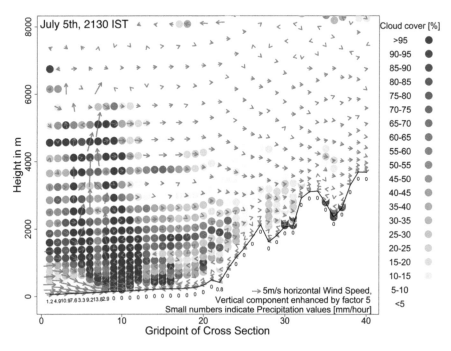

Fig. 5.12 Cloud cover, wind speed, and precipitation diurnal cycle for August 2nd–3rd along the cross-section in region UT (snapshots at Aug. 2nd, 1720 IST (left) and Aug. 3rd, 0330 IST (right)). Color coding is the same as in Fig. 5.11

5.4 Conclusions

The investigation of the satellite-based CMORPH data confirmed previous findings about the diurnal cycle of precipitation with maximum rainfall during the night along the whole Himalayan foothill chain. Further south, for example in Central India, the rainfall maximum is in the late afternoon. The regional climate model COSMO-CLM with grid-spacing of 25 km, i.e., a model unable to resolve deep convection and thus needs parameterization of deep convection, simulated a midday rainfall maximum in the Himalayan foothills and in the flatlands further in the South. This too early midday maximum is an often-discussed challenge of models with parameterized convection (e.g., Bechtold et al. 2004) which improves in simulations with convection-permitting grid-spacing (e.g., Prein et al. 2015; Kendon et al. 2017).

Here, we focused on the observed nighttime precipitation maximum in the Himalayan foothills, which was not simulated in the 25-km COSMO-CLM simulation. Convection-permitting simulations (grid-spacing of 2.8 km) with COSMO-CLM were done in two regions of the Himalayan forelands and foothills: in a region called Bhutan (BT) and in a region called Uttarakhand (UT) (east and west of Nepal, respectively).

Our simulations with a convection-permitting set-up of COSMO-CLM with grid-spacing of 2.8 km improved the diurnal cycle of precipitation substantially in both regions. This improvement was partly due to the no longer necessary parameterization of deep convection in the simulations. However, in the Himalayan foothills, our results suggest that the improved description of orography allows for an importantly more realistic representation of precipitation forcing through mountain-valley wind systems and orographic lifting than in the 25 km grid-spacing simulation. Indeed, the fine-grid simulations showed that the interplay of mountainous wind systems and monsoonal flow is controlling the diurnal cycle of simulated precipitation in this area.

In addition, we further investigated some hypotheses formulated in previous studies to explain nighttime maximum precipitation. Triggering of convection via nocturnal low-level jet (as in the foothills of the Central Andes, Romatschke and Houze 2010) was not observed in our simulations. Long-living organized convection as hypothesized in Romatschke and Houze (2011) was of importance for the diurnal cycle in the UT region, but of less importance in the BT region with convection along the convergence of prevailing low-level monsoon southwesterlies and downslope katabatic flow (as previously discussed in Sahany et al. 2010, Rüthrich et al. 2013).

High-resolution convection-permitting atmospheric simulations are computationally expensive and challenging (e.g., for climate projections). However, they are necessary for providing realistic representations of precipitation regimes in complex areas and for improving our understanding of atmospheric processes. In the near future, convection-permitting-model-based climate projections may be feasible and more often used given the ongoing developments in computing (Fuhrer et al. 2018).

References

Albrecht RI, Goodman SJ, Buechler DE, Blakeslee RJ, Christian HJ, Albrecht RI, Goodman SJ, Buechler DE, Blakeslee RJ, Christian HJ (2016) Where are the lightning hotspots on Earth? Bull Am Meteorol Soc 97(11):2051–2068. https://doi.org/10.1175/BAMS-D-14-00193.1

Asharaf S, Ahrens B (2015) Indian summer monsoon rainfall feedback processes in climate change scenarios. J Clim 28(13):5414–5429. https://doi.org/10.1175/JCLI-D-14-00233.1

Barros AP, Lang TJ (2003) Monitoring the monsoon in the Himalayas: observations in central Nepal, June 2001. Mon Weather Rev 131(7):1408–1427

Barros AP, Joshi M, Putkonen J, Burbank DW (2000) A study of the 1999 monsoon rainfall in a mountainous region in central Nepal using TRMM products and rain gauge observations. Geophys Res Lett 27:3683–3686. https://doi.org/10.1029/2000GL011827

Barros AP, Kim G, Williams E, Nesbitt SW (2004) Probing orographic controls in the Himalayas during the monsoon using satellite imagery. Nat Hazards Earth Syst Sci 4:29–51. https://doi.org/10.5194/nhess-4-29-2004

Bechtold P, Chaboureau JP, Beljaars A, Betts AK, Köhler M, Miller M, Redelsperger J-L (2004) The simulation of the diurnal cycle of convective precipitation over land in a global model. Q J Roy Meteorol Soc 130(604):3119–3137. https://doi.org/10.1256/qj.03.103

Bhatt BC, Nakamura K (2006) A climatological-dynamical analysis associated with precipitation around the southern part of the Himalayas. J Geophys Res 111:D02115. https://doi.org/10.1029/2005JD006197

Brisson E, Van Weverberg K, Demuzere M, Devis A, Saeed S, Stengel M, van Lipzig NPM (2016) How well can a convection-permitting climate model reproduce decadal statistics of precipitation, temperature and cloud characteristics? Clim Dyn 47(9–10):3043–3061. https://doi.org/10.1007/s00382-016-3012-z

Brisson E, Brendel C, Herzog S, Ahrens B (2018) Lagrangian evaluation of convective shower characteristics in a convection permitting model. Meteorol Zeitschrift 27(1):59–66. https://doi.org/10.1127/metz/2017/0817

Dai A, Giorgi F, Trenberth KE (1999) Observed and model-simulated diurnal cycles of precipitation over the contiguous United States. J Geophys Res 104(D6):6377–6402. https://doi.org/10.1029/98JD02720

Dobler A, Ahrens B (2008) Precipitation by a regional climate model and bias correction in Europe and South-Asia. Meteorol Zeitschrift 17(4):499–509. https://doi.org/10.1127/0941-2948/2008/0306

Dobler A, Ahrens B (2011) Four climate change scenarios for the Indian summer monsoon by the regional climate model COSMO-CLM. J Geophys Res 116:D24104. https://doi.org/10.1029/2011JD016329

Egger J, Bajrachaya S, Egger U, Heinrich R, Reuder J, Shayka P, Wendt H, Wirth V (2000) Diurnal winds in the Himalayan Kali Gandaki Valley. Part I: observations. Mon Weather Rev 128(4):1106–1122. https://doi.org/10.1175/1520-0493(2000)128<1106:DWITHK>2.0.CO;2

Fuhrer O, Chadha T, Hoefler T, Kwasniewski G, Lapillonne X, Leutwyler D, Luthi D, Osuna C, Schär C, Schulthess TC, Vogt H (2018) Near-global climate simulation at 1 Km resolution: establishing a performance baseline on 4888 GPUs with COSMO 5.0. Geosci Model Dev 11:1665–1681

Geerts B et al (2017) The 2015 plains elevated convection at night field project. Bull Am Meteorol Soc 98:67–786

Houze RA Jr (2012) Orographic effects on precipitating clouds. Rev Geophys 50(1):1–47. https://doi.org/10.1029/2011RG000365

Houze RA, Rasmussen KL, Zuluaga MD, Brodzik SR (2015) The variable nature of convection in the tropics and subtropics: a legacy of 16 years of the tropical rainfall measuring mission satellite. Rev Geophys 53(3):994–1021. https://doi.org/10.1002/2015RG000488

Joyce R, Janowiak J, Arkin PA, Xie P (2004) CMORPH: a method that produces global precipitation estimates from passive microwave and infrared data at high spatial and temporal resolution. J Hydrometeorol 5:487–503. https://doi.org/10.1175/1525-7541(2004)005<0487:CAMTPG>2.0.CO;2

Jury MR (2016) Large-scale features of Africa's diurnal climate. Phys Geogr 37(2):1–12. https://doi.org/10.1080/02723646.2016.1163004

Kendon EJ, Ban N, Roberts NM, Fowler HJ, Roberts MJ, Chan SC, Evans JP, Fosser G, Wilkinson JM (2017) Do convection-permitting regional climate models improve projections of future precipitation change? Bull Am Meteorol Soc 98(1):79–93. https://doi.org/10.1175/BAMS-D-15-0004.1

Kikuchi K, Wang B (2008) Diurnal precipitation regimes in the global tropics. J Clim 21(11):2680–2696. https://doi.org/10.1175/2007JCLI2051.1

Kraus EB (1963) The diurnal precipitation change over the sea. J Atmos Sci 20:551–556

Kumar P, Wiltshire A, Mathison C, Asharaf S, Ahrens B, Lucas-Picher P, Christensen JH, Gobiet A, Saeed F, Hagemann S, Jacob D (2013) Downscaled climate change projections with uncertainty assessment over India using high resolution multi model approach. Sci Total Environ 468:18–30. https://doi.org/10.1016/j.scitotenv.2013.01.051

Lucas-Picher P, Christensen JH, Saeed F, Kumar P, Asharaf S, Ahrens B, Wiltshire A, Jacob D, Hagemann S (2011) Can regional climate models represent the Indian monsoon? J Hydrometeorol 12:849–868. https://doi.org/10.1175/2011JHM1327.1

Maddox RA (1980) Mesoscale convective complexes. Bull Am Meteorol Soc 61(11):1374–1387

Meier T (2017) Diurnal cycle of precipitation in the Himalayan foothills. MSc thesis, Goethe University Frankfurt am Main

Pfeifroth U, Trentmann J, Fink AH, Ahrens B (2015) Evaluating satellite-based diurnal cycles of precipitation in the African tropics. J Appl Meteorol Climatol 55(1):23–39. https://doi.org/10.1175/JAMC-D-15-0065.1

Prein AF, Langhans W, Fosser G, Ferrone A, Ban N, Goergen K, Keller M, Tölle M, Gutjahr O, Feser F, Brisson E, Kollet S, Schmidli J, van Lipzig NPM, Leung R (2015) A review on regional convection-permitting climate modeling: demonstrations, prospects, and challenges. Rev Geophys 53(2):323–361. https://doi.org/10.1002/2014RG000475

Rasmussen KL, Houze RA Jr (2012) A flash-flooding storm at the steep edge of high terrain – disaster in the Himalayas. Bull Am Meteorol Soc 93(11):1713–1724. https://doi.org/10.1175/BAMS-D-11-00236.1

Romatschke U, Houze RA Jr (2010) Extreme summer convection in South America. J Clim 23(14):3761–3791. https://doi.org/10.1175/2010JCLI3465.1

Romatschke U, Houze RA Jr (2011) Characteristics of precipitating convective systems in the South Asian monsoon. J Hydrometeorol 12:3–26. https://doi.org/10.1175/2010JHM1289.1

Romatschke U, Medina S, Houze RA Jr (2010) Regional, seasonal, and diurnal variations of extreme convection in the South Asian region. J Clim 23:419–439. https://doi.org/10.1175/2009JCLI3140.1

Rüthrich F, Thies B, Reudenbach C, Bendix J (2013) Cloud detection and analysis on the Tibetan Plateau using Meteosat and Cloudsat. J Geophys Res 118(17):10082–10099. https://doi.org/10.1002/jgrd.50790

Sahany S, Venugopal V, Nanjundiah RS (2010) Diurnal-scale signatures of monsoon rainfall over the Indian Region from TRMM satellite observations. J Geophys Res 115:D02103. https://doi.org/10.1029/2009JD012644

Sen Roy S (2008) Spatial variations in the diurnal patterns of winter precipitation in India. Theor Appl Climatol 96:347–356. https://doi.org/10.1007/s00704-008-0045-1

Simmons AJ, Uppala S, Dee D, Kobayashi S (2007) ERA-interim: new ECMWF reanalysis products from 1989 onwards. ECMWF Newsl 110:25–35

Thiery W, Davin EL, Seneviratne SI, Bedka K, Lhermitte S, van Lipzig NPM (2016) Hazardous thunderstorm intensification over Lake Victoria. Nat Commun 7:12786. https://doi.org/10.1038/ncomms12786

Tiedtke M (1989) A comprehensive mass flux scheme for cumulus parameterization in large-scale models. Mon Weather Rev 117:1779–1800

Trachte K, Bendix J (2012) Katabatic flows and their relation to the formation of convective clouds-idealized case studies. J Appl Meteorol Climatol 51(8):1531–1546. https://doi.org/10.1175/JAMC-D-11-0184.1

Chapter 6
Impact of Cloud Microphysical Processes on the Dynamic Downscaling for Western Himalayas Using the WRF Model

S. C. Kar and Sarita Tiwari

Abstract Dynamic downscaling of climate is a useful procedure to downscale the climate especially over the data sparse regions of the Himalayas. The global reanalysis data are too coarse to represent the hydroclimate over the regions with sharp orography gradient in the western Himalayas. The present study attempts to carry out dynamic downscaling of ERA-Interim dataset (January to May) over the western Himalayas using the weather research and forecasting (WRF) model. Sensitivity studies have been carried out using four microphysics parameterization schemes (namely WSM3, WSM6, Morrison and Thompson schemes). It is seen that the model is able to simulate large scale patterns of precipitation, temperature and winds reasonably well. The impact of the Morrison and Thompson schemes is to shift the zone of maximum precipitation more downwind as compared to WSM6 during winter. The WSM6 favors precipitation on the slopes of the terrain, Morrison and Thompson schemes simulate more precipitation on the mountain top (more snow) as the snow particles get advected more downwind. The Morrison scheme simulates less amount of graupels over the region than the WSM6. The narrow zone of sharply rising orography is the area where the WSM6 scheme simulates more rain than the Morrison scheme. This study emphasizes that a correct representation of the microphysical processes in the models is crucial for long-term climate simulations for correct representation of partitioning atmospheric water into vapor, cloud liquid water, cloud ice etc. leading either to solid or liquid precipitation.

S. C. Kar (✉)
National Centre for Medium Range Weather Forecasting, Noida, India
e-mail: sckar@ncmrwf.gov.in

S. Tiwari
National Centre for Medium Range Weather Forecasting, Noida, India

Geological Survey of India, Hyderabad, India

© Springer Nature Switzerland AG 2020
A. P. Dimri et al. (eds.), *Himalayan Weather and Climate and their Impact on the Environment*, https://doi.org/10.1007/978-3-030-29684-1_6

6.1 Introduction

Dynamic downscaling studies are mostly carried out using either Regional Climate Model (RegCM) or the Weather Research and Forecast (WRF) model. Several sensitivity studies using the WRF model with different microphysics options, land surface schemes and convection schemes have been performed over many regions including over India (Flaounas et al. 2010; Kim and Wang 2011; Raju et al. 2011; Crâetat et al. 2012). However, most of the sensitivity studies using high-resolution regional models are to simulate the extreme events over the Indian region. Patil and Kumar (2016) examined the sensitivity of the WRF model to different physical parameterization schemes to identify a combination of the best physics options suited during the passage of a western disturbance (WD) over northwest India. Rajeevan et al. (2010) found that there are major differences in the simulations of a severe thunderstorm in southwest India among the cloud microphysics schemes, in spite of using the same initial and boundary conditions and model configuration. Kar and Tiwari et al. (2016) have simulated the extreme rainfall events of September 2014 over Jammu and Kashmir; India using the WRF model. Dimri and Chevuturi (2014) have studied the model sensitivity with five microphysics options during western disturbance over northwest India. The above studies refer to model simulation in the scale of days and pertain only to the specific events considered.

Climate of the Himalayan region has large spatial variations based on its altitude. While the southern part is predominantly wet due to monsoon rains, the northern part is relatively dry as the high mountain range obstructs the passage of the monsoon systems to the northern part. The western Himalayas region receives substantial amount of precipitation in the form of snow during winter months. The precipitation over this region is vital for several sectors such as agriculture/horticulture, transportation, tourism, hydropower projects and water resource management. Due to the complex orography, nonlinear interaction of land-atmosphere process and insufficient observed datasets, seasonal-scale prediction of precipitation over Himalayan region is quite challenging. Precipitation over this region is mainly associated with western disturbances, which bring heavy bursts of rain and snow. Dimri et al. (2015) have provided an excellent review of the western disturbances. The frequency and amplitude of these westerly disturbances (WDs) in a given month or season decide if the winter season will experience above normal or below normal precipitation (Kar and Rana 2014). Systematic improvement in physical parameterization schemes have been made to utilize the WRF model for climate simulation (e.g. Givati et al. (2012); Liang et al. 2012). Lo et al. (2008) have proposed few strategies for dynamic downscaling using the WRF model which include (i) long-term continuous integration and (ii) consecutive integration with frequent re-initialization. Cossu and Hocke (2014) have examined the impact of different microphysical parameterization schemes in the WRF model on orography induced precipitation and the distributions of hydrometeors and water vapour. Tiwari et al. (2015) used a regional climate model and examined the influence of land surface processes on the simulated climate of the western Himalayas. Sinha et al. (2013) and Sinha et al.

(2015) have carried out sensitivity studies of convective schemes in simulations of summer and wintertime precipitation over western Himalayas respectively using RegCM. Dimri (2009) used a mosaic type parameterization of sub-grid scale topography and land use within a framework of a regional climate model (RegCM3) and studied the inter-seasonal variability of surface climate during a winter season over western Himalayas.

There is a need to make systematic dynamic downscaling studies of the Himalayan hydroclimate using high-resolution regional models. Tiwari et al. (2018) have studied he impacts of convection and cloud microphysics schemes on the precipitation simulation over the western Himalayas using the WRF model. However, the simulations were carried out as short-term forecasts such that the WRF model was run for 7-days each starting from initial conditions spaced every 5 days. The first 2-days of simulation outputs from every run are not used in the analysis as initial 48-h period is considered as a spin-up time of the model. Such simulations are not able to provide strength and weaknesses of the model for climate downscaling at seasonal scale. Therefore, the main objectives of the study is to carry out the WRF model simulations from January to May as single long-term integration and examine the skill of the model. Sensitivity experiments have been carried out using several microphysics options available in the WRF model. Section 6.2 of the paper explains data used, the model configuration and sensitivity experiments. Simulations of hydroclimate and the results of the sensitivity experiments are described in Sect. 6.3. The study is concluded in Sect. 6.4.

6.2   Data, Model and Methodology

6.2.1   Data

Six hourly high resolutions ($0.75° \times 0.75°$) European Centre for Medium Range Weather Forecast (ECMWF) Reanalysis-Interim (ERA-I) data (Dee et al. 2011) have been used as initial and boundary conditions of the WRF model as well as for model diagnostics. The ERA–I data used in the present study are from January 01 2003 to June 01 2003 covering winter and spring periods in the western Himalayan region. This study also utilizes gridded temperature data ($1° \times 1°$) from IMD (Srivastava et al. 2009) over the Indian region.

6.2.2   WRF Model Configuration

The model used for the present study is the WRF model version 3.7.1 (Skamaraock et al. 2005). The microphysics scheme of the model controls the formation of cloud droplets and ice crystals, their growth and fall down as precipitation. Out of many microphysics schemes available with the WRF model, four schemes have been used in the present study. These are WSM6-class scheme (Hong and Lim 2006), WSM3-class scheme (Hong et al. 2004), Morrison 2-moment scheme (Hong et al. 2004) and Thompson 6-class microphysics scheme with graupel (Thompson et al. 2008). WSM 3-class scheme deals with ice water content and rain/snow. The WSM-6 class microphysics scheme involves the development of graupel. Similar to WSM-6 class scheme, Morrison 2-moment scheme involves the development of graupel in generation of precipitation. In this scheme number concentrations are also predicted for ice, snow, rain and graupel. Thompson scheme deals with mixed phase processes similar to the WSM6 and Morrison schemes. This scheme explicitly predicts the mixing ratios of cloud water, rain, cloud ice, snow, and graupel. This scheme has the double-moment cloud ice variable and predicts the number concentration of cloud ice. The cloud droplet size distribution assumed in the scheme has a variable gamma shape distribution that shifts according to the assumed droplet number concentration. It may be noted that the WSM3 is a simple ice scheme and doesn't support mixed-phase processes like other schemes mentioned above. However, the simulations with WSM3 scheme were included in the present study for the sake of completeness.

All the experiments in the present study used the unified NOAH land-surface model (Chen and Dudhia 2001) with 4-layer soil and Yonsei University (YSU) planetary boundary layer (PBL) parameterization schemes (Hong and Pan 1996). The Grell-Devenyi convection scheme (Grell and Devenyi 2002) has been used in the present work in which the convective trigger function is based on grid-resolved vertical velocity. Grell-Devenyi (GD) scheme is formulated as a multi-closure (16 mass flux closure) and it considers ensemble average of more than 100 types of clouds.

6.2.3   Design of Experiments

Four sensitivity experiments have been carried out using different microphysics while keeping the other scheme same. The experiments are

(i)   CNTL: uses WSM6 scheme
(ii) WSM3: uses WSM3 scheme
(iii) Morrison: uses Morrison scheme
(iv) Thompson: uses Thompson scheme

In the present study, two nested domains of resolution 24 and 8 Km as shown in Fig. 6.1 have been used. For every experiment, the WRF model has been run using initial conditions of 00 UTC of 1 January 2003. All the runs were made till June 01 2003.

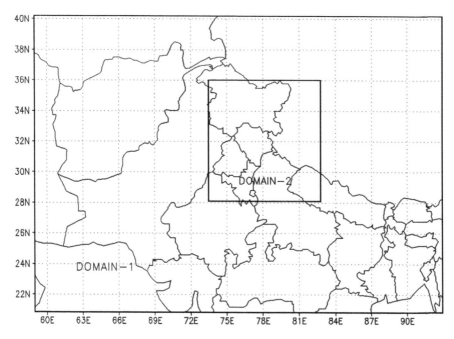

Fig. 6.1 Two nested domains (Domain-1 and Domain-2) of the WRF model used in the present study

6.3   Results and Discussion

6.3.1   CNTL Simulation

Large scale circulation, precipitation patterns including hydroclimate features are first analysed for the bigger domain i.e. Domain-1. The simulations are compared with ERA-I data for mean of January and February (JF) and March, April and May (MAM). Figure 6.2 shows the mean precipitation from WRF model (CNTL) simulations and ERA-I for the Domain-1. During January and February precipitation occurs mostly over Jammu and Kashmir (J&K), Pir Panjal range and Hindu Kush region as seen in ERA-I (Fig. 6.2c). There is a northwestsoutheast oriented precipitation patch extending from northern parts of Pakistan, Afghanistan up to some parts of Jammu and Kashmir (J&K), Himachal Pradesh (HP) and Uttarakhand (UK). It is noticed that while the precipitation pattern remains the same in MAM as in JF, mean precipitation in MAM is less than that in JF (Fig. 6.2d). The precipitation patch extends more to the east in MAM as seen in Tiwari et al. (2016) including snowfall that continues in March and April. Precipitation from gridded observed data from IMD has also been examined over the Indian part of the domain (figure not shown). It is seen from the IMD and reanalysis (ERA–I) datasets that precipitation is mostly received over the western parts of J&K and the precipitation band

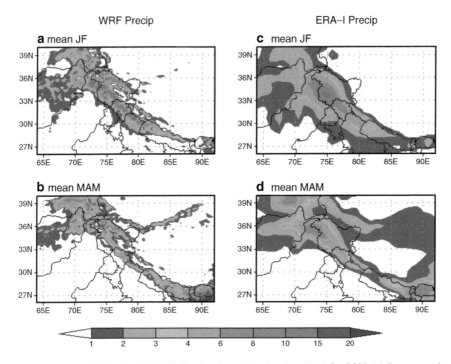

Fig. 6.2 Mean WRF model (CNTL) simulated precipitation (mm/day) for 2003 (**a**) January and February (JF); (**b**) March to May (MAM);). Mean of ERA-I total precipitation (mm/day) for (**c**) JF; (**d**) MAM for the same year

orients south-eastwards towards Himachal with reduced magnitude. Some amount of precipitation is received over the slope of mountains in HP and UK. However, eastern parts of J&K (the Ladakh region) as well as Tibet receives very less or no precipitation. The model simulated (CNTL) precipitation for JF and MAM are shown in Fig. 6.2a, b respectively. It is seen that the high-resolution model simulations have sharp features in precipitation in JF as well as in MAM. The ERA-I precipitation is from a coarse resolution model as compared to the WRF (24 km resolution for Domain-1). The model simulated precipitation over J&K and Hindukush regions does not spread horizontally over to other regions as in ERA-I. In JF (Fig. 6.2a), the precipitation amount is more (about 10–15 mm/day) along the mountain range of J&K while it is only 5–8 mm/day in ERA-I. Moreover, the amount of precipitation over HP is more in JF in the WRF model as compared to ERA-I. In MAM, the model simulates less precipitation as compared to reanalysis data (Fig. 6.2b). The region with precipitation between 6 and 8 mm/day is confined to smaller regions in J&K as compared to a broad region in reanalysis. Therefore, it is found that spatial distribution of precipitation over western Himalayas in the WRF model CNTL simulation is similar to the reanalysis products in JF and MAM. Therefore this model product can be used for further study for the region.

The mean winds at 500 hpa for JF and MAM have been examined from reanalysis and WRF model CNTL simulation to establish the consistencies between circulation and precipitation. During winter and spring seasons (JF and MAM), westerly flow dominates the study region and a large number of cyclonic systems (western disturbances) travel across the Himalayan region from west to east. In ERA-I, in JF and MAM a trough is seen in the westerlies around 70°E–80°E and 30°N to 36°N (Fig. 6.3a, b). The westerly trough is oriented northwest-southeastward. In the model simulations (Fig. 6.3c, d), all the essential features of the circulation are seen indicating that the model is able to simulate consistent circulation and precipitation over the Domain-1 of the present study. However, the westerly strength of the WRF model in JF and MAM is stronger (by about 5 m/s) over region extending from 60°E to 75°E and 33°N to 36°N.

Surface air temperature (T2m, °K) is a key parameter in the western Himalayas as important hydrological process such as snowmelt and evapotranspiration are functions of surface temperature. In order to examine the model simulated temperature over the area of interest, T2m from ERA-I has been compared with T2m from the CNTL run of the model during JF and MAM as shown in Fig. 6.4. Seasonal averages have been computed using daily mean data (average of 06, 12, 18 and 00 h)

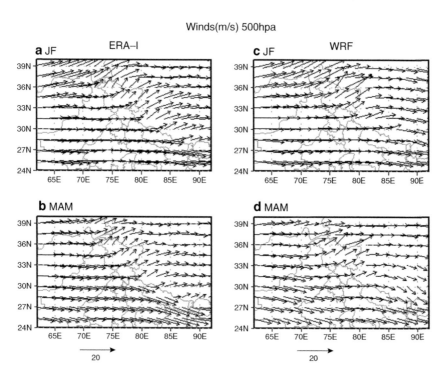

Fig. 6.3 Mean ERA-I winds (m/s) at 500 hPa for (**a**) January and February (JF); (**b**) March, April and May (MAM) for 2003. Mean of WRF model (CNTL) winds (m/s) at 500 hPa for (**c**) JF; (**d**) MAM for the same year

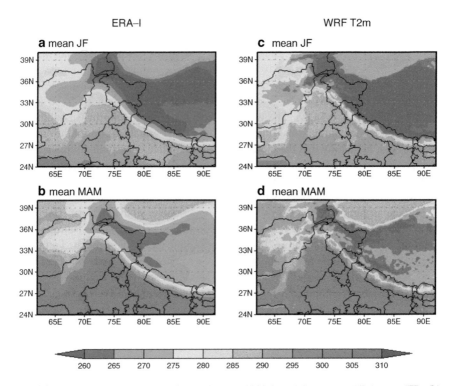

Fig. 6.4 Mean ERA-I temperature (°K) at 2 m for 2003 for (**a**) January and February (JF); (**b**) March, April and May (MAM) Mean of WRF model (CNTL) temperature (°K) at 2 m for (**c**) JF; (**d**) MAM for the same year

for both ERA-I and model simulations. It is found that the model simulated temperature values are comparable to the reanalysis temperatures. Sharp gradient in surface air temperature in the foothills of Himalayas is clearly brought out by the model, especially over the northwest India. During winter and spring, the model temperature values are less as compared to the reanalysis over higher elevation areas that covers western Himalayas and Tibet. It may be noted that observations of T2m available for assimilation in ERA-I are too few over the data sparse region such as the Himalayas. In order to examine the bias in model simulated temperatures, these values over an area over the plains of north India (75°E–80°E, 27°N–31°N) have been compared against IMD daily mean gridded temperatures (Srivastava et al. 2009). During JF and MAM, all the data have similar values as shown in Fig. 6.5.

In order to further downscale the ERA-I data to a smaller domain, the WRF model was configured for the domain-2 as shown in Fig. 6.1. The horizontal resolution of the model for this nested domain is set at 8 km as compared to 24 km set for domain-1. All other dynamics and physics options were set same as the runs for domain-1. Figure 6.6a shows the mean of accumulated snow water equivalent (SWE, mm) simulated using WRF model (CNTL run) at the end of February 2003.

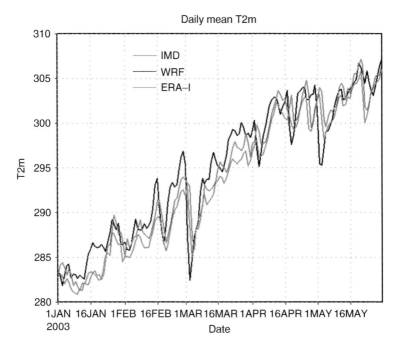

Fig. 6.5 Daily mean T2m (°K) from IMD observations, WRF model and ERA-I over the (75–80°E, 27–31°N) for January to May 2003

SWE describes the amount of water stored within the snow-pack that would be available upon melting and is a major driver of local snowmelt release and hydrological cycles (Derksen et al. 1998). The availability of spatially and temporally extensive SWE data enables a better understanding of space-time trends in snow cover. In situ observations of SWE are made by measuring the snow depth and its density. Observations of microwave sensors are also used to estimate SWE magnitude in large-scale. Tiwari et al. (2016) have examined the large-scale characteristics of SWE using satellite remote sensing data over the western Himalayan region. In the WRF model, precipitation amount is partitioned in to liquid precipitation (rainfall) and solid precipitation (snowfall). Accumulation and ablation of the snowfall occurs based on surface temperature and other meteorological conditions. The model then provides an estimate of accumulated SWE. In the WRF model simulations in the present study, the SWE pattern is characterized by a northwest-southeast band across the Satluj basin. In the Domain-2 region (Satluj basin), maximum snowfall occurs in March and April. Figure 6.6b shows the difference of SWE (mm) between that at the end of May and February 2003. The SWE accumulated on the surface in model simulations at the end of May is more than about 400 mm over most parts of the Satluj basin. The zone of Satluj basin where maximum snowfall and accumulation occurs is the major catchment area of the Satluj river. This area has maximum elevation (> 6000 m) and the Spiti river originating from this area

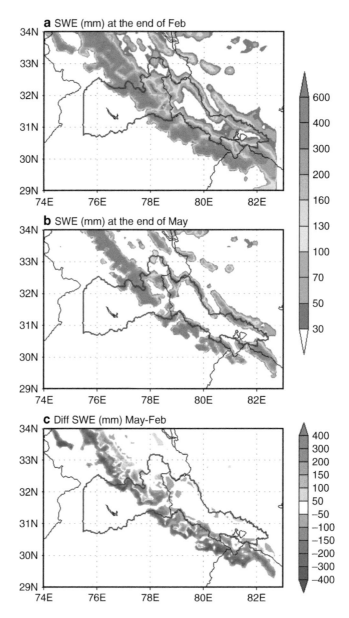

Fig. 6.6 WRF model (CNTl) simulated snow water equivalent (SWE, mm) accumulation (**a**) at the end of February 2003; (**b**) difference between SWE (mm) at the end of May and end of February 2003

contribute most of the snowmelt water to Satluj River. The accumulated snow at the end of May becomes available for snowmelt runoff in the months of June, July and august.

6.3.2 Sensitivity Studies

As mentioned earlier, a set of sensitivity experiments were carried out using various cloud microphysics schemes in the WRF model and the precipitation simulations over the western Himalayas from these experiments were compared among each other. A composite analysis was made by considering the peak precipitation period during winter (for 18 days in February 2003) over Satluj basin and adjoining areas.

Figure 6.7a shows total precipitation from CNTL experiment for the study period. It is seen that the higher elevation region receive >20 mm/day of precipitation. In contrast, the plain regions in the Satluj basin get about 5 mm/day. The precipitation amount received over J&K is more than 20 mm/day in these 18 days of composite analysis. On the slope of higher elevation areas of Satluj basin, it is seen that WSM6 produces more precipitation than other microphysics schemes (Fig. 6.7b, c). Moreover, the WSM6 scheme has more precipitation than the Morrison scheme over the region of SE to NW precipitation band. Very less difference (between 1 and 2 mm/day) of precipitation is found between WSM6 and WSM3 schemes (Fig. 6.7d)

Differences in non-convective precipitation in WRF model simulations come from the differences in the treatment of phase change of water in the cloud microphysics schemes. Accumulated solid precipitation (snow) and liquid precipitation (rain) obtained from the composite analysis of the study period in February 2003 are shown respectively in Fig. 6.8a, b. It may be noted that during this 18-day period, >500 mm of snowfall was simulated over the over the higher reaches of the Satluj basin. Rainfall occurred mostly over the plain areas and lower reaches (Fig. 6.8b). The differences in total solid and liquid precipitation between CNTL and Morrison scheme are shown in Fig. 6.8c, d respectively. From the difference plot, it can be seen that the WSM6 (CNTL) scheme simulates more snowfall (> 100 mm/day) than Morrison in the northern part of the domain. However, the Morrison scheme simulates about 100 mm more rain over western and northern region than that in the WSM6 scheme. The differences are similar between CNTL and Thompson scheme as seen in Fig. 6.8e, f as the Thompson scheme also simulates more snow on the mountain top than CNTL (Tiwari et al. 2018). No large differences are seen between simulated snow or liquid rainfall in WSM3 and CNTL simulations (Fig. 6.8g, h). It may be noted here that in the double moment schemes (e.g. Morrison scheme) particle type vary as a function of the time and the ice mixing ratio is diagnosed separately. As a result of additional moment computation, the double moment scheme (Morrison scheme here) has greater flexibility in representing size distributions impacting simulation of ice, snow, graupels or hails in the present study.

In the WRF model simulations, the hydrometeors are separated into non-precipitating particles and precipitating particles. The mass mixing ratios of cloud

Fig. 6.7 Composite analysis of total precipitation (mm/day) for excess precipitation days in February 2003 for (**a**) total precipitation in CNTL; difference in precipitation (mm/day) between CNTL and Morrison; difference between CNTL and Thompson and (**d**) difference between CNTL and WSM3 experiments

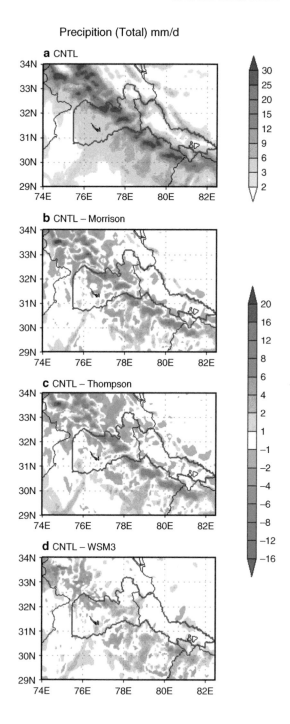

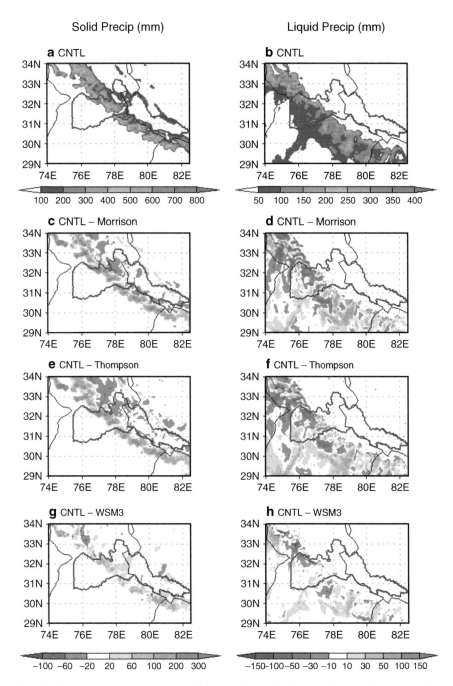

Fig. 6.8 Composite analysis of total solid (snowfall) and liquid (rainfall) precipitation (mm/day) for excess precipitation days in February 2003 for (**a**) total solid precipitation in CNTL; (**b**) total liquid precipitation in CNTL; (**c**) and (**d**) difference in solid and liquid precipitation (mm/day) between CNTL and Morrison respectively; (**e**) and (**f**) difference between CNTL and Thompson; (**g**) and (**h**) difference between CNTL and WSM3 experiments

and ice (in kg/kg) are referred to as QCLOUD and QICE respectively in the present study. The mass mixing ratios of precipitating particles such as rain, snow and graupels are referred to as QRAIN, QSNOW and QGRAUP respectively. Spatial pattern of QSNOW at 500 hPa from CNTL simulation and its difference from the Morrison scheme are depicted in Fig. 6.9a, b respectively. Magnitude of QSNOW (> 12 kg/kg) is seen over the region of maximum precipitation. The orientation of QSNOW band is same as the precipitation band seen earlier. QSNOW has noticeable magnitude between 700 hPa and 300 hPa (not shown in figure). Difference between single moment scheme (WSM6) and double moment scheme (Morrison) indicates that the WSM6 scheme simulates less snow (about 3–4 kg/kg) over the peak mountainous region as compared to Morrison which has more QSNOW to the downwind. Similar to QSNOW, QICE also forms only above 700 hPa with a maximum (3–4 kg/kg) reaching at about 400 hPa (figure not shown). Figure 6.9c, d has the plots of QICE at 500 hPa in the study domain from CNTL and its difference between the Morrison scheme. The single moment scheme (WSM6) simulates more amount of ice particles at this altitude than the double moment Morrison scheme. Graupel particle (QGRAUPEL) forms only on the windward side of the high mountain at 600 hPa with a maximum of about 6–8 kg/kg in CNTL simulations (Fig. 6.9e). It is also seen that in WSM6 scheme, preferred locations of snow and graupel formation is on the upwind side on the mountain slope with the Morrison scheme simulating less graupel The WSM6 (CNTL) has more rain over the narrow zone of sharply rising orography. However, the Morrison scheme has more precipitation downstream at higher elevations.

6.4 Conclusion

The climate system over the Himalayas is vulnerable to global warming and climate change. The inhomogeneous terrain and high altitude of the Himalayas has restricted observational data of the climate system in the region. Therefore, detailed climate of Himalayas is still unknown. Moreover the western Himalayan region has sharp orography gradient, therefore, the global reanalysis data are too coarse to represent the hydroclimate over the region. As a result, in order to have a high resolution dataset for the region, it is necessary to employ dynamic downscaling method. In this study, the ERA-Interim dataset (January to May) over the western Himalayas has been downscaled using the high-resolution (8 km) WRF model and several sensitivity studies have been carried out using microphysics schemes.

Examination of the WRF model simulated surface temperature over the western Himalayas indicate that the model has been successful in bringing out the sharp temperature gradient between plains of northwest India and the Himalayas. The downscaled precipitation pattern has sharp features unlike that of ERA-Interim. The WRF model simulated precipitation from various sensitivity experiments have been compared. It is found that there is a shift in the location of maximum precipitation zone in Morrison and Thompson schemes as compared to WSM6 during winter.

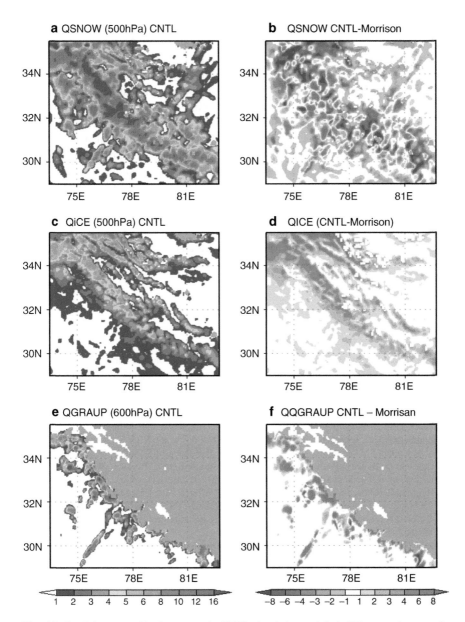

Fig. 6.9 Spatial pattern of hydrometeors in CNTL simulation and their differences between the simulations with the Morrison scheme. (**a**) QSNOW (kg/kg) at 500 hPa; (**c**) QICE (kg/kg) at 500 hPa and (**e**) QGRAUP (Kg/kg) at 600 hPa from CNTL simulations. Differences between CNTL and Morrison scheme (**b**) QSNOW; (**d**) QICE and (**f**) QGRAUP

The WSM6 scheme has maximum precipitation more upwind as compared to other schemes as the WSM6 favors precipitation on the slopes of the terrain whereas the Morrison and Thompson schemes produce more snow on the mountain top.

This study emphasizes that a correct representation of the microphysical processes in the models is crucial for long-term climate simulations as these microphysical schemes are responsible for partitioning atmospheric water into vapor, cloud liquid water, cloud ice etc. leading either to solid or liquid precipitation. One of the limiting factor of such studies is the lack of observational data over the Himalayan region to carry out proper verification studies. Moreover, this study has been carried out using simulations of 1 year only. Further simulations are being carried out for other years for developing a regional forecasting system for the region using dynamic downscaling strategy.

Acknowledgements This work has been carried out as a part the project "Dynamics of Himalayan ecosystem and its impact under changing climate scenario in Western Himalaya" under the National Mission on Himalayan Studies (NMHS) of the Ministry of Environment, Forest & Climate Change, Government of India.

References

Chen F, Dudhia J (2001) Coupling an advanced land-surface/hydrology model with the Penn State/ NCAR MM5 modeling system. Part I: model description and implementation. Mon Weather Rev 129:569–585

Cossu F, Hocke K (2014) Influence of microphysical schemes on atmospheric water in the Weather Research and Forecasting model. Geosci Model Dev 7:147–160

Crâetat J, Pohl B, Richard Y, Drobinski P (2012) Uncertainties in simulating regional climate of Southern Africa: sensitivity to physical parameterizations using WRF. Clim Dyn 38:613–634

Dee DP, Uppala SM, Simmons AJ, Berrisford P, Poli P, Kobayashi S, Andrae U, Balmaseda MA, Balsamo G, Bauer P, Bechtold P, Beljaars ACM, van de Berg L, Bidlot J, Bormann N, Delsol C, Dragani R, Fuentes M, Geer AJ, Haimberger L, Healy SB, Hersbach H, Hólm EV, Isaksen L, Kållberg P, Köhler M, Matricardi M, McNally AP, Monge-Sanz BM, Morcrette J-J, Park B-K, Peubey C, de Rosnay P, Tavolato C, Thépaut J-N, Vitart F (2011) The ERA-Interim reanalysis: configuration and performance of the data assimilation system. Q J Roy Meteorol Soc 137:553–597

Derksen C, LeDrew E, Goodison B (1998) SSM/I derived snow water equivalent data: the potential for investigating linkages between snow cover and atmospheric circulation. Atmosphere-Ocean 36(2):95–117

Dimri AP (2009) Impact of subgrid scale scheme on topography and landuse for better regional scale simulation of meteorological variables over the western Himalayas. Clim Dyn 32(4):565–574

Dimri AP, Chevuturi A (2014) Model sensitivity analysis study for western disturbances over the Himalayas. Meteorog Atmos Phys 123:155–180

Dimri AP, Niyogi D, Barros AP, Ridley J, Mohanty UC, Yasunari T, Sikka DR (2015) Western disturbances: a review. Rev Geophys 53(2):225–246

Flaounas F, Bastin S, Janicot S (2010) Regional climate modelling of the 2006 West African monsoon: sensitivity to Cumulus and planetary boundary layer parameterisation using WRF. Clim Dyn 36:1083–1105

Givati A, Lynn B, Liu Y, Rimmer A (2012) Using the WRF model in an operational streamflow forecast system for the Jordan River. J Appl Meteorol Climatol 51:285–299. https://doi.org/10.1175/JAMC-D-11-082.1

Grell G, Devenyi D (2002) A generalized approach to parameterizing convection combining ensemble and data assimilation techniques. Geophys Res Lett 29(14):38–31

Hong SY, Lim J (2006) The WRF single-moment 6-class microphysics scheme (WSM6). J Korean Meteorol Soc 42:129–151

Hong SY, Pan HL (1996) Nonlocal boundary layer vertical diffusion in a medium range forecast model. Mon Weather Rev 124(10):2322–2339

Hong SY, Dudhia J, Chen SH (2004) A revised approach to ice microphysical processes for the bulk parameterization of clouds and precipitation. Mon Weather Rev 132:103–120

Kar SC, Rana S (2014) Interannual variability of winter precipitation over northwest India and adjoining region: impact of global forcings. Theor Appl Climatol 116(3–4):609–623

Kar SC, Tiwari S (2016) Model simulations of heavy precipitation in Kashmir, India, in September 2014. Nat Hazards 81:167–188. https://doi.org/10.1007/s11069-015-2073-3

Kim HJ, Wang B (2011) Sensitivity of the WRF model simulation of the East Asian summer monsoon in 1993 to shortwave radiation schemes and ozone absorption. Asia-Pac J Atmos Sci 47:167–180

Liang X-Z, Xu M, Yuan X, Ling T, Choi HI, Zhang F, Chen L, Liu S, Su S, Qiao F, He Y, Wang JL, Kunkel KE, Gao W, Joseph E, Morris V, Yu T-W, Dudhia J, Michalakes J (2012) Regional climate–weather research and forecasting model. B Am Meteorol Soc 93:1363–1387. https://doi.org/10.1175/BAMS-D-11-00180.1

Lo JC-F, Yang Z-L, Pielke RA Sr (2008) Assessment of three dynamical climate downscaling methods using the Weather Research and Forecasting (WRF) model. J Geophys Res 113:D09112. https://doi.org/10.1029/2007JD009216

Patil R, Kumar PP (2016) WRF model sensitivity for simulating intense western disturbances over North West India. Model Earth Syst Environ 2:82. https://doi.org/10.1007/s40808-016-0137-3

Rajeevan M, Kesarkar A, Thampi SB, Rao TN, Radhakrishna B, Rajasekhar M (2010) Sensitivity of WRF cloud microphysics to simulations of a severe thunderstorm event over Southeast India. Ann Geophys 28:603–619

Raju PV, Potty J, Mohanty UC (2011) Sensitivity of physical parameterizations on prediction of tropical cyclone Nargis over the Bay of Bengal using WRF model. Meteorog Atmos Phys 113:125–137

Sinha P, Mohanty UC, Kar SC, Dash SK, Kumari S (2013) Sensitivity of the GCM driven summer monsoon simulations to cumulus parameterization schemes in nested RegCM3. Theor Appl Climatol 112(1–2):285–306

Sinha P, Tiwari PR, Kar SC, Mohanty UC, Raju PVS, Dey S, Shekhar MS (2015) Sensitivity studies of convective schemes and model resolutions in simulations of wintertime circulation and precipitation over the Western Himalayas. Pure Appl Geophys 172(2):503–530

Skamaraock WC, Klemp JB, Dudhia J, Gill DO, Barker DM, Wang W, Powers JG (2005) A description of the advanced research WRF version 2, NCAR Technical Note. www.wrf-model.org

Srivastava AK, Rajeevan M, Kshirsagar SR (2009) Development of a high resolution daily gridded temperature data set (1969–2005) for the Indian region. Atmos Sci Lett. https://doi.org/10.1002/asl.232

Thompson G, Field PR, Rasmussen RM, Hall WD (2008) Explicit forecasts of winter precipitation using an improved bulk microphysics scheme. Part II: implementation of a new snow parameterization. Mon Weather Rev 136(12):5095–5115

Tiwari PR, Kar SC, Mohanty UC, Dey S, Sinha P, Raju PVS (2015) The role of land surface schemes in the regional climate model (RegCM) for seasonal scale simulations over Western Himalaya. Atmosfera 28(2):129–142

Tiwari S, Kar SC, Bhatla R (2016) Interannual variability of snow water equivalent (SWE) over Western Himalayas. Pure Appl Geophys 173(4):1317–1335

Tiwari S, Kar SC, Bhatla R (2018) Dynamic downscaling over western Himalayas: Impact of cloud microphysics schemes. Atmos Res 201:1–16. https://doi.org/10.1016/j.atmosres.2017.10.007

Chapter 7
Climate Variability and Extreme Weather in High Mountain Asia: Observation and Modelling

Leila M. V. Carvalho, Jesse Norris, Forest Cannon, and Charles Jones

Abstract The hydrological cycle in High Mountain Asia is of critical importance for water security, agriculture and power generation for one of the most densely populated regions in the world. Glaciers over the central Himalaya have retreated at particularly rapid rates in recent decades, while glacier mass in the Karakoram appears stable. This chapter focuses on the climate of recent decades and discusses how the unique orography and complex terrain of this area of the planet influence weather patterns and the climate of the region and implications of these interactions for the future of high-elevation freshwater reservoirs.

7.1 High Mountain Asia Climate

High Mountain Asia (HMA) is the definition of the area that extends from the Hindu Kush and Tien Shan in the west to the Eastern Himalaya in the east. The Lesser Himalayas (average height 1200–3000 m) and the Great Himalayas (snow-covered mountains having an average height of 3000–7000 m above Mean Sea Level) are also part of HMA (Joshi 2004).

The regional climate of the HMA is influenced by two predominant systems: the Indian Summer Monsoon that extends approximately from June to September and winter western disturbances (WWD) from December to March. The interaction between these systems results in two distinct climatic regimes with different precipitation and water storage characteristics (Bookhagen and Burbank 2010; Cannon

L. M. V. Carvalho (✉) · C. Jones
Department of Geography, UC Santa Barbara, Santa Barbara, CA, USA
e-mail: leila@eri.ucsb.edu; cjones@eri.ucsb.edu

J. Norris
Department of Ocean and Atmospheric Sciences, UC Los Angeles, Los Angeles, CA, USA
e-mail: jessenorris@ucla.edu

F. Cannon
Scripps Institution of Oceanography, UC San Diego, San Diego, CA, USA
e-mail: fcannon@ucsd.edu

© Springer Nature Switzerland AG 2020
A. P. Dimri et al. (eds.), *Himalayan Weather and Climate and their Impact on the Environment*, https://doi.org/10.1007/978-3-030-29684-1_7

et al. 2016). Consequently, there is an east-west gradient in monsoonal influence across the HMA (Fig. 7.1) with the Central Himalaya (CH) receiving up to 80% of its annual precipitation during the monsoon months. In contrast, the Karakoram-Hidu Kush region (KH) receives more than 50% of its precipitation during the winter months by fewer than ten WWD (Lang and Barros 2004). This chapter discusses the influence of both systems and their variability in the hydrology of HMA and ends showing recent climatic trends affecting the region.

Synoptic scale systems associated with significant snowstorms in the Himalaya and KH during winter have been associated with the propagation of midlatitude wave trains and respective extratropical cyclones that disturb circulation in the lower and upper atmosphere (Singh et al. 1995; Lang and Barros 2004; Barlow et al. 2005; Dimri et al. 2015). These systems affect weather in the KH and adjacent regions (Ridley et al. 2013; Cannon et al. 2015) and although they are more common during winter (December–February) they are often observed during Spring (March–April) (Cannon et al. 2015, 2016; Norris et al. 2015). When cyclones associated with WWD are terrain-locked, they often result in extreme precipitation and snowfall caused by topographic uplift (Dimri 2006; Norris et al. 2015). Snowfall generated by WWD maintains regional snowpack and glaciers (Anders et al. 2006; Tahir et al. 2011; Ridley et al. 2013; Cannon et al. 2015), and is essential to water resources for hundreds of millions of people (Hewitt 2005; Immerzeel and Bierkens 2010; Miller et al. 2012). Nonetheless, while precipitation is primarily orographically forced during intense WWD with strong cross-barrier winds (Braun et al. 1997; Medina et al. 2005; Norris et al. 2017), weaker WWD with similar precipitation totals may occur when enhanced instability and high moisture content are observed at low levels in late winter and spring (Cannon et al. 2016).

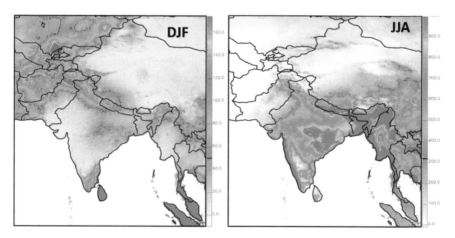

Fig. 7.1 Total seasonal mean precipitation over HMA, Subcontinental India, Southwestern China, Pakistan, Afghanistan, Kyrgyzstan, Uzbekistan, Kazakhstan, Burma, Thailand and Malaysia during December-to-February (left) and July-August (right). Note the difference in scale ranges. Data source: CHIRPS. (Funk et al. 2015)

Cannon et al. (2015) examined variations and changes in the WWD and relationships with extreme precipitation in the KH and central Himalaya (CH) regions from 1979 to 2010 using multiple data sets. They showed that extreme events occurring in these two regions are often independent. Additionally, they used the wavelet-power spectrum of 200 hPa geopotential anomalies on synoptic time-scales as an index for WWD and investigated trends in the amplitude of the signal affecting KH and CH separately. They showed evidence that the frequency and strength of WWD affecting KH have increased, possibly increasing local extreme events. Contrasting trends were observed for the CH domain. Although teleconnections with known modes of climate variability that affect central Asia (e.g., Arctic Oscillation, Eurasian/Polar Pattern, the El Niño Southern Oscillation, and the Siberian High) may influence the interannual variability of WWD, they do not explain the observed trends.

Central Himalaya (CH) is also deeply affected by the Indian Summer Monsoon (Fig. 7.1) (Parthasarathy et al. 1994; Bookhagen and Burbank 2006; Guhathakurta and Rajeevan 2008; Ueno et al. 2008; Mukherjee et al. 2015; Carvalho et al. 2016). In Nepal, the onset of precipitation occurs in the middle of June. Monsoon precipitation generally accounts for 80% of annual precipitation (Ueno et al. 2008), but this amount vary during the season depending on intraseasonal variations of the Indian Summer Monsoon (Carvalho et al. 2016). The El Nino/Southern Oscillation (ENSO) plays a primary role in modulating the interannual variability of precipitation in the CH. Numerous studies have shown that precipitation in the CH exhibits a strong diurnal cycle, with prevalence of daytime precipitation on ridges and night time precipitation in valleys, with the main mechanism being the development of orographic convection (Higuchi 1977; Bhatt and Nakamura 2005). Nonetheless, precipitation patterns are heterogeneous in this region and depend on the steepness and orientation of the slopes relative to the monsoon circulation (Yasunari and Inoue 1978; Lang and Barros 2004).

7.2 Intraseasonal to Interannual Variability

The WWD activity over southwest Asia (25–40°N; 40–80°E) is influenced by global modes of climate variability including the Madden-Julian Oscillation (MJO) (Jones et al. 2004b; Barlow et al. 2005; Dimri 2013b; Hoell et al. 2014), El Niño Southern Oscillation (ENSO) (Syed et al. 2006; Yadav et al. 2010, 2013; Dimri 2013a; Hoell et al. 2013), Arctic Oscillation/North Atlantic Oscillation (NAO) (Gong et al. 2001; Syed et al. 2006; Yadav et al. 2010; Filippi et al. 2014), and the Polar Eurasia Pattern (Lang and Barros 2004).

During the boreal winter, southwest Asia, and particularly the KH, are under influence of upper-level westerlies (Krishnamurti 1961). Variability in the upper-level jet over southwest Asia (i.e, shear, maximum wind speed, and deformation) has a complex relationship with WWD activity (Barlow et al. 2005). Teleconnections with ENSO (Rasmusson and Carpenter 1982) and NAO (Barnston and Livezey

1987) on interannual time-scales and the MJO (Madden and Julian 1971, 1972, 1994; Jones et al. 2004a) on intraseasonal timescales modify circulation in the lower and upper troposphere and the characteristics of the subtropical jet, significantly influencing the frequency and intensity of WWD with implications to the seasonal precipitation totals in HMA (Yadav et al. 2010; Filippi et al. 2014; Dimri et al. 2015). However, Cannon et al. (2015) showed that the characteristics and development of WWD and orographic precipitation affecting KH depend on how these global modes of variability interact and modify the atmospheric dynamic and thermodynamic conditions.

Cannon et al. (2017) examined the influence of ENSO and the MJO on the characteristics and frequency of WWD. They showed that on interannual time-scales, El Niño related changes in tropical diabatic heating induce a Rossby wave response over southwest Asia that enhances the dynamical forcing of WWD and increases available moisture. This combined effect enhances the frequency of extreme orographic precipitation in the KH during El Niño compared to La Niña or neutral conditions. Nonetheless, the Rossby wave response associated with the MJO activity is less spatially uniform over southwest Asia and varies on intraseasonal timescales. Therefore, the relationships between MJO activity and WWD, and how the MJO phases affect KH precipitation, are more complex and depend on numerous factors. For instance, some phases of the MJO favor the dynamical enhancement of WWD and at the same time suppress available moisture over southwest Asia, whereas other phases favor exactly opposite conditions. Because moisture availability and convective instability (thermodynamic factors) and strong cross-barrier winds (dynamical enhancement) may equally influence orographic precipitation (Cannon et al. 2015; Norris et al. 2015), most MJO phases can induce extreme precipitation in the KH as the oscillation evolves. Therefore, understanding and predicting independent and combined effects in tropical forcing by ENSO and MJO must be considered for long-term evaluations of the KH hydrology.

7.3   Recent Climatic Trends Affecting Glaciers in HMA

Most of Earth's Alpine glaciers have generally retreated in recent decades in response to global warming. However, evidence exists that in some areas increasing glacier melt may be offset by regional snowfall and/or temperature trends. This intriguing behavior seems particularly evident in glaciers over HMA (Hewitt 2005; Scherler et al. 2011; Bolch et al. 2012; Kaab et al. 2012). While glaciers in the CH exhibit some of the fastest retreat rates on Earth, many glaciers in the Karakoram appear to be stable or even advancing (Scherler et al. 2011; Bolch et al. 2012; Gardelle et al. 2012). Glaciers are reservoirs of moisture from precipitation for populations and ecosystems. Therefore, understanding how recent changes in atmosphere conditions may have contributed to the observed glacial melting rates in recent decades and whether these changes will remain, cease, or intensify is crucial in predicting the future of freshwater reservoirs for millions of people as the planet warms.

Since precipitation over the CH predominantly occur during summer (Fig. 7.1), the rapid retreat of central Himalayan glaciers has been attributed to increasing summer temperatures at high elevations (Shrestha et al. 1999). Therefore, changes to the Indian summer monsoon (ISM) precipitation are particularly relevant for CH glaciers. In fact, many studies have shown that precipitation associated with the ISM has decreased over the twentieth century because of the weakening of the over-turning monsoonal circulation caused by the great warming of the Indian Ocean (Krishnan et al. 2013; Zhao et al. 2014; Roxy et al. 2015). To investigate the influence of these circulations on the CH, Duan et al. (2006) analyzed ice cores from a central-Himalayan glacier and found significant decreases in orographic monsoonal precipitation during the twentieth century. Therefore, there is mounting evidence that increasing temperatures and decreasing precipitation in summer have both contributed to glacier retreat over the CH. Contrastingly, the advance of some glaciers over the KH has been attributed to increasing precipitation in winter and summer (Archer and Fowler 2004), as well as decreasing summer temperature (Fowler and Archer 2006; Forsythe et al. 2017). These observations are consistent with the increase in the frequency and intensity of synoptic-scale activity in the KH together with the decrease of troughs affecting CH in the winter shown in Cannon et al. (2015).

Forsythe et al. (2017) proposed an index to quantify large-scale variations in circulation affecting the Karakoram (Karakoram Zonal Index – KZI). When KZI is positive (negative) there is an anticyclonic (cyclonic) anomaly in the Karakoram. They found a significant negative trend in the KZI over recent decades in summer months correlated with a cooling trend in the upper Indus basin, near the Karakoram. They attributed the negative trend to adiabatic cooling associated with a cyclonic trend, increased cloudiness, and decreased insolation. Contrastingly, they found that CH is under the influence of a vortex over the Indian subcontinent associated with the monsoon that is anomalously anticyclonic when the KZI is anomalously cyclonic. Therefore, these results provide additional evidence that the Karakoram and central Himalayan glaciers appear to have been under the influence of contrasting climatic trends in both winter and summer over the last few decades.

Norris et al. (2018) deciphered some of the regional mechanisms explaining the observed dichotomy between climatic regimes affecting the KH and the CH. They analyzed 36 years (1979–2015) of the Climate Forecast System Reanalysis (CFSR) (Saha et al. 2010) dynamically downscaled with the Weather Research and Forecasting (WRF) model over High Mountain Asia at 6.7 km resolution. They showed that an anticyclonic warming trend is observed over the majority of High Mountain Asia in all seasons, with distinctive differences observed between the CH and KH in winter and summer (Fig. 7.2).

Although the CH has been under the influence of an anticyclonic trend in winter and summer, the WRF simulations show reduced cloud cover in the summer leading to significant warming and reduced snowfall in recent years. Contrastingly, the KH is at the boundary between large-scale cyclonic and anticyclonic trends (Fig. 7.2). WRF downscaling shows that the KH has not experienced significant snowfall or temperature changes in the winter or summer, despite significant positive trends in cloud cover and consequent negative trends in shortwave radiation. The downscaling at 6.7 km resolution did not identify any significant trends over glaciers or in

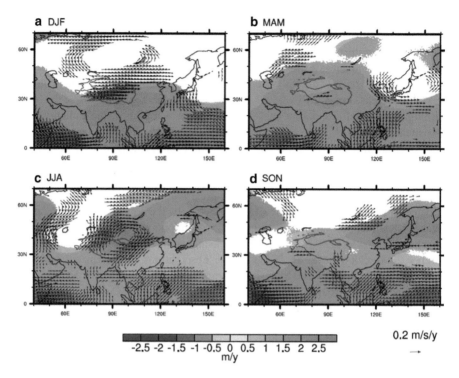

Fig. 7.2 Seasonal trends over Asia (1979/80 to 2014/15) obtained with CFSR of 200 hPa geopotential heights (colours, m/year) and winds (ms^{-1}/year). Only statistically significant trends (at 5% significance level) are plotted. Blue contour indicates the 3-km elevation to locate HMA. Reproduced from Norris et al. (2017)

neighboring regions to the KH or Spiti Lahaul, where glaciers have retreated as over the CH (Gardelle et al. 2012; Kaab et al. 2012). Although the reasons for the KH anomaly and the fast melting rate of the CH cannot be totally explained by atmospheric mechanisms, this study showed the complexity of the problem and the need of assessing variations in local circulations in regions with complex terrain such as HMA under conditions of a changing climate.

7.4 Conclusions

HMA is an important reservoir of fresh water for millions of people in the form of perennial glaciers and snow. The hydrological cycle in HMA is influenced by the Summer Indian Monsoon and the WWD. These two systems distinctly affect circulation and diurnal cycles of precipitation across the steep terrain and complex network of mountains and valleys that characterize HMA. Both the Summer Indian Monsoon and the WWD exhibit variations on multiple time-scales that modulate circulation and precipitation in HMA. These distinct systems and respective driving mechanisms play significant role in glacial growth rates and in the observed dichotomy between glaciers in the western HMA (mostly stable or growing) and central HMA (where glaciers exhibit fast melting rates) in recent decades. This apparent paradox appears to be associated with an anomalous cyclonic trend affecting the upper Indus basin near the Karakoram, whereas an anomalous anticyclonic trend is influencing Central Himalaya. These trends have contributed to variations in cloudiness, temperature, radiation and circulation across HMA that are relevant to the regional hydrological cycle. Additional observational and modeling studies are necessary to investigate the impacts of these trends, how they have contributed to glacial balance in these distinct regions and whether these conditions will persist in the future.

Acknowledgments We appreciate the contribution of Pete Peterson in providing CHIRPS data. We acknowledge NSF AG AGS-1116105 support.

References

Anders AM, Roe GH, Hallet B, Montgomery DR, Finnegan NJ, Putkonen J (2006) Spatial patterns of precipitation and topography in the Himalaya. Geol Soc Am Spec Pap 398:39–53

Archer DR, Fowler HJ (2004) Spatial and temporal variations in precipitation in the Upper Indus Basin, global teleconnections and hydrological implications. Hydrol Earth Syst Sci 8:47–61

Barlow M, Wheeler M, Lyon B, Cullen H (2005) Modulation of daily precipitation over Southwest Asia by the Madden-Julian Oscillation. Mon Weather Rev 133:3579–3594

Barnston AG, Livezey RE (1987) Classification, seasonality and persistence of low-frequency atmospheric circulation patterns. Mon Weather Rev 115:1083–1126

Bhatt BC, Nakamura K (2005) Characteristics of monsoon rainfall around the Himalayas revealed by TRMM precipitation radar. Mon Weather Rev 133:149–165

Bolch T et al (2012) The state and fate of Himalayan Glaciers. Science 336:310–314

Bookhagen B, Burbank DW (2006) Topography, relief, and TRMM-derived rainfall variations along the Himalaya (vol 33, art no L08405, 2006). Geophys Res Lett 33

Bookhagen B, Burbank DW (2010) Towards a complete Himalayan hydrologic budget: the spatiotemporal distribution of snow melt and rainfall and their impact on river discharge. J Geophys Res Earth Surface. https://doi.org/10.1029/2009JF001426

Braun SA, Houze RA, Smull BF (1997) Airborne dual-Doppler observations of an intense frontal system approaching the Pacific Northwest coast. Mon Weather Rev 125:3131–3156

Cannon F, Carvalho LMV, Jones C, Bookhagen B (2015) Multi-annual variations in winter westerly disturbance activity affecting the Himalaya. Clim Dyn 44:441–455

Cannon F, Carvalho LMV, Jones C, Norris J (2016) Winter westerly disturbance dynamics and precipitation in the western Himalaya and Karakoram: a wave-tracking approach. Theor Appl Climatol 125:27–44

Cannon F, Carvalho LMV, Jones C, Hoell A, Norris J, Kiladis GN, Tahir AA (2017) The influence of tropical forcing on extreme winter precipitation in the western Himalaya. Clim Dyn 48:1213–1232

Carvalho LMV, Jones C, Cannon F, Norris J (2016) Intraseasonal-to-interannual variability of the Indian Monsoon identified with the Large-Scale Index for the Indian Monsoon System (LIMS). J Clim 29:2941–2962

Dimri AP (2006) Surface and upper air fields during extreme winter precipitation over the western Himalayas. Pure Appl Geophys 163:1679–1698

Dimri AP (2013a) Relationship between ENSO phases with Northwest India winter precipitation. Int J Climatol 33:1917–1923

Dimri AP (2013b) Intraseasonal oscillation associated with the Indian winter monsoon. J Geophys Res-Atmos 118:1189–1198

Dimri AP, Niyogi D, Barros AP, Ridley J, Mohanty UC, Yasunari T, Sikka DR (2015) Western disturbances: a review. Rev Geophys 53:225–246

Duan K, Yao T, Thompson LG (2006) Response of monsoon precipitation in the Himalayas to global warming. J Geophys Res Atmos 111

Filippi L, Palazzi E, von Hardenberg J, Provenzale A (2014) Multidecadal variations in the relationship between the NAO and winter precipitation in the Hindu Kush-Karakoram. J Clim 27:7890–7902

Forsythe N, Fowler HJ, Li XF, Blenkinsop S, Pritchard D (2017) Karakoram temperature and glacial melt driven by regional atmospheric circulation variability. Nat Clim Chang 7:664

Fowler HJ, Archer DR (2006) Conflicting signals of climatic change in the Upper Indus Basin. J Clim 19:4276–4293

Funk C et al (2015) The climate hazards infrared precipitation with stations-a new environmental record for monitoring extremes. Sci Data 2:150066

Gardelle J, Berthier E, Arnaud Y (2012) Slight mass gain of Karakoram glaciers in the early twenty-first century. Nat Geosci 5:322–325

Gong DY, Wang SW, Zhu JH (2001) East Asian winter monsoon and Arctic Oscillation. Geophys Res Lett 28:2073–2076

Guhathakurta P, Rajeevan M (2008) Trends in the rainfall pattern over India. Int J Climatol 28:1453–1469

Hewitt K (2005) The Karakoram anomaly? Glacier expansion and the 'elevation effect,' Karakoram Himalaya. Mt Res Dev 25:332–340

Higuchi K (1977) Effects of nocturnal precipitation on the mass balance of the Rikha Sambe glacier, Hidden valley, Nepal. Seppyo 39:43–49

Hoell A, Barlow M, Saini R (2013) Intraseasonal and seasonal-to-interannual Indian Ocean convection and hemispheric teleconnections. J Clim 26:8850–8867

Hoell A, Barlow M, Wheeler MC, Funk C (2014) Disruptions of El Nino-Southern Oscillation teleconnections by the MaddenJulian Oscillation. Geophys Res Lett 41:998–1004

Immerzeel WW, Bierkens MFP (2010) Asian water towers: more on monsoons response. Science 330:585–585

Jones C, Waliser DE, Lau KM, Stern W (2004a) Global occurrences of extreme precipitation and the Madden-Julian oscillation: observations and predictability. J Clim 17:4575–4589

Jones C, Carvalho LMV, Higgins RW, Waliser DE, Schemm JKE (2004b) Climatology of tropical intraseasonal convective anomalies: 1979-2002. J Clim 17:523–539

Joshi SC (2004) Uttaranchal: environment and development – a geo-ecological overview. Gyanodaya Prakashan, Nainital, p 426

Kaab A, Berthier E, Nuth C, Gardelle J, Arnaud Y (2012) Contrasting patterns of early twenty-first-century glacier mass change in the Himalayas. Nature 488:495–498

Krishnamurti TN (1961) The subtropical jet stream of winter. J Meteorol 18:172–191

Krishnan R et al (2013) Will the South Asian monsoon overturning circulation stabilize any further? Clim Dyn 40:187–211

Lang TJ, Barros AP (2004) Winter storms in the central Himalayas. J Meteorol Soc Jpn 82:829–844

Madden RA, Julian PR (1971) Detection of a 40-50 day oscillation in the zonal wind in the tropical Pacific. J Atmos Sci 28:702–708

Madden RA, Julian PR (1972) Description of global-scale circulation cells in the tropics with a 40-50 day period. J Atmos Sci 29:1109–1123

Madden RA, Julian PR (1994) Observations of the 40-50-day tropical oscillation-a review. Mon Weather Rev 122:814–837

Medina S, Smull BF, Houze RA, Steiner M (2005) Cross-barrier flow during orographic precipitation events: results from MAP and IMPROVE. J Atmos Sci 62:3580–3598

Miller JD, Immerzeel WW, Rees G (2012) Climate change impacts on glacier hydrology and river discharge in the Hindu Kush-Himalayas. A synthesis of the scientific basis. Mt Res Dev 32:461–467

Mukherjee S, Joshi R, Prasad RC, Vishvakarma SCR, Kumar K (2015) Summer monsoon rainfall trends in the Indian Himalayan region (vol 121, pg 789, 2015). Theor Appl Climatol 121:803–805

Norris J, Carvalho LMV, Jones C, Cannon F (2015) WRF simulations of two extreme snowfall events associated with contrasting extratropical cyclones over the western and central Himalaya. J Geophys Res-Atmos 120:3114–3138

Norris J, Carvalho LMV, Jones C, Cannon F, Bookhagen B, Palazzi E, Tahir AA (2017) The spatiotemporal variability of precipitation over the Himalaya: evaluation of one-year WRF model simulation. Clim Dyn 49:2179–2204

Norris J, Carvalho LMV, Jones C, Cannon F (2018) Deciphering the contrasting climatic trends between the Central Himalaya and Karakoram with 34 years of WRF simulations. Clim Dyn. (In Press)

Parthasarathy B, Munot AA, Kothawale DR (1994) All-India monthly and seasonal rainfall series – 1871-1993. Theor Appl Climatol 49:217–224

Rasmusson EM, Carpenter TH (1982) Variations in tropical sea surface temperature and surface wind fields associated with the Southern Oscillation/El Niño. Mon Weather Rev 110:354–384

Ridley J, Wiltshire A, Mathison C (2013) More frequent occurrence of westerly disturbances in Karakoram up to 2100. Sci Total Environ 468:S31–S35

Roxy MK, Ritika K, Terray P, Murtugudde R, Ashok K, Goswami BN (2015) Drying of Indian subcontinent by rapid Indian Ocean warming and a weakening land-sea thermal gradient. Nat Commun 6

Saha S et al (2010) The NCEP climate forecast system reanalysis. Bull Am Meteorol Soc 91:1015–1057

Scherler D, Bookhagen B, Strecker MR (2011) Spatially variable response of Himalayan glaciers to climate change affected by debris cover. Nat Geosci 4:156–159

Shrestha AB, Wake CP, Mayewski PA, Dibb JE (1999) Maximum temperature trends in the Himalaya and its vicinity: an analysis based on temperature records from Nepal for the period 1971-94. J Clim 12:2775–2786

Singh P, Ramasastri KS, Kumar N (1995) Topographical influence on precipitation distribution in different ranges of western Himalayas. Nord Hydrol 26:259–284

Syed FS, Giorgi F, Pal JS, King MP (2006) Effect of remote forcings on the winter precipitation of central southwest Asia part 1: observations. Theor Appl Climatol 86:147–160

Tahir AA, Chevallier P, Arnaud Y, Ahmad B (2011) Snow cover dynamics and hydrological regime of the Hunza River basin, Karakoram Range, Northern Pakistan. Hydrol Earth Syst Sci 15:2275–2290

Ueno K, Toyotsu K, Bertolani L, Tartari G (2008) Stepwise onset of monsoon weather observed in the Nepal Himalaya. Mon Weather Rev 136:2507–2522

Yadav RK, Yoo JH, Kucharski F, Abid MA (2010) Why is ENSO influencing Northwest India winter precipitation in recent decades? J Clim 23:1979–1993

Yadav RK, Ramu DA, Dimri AP (2013) On the relationship between ENSO patterns and winter precipitation over North and Central India. Glob Planet Chang 107:50–58

Yasunari T, Inoue J (1978) Characteristics of monsoon precipitation around peaks and ridges in Shorong and Khumbu Himal. Seppyo 40

Zhao Y et al (2014) Impact of the middle and upper tropospheric cooling over Central Asia on the summer rainfall in the Tarim Basin, China. J Climate 27:4721–4732

Chapter 8
Remotely Sensed Rain and Snowfall in the Himalaya

Taylor Smith and Bodo Bookhagen

Abstract Waters sourced in the Himalaya flow through the Ganges and Indus basins, which are some of the most densely populated regions of the world. Both communities in the mountains and those downstream are highly dependent on the volume and consistency of runoff. A growing body of research has pointed towards changes in the timing, volume, and spatial distribution of precipitation in the region over the past decades, but our understanding of the magnitude and direction of these trends is limited by lack of in-situ data availability, complex terrain, and poor process understanding.

Remote sensing provides long-term and spatially-extensive climate data over the entire Himalayan region, and allows for detailed analysis of large-scale environmental change. Here we use several complimentary datasets to explore recent changes in both liquid and solid precipitation, and the knock-on impacts on the Himalayan cryosphere. We find that the spatial and temporal distribution of water resources has shifted, with potentially significant consequences for downstream water provision. In particular, we find that there has been less water stored in snowpack over the past decades, and that the timing of the snowmelt season has shifted earlier in the year. The length of the snowmelt season has also been compressed in much of the region. Rainfall trends can also be detected in the time series; however, multi-annual oscillations and intra-seasonal variations make it difficult to obtain statistically significant trends. Continued exploration of these time series and their associated trends will be essential for understanding hydro-meteorologic processes and improving future regional water planning.

T. Smith · B. Bookhagen (✉)
Institute of Geosciences, University of Potsdam, Potsdam, Germany
e-mail: bodo.bookhagen@uni-potsdam.de

© Springer Nature Switzerland AG 2020
A. P. Dimri et al. (eds.), *Himalayan Weather and Climate and their Impact on the Environment*, https://doi.org/10.1007/978-3-030-29684-1_8

8.1   Introduction

More than a billion people rely on water sourced in the Himalayan region for hydro-power, agriculture, and household water resources (Bolch et al. 2012; Immerzeel et al. 2010). Both small and large communities are often highly dependent on the consistency of precipitation; many lack the resources to respond to rapid changes in water availability. Over the past decades, significant changes in the hydrological cycle of the Himalaya have been observed.

The study region runs from 25–40 N to 65–95 E, covering several large and gla-ciated watersheds (Fig. 8.1a). There exists a distinct precipitation gradient from east to west and north to south, which generally follows the track of the Indian Summer Monsoon (ISM) along the front of the Himalaya (Fig. 8.1b, cf. Bookhagen and Burbank 2012). Large regions of the Himalaya, particularly those above 4000 m asl, also receive significant amounts of precipitation in the form of snow throughout the year (Fig. 8.2).

In the eastern regions of the Himalaya, there is significantly more rainfall than snowfall. The western reaches of the Himalaya, however, receive a large percentage of their water budget in the form of snow, particularly in the coldest regions of the Himalaya (Fig. 8.3). The western parts of the Himalaya are generally colder than the eastern regions, and maintain much larger concentrations of glaciers (cf. Figure 8.1a, blue regions). However, there exists a very high elevation band – along the highest peaks of the world – that is well below freezing for much of the year and receives significant snowfall.

Throughout the region, and in particular in the very cold parts of the study area, snow remains on the ground through a large portion of the year (Fig. 8.4). These snow-covered regions are not confined to areas surrounding glaciers, but also extend into the high Tibetan Plateau – although this snow-cover is often quite shallow.

8.1.1   Recent Climate Changes in the Himalaya

There have been substantial changes in the regional climate, including increased temperatures (Vaughan et al. 2013), increased storm intensity (Singh et al. 2014; Yao et al. 2012; Bookhagen and Burbank 2012; Malik et al. 2016; Fu 2003; Palazzi et al. 2013), changes in the ISM (Gautam et al. 2009; Menon et al. 2013; Kitoh et al. 2013), intensification of the Winter Westerly Disturbances (WWD) (Cannon et al. 2014, 2015), and substantial changes in glaciers throughout the region (Bolch et al. 2012; Kääb et al. 2012, 2015; Kapnick et al. 2014; Gardner et al. 2013; Yao et al. 2012; Gardelle et al. 2012; Scherler et al. 2011; Frey et al. 2014).

The shrinking of Himalayan glaciers over the past decades is a highly visible sign of regional climate change (Bolch et al. 2012). As water stored in these glaciers is an important part of the hydrological budget of many Himalayan catchments, any changes in glacier water storage capacity will have significant downstream impacts

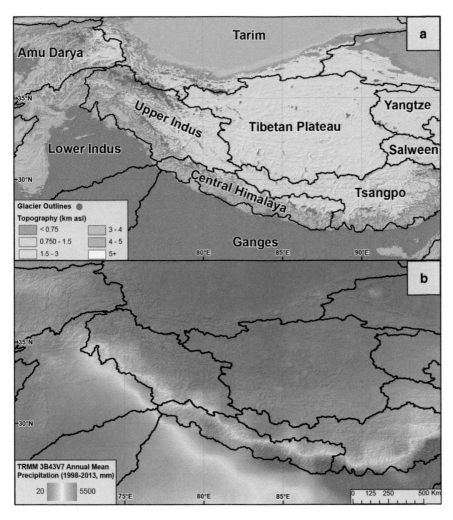

Fig. 8.1 (a) Topography of the Himalaya region, showing major watershed boundaries and glacier outlines in blue based on the Randolph Glacier Inventory (V6) (RGI Consortium 2017). (b) Mean annual Tropical Rainfall Measurement Mission (TRMM) 3B43 rainfall (Huffman et al. 2007), showing distinctive precipitation gradient along and across strike. (cf. Bookhagen and Burbank 2012)

(Vaughan et al. 2013). Historically, water has been slowly released from snow and glaciers throughout the spring and summer, providing year-round water to both high-elevation and downstream communities. However, recent climate changes have reduced the dependability of the yearly hydrological cycle. Many regions have seen increases in early season runoff alongside drastic decreases in late season run-off – particularly in those regions where glaciers have substantially retreated or disappeared (Lutz et al. 2014). The Intergovernmental Panel on Climate Change (IPCC) forecasts that runoff in major river basins of the Himalaya will increase

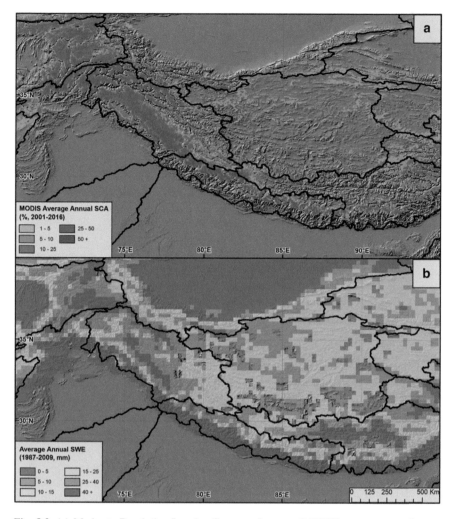

Fig. 8.2 (**a**) Moderate Resolution Imaging Spectroradiometer (MODIS) average annual snow covered area (Hall et al. 2006), showing distinctly higher snow cover in the west and at high elevations. (**b**) Average annual Snow-Water Equivalent (SWE) from Special Sensor Microwave/Imager (SSMI) data (1987–2009), showing similar but not identical spatial patterns (for data processing information see Smith and Bookhagen 2018). The largest snow-water volumes are stored in the western regions of the Himalaya

through 2100 due to both changes in glaciers and large-scale precipitation patterns, and then decrease after the end of the century (Vaughan et al. 2013). However, these large-scale changes conceal regional, small-scale, and seasonal variation in climate change impacts which are already being felt in the region.

Recent work has noted that the spatial and temporal patterns of snow-water storage have shifted over the past decades (Smith et al. 2017; Smith and Bookhagen 2018). In many regions, snow-water storage has peaked earlier in the year, and

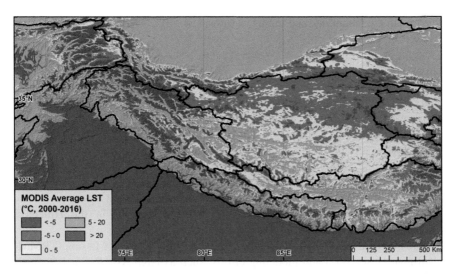

Fig. 8.3 MODIS average annual land-surface temperature (LST, product MOD11A2 V006, Wan et al. 2015). Significant regions of the Himalaya maintain average annual temperatures well below zero

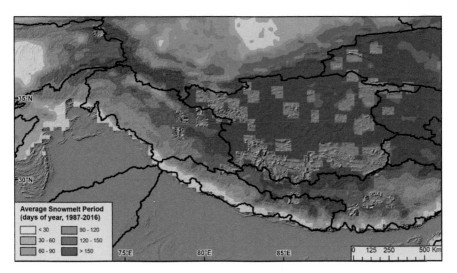

Fig. 8.4 Average annual snowmelt period based on passive microwave data (modified after Smith et al. 2017). Cold regions, particularly around the highest peaks and through the central Tibetan Plateau, maintain at least sparse snow cover for large portions of the year

melted more rapidly during the spring (Smith et al. 2017). Annual trends in snow-water storage throughout the region are also majority negative, implying that changes in temperature and precipitation patterns have already had strong impacts on the cryosphere (Immerzeel et al. 2010, 2012; Smith and Bookhagen 2018). As snow meltwaters account for a large portion of many hydrological budgets, any changes in the magnitude and timing of snowmelt will have significant downstream impacts.

Large-scale climate patterns – such as the ISM and WWD – have changed in strength and timing (e.g., Cannon et al. 2014, 2015; Gautam et al. 2009; Menon et al. 2013; Kitoh et al. 2013; Fu 2003; Palazzi et al. 2013; Ramanathan et al. 2005; Lau et al. 2010). For example, the ISM has increased in strength since the 1950s due to increases in moisture availability (Menon et al. 2013; Kitoh et al. 2013) and increased regional heat-trapping potential due to air pollution (Ramanathan et al. 2005; Lau et al. 2010). In addition to climate change, shifts in Southeast Asia's landcover (i.e., deforestation, extensive irrigation and agriculture, urbanization) have modified regional weather and climate patterns (Fu 2003; Gautam et al., 2009; Bookhagen and Burbank 2012). Large-scale changes in ENSO patterns due to warming oceans have also impacted precipitation patterns in the Himalaya (Bookhagen et al. 2005).

Satellite datasets, such as the Tropical Rainfall Measurement Mission (TRMM) (Huffman et al. 2007) and modeling efforts, such as High Asia Refined Analysis (Maussion et al. 2014) and Asian Precipitation – Highly-Resolved Observational Data Integration Towards Evaluation (APHRODITE) (Yatagai et al. 2012), rarely agree on the magnitude and direction of changes in temperature and precipitation (Malik et al. 2016), and have trouble correctly quantifying high-elevation precipitation (Li et al. 2017; Immerzeel et al. 2015). A lack of long-term and spatially extensive in-situ empirical observations (Fig. 8.5) limits our process understanding (Bookhagen and Burbank 2012; Sorg et al. 2012), and leads to large disagreements

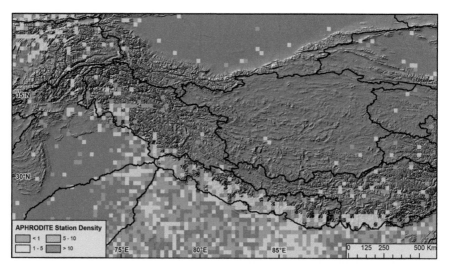

Fig. 8.5 APHRODITE climate station density (1961–2007) (Yatagai et al. 2012, figure modified after Smith and Bookhagen 2018). While parts of the Himalayan foreland are well monitored, much of the Tibetan interior lacks in-situ climate measurements. The majority of these stations measure rainfall and temperature, and very few measure snow depth or snow-water equivalent

in climate projections in global and regional models (Kapnick et al. 2014; Vaughan et al. 2013).

While satellite datasets each have their own strengths and weaknesses, they provide the only long-term, spatially continuous, empirical set of observations of the remote and rugged Himalayan region. Several different satellite platforms have been used to great effect in the Himalaya, and provide optical, active radar, passive microwave, gravimetry, and precipitation measurements across the entire region with high temporal frequency. This chapter explores changes in rain and snowfall in the Himalaya using a set of complimentary remote sensing datasets.

8.2 Data and Methods

Rainfall data are derived from the TRMM product 3B42 (Huffman et al. 2007), at a 3-hour temporal resolution. The spatial resolution of this data is $0.25° \times 0.25°$ (~25 × 25 km²), with data available from 1998 to 2014. As short-duration storms may be missed during satellite overpasses, we aggregate data here to daily and monthly means before performing precipitation analyses.

The Moderate Resolution Imaging Spectroradiometer (MODIS) land-surface temperature (LST) product MOD11A2 (V006) dataset was used to explore regional temperature (Wan et al. 2015). The data has an 8-day temporal frequency and a 1-km spatial resolution. In this study, we use data from 2000 until 2017. We also leverage MODIS maximum mean monthly snow cover (product MOD10C1, V005) at $0.05° \times 0.05°$ (~5 × 5 km²) spatial resolution over the same study period (Hall et al. 2006).

We use reprocessed Satellite Pour l'Observation de la Terre (SPOT) data from 1998 to 2014 (product VGT-S10, Deronde et al. 2014) to generate long-term average normalized difference vegetation index (NDVI) values. This data has a 10-day temporal and 1-km spatial resolution.

Several passive microwave sensors were used to examine changes in snow-water storage and snowmelt timing. The five sensors used are the Special Sensor Microwave/Imager (SSMI, 1987–2009) (Wentz 2013), Special Sensor Microwave Imager/Sounder (SSMIS, 2003–2017) (Sun and Weng 2008), Advanced Microwave Scanning Radiometer – Earth Observing System (AMSR-E, 2002–2011) (Ashcroft and Wentz 2013), AMSR2 (2012–2017) (Imaoka et al. 2010) and the Global Precipitation Measurement Core Observatory (GPM, 2014–2017) (GPM Science Team 2014). These satellites each carry slightly different microwave frequencies, but can each be used to derive snow properties.

Snow-water equivalent (SWE) is derived from the passive microwave data using the Chang equation (Chang et al. 1987), with modifications for non-SSMI platforms as proposed by Armstrong and Brodzik (2001), and a constant snow density of 0.24 g/cm³ as proposed by Takala et al. (2011). This data is used to both derive annual and seasonal trends (after Smith and Bookhagen 2018) and to assess changes in the timing and length of the snowmelt season (after Smith et al. 2017). For a more detailed discussion of passive microwave data processing methodologies, please see Smith and Bookhagen (2018), Smith et al. (2017), and Smith and Bookhagen (2016).

8.3 Results

8.3.1 Along and Across-Strike Rainfall Gradients

There exists a clear rainfall gradient running from east to west along the front of the Himalaya (cf. Fig. 8.1). This precipitation gradient also engenders a vegetation gradient (Fig. 8.6). In particular, the interior of the Tibetan Plateau is sparsely vegetated, especially at high elevations and in the west. In addition to the along-strike gradients, there exist clear differences in the across-strike topographic and precipitation patterns throughout the Himalaya.

In the eastern and western reaches of the Himalaya, along the syntaxes (cf. Fig. 8.7a), topography rises smoothly towards 5000 m asl. Precipitation mirrors this smooth rise with a single precipitation maximum along the topographic break. In the central portion of the Himalaya, however, there are two topographic steps, along the lesser and greater Himalaya (cf. Fig. 8.7b). These correspond to two distinct rainfall peaks – one at each topographic break.

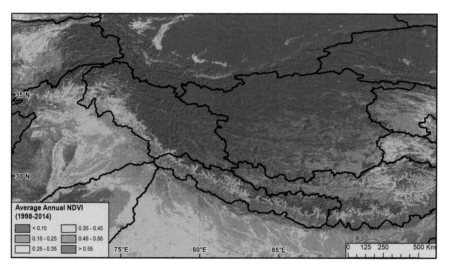

Fig. 8.6 Annual average NDVI values (1998–2014), from the SPOT (VGT-S10) dataset (Deronde et al. 2014). There exists a gradient running from denser vegetation in the east to sparse vegetation in the west and in the Tibetan interior (cf. Olen et al. 2016)

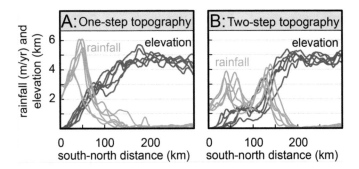

Fig. 8.7 Rainfall and elevation profiles perpendicular to the strike of the Himalaya in the (**a**) western and eastern ends and (**b**) central portion of the study area. The edges have a single topographic step and rainfall peak, whereas the central Himalaya has two topographic breaks and rainfall peaks. Modified from Bookhagen and Burbank (2012)

8.3.2   Along-Strike Snow and Glacier Distribution

The distribution of snow-water resources also follows a similar east-west pattern (cf. Fig. 8.2). In particular, there is more extensive snow cover in the western regions of the Himalaya, and higher SWE storage. There is also significantly more glaciated area in the west (cf. Fig. 8.1). This can be easily seen when examining the aggregated snow, glacier, and topographic profiles of three catchments moving from west to east (Fig. 8.8). Both the spatial and elevation distributions of snow change along this east-west gradient. While the general topographic distribution of the three examined catchments (Upper Indus, Central Himalaya, Tsangpo) remains similar, snow and glaciers occur at distinctly higher elevations moving towards the east.

8.3.3   Along-Strike Climate Analysis

In order to examine east-west climate and glacier coverage gradients, we construct a swath profile running along the length of the Himalaya from the Indus catchment through to the Tsangpo (Fig. 8.9). We focus on areas along the main Himalayan arc, above 500 m asl. We find that maximal elevations along strike of the Himalaya remain roughly equivalent, at around 6000 to 8000 m asl, but that the area above 5000 m asl varies significantly. There exist clear opposing east-west trends in frozen and liquid precipitation, where the eastern regions receive more rainfall and the western regions maintain higher SWE and snow-covered area. This is also reflected in the atmospheric lapse rate, which increases from (negative) 5–6 °C/km in the east to (negative) 3.5–4.5 °C/km in the west. We measure the lapse rate here not vertically in the atmosphere but along a N-S swath from the low-elevation foreland to the high-elevation Himalaya. The eastern Himalaya have a lower lapse rate than the western regions – particularly during the monsoon season – due to the relatively smaller temperature difference between the foreland and high-elevation areas.

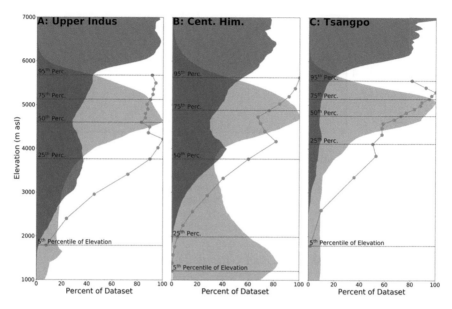

Fig. 8.8 (**a**) Upper Indus, (**b**) Central Himalaya, and (**c**) Tsangpo catchment hypsometries (grey), percentage glaciated area (blue) and normalized SWE distribution (red). All three values displayed as percentage of dataset maximum. While all three catchments have similar elevation distributions, SWE is stored at generally higher elevations moving from west to east. Modified after Smith and Bookhagen (2018)

Further west, however, there exist much larger temperature gradients between the warmer foreland and extremely cold high-elevation zones. This effect is particularly strong during the winter months.

8.4 Discussion

8.4.1 Spatial and Temporal Patterns of Rainfall Changes in the Himalaya

The spatial east-west rainfall gradient has been used in previous research to explain differences in sediment fluxes and erosion (Olen et al. 2016; Thiede et al. 2009; Dey et al. 2016; Hirschmiller et al. 2014). Similarly, the north-south rainfall gradient and the strike-parallel bands of orographic rainfall (cf. Fig. 8.7) have been used to explain spatially-focused erosion and landslide activity (e.g., Bookhagen et al. 2005). Recent research also has investigated the differences in rainfall dynamics between areas of high annual precipitation (orographic rainfall) and regions in the luv and lee of the rainfall bands (e.g., Wulf et al. 2016). Time series analysis shows that rainfall magnitude-frequency distributions differ significantly on short spatial distances.

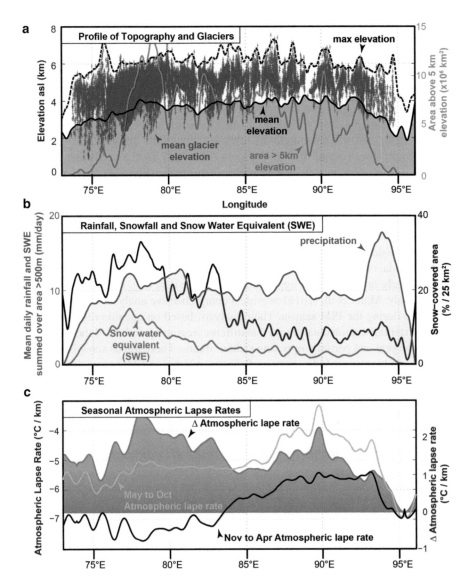

Fig. 8.9 West-to-east swaths with averaged profiles showing topography, climate, and atmospheric lapse rates (modified from Bookhagen 2017), with values averaged north-to-south perpendicular to the strike of the Himalaya. (**a**) Maximal (black dashed) and mean (black solid) elevation, and area above 5000 m asl. Blue crosses represent mean glacier elevation. (**b**) TRMM 3B42-derived precipitation (blue), SSMI SWE (purple) and MODIS snow-covered area (black). These profiles show strong west-east gradients, with more snow in the western reaches of the Himalaya. (**c**) Atmospheric lapse rates (summer, gold; winter, black; annual, shaded). Lapse rates in the eastern Himalaya have higher negative values during the summer due to the heating of the Tibetan Plateau and the increase of the temperature gradient

Rainfall during the monsoon season is the main driver of hydrological processes in the central Himalaya and contributes significantly to the annual precipitation budget. In most parts of the central Himalaya, rainfall during the monsoon season – between June and September – contributes on average more than 80% of the annual rainfall (Bookhagen and Burbank 2012). The eastern and western parts of the Himalaya (syntaxes) receive about half of their annual precipitation during the monsoon season; high-elevation regions of the Himalaya, especially in the syntaxes, receive significant precipitation in the form of snow. While this general spatiotemporal pattern is well established (Bookhagen 2016, 2017), patterns of rainfall trends are less well-studied and generally less reliable due to the short time series and large rainfall heterogeneities in mountainous terrain. Reliable and high-spatial resolution rainfall measurements from satellites started with the TRMM mission (Kummerow et al. 1998) and continue today with the Global Precipitation Measurement Mission (Hou et al. 2014). The ISM exhibits strong decadal and longer-timescale oscillations; trends determined from short timescales do not provide meaningful results for multi-decadal or centennial studies. In order to study multi-decadal trends, researchers often rely on gridded and aggregated rain-gauge data (e.g. APHRODITE). In a recent study, Malik et al. (2016) provide a comprehensive analysis of trends in the extremes during the ISM season. Their analysis, based on quantile regression and gridded rainfall-station data, shows that different regions in India and the Himalaya have divergent and partially opposing rainfall trends. These trends show intensified droughts in Northwest India, parts of Peninsular India, and Myanmar; in contrast, parts of Pakistan, Northwest Himalaya, and Central India show increased extreme daily rain intensity leading to higher flood vulnerability.

8.4.2 Spatial Patterns of Changes in the Himalayan Cryosphere

Spatially and temporally extensive passive microwave SWE estimates can be used to examine decadal trends in snow-water storage throughout the Himalaya. As can be seen in Fig. 8.10, the majority of annual SWE trends are negative throughout the Himalaya, with the exception of parts of the Pamir, the eastern edge of the Tibetan Plateau, and a region along the edge of the Tarim Basin. This implies that over the period 1987 to 2009, less water was stored (on average) in snowpack throughout the Himalaya. To assess the impact of these changes at a watershed scale, we aggregate both SWE volumes and SWE trends across elevations within selected Himalayan catchments to identify regions where changes in SWE will have the strongest downstream effects (Fig. 8.11).

Across all catchments, there is a strong and non-linear elevation-SWE relationship. In the majority of catchments examined here, the highest-SWE elevation slice occurs below the maximum catchment elevation. The majority of SWE in these catchments is stored in their mid-to-high elevation regions (above 3500 m asl).

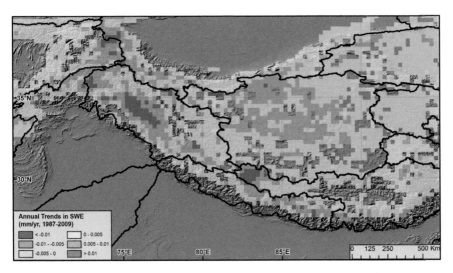

Fig. 8.10 Annual trends in SWE volume (data derived from SSMI, 1987–2009). While the majority of SWE trends are negative, there exist positive SWE storage trends in the Pamir, parts of the Kunlun Shan, and in parts of Eastern Tibet. Modified from Smith and Bookhagen (2018)

These mid-elevation, high-SWE zones also have some of the most negative SWE trends, implying that the negative trends in SWE will likely have a strong impact on downstream water provision. This is in line with increased temperatures in low-precipitation, high-elevation zones of the Himalaya (Bolch et al. 2012; Pepin et al. 2015), and observed changes in Himalayan runoff (Lau et al. 2010; Panday et al. 2011; Lutz et al. 2014).

In addition to changes in the volume of water stored in snowpack, there have been measurable changes in the timing of the snowmelt season (Smith et al. 2017; Xiong et al. 2017; Panday et al. 2011). In particular, the length of the snowmelt season (time from the onset of the main melt phase until the clearance of snow), has been shrinking over the period 1987 to 2016 (Fig. 8.12). The snowmelt season has also tended to both start and end earlier in the year.

8.4.3 Dynamics of Snow-Water Storage and Snowmelt

Temperatures in the Himalaya are increasing faster than the global average (Vaughan et al. 2013; Lau et al. 2010). These temperature increases are likely the driver of changes in snow-water storage and snowmelt, due to changes in the timing of snowfall, precipitation phase, and the spatial distribution of precipitation.

There has been an overall decrease in SWE storage in the Himalaya, as well as shifts in the seasonality of SWE buildup and melt (Lau et al. 2010; Panday et al. 2011; Smith and Bookhagen 2018; Smith et al. 2017). The mechanism behind these

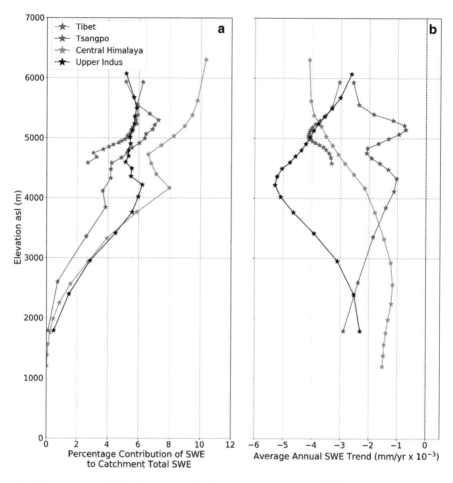

Fig. 8.11 Elevation SWE relationships. (**a**) Elevation distribution of SWE in each catchment, at each fifth percentile elevation slice. (**b**) Average SWE trend at the same fifth percentile elevation slices. SWE trends tend to be the most negative where the highest SWE volumes are stored. This indicates that SWE changes are not only impacting low-elevation, shallow-snow areas, but have also impacted medium-elevation zones where there is high SWE storage. See Fig. 8.1 for watershed boundaries. Modified from Smith and Bookhagen (2018)

SWE changes is not well defined, but likely includes contributions from aerosol contamination (Lau et al. 2010), changes in precipitation phase (Lutz et al. 2014), changes in the strengths of the WWD (Cannon et al. 2014, 2015) and ISM (Singh et al. 2014; Palazzi et al. 2013), and increases in regional temperatures which lead to both more atmospheric water storage and decreased SWE persistence (Vaughan et al. 2013; Yao et al. 2012; Trenberth 2011). Changes in snowmelt seasonality has been shown to modify downstream water availability (Barnett et al. 2005; Berghuijs et al. 2014).

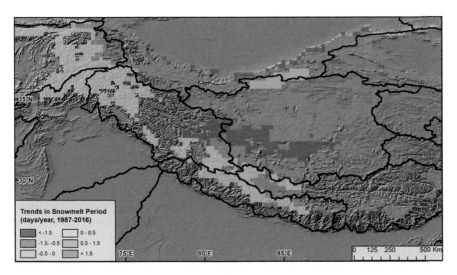

Fig. 8.12 Trends in snowmelt period (length of time between onset and end of snowmelt), 1987–2016. The majority of the study region has experienced compressed snowmelt seasons over the past decades. Modified from Smith et al. (2017)

Glaciers in the Himalaya are generally retreating (Bolch et al. 2012; Gardner et al. 2013; Kääb et al. 2012, 2015); in many cases, retreat is accelerating and small glaciers are disappearing (Armstrong et al. 2010). The reasons behind these changes are multi-faceted and poorly constrained, although debris cover, topography, and precipitation seasonality are factors known to impact glacier stability.

While much of the Himalaya's water budget is monsoon-driven, there exists a precipitation gradient moving west along the front of the Himalaya, where the western reaches of the Himalaya region have a much higher snowmelt and glacier contribution to their water budgets (cf. Fig. 8.9) (Bookhagen and Burbank 2012). Even in those regions where rainfall is primary, seasonal snowmelt is an important water source for mountain communities and local ecologies.

Snow and glacier melt are primary sources of water at different times of the year; snowmelt generally peaks in the spring before the monsoon, and glacier melt is primary in the post-monsoon season. Both of these segments of the hydrosphere are essential for maintaining consistent and reliable water flow in both natural areas, such as wetlands, and in developed areas, such as hydropower, irrigation, and municipal water systems. Any changes in the temporal distribution of these water resources can increase the frequency of short-term water surpluses and droughts, particularly in the western and northwestern Himalaya where snow-water resources form a large part of the yearly hydrological budget (Vaughan et al. 2013).

8.4.4 Limitations and Caveats of Remote-Sensing Datasets

While remote sensing datasets provide a quasi-continuous and long-term record of earth surface processes, there are several important caveats to consider when interpreting these data records. Each satellite dataset has its own set of strengths and weaknesses that impact the reliability of derived environmental analyses.

TRMM data is limited by its temporal resolution with respect to the often-short duration of precipitation events. In particular, intense monsoonal rainstorms, which account for a significant portion of the regional water budget, are often missed in the gaps between satellite overpasses. In this analysis, we average our high-temporal resolution data to monthly averages to limit the impacts of gaps in the TRMM data record. A second caveat of the TRMM dataset is the well-documented elevation-dependent error, where high-elevation precipitation is underestimated in the region (Bharti and Singh 2015; Wulf et al. 2016). Thus, precipitation estimates in some of the poorly-monitored, high-elevation regions of the study area should be considered with caution. Furthermore, the low spatial resolution of the TRMM 3B42 datasets (and similar datasets, e.g., APHRODITE) do not capture distinctive orographic rainfall peaks, but provide average measurements for ~25 × 25 km areas. Importantly, the previously identified orographic rainfall peak (cf. Fig. 8.7, Bookhagen and Burbank 2006; Bookhagen and Burbank 2012) is not well captured in lower-resolution TRMM data.

Both MODIS and SPOT data are limited by both the temporal and radiometric resolution of data collection. In much of the study area, long-duration cloud cover and storm systems can prevent optical measurement of earth-surface characteristics for several days at a time. We thus rely on lower temporal resolution products in this study. A second caveat of the MODIS products is the poor performance of LST estimates over snow-covered terrain (Wan 2008). The MODIS LST algorithm is based on radiation balances, and is adversely impacted by high reflectivity values in the visible and near-infrared spectra over snow.

While passive microwave data provides the most globally extensive means of measuring snow buildup and melt, there are several important sources of error in SWE estimation. The most commonly used SWE estimation algorithms assume that the snowpack is comprised of dry, evenly sized, snow crystals at a constant density. While fresh snow in cold regions often satisfies these conditions, in complex and mountainous terrain, snowpack undergoes progressive metamorphism throughout a given snow season. This changes both the size and density distribution of snowpack, and leads to errors in SWE estimation (Kelly et al. 2003).

There is a well-documented signal saturation in passive microwave data over glaciers and in deep snowpacks (Takala et al. 2011; Tedesco and Narvekar 2010). As the estimated SWE in a passive microwave pixel is sensitive to the depth of snow throughout the pixel, this saturation can occur even in regions where only a small portion of the pixel overlaps with deep snow (Vander Jagt et al. 2013). In our study area, this mostly impacts glaciated regions, and thus SWE estimates close to glaciers (cf. Fig. 8.1) should be considered more error-prone. A final caveat of passive microwave data is that SWE estimates in the presence of liquid water are highly biased, due to large differences in microwave signal strength between liquid and frozen water. Thus, SWE cannot be estimated near large bodies of water, and SWE estimates during the spring snowmelt season are less reliable than those earlier in the season.

Despite these caveats, passive microwave data remains the only empirical method to estimate SWE over large areas of complex and inhospitable terrain (Chang et al. 1982, 1987; Kelly et al. 2003; Abdalati and Steffen 1995; Drobot and Anderson 2001; Takala et al. 2011). The lack of an extensive in-situ sensor network in the region (cf. Fig. 8.5), as well as the difficulties associated with ground-data collection, mean that passive microwave SWE estimates remain the best option for monitoring large-scale snow patterns.

8.5   Conclusions

Remote sensing provides the only long-term and spatially-extensive climatic datasets across the diverse terrain of the Himalaya. These data are essential for improving our understanding of environmental changes and our continued prediction of their future magnitudes and impacts. They also provide the basis for calibrating large-scale climate models, which are important tools in predicting the future impacts of climate change.

While remotely sensed data provides valuable insight into large-scale environmental change, the uncertainties in the data must be carefully considered in combination with calibration and field observations – especially when high-resolution datasets are used.

There have already been significant changes to the precipitation regime across the Himalaya, which have impacted not only liquid precipitation but also the buildup and melt of frozen water. Long-term trends indicate a shift towards a shorter and earlier snowmelt season. This has already had, and will continue to have, strong impacts upon both the natural environment and communities which rely on consistency in the volume and timing of snowmelt for year-round water provision. Information on the magnitude and direction of changes will be essential for future water planning.

References

Abdalati W, Steffen K (1995) Passive microwave-derived snow melt regions on the Greenland Ice Sheet. Geophys Res Lett 22:787–790

Armstrong R, Brodzik M (2001) Recent Northern Hemisphere snow extent: a comparison of data derived from visible and microwave satellite sensors. Geophys Res Lett 28:3673–3676

Armstrong RL et al (2010) The glaciers of the Hindu Kush-Himalayan region: a summary of the science regarding glacier melt/retreat in the Himalayan, Hindu Kush, Karakoram, Pamir, and Tien Shan mountain ranges. International Centre for Integrated Mountain Development (ICIMOD)

Ashcroft P, Wentz F(2013) AMSR-E/aqua L2A global swath spatially-resampled brightness temperatures V003 [2002–2010]. National Snow and Ice Data Center, Boulder, Colorado, USA

Barnett TP, Adam JC, Lettenmaier DP (2005) Potential impacts of a warming climate on water availability in snow-dominated regions. Nature 438:303–309

Berghuijs W, Woods R, Hrachowitz M (2014) A precipitation shift from snow towards rain leads to a decrease in stream-ow. Nat Clim Chang 4:583–586

Bharti V, Singh C (2015) Evaluation of error in TRMM 3B42V7 precipitation estimates over the Himalayan region. J Geophys Res Atmos 120:12458–12473

Bolch T, Kulkarni A, Kääb A, Huggel C, Paul F, Cogley J, Frey H, Kargel J, Fujita K, Scheel M et al (2012) The state and fate of Himalayan glaciers. Science 336:310–314

Bookhagen B (2016) Chapter 11: Glaciers and monsoon systems. In: Carvalho L, Jones C (eds) The monsoons and climate change: observations and modeling. Springer Climate, Cham

Bookhagen B (2017) Chapter 11: The influence of hydrology and glaciology on wetlands in the Himalaya. In: Prins H, Namgail T (eds) Bird migration across the Himalayas: wetland functioning amidst mountains and glaciers. Cambridge University Press, Cambridge

Bookhagen B, Burbank DW (2006) Topography, relief, and TRMM-derived rainfall variations along the Himalaya. Geophys Res Lett 33(8)

Bookhagen B, Burbank DW (2003–2012) Toward a complete Himalayan hydrological budget: Spatiotemporal distribution of snowmelt and rainfall and their impact on river discharge, J Geophys Res Earth 115: 2010

Bookhagen B, Thiede RC, Strecker MR (2005) Abnormal monsoon years and their control on erosion and sediment flux in the high, arid northwest Himalaya. Earth Planet Sci Lett 231:131–146

Cannon F, Carvalho L, Jones C, Bookhagen B (2014) Multi-annual variations in winter westerly disturbance activity affecting the Himalaya. Clim Dyn:1–15

Cannon F, Carvalho LM, Jones C, Norris J (2015) Winter westerly disturbance dynamics and precipitation in the western Himalaya and Karakoram: a wave-tracking approach. Theor Appl Climatol:1–18

Chang A, Foster J, Hall D, Rango A, Hartline B (1982) Snow water equivalent estimation by microwave radiometry. Cold Reg Sci Technol 5:259–267

Chang A, Foster J, Hall D (1987) Nimbus-7 SMMR derived global snow cover parameters. Ann Glaciol 9:39–44

Deronde B, Debruyn W, Gontier E, Goor E, Jacobs T, Verbeiren S, Vereecken J (2014) 15 years of processing and dissemination of SPOT-VEGETATION products. Int J Remote Sens 35(7):2402–2420

Dey S, Thiede RC, Schildgen TF, Wittmann H, Bookhagen B, Scherler D, Vikrant J, Strecker MR (2016) Climate-driven sediment aggradation and incision phases since the Late Pleistocene in the NW Himalaya, India. Earth Planet Sci Lett 449:321–331

Drobot SD, Anderson MR (2001) An improved method for determining snowmelt onset dates over Arctic sea ice using scanning multichannel microwave radiometer and Special Sensor Microwave/Imager data. J Geophys Res 106:24033–24049

Frey H, Machguth H, Huss M, Huggel C, Bajracharya S, Bolch T, Kulkarni A, Linsbauer A, Salzmann N, Stoffel M (2014) Estimating the volume of glaciers in the Himalayan–Karakoram region using different methods. Cryosphere 8:2313–2333

Fu C (2003) Potential impacts of human-induced land cover change on East Asia monsoon. Glob Planet Chang 37:219–229

Gardelle J, Berthier E, Arnaud Y (2012) Slight mass gain of Karakoram glaciers in the early twenty-first century. Nat Geosci 5:322–325

Gardner AS, Moholdt G, Cogley JG, Wouters B, Arendt AA, Wahr J, Berthier E, Hock R, Pfeffer WT, Kaser G et al (2013) A reconciled estimate of glacier contributions to sea level rise: 2003 to 2009. Science 340:852–857

Gautam R, Hsu N, Lau K-M, Kafatos M (2009) Aerosol and rainfall variability over the Indian monsoon region: distributions, trends and coupling. Ann Geophys 27:3691–3703

GPM Science Team (2014) GPMGMI Level 1B Brightness Temperatures, version 03. NASA Goddard Earth Science Data and Information Services Center (GES DISC), Greenbelt, MD, USA

Hall DK, Salomonson VV, Riggs GA (2006) MODIS/Terra Snow Cover Daily L3 Global 0.05Deg CMG, Version 5. NASA National Snow and Ice Data Center Distributed Active Archive Center, Boulder, Colorado USA. https://doi.org/10.5067/EI5HGLM2NNHN

Hirschmiller J, Grujic D, Bookhagen B, Coutand I, Huyghe P, Mugnier J-L, Ojha T (2014) What controls the growth of the Himalayan foreland fold-and-thrust belt? Geology 42(3):247–250

Hou AY, Kakar RK, Neeck S, Azarbarzin AA, Kummerow CD, Kojima M, Oki R, Nakamura K, Iguchi T (2014) The Global Precipitation Measurement Mission. Bull Am Meteorol Soc 95:701–722

Huffman GJ, Bolvin DT, Nelkin EJ, Wolff DB, Adler RF, Gu G, Hong Y, Bowman KP, Stocker EF (2007) The TRMM multisatellite precipitation analysis (TMPA): Quasi-global, multiyear, combined-sensor precipitation estimates at fine scales. J Hydrometeorol 8:38–55

Imaoka K, Kachi M, Kasahara M, Ito N, Nakagawa K, Oki T (2010) Instrument performance and calibration of AMSR-E and AMSR2. Int Arch Photogramm Remote Sens Spat Inf Sci 38:13–18

Immerzeel WW, Van Beek LP, Bierkens MF (2010) Climate change will affect the Asian water towers. Science 328:1382–1385

Immerzeel WW, van Beek LPH, Konz M et al (2012) Clim Chang 110:721. https://doi.org/10.1007/s10584-011-0143-4

Immerzeel W, Wanders N, Lutz A, Shea J, Bierkens M (2015) Reconciling high-altitude precipitation in the upper Indus basin with glacier mass balances and runoff. Hydrol Earth Syst Sci 19:4673

Kääb A, Berthier E, Nuth C, Gardelle J, Arnaud Y (2012) Contrasting patterns of early twenty-first-century glacier mass change in the Himalayas. Nature 488:495–498

Kääb A, Treichler D, Nuth C, Berthier E (2015) Brief communication: contending estimates of 2003–2008 glacier mass balance over the Pamir–Karakoram–Himalaya. Cryosphere 9:557–564

Kapnick SB, Delworth TL, Ashfaq M, Malyshev S, Milly P (2014) Snowfall less sensitive to warming in Karakoram than in Himalayas due to a unique seasonal cycle. Nat Geosci 7:834–840

Kelly RE, Chang AT, Tsang L, Foster JL (2003) A prototype AMSR-E global snow area and snow depth algorithm. IEEE Trans Geosci Remote Sens 41:230–242

Kitoh A, Endo H, Krishna Kumar K, Cavalcanti IF, Goswami P, Zhou T (2013) Monsoons in a changing world: a regional perspective in a global context. J Geophys Res Atmos 118:3053–3065

Kummerow C, Barnes W, Kozu T, Shiue J, Simpson J (1998) The tropical rainfall measuring mission (TRMM) sensor package. J Atmos Ocean Technol 15(3):809–817

Lau WK, Kim M-K, Kim K-M, Lee W-S (2010) Enhanced surface warming and accelerated snow melt in the Himalayas and Tibetan Plateau induced by absorbing aerosols. Environ Res Lett 5:025–204

Li L, Gochis DJ, Sobolowksi S, Mesquita M d S (2017) Evaluating the present annual water budget of a Himalayan headwater river basin using a high-resolution atmosphere-hydrology model. J Geophys Res Atmos

Lutz A, Immerzeel W, Shrestha A, Bierkens M (2014) Consistent increase in High Asia's runoff due to increasing glacier melt and precipitation. Nat Clim Chang 4:587–592

Malik N, Bookhagen B, Mucha PJ (2016) Spatiotemporal patterns and trends of Indian monsoonal rainfall extremes. Geophys Res Lett 43:1710–1717

Maussion F, Scherer D, Mölg T, Collier E, Curio J, Finkelnburg R (2014) Precipitation seasonality and variability over the Tibetan Plateau as resolved by the high Asia reanalysis. J Clim 27:1910–1927

Menon A, Levermann A, Schewe J (2013) Enhanced future variability during India's rainy season. Geophys Res Lett 40:3242–3247

Olen SM, Bookhagen B, Strecker MR (2016) Role of climate and vegetation density in modulating denudation rates in the Himalaya. Earth Planet Sci Lett 445:57–67

Palazzi E, Hardenberg J, Provenzale A (2013) Precipitation in the Hindu-Kush Karakoram Himalaya: observations and future scenarios. J Geophys Res Atmos 118:85–100

Panday PK, Frey KE, Ghimire B (2011) Detection of the timing and duration of snowmelt in the Hindu Kush-Himalaya using QuikSCAT, 2000–2008. Environ Res Lett 6:024007

Pepin N, Bradley RS, Diaz HF, Baraer M, Caceres EB, Forsythe N, Fowler H, Greenwood G, Hashmi MZ, Liu XD, Miller JR, Ning L, Ohmura A, Palazzi E, Rangwala I, Schöner W, Severskiy I, Shahgedanova M, Wang MB, Williamson SN, Yang DQ (2015) Elevation-dependent warming in mountain regions of the world. Nat Clim Chang 5:424–430

Ramanathan V, Chung C, Kim D, Bettge T, Buja L, Kiehl J, Washington W, Fu Q, Sikka D, Wild M (2005) Atmospheric brown clouds: Impacts on South Asian climate and hydrological cycle. Proc Natl Acad Sci USA 102:5326–5333

RGI Consortium (2017) Randolph glacier inventory – a dataset of global glacier outlines: version 6.0: technical report. Global Land Ice Measurements from Space, Colorado, USA. Digital Media. https://doi.org/10.7265/N5-RGI-60

Scherler D, Bookhagen B, Strecker MR (2011) Spatially variable response of Himalayan glaciers to climate change affected by debris cover. Nat Geosci 4:156–159

Singh D, Tsiang M, Rajaratnam B, Diffenbaugh NS (2014) Observed changes in extreme wet and dry spells during the South Asian summer monsoon season. Nat Clim Chang 4:456–461

Smith T, Bookhagen B (2016) Assessing uncertainty and sensor biases in passive microwave data across High Mountain Asia. Remote Sens Environ 181:174–185

Smith T, Bookhagen B (2018) Changes in seasonal snow- water equivalent distribution in High Mountain Asia (1987 to 2009). Sci Adv 4(1)

Smith T, Bookhagen B, Rheinwalt A (2017) Spatiotemporal patterns of High Mountain Asia's snowmelt season identified with an automated snowmelt detection algorithm, 1987–2016. Cryosphere

Sorg A, Bolch T, Stoffel M, Solomina O, Beniston M (2012) Climate change impacts on glaciers and runoff in Tien Shan (Central Asia). Nat Clim Chang 2:725–731

Sun N, Weng F (2008) Evaluation of special sensor microwave imager/sounder (SSMIS) environmental data records. IEEE Trans Geosci Remote Sens 46:1006–1016

Takala M, Luojus K, Pulliainen J, Derksen C, Lemmetyinen J, Kärnä J-P, Koskinen J, Bojkov B (2011) Estimating northern hemisphere snow water equivalent for climate research through assimilation of space-borne radiometer data and ground-based measurements. Remote Sens Environ 115:3517–3529

Tedesco M, Narvekar PS (2010) Assessment of the NASA AMSR-E SWE product. IEEE J Sel Top Appl Earth Obs Remote Sens 3:141–159

Thiede RC, Ehlers TA, Bookhagen B, Strecker MR (2009) Erosional variability along the north-west Himalaya. J Geophys Res 114:F01015. https://doi.org/10.1029/2008JF001010

Trenberth KE (2011) Changes in precipitation with climate change. Clim Res 47:123–138

Vander Jagt BJ, Durand MT, Margulis SA, Kim EJ, Molotch NP (2013) The effect of spatial variability on the sensitivity of passive microwave measurements to snow water equivalent. Remote Sens Environ 136:163–179

Vaughan D, Comiso J, Allison I, Carrasco J, Kaser G, Kwok R, Mote P, Murray T, Paul F, Ren J, Rignot E, Solomina O, Steffen K, Zhang T (2013) Observations: cryosphere. In: Climate change 2013: the physical science basis. Contribution of working group I to the Fifth Assessment Report of the IPCC

Wan Z (2008) New refinements and validation of the MODIS land-surface temperature/emissivity products. Remote Sens Environ 112(1):59–74

Wan Z, Hook S, Hulley G (2015) MOD11A2 MODIS/Terra Land Surface Temperature/Emissivity 8-Day L3 Global 1km SIN Grid V006

Wentz FJ (2013) SSM/I version-7 calibration report. Remote Sensing Systems Rep 11012:46

Wulf H, Bookhagen B, Scherler D (2016) Differentiating between rain, snow, and glacier contributions to river discharge in the western Himalaya using remote-sensing data and distributed hydrological modeling. Adv Water Resour 88:152–169

Xiong C, Shi J, Cui Y, Peng B (2017) Snowmelt pattern over High-Mountain Asia detected from active and passive microwave remote sensing. IEEE Geosci Remote Sens Lett

Yao T, Thompson L, Yang W, Yu W, Gao Y, Guo X, Yang X, Duan K, Zhao H, Xu B et al (2012) Different glacier status with atmospheric circulations in Tibetan Plateau and surroundings. Nat Clim Chang 2:663–667

Yatagai A, Kamiguchi K, Arakawa O, Hamada A, Yasutomi N, Kitoh A (2012) APHRODITE: constructing a long-term daily gridded precipitation dataset for Asia based on a dense network of rain gauges. Bull Am Meteorol Soc 93:1401–1415

Chapter 9
Elevation Dependent Warming over Indian Himalayan Region

A. P. Dimri, A. Choudhary, and D. Kumar

Abstract Recent studies reported an elevation dependent signal of warming in mountainous regions of the world including the Himalayas. Various mechanisms are proposed to link this phenomenon with other atmospheric variables. In the present study, long-term (1970–2099) trend of near-surface air temperature at different elevations in the Indian Himalayan region (IHR) is assessed from Regional Climate Model (REMO) simulations. This is done for four different seasons- winter, pre-monsoon, monsoon and post-monsoon – to detect any signal of elevation dependency in the rate of warming and its seasonal response. Our results show enhanced trends in temperature during post-monsoon and winter season at higher elevations, which is concurrent with increased trends in surface downwelling longwave radiation (DLR) at higher elevations. Further, the elevation dependency of other climatic variables like – soil moisture, surface snow amount, cloud fraction etc. are studied to understand the possible factors behind higher DLR trend at higher altitudes in specific seasons.

9.1 Introduction

Under various geographical situations climate change signals are more evident in regions where they are emphasized or occur at a rapid pace and thus act like sentinels of climate and associated environmental changes. Mountains and highland regions are among the most vulnerable areas to climate change and to its impacts over their surroundings (Xu et al. 2009). In this regard, Messerli and Ives (1997) highlighted the major problems related with changing climate in mountains on sustainable development with an interdisciplinary approach where questions concerning mountain cultures, water, energy, biodiversity, environment and socio-economic issues are documented. Barry (1992) discussed on pronounced amplitude of climate variability and change at various scales in several mountainous regions across the globe and the study limitations that exist therein which are associated with scarcity

A. P. Dimri (✉) · A. Choudhary · D. Kumar
School of Environmental Sciences, Jawaharlal Nehru University, New Delhi, India

© Springer Nature Switzerland AG 2020   141
A. P. Dimri et al. (eds.), *Himalayan Weather and Climate and their Impact on the Environment*, https://doi.org/10.1007/978-3-030-29684-1_9

of observations and lack of theoretical understanding of physical processes related with mountain climate. Using station observations over various mountain ranges, Diaz and Bradley (1997) provided a comprehensive survey of differential temperature changes with altitude and found strong evidences of high altitude warming in parts of Asia and Europe. Liu and Chen (2000) illustrated elevation dependent warming, i.e., significant amplification of warming rates with elevation, analysing temporal trends of temperature measured at 197 in situ stations over the Tibetan Plateau. Thompson et al. (2003) showed elevation dependency in millennium scale temperature trend record from Tibet. Similar studies over the Alps (Giorgi et al. 1997) and the Rocky mountains (Fyfe and Flato 1999; Snyder et al. 2002) were carried out. Nogues-Bravo et al. (2007) using 5 AOGCMs to simulate future climate under IPCC scenarios for major mountain regions did not find any consistent difference in warming between the high elevations and low-lying areas at the same corresponding latitudinal belt. In a study based on 1000 high elevation stations across the globe, Pepin and Lundquist (2008) found no globally concurrent relationships between warming rates and elevation. However, they found strongest warming trends near 0° isotherm due to snow-ice feedback and also concluded that mountain summits and free draining slopes are better indicators of global warming as they are more exposed to free-air advection. Rangwala et al. (2009) studied the influence of changes in surface specific humidity on downwelling longwave radiation (DLR) which is responsible for pronounced warming during winter season over higher altitudes in Tibetan Plateau. In their study over high altitude stations in Alps, Ruckstuhl et al. (2007) noted elevation dependent warming and found it to be related with enhanced DLR at high elevations due to its increased sensitivity to surface water vapour. Liu et al. (2009) reported elevation dependent changes over most of the mountain ranges across the globe, including Tibetan plateau. They have shown it in the instrumentation records and NCAR CCSM3 based future projections in different elevation zones. During winter and spring, warming is more pronounced at higher elevation ranges than over lower elevation ranges with similar tendency in future. Qin et al. (2009) have shown higher warming over 2000 to 4800 m asl in Tibetan Plateau using satellite information of Moderate Resolution Imaging Spectrometer (MODIS). In a study carried out over 10 major mountain ranges across the world, Ohmura (2012) found temperature variability and trend to increase with elevation. He linked this elevation dependent warming to enhanced diabatic processes in the middle to high troposphere as a result of the cloud condensation. Rangwala and Miller (2012) have given a comprehensive review over the global mountain ranges and provided corresponding four mechanisms of warming in details due to (1) snow/ice albedo feedback, (2) cloud cover, (3) water vapour modulation of longwave heating and (4) aerosol impact. Further, using a 1-D radiative transfer model, Rangwala (2013) has shown the possibility of strong modulation of surface DLR caused by increase in atmospheric moisture in high altitudes (>3000 m) during winter which is responsible for winter warming. A review of studies from different mountain regions of the world by Pepin et al. (2015) finds strong and emerging evidence of higher rate of warming over elevated regions, thus leading to impact on the adjacent environment. Based on global climate simulations from

CMIP5, Rangwala et al. (2016) have shown that amplified warming during winter season in higher elevation regions of boreal midlatitudes is strongly correlated with elevation dependent increase in water vapour. In another model based study, carried out by Palazzi et al. (2018), where the GCM- EC-EARTH is applied at different resolutions, it is found that the most significant drivers· of elevation dependent warming in different mountain regions across the globe are changes in surface albedo and DLR. However, the same study shows that over Himalayan region an additional key driver is the change in surface specific humidity with elevation. In view of such findings, altitude dependent warming/cooling or climate signal thereof is one of the prime research interests for mountain researchers.

The Himalayas are identified as a climate and climate change hot-spot, especially since they host numerous glaciers representing a water source for northern Indian rivers, as well as a hotspot of biodiversity. The progress and interpretation on this subject was limited due to the paucity of the observations. Most of the studies discussed above and others have shown warming over the Tibetan Plateau in the recent decades. There are very few studies which are focused on Indian Himalayan regions (IHR, Fig. 9.1). This region of the Himalayas is much less monitored and plays a crucial role in defining hydro-climatic regime of the Indian sub-continent. Debate on disappearing glacier (Bolch et al. 2012), snow pack and cover, permafrost, acceding snowline, receding treeline etc. in this region is looming large as it can have significant consequences for the hundreds of millions of people living in Indian sub-continent.

Keeping these in mind and with few available researches over IHR, this study examines mechanism for elevation dependent warming over the IHR using recently

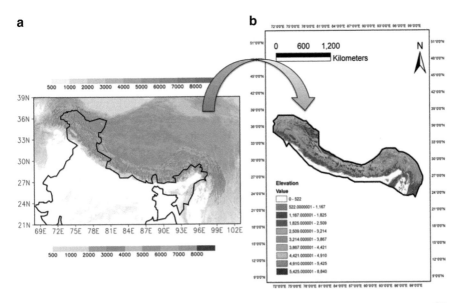

Fig. 9.1 Topography (metres) over (**a**) Himalayan and Tibetan region (grey shaded) and over (**b**) study area (color shaded). (Reproduced from Ghimire et al. 2015)

available regional model simulations (1970–2099). The primary objective of this study is to understand the mechanism behind elevation dependent warming in future climatic projections over Himalayan region and to investigate its seasonal response.

9.2 Study Area, Data and Methods

9.2.1 Study Area

The study region considered here is the great Himalayan range along with Hindu-Kush and Karakoram region (Fig. 9.1). Considering the resolution of the model and the high spatial variability of climate in this region, the domain of study is selected wide enough in order to include the entire stretch of Himalayan orography. This also ensures a sampling of a large number of grid cells for a comprehensive assessment of the elevation based response of climate in the region. The other reason to select this area is to maintain coherency with the recent model based studies over Himalayan region where the same region is considered (see Ghimire et al. 2015; Nengker et al. 2017; Choudhary and Dimri 2017).

9.2.2 Data and Methods

The required climatic information for the study area is obtained from the regional climate model (RCM) REMO (Teichmann et al. 2013). REMO regional climate simulations considered here are performed within the framework of Coordinated Regional Climate Downscaling Experiment-South Asia (CORDEX-SA) where the necessary forcings are provided by global climate simulations from- MPI-ESM-LR (Giorgetta et al. 2013). CORDEX-SA is a part of larger regional climate modeling initiative called CORDEX (Giorgi et al. 2009; Lake et al. 2017) which is coordinated by World Climate Research Programme. The horizontal resolution of the model simulation in the present study is 0.44° (approximately 50 km). The REMO model output is provided by Climate Service Center (CSC), Germany, and is retrieved from CORDEX-SA portal of CCCR, IITM, India. For further information on the model configuration and experimental design, the readers can refer to Teichmann et al. (2013). For information on assessment and validation of REMO over the present study region the readers are referred to work by Nengker et al. (2017) where the inherent cold bias in the model and the skill in capturing the spatial distribution of temperature is thoroughly discussed. For elevation of the Himalayan region based on the present REMO grid at 50 km, the readers can refer to Fig. 9.1 in Kumar et al. (2013). The study period considered here is 1970–2099 to assess the long-term changes in climate over the study region beginning from the recent past. Here, the regional climate data for the period 1970–2005 is obtained by GCM forcings under historical emissions whereas that for the period 2006–2099 is obtained by GCM forcings under projected emissions based on the representative concentration pathway RCP2.6 scenario. The RCP2.6 was formulated by the modeling team- IMAGE from Environmental Assessment Agency of Netherlands. This emission pathway represents the trajectory of achieving the least greenhouse gas concentration levels in future through a stringent climate policy (Van Vuuren et al. 2011). Under this mitigation scenario, the radiative forcing level first rises up to a value of around 3.1 W/m^2 by mid-century, and then comes down to 2.6 W/m^2 by 2100 (Van Vuuren et al. 2006). A single pathway RCP2.6 is chosen here to see specifically how Himalayan region, which is considered to be sensitive to even a small scale change in the global climate, responds to the most conservative of all emission scenarios.

First, the long-term linear trend (1970–2099) of near surface mean air temperature and its altitudinal distribution over Himalayan region is examined for identifying signals of elevation dependency of warming rate. This is done for four seasons- December–January–February (DJF), March–April–May (MAM), June–July–August–September (JJAS) and October–November (ON) to assess the seasonal response of elevation dependent warming. Next, to understand the importance of different climatic drivers in contributing to this elevation dependent warming, long-term trends in other variables for the same period and RCP scenario is studied with respect to its elevation dependent response. These variables are- DLR, total cloud fraction, total soil moisture, near surface specific humidity and surface snow melt along with surface albedo as an indirect measure of snow cover. The surface albedo is calculated here as the ratio (in %) of reflected to incident shortwave radiation. Further, the elevation dependency of the sensitivity of warming rate to that of moisture is examined. For this purpose, the ratio of the trend of temperature with that of near-surface specific humidity is studied with respect to its altitudinal distribution.

9.3 Results and Discussion

In the following sections, evaluation of seasonal (winter: DJF; pre-monsoon: MAM; monsoon: JJAS and post-monsoon: ON) response is provided. Distribution of discussed variables are depicted in vertical elevations ranges averaged over the number grids in that particular elevation range.

9.3.1 DJF

Figure 9.2 shows the altitudinal distribution of trend of various climatic variables over the study region during winter season. The increase in the rate of warming with elevation beyond a certain altitude is clearly reflected in Fig. 9.2a. In case of DLR, Fig. 9.2b, similar increasing trend is seen in higher elevation, i.e., beyond 4000

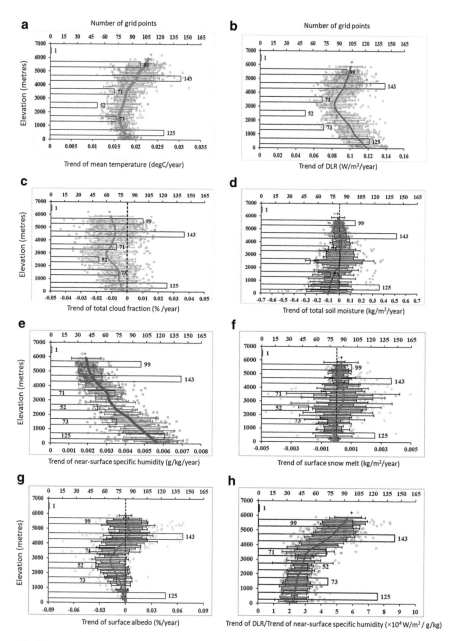

Fig. 9.2 Trends of (**a**) mean near surface air temperature (°C) (**b**) downwelling longwave radiation (DLR) (W/m 2/year), (**c**) total cloud fraction (%/year), (**d**) total soil moisture (kg/m 2/year), (**e**) near surface specific humidity (g/kg/year), (**f**) surface snow melt (kg/m 2/year), (**g**) surface albedo (%/year) and (**h**) ratio of the trend of DLR and near-surface specific humidity (×10 4 W/m 2/g/kg) during DJF under RCP2.6 scenario from REMO at every grid point over the Himalayan region plotted against surface elevation (metres) for the period (1970–2099) including present (1970–2005) and future climate (2006–2100). The scatter plot of respective trends from REMO has been shown as background dots representing grid points; the curves in same color as their corresponding dots represent the mean in 100 m- thick elevational bins smoothed by LOWESS method (Cleveland 1979). The error bar in red shows the spatial variability within each 100 m class. The rectangular bars with numbers indicate the number of grid points falling within each 1000 m altitude range

meters, while trend decreases with elevation in lower elevation areas. A reversal of the trend is noticed near 3500 m. Total cloud fraction trend (Fig. 9.2c) shows an increase with elevation between 3000 m and 5000 m with elevation (although the trend values themselves are negative). In case of total soil moisture, Fig. 9.2d, increasing trends from the surface up to 2000 m dominates, which becomes almost constant beyond this elevation. Near-surface specific humidity (Fig. 9.2e), shows decreasing trends with elevation. The increase in DLR with elevation leading to enhanced surface heat storage is one of the primary mechanisms responsible for high altitude warming (e.g. Rangwala et al. 2009, 2010, 2016; Rangwala 2013; Ruckstuhl 2007) and could be explained by the decrease of humidity/water-vapor with elevation and its feedback to the DLR. During winter, there is a large change in DLR which is associated with a greater sensitivity of DLR to absorption by water vapor at lower levels of atmospheric moisture content (typically <2.5 g/kg; Rangwala et al. 2009) – conditions which exist during the cold season at high altitudes. Further, decreasing value of humidity trends with altitude also indicate higher convective loss of moisture further leading to a higher sensible heat flux and amplification of near surface temperature. Altitudinal variation of the trend of surface snow melt (Fig. 9.2f), appears to regulate surface albedo (Fig. 9.2g), where the former seems to be clearly imprinted as a mirror image on that of latter. The decrease in the rate of change of snow melt and dependent increase in that of surface albedo beyond 3000 m could subdue the DLR-moisture positive feedback effect on surface heating. However, latter feedback plays a much larger dominating role in elevation dependent warming as the ratio of the rate of change of DLR and that of near-surface specific humidity (Fig. 9.2h) increases with elevation. This implies that at a higher elevation there is an enhanced increase in DLR with a certain increase in moisture compared with lower elevation where the sensitivity of DLR on moisture content is less. This case is typical to winter season as the moisture is within a threshold In addition to this, the total cloud fraction change will also control the DLR which, thus, is increasing over higher elevation. An increase in daytime cloud cover could reduce surface insolation which could lead to decrease in temperature thus counteracting the DLR-moisture feedback and subduing the elevation dependent rise in warming rate. In winter, a distinct kink partitioning the trend reversal at ~3500 m is seen in most of the variables. It suggests the existence of an altitude threshold, beyond which the mechanisms associated with elevation dependent warming changes.

9.3.2 MAM

The variation of the trend of different climatic variables with elevation during MAM is shown in Fig. 9.3. The entire elevations, during MAM Fig. 9.3a, show warming trends. However, DLR trend, Fig. 9.3b, decreases throughout with elevation, unlike during DJF, which corresponds to reduced increase in temperature trend values with elevation (Fig. 9.3a). It could be due to atmospheric dryness and stability at high altitudes in this season. A distinct and peculiar trend – first with steady reduction till 4000 m, followed by sudden reduction up to 5000 m and then again steady

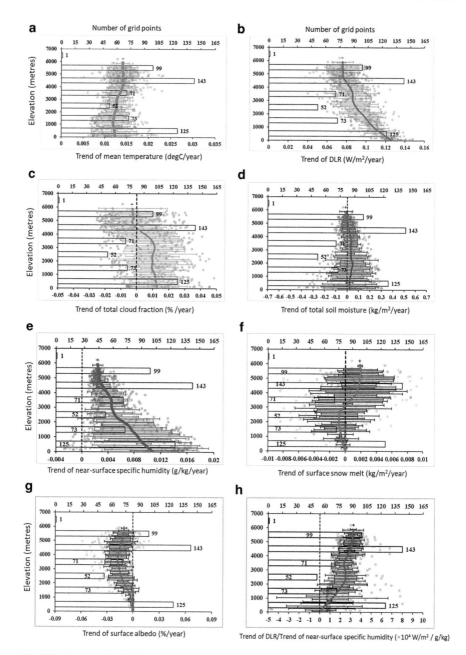

Fig. 9.3 Same as Fig. 9.2, but for MAM

reduction above – is seen in total cloud fraction with elevation, Fig. 9.3c. However up to mid elevation trends of total cloud fraction remained positive and there after negative. It indicates stable atmosphere in mid elevation ranges which in upper elevation becomes more stable by bring the lifting condensation level down. The reduction in cloud fraction trend values above 3000 m could favor the enhancement in net solar radiation received at the surface, with further increase in snow melt thus allowing for increased absorption of solar radiation at higher elevations, implying more storage of heat at the higher elevation surface and thereby amplifying the temperature (Yan et al. 2016). No significant change in total soil moisture trend with elevation is depicted, Fig. 9.3d. However, in case of near-surface specific humidity decreasing trend with elevation is seen, Fig. 9.3e. In case of trend of surface snow melt, Fig. 9.3f, decreasing negative trend up to 3000 m, followed by increasing snow melt trend thereafter is depicted which could be linked with the trend pattern of cloud fraction as explained earlier. Further, it is found that at lower elevations there will be more snow conserved as compared to that in higher elevations. This is distinctly different from results found for DJF. It could be attributed to the fact that more snow is deposited over higher elevations due to winter time conditions. Clear impact of the snow melt, Fig. 9.3f, is seen in associated trends of surface albedo, Fig. 9.3g, where snowmelt trends show higher (lower) values over the lower and higher (mid) elevations. A decreased surface albedo/snow could lead to increase in surface absorption of insolation which primarily occurs in summer season at high altitudes in association with 0 °C isotherm (Pepin and Lundquist 2008). This may be responsible for enhanced temperature trends. Trends in DLR to near-surface specific humidity ratio, Fig. 9.3h, shows increasing trends from lower elevations to higher elevations with stable trend in mid-elevation.

9.3.3   JJAS

The distribution of trend of seasonal mean values of different variables is depicted in Fig. 9.4. In case of near surface temperature during JJAS, slight reduction in trend values with elevation is visible, Fig. 9.4a. Whereas, DLR trend values decreases with elevation up to 3000 m, increases beyond till 4000 m and again decreases above it, Fig. 9.4b. Corresponding total cloud fraction trends show almost reverse mirror image of DLR, Fig. 9.4c, with increase in trend values along the elevation. It is important to mention here that JJAS is monsoonal season with increased moisture in the free atmosphere and hence play controlling role due to cloud formation. An increased daytime cloud cover decreases surface insolation which has a profound influence on near-surface temperature during this time of the year and could explain the reduced trend values during JJAS. Trend of cloud-fraction increases with elevation indicative of an increased availability of moisture which enhances the DLR but possibly the moisture level is beyond the threshold limit (2.5 g/kg; Rangwala et al. 2009) unlike the case in DJF season when the atmosphere is relatively drier. Further, trend of total soil moisture as well does not show changes with elevations, Fig. 9.4d.

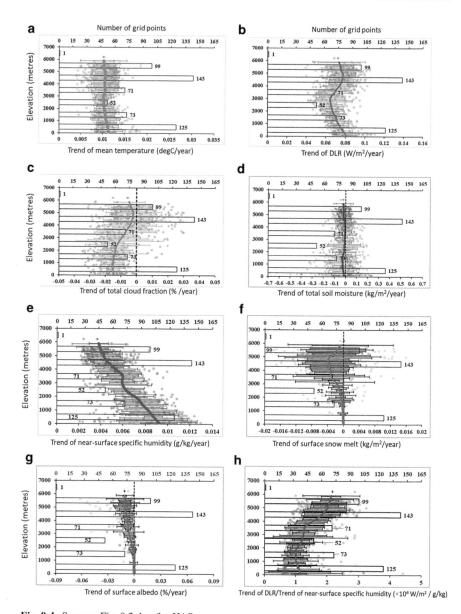

Fig. 9.4 Same as Fig. 9.2, but for JJAS

Trends of near-surface specific humidity decreases with elevation, Fig. 9.4e. This is always the case in other seasons as well. It suggests that lower elevation ranges have higher moisture which decreases as we move to higher elevations. Higher elevations will retain more snow as snow melt trends over these regions are decreasing as compared to lower elevation regions, Fig. 9.4f. At around 5000 m, trends show distinctly decreasing trends. Corresponding trends in surface albedo distinctly show these changes in their trends as well, Fig. 9.4g. The increased surface absorption of insolation is important mechanism as higher elevations show lesser trends than lower elevations. Trends of ratio of DLR to the near-surface specific humidity show variable trends but with a general increase with elevation.

9.3.4 ON

The altitudinal distribution of trend of different variables during ON is presented in Fig. 9.5. In regions below 2000 m, temperature trend remains almost unchanged with elevation, while in mid elevations (2000–3500 m) it follows a curvilinear path with first increase and then decrease and further beyond 4000 m it increases all along (Fig. 9.5a). In general, after DJF, the most significant dependency of warming rate with elevation with an increase in the trend with altitude is clearly seen during this season. Previous studies have also reported that autumn season show strongest signal of elevation dependent warming after winter (e.g. Liu et al. 2006; Rangwala et al. 2009). DLR reflects the altitudinal distribution of temperature (Fig. 9.5b) with a sharp increase in trend beyond 3500 m. In mid elevation region, there is cascading increased and decrease dominates. In lower elevation, no significant change in trends exist. The trend of cloud fraction increases with elevation between 3500 m and 5000 m explaining the increased DLR on surface, Fig. 9.5c. Trends in total soil moisture show decrease in lower elevations than that in the upper elevations, Fig. 9.5d. A decrease in soil moisture implies a reduction (increase) in latent (sensible) heat fluxes which strongly affects the snowmelt activities at the surface. Near-surface humidity shows consistent decrease in trend values with elevation corresponding to convective loss of moisture due to near surface heating leading to a higher sensible heat flux, Fig. 9.5e. In case of surface snow melt, upper elevations indicate higher snow melt then the lower elevations, Fig. 9.5f. It corresponds to similar mirror image in trends of surface albedo, Fig. 9.5g. Upper elevation has lower albedo than the lower elevations. In case of ratio of trend DLR to the trend of near-specific humidity, it increases with elevations, Fig. 9.5h.

9.4 Summary and Conclusions

Previous studies showed that among the possible mechanisms behind amplified warming at higher elevations are several feedbacks acting in the climate system, like snow-albedo (Giorgi et al. 1997; Fyfe and Flato 1999; Rangwala et al. 2010); cloud-radiation (Liu et al. 2009); humidity-DLR (Rangwala et al. 2009; Rangwala 2013; Naud et al. 2013) feedbacks. These are associated with changes in a number of relevant variables such as- soil moisture (Liu et al. 2009; Naud et al. 2013), aerosols (Lau et al. 2010), clouds and their coverage (Sun et al. 2000) all of them contributing to variations in the energy balance at the surface at various scales. In the present

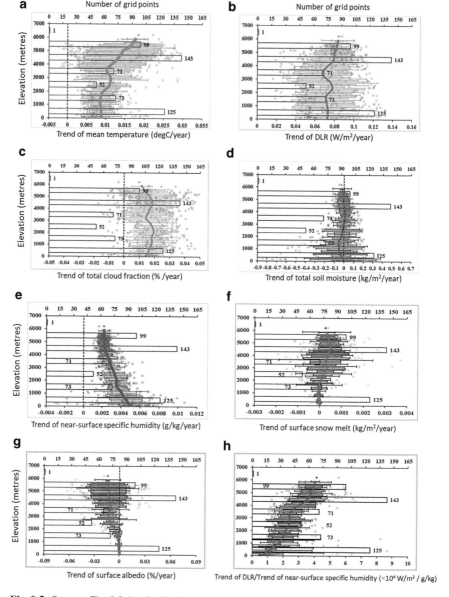

Fig. 9.5 Same as Fig. 9.2, but for ON

study, a high resolution long-term climate simulation of climate over IHR was analyzed to study elevation-dependent warming and its mechanisms over the area.
Results indicate that enhanced increase in DLR flux at the higher elevation surface
during winter is primarily responsible for high altitude warming amplification.
Possible coupling between multiple land-atmosphere feedbacks could explain the
magnitude and peculiar pattern of DLR variation during this season characterized

by trend amplification above a certain altitude. The primary feedback which is responsible for higher trend of DLR beyond a certain altitude is the humidity- surface DLR feedback which is a significant player during winter season. However, the decrease in the rate of change of snow melt and dependent increase in that of surface albedo beyond 3000 m could subdue the DLR-moisture positive feedback effect on surface heating. On the other hand, there are counter acting mechanisms existing to this process. The reduction in cloud fraction trend values above 3000 m favors the enhancement in net solar radiation received at the surface, with further increase in snow melt/decrease in snow depth thus leading to the reduced surface albedo. This further allows the absorption of solar radiation at higher elevations implying more storage of heat at the higher elevation surface and thereby amplifying the temperature (Yan et al. 2016).

Although the increase in DLR with increase in specific humidity occurs globally, the sensitivity of former to latter follows a non-linear relationship (Ruckstuhl et al. 2007; Rangwala and Miller 2012) and is particularly high when the humidity levels are low which exists typically at high elevations during winter. In other words, the drier the atmosphere, magnified will be the impact of even smaller changes in humidity on the DLR (Ruckstuhl et al. 2007; Rangwala et al. 2010; Naud et al. 2013). Changes in DLR are more sensitive to changes in humidity when the latter is less than 2.5 g/kg i.e. when the atmosphere is dry (Rangwala et al. 2009) a condition which is more prevalent during winter in the elevated regions. Instead, this phenomenon does not occur during summer season since, as background humidity values are already very high, the sensitivity of surface DLR to any further increase of atmospheric moisture is much less (e.g. Ruckstuhl et al. 2007). Also, as shown in the present study the sensitivity of longwave radiation to surface air humidity increases with altitude above a certain threshold (3000 m) corroborating the results found by Ruckstuhl et al. (2007). This means that, the same amount of changes in the surface air humidity will cause higher amount of changes in DLR at higher elevation sites in comparison to the lower elevation locations (Rangwala 2013). Increased DLR at the surface in higher elevations or above a critical altitude plays significant role in elevation dependent warming during winter through coupled feedbacks of moisture, cloud and snow cover with radiation.

Since the simulation used in this study did not include any aerosol component, the role of this variable in influencing high elevation temperature changes could not be assessed. Incorporating aerosol feedbacks in climate model would imply nesting an aerosol component through parametrization of the related forcings or processes. Further, to properly represent the relevant mechanisms and provide a more realistic simulation of the changes in the cryosphere system of high elevation regions an interactive snow/glacier model feedback into a high resolution regional climate model is required. There is also a need for increasing climate monitoring programmes at high elevation regions with greater number of climatic variables. This will aid in better understanding of present trends and processes that are affecting the state of climate in IHR as well as for validating the model generated information.

Acknowledgements APD acknowledges financial support by MoEFandCC under NMHS scheme. Authors also acknowledge the Earth System Grid Federation (ESGF) infrastructure and the Climate Data Portal at Centre for Climate Change Research (CCCR), Indian Institute of Tropical Meteorology, India for provision of REMO data under CORDEX-SA.

Author Contributions A.P.D. conceived the idea. A.P.D. and A.C. designed the study. D.K. did the major part of analysis and prepared the plots with ideas from A.P.D. All the three authors discussed and interpreted the results. A.P.D wrote the paper.

Declaration of Interests None.

References

Barry RG (1992) Mountain climatology and past and potential future changes in mountain regions: a review. Mt Res Dev 12:71–86. https://doi.org/10.2307/3673749

Bolch T, Kulkarni A, Kääb A, Huggel C, Paul F, Cogley JG, Frey H, Kargel JS, Fujita K, Scheel M, Bajracharya S (2012) The state and fate of Himalayan glaciers. Sci 336(6079):310–314. https://doi.org/10.1007/s10584-016-1599-z

Choudhary A, Dimri AP (2017) Assessment of CORDEX-South Asia experiments for monsoonal precipitation over Himalayan region for future climate. Clim Dyn 1–22. https://doi.org/10.1007/s00382-017-3789-4

Cleveland WS (1979) Robust locally weighted regression and smoothing scatterplots. J Am Stat Assoc 74(368):829–836

Diaz HF, Bradley RS (1997) Temperature variations during the last century at high elevation sites. In: Climatic Change at High Elevation Sites, pp 21–47. https://doi.org/10.1007/978-94-015-8905-5_2

Fyfe JC, Flato GM (1999) Enhanced climate change and its detection over the Rocky Mountains. J Clim 12:230–243. https://doi.org/10.1175/1520-0442-12.1.230

Ghimire S, Choudhary A, Dimri AP (2015) Assessment of the performance of CORDEX-South Asia experiments for monsoonal precipitation over the Himalayan region during present climate: part I. Clim Dyn 2–4. https://doi.org/10.1007/s00382-015-2747-2

Giorgetta MA, Jungclaus J, Reick CH, Legutke S, Bader J, Böttinger M, Brovkin V, Crueger T, Esch M, Fieg K, Glushak K (2013) Climate and carbon cycle changes from 1850 to 2100 in MPI-ESM simulations for the Coupled Model Intercomparison Project phase 5. J Adv Model Earth Syst 5:572–597. https://doi.org/10.1002/jame.20038

Giorgi F, Hurrell JW, Marinucci MR, Beniston M (1997) Elevation dependency of the surface climate change signal: a model study. J Clim 10:288–296. https://doi.org/10.1175/1520-0442(1997)010<0288:EDOTSC>2.0.CO2

Giorgi F, Jones C, Asrar GR (2009) Addressing climate information needs at the regional level: the CORDEX framework. WMO Bull 58:175–183. https://doi.org/10.1016/j.jjcc.2009.02.006

Kumar P, Wiltshire A, Mathison C, Asharaf S, Ahrens B, Lucas-Picher P et al (2013) Downscaled climate change projections with uncertainty assessment over India using a high resolution multi-model approach. Sci Total Environ 468:S18–S30. https://doi.org/10.1016/j.scitotenv.2013.01.051

Lake I, Gutowski W, Giorgi F, Lee B (2017) CORDEX: Climate research and information for regions. Bull Am Meteorol Soc 98:ES189–ES192. https://doi.org/10.1175/BAMS-D-17-0042.1

Lau WKM, Kim MK, Kim KM, Lee WS (2010) Enhanced surface warming and accelerated snow melt in the Himalayas and Tibetan Plateau induced by absorbing aerosols. Environ Res Lett 5. https://doi.org/10.1088/1748-9326/5/2/025204

Liu X, Chen B (2000) Climatic warming in the Tibetan Plateau during recent decades. Int J Climatol 20:1729–1742. https://doi.org/10.1002/1097-0088(20001130)20:14<1729::AID-JOC556>3.0.CO;2-Y

Liu X, Yin ZY, Shao X, Qin N (2006) Temporal trends and variability of daily maximum and minimum, extreme temperature events, and growing season length over the eastern and central Tibetan Plateau during 1961–2003. J Geophys Res Atmos 111(D19). https://doi.org/10.1029/2005JD006915

Liu X, Cheng Z, Yan L, Yin ZY (2009) Elevation dependency of recent and future minimum surface air temperature trends in the Tibetan Plateau and its surroundings. Glob Planet Change 68:164–174. https://doi.org/10.1016/j.gloplacha.2009.03.017

Messerli B, Ives JD (eds) (1997) Mountains of the world: a global priority. Parthenon Publishing, New York/Carnforth, 495pp. ISBN 1850707812 9781850707813

Naud CM, Chen Y, Rangwala I, Miller JR (2013) Sensitivity of downward longwave surface radiation to moisture and cloud changes in a high-elevation region. J Geophys Res Atmos 118:10072–10081. https://doi.org/10.1002/jgrd.50644

Nengker T, Choudhary A, Dimri AP (2017) Assessment of the performance of CORDEX-SA experiments in simulating seasonal mean temperature over the Himalayan region for the present climate: Part I. Clim Dyn 1–31. https://doi.org/10.1007/s00382-017-3597-x

Nogués-Bravo D, Araújo MB, Errea MP, Martínez-Rica JP (2007) Exposure of global mountain systems to climate warming during the 21st Century. Glob Environ Chang 17:420–428. https://doi.org/10.1016/j.gloenvcha.2006.11.007

Ohmura A (2012) Enhanced temperature variability in high-altitude climate change. Theor Appl Climatol 110:499–508. https://doi.org/10.1007/s00704-012-0687-x

Palazzi E, Mortarini L, Terzago S, von Hardenberg J (2018) Elevation-dependent warming in global climate model simulations at high spatial resolution. Clim Dyn:1–18. https://doi.org/10.1007/s00382-018-4287-z

Pepin NC, Lundquist JD (2008) Temperature trends at high elevations: Patterns across the globe. Geophys Res Lett 35. https://doi.org/10.1029/2008GL034026

Pepin N, Bradley RS, Diaz HF, Baraer M, Caceres EB, Forsythe N, Fowler H, Greenwood G, Hashmi MZ, Liu XD, Miller JR, Ning L, Ohmura A, Palazzi E, Rangwala I, Schöner W, Severskiy I, Shahgedanova M, Wang MB, Williamson SN, Yang DQ (2015) Elevation-dependent warming in mountain regions of the world. Nat Clim Chang 5:424–430. https://doi.org/10.1038/nclimate2563

Qin J, Yang K, Liang S, Guo X (2009) The altitudinal dependence of recent rapid warming over the Tibetan Plateau. Clim Change 97:321–327. https://doi.org/10.1007/s10584-009-9733-9

Rangwala I (2013) Amplified water vapour feedback at high altitudes during winter. Int J Climatol 33:897–903. https://doi.org/10.1002/joc.3477

Rangwala I, Miller JR (2012) Climate change in mountains: A review of elevation-dependent warming and its possible causes. Clim Chang 114:527–547. https://doi.org/10.1007/s10584-012-0419-3

Rangwala I, Miller JR, Xu M (2009) Warming in the Tibetan Plateau: possible influences of the changes in surface water vapor. Geophys Res Lett 36 https://doi.org/10.1029/2009GL037245

Rangwala I, Miller JR, Russell GL, Xu M (2010) Using a global climate model to evaluate the influences of water vapor, snow cover and atmospheric aerosol on warming in the Tibetan Plateau during the twenty-first century. Clim Dyn 34:859–872. https://doi.org/10.1007/s00382-009-0564-1

Rangwala I, Sinsky E, Miller JR (2016) Variability in projected elevation dependent warming in boreal midlatitude winter in CMIP5 climate models and its potential drivers. Clim Dyn 46:2115–2122. https://doi.org/10.1007/s00382-015-2692-0

Ruckstuhl C, Philipona R, Morland J, Ohmura A (2007) Observed relationship between surface specific humidity, integrated water vapor, and longwave downward radiation at different altitudes. J Geophys Res Atmos 112. https://doi.org/10.1029/2006JD007850

Sun B, Groisman PY, Bradley RS, Keimig FT (2000) Temporal changes in the observed relationship between cloud cover and surface air temperature. J Clim 13:4341–4357. https://doi.org/10.1175/1520-0442(2000)013<4341:TCITOR>2.0.CO2.2

Snyder MA, Bell JL, Sloan LC, Duffy PB, Govindasamy B (2002) Climate responses to a doubling of atmospheric carbon dioxide for a climatically vulnerable region. Geophys Res Lett 29(11):9–1

Teichmann C, Eggert B, Elizalde A, Haensler A, Jacob D, Kumar P, Moseley C, Pfeifer S, Rechid D, Remedio AR, Ries H (2013) How does a regional climate model modify the projected climate change signal of the driving GCM: a study over different CORDEX regions using REMO. Atmosphere (Basel) 4:214–236. https://doi.org/10.3390/atmos4020214

Thompson LG, Mosley-Thompson E, Davis ME, Lin PN, Henderson K, Mashiotta TA (2003) Tropical glacier and ice core evidence of climate change on annual to millennial time scales. Clim Chang 59:137–155. https://doi.org/10.1023/A:1024472313775

Van Vuuren DP, Eickhout B, Lucas PL, den Elzen MGJ (2006) Long-term multi-gas scenarios to stabilise radiative forcing – exploring costs and benefits within an integrated assessment framework. Energy J 27:201–233. https://doi.org/10.5547/ISSN0195-6574-EJ-VolSI2006-NoSI3-10

Van Vuuren DP, Edmonds J, Kainuma M, Riahi K, Thomson A, Hibbard K, Hurtt GC, Kram T, Krey V, Lamarque JF, Masui T (2011) The representative concentration pathways: an overview. Clim Chang 109:5–31. https://doi.org/10.1007/s10584-011-0148-z

Xu J, Grumbine RE, Shrestha A, Eriksson M, Yang X, Wang YU, Wilkes A (2009) The melting Himalayas: cascading effects of climate change on water, biodiversity, and livelihoods. Conserv Biol 23:520–530. https://doi.org/10.1111/j.1523-1739.2009.01237.x

Yan L, Liu Z, Chen G, Kutzbach JE, Liu X (2016) Mechanisms of elevation-dependent warming over the Tibetan plateau in quadrupled CO2 experiments. Clim Chang 135(3–4):509–519. https://doi.org/10.1007/s10584-016-1599-z

Part II
Paleoclimate

Chapter 10
Geomorphological Changes During Quaternary Period Vis a Vis Role of Climate and Tectonics in Ladakh, Trans-Himalaya

Anupam Sharma and Binita Phartiyal

Abstract The Ladakh region of NW India is relatively unexplored as compared to its counterpart, the Tibetan Plateau. The region provides ample opportunity to understand the dynamic relationship of the Indian and Eurasian plates, role in governing the global climate and controlling the earth surface temperature by carbon sequestration through weathering of rocks. The differential uplift along several thrust planes and associated climatic conditions are primarily governing the geomorphic evolution of the Ladakh, and amongst several agents of material movement, glaciers are the most effective agent. Overall ten glacial stages (~430–0.4 ka) are recognized from Ladakh. Additionally, the major sources of moisture to the Ladakh region are the Indian Summer monsoon (ISM) and the Central Asian westerlies contributing in subequal amounts. Several morphometric parameters suggest that tectonics is governing the topography. It was most pronounced at 27 ka, 23 ka, 17–19 ka, 11–10 ka and 6 ka along the Karakorum Fault and Indus Suture Zone. The climate studies suggest that during around 35–25 ka, the climate was cold and humid followed by the relatively dry at the Last Glacial Maximum (LGM). A rise in lakes during ~17–5 ka was observed. However, major fluctuations in limiting the lake levels during the Older Dryas, Younger Dryas and 8.2 ka were also noticeable. The variation in different chronological technique, however, poses a serious challenge in determining the geomorphic evolution of the landscape in the region.

10.1 Introduction

The rifting in Gondwanaland resulted in the birth of several continents distributed across the globe (Miashita and Yamamoto 1996; Li et al. 2008; Seton et al. 2012; Torsvik and Cocks 2013). Among all landmasses, the Indian subcontinent not only

A. Sharma (✉) · B. Phartiyal
Birbal Sahni Institute of Palaeosciences, Lucknow, India

© Springer Nature Switzerland AG 2020
A. P. Dimri et al. (eds.), *Himalayan Weather and Climate and their Impact on the Environment*, https://doi.org/10.1007/978-3-030-29684-1_10

traveled very fast but also covered maximum distance before colliding with the Eurasian plate (Klootwijk et al. 1992; Gaina et al. 2007). The collision of the Indian plate with the Eurasian plate is responsible for the uplift of the mighty and young Himalayan mountain range (Heim and Gansser 1939; Besse et al. 1984; Klootwijk et al. 1985; Yin 2006; https://pubs.usgs.gov/publications/text/himalaya.html). The structural and tectonic history of the Himalaya reveals that differential uplift along several thrust planes and associated climatic conditions are primarily governing the geomorphic evolution of the region. The higher Himalayan ranges, comprised largely of mechanically resistant, massive and hard crystalline rocks, differentiate the Indian landmass into two climatically contrasting regions. The region lying to the south of the higher Himalaya receives significant rainfall from SW Monsoon; however, the northern region that lies in the rain shadow of SW Monsoon remains more or less dry (Pande et al. 2000; Thompson et al. 2000; Dalai et al. 2002a; Karim and Veizer 2002; Bookhagen and Burbank 2006; Sharma et al. 2017).

The region lying north to the higher Himalaya and Tethyan Himalaya is also named as Trans Himalaya (Fig. 10.1a). Since the larger part of the Trans Himalaya lies in the Tibet, it is often called as Tibetan Himalaya. Spatially, the Trans Himalayan region is extended to ~1000 km in the east-west direction having an average elevation of 3000 m above mean sea level. However, its width varies from 40 km in the eastern and western extremities to ~250 km in the central part (Kumar et al. 2010). Ladakh sector of the Tibetan Himalaya is the most important center in the Indian Territory and known for its rugged topography with almost barren mountain ranges.

Ladakh, literary means land of High passes, also called little Tibet, is one of the three administrative divisions of Jammu and Kashmir State of India. Among the three divisions, Ladakh is the largest division having an area of 59,146 km^2 (58.33%) and most of its region is lying 3000 m above mean sea level. Because of its general

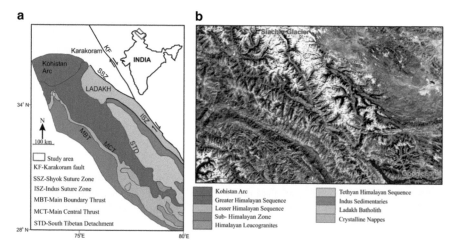

Fig. 10.1 (**a**) Gelogical map of Ladakh showing major geological formations. The location of the study area is marked by red rectangle in the sketch map of india (inset); (**b**) Google Earth Pro-image of Ladakh region showing snow cover and glacier distribution

elevation and being in the rain shadow for monsoon clouds, having very sparse vegetation and an extremely cold climate, Ladakh is also known as the cold desert. Earlier workers largely believed that the major source of moisture in Ladakh is mid-latitude westerlies, which brings the moisture in the form of snow during the winter months (Benn and Owen 1998; Bookhagen and Burbank 2010; Bolch et al. 2012). However, the recent study of Sharma et al. (2017) adequately described, using mass balance of Indus and its tributary river water isotopic composition, that the major (~74%) source of moisture is supplied from the Bay of Bengal, Arabian Sea and the Indian Ocean and remaining ~26% is supplied through the Central Asian region to the Indus catchment in Ladakh.

In Ladakh, the Indus and its tributary rivers are the backbone of the agriculture, livelihood as well as the socio-economic fabric of the region (Fig. 10.2). In a larger perspective, the Indus River System comprises of several west-flowing rivers of the Indian subcontinent and also forms one of the world's largest freshwater systems. All along the Indus and its tributary river basins in Ladakh, a variety of collision related rock types, weathering and erosion linked glacial, fluvial, lacustrine and aeolian sediments and landforms are ubiquitous. These characteristics make it an open laboratory, particularly for the geologists, to understand the effects of earth surface and subsurface processes in geomorphic evolution of the region.

Ever since (>50 Ma) the northward-moving Indian plate collided with the Eurasian plate, there has been a continuous horizontal push, which not only closed the Tethys basin but also uplifted and exhumed its sediment to greater heights. The processes, which were associated with magmatism and crustal deformation initially, shaped the geological framework (Gansser 1964; Yin 2006). Soon as the material attained topographic relief, the earth surface processes such as weathering and erosion started acting upon. The present landscape of the Ladakh region is an outcome of Quaternary processes that includes several glacial and interglacial intervals and an on-going tectonics associated with the build-up of Himalaya. In all large and small river valleys, the signatures of these in the form of various landforms (glacial,

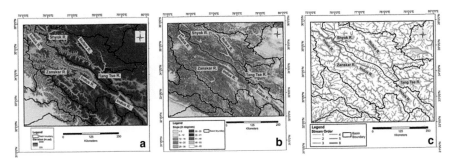

Fig. 10.2 SRTM maps showing (**a**) Basin boundary of Indus and its major tributary rivers with elevation; (**b**) Slope map of different rivers catchment area. It is interesting to note that in a short distance the elevation gradient varies significantly indicating tectonic control in generating the observed relief variation in the Ladakh region; (**c**) Map showing the seven different stream orders in the Indus River catchment

fluvial, lacustrine and aeolian) (Fig. 10.3) are easily noticeable. These landforms include moraines ridges, cirques, drumlin, Roche montonnees, glacial troughs, amphitheater valleys (glacial); alluvial fans, channel bars, outwash plains, gorges, valley fill and strath terraces (fluvial); clay-silt-sand intercalated and usually horizontal with occasional varve deposits (lacustrine) and sand dunes, barchans, sand ramps (aeolian) etc. Macro to microscale faults, fault gouge, sand dykes, broken clasts and multiple types of deformation structures are easily noticeable in these sediments, which are the examples of seismic/tectonic activity in the region (Phartiyal et al. 2013, 2015; Juyal 2014; Nag and Phartiyal 2015).

Fig. 10.3 Field photographs showing general geomorphological characteristics including major agencies responsible for generation and distribution of sediments, (**a**) Palaeo-lacustrine sediments exposed at Spituk-Leh (Indus River Valley); (**b**) Meandering Indus River with fluvial terrace on its left bank; (**c**) Glacial features at the Khardungla Pass northern slope (Ladakh Range); (**d**) View of Zanskar River near Sani village; (**e**–**f**) Aeolian Deposits (**e**) Sand ramp near Leh; (**f**) Sand dunes at Hunder in Shyok River Valley, a tributary of the Indus River

10.2   Important Features of the Study Area

10.2.1   Indus and Its Tributary Rivers

The Indus and its tributary rivers are the lifeline of Ladakh. The Indus River originates at an elevation of 6714 m in the vicinity of Mansarovar Lake and Mount Kailash in Tibet. In it's ~422 km northwest-southeast long course in Ladakh, the Indus flows in a well-defined valley parallel to the Indus Suture Zone (ISZ) and forms one of the world's largest freshwater systems. During its journey the Indus follows a narrow, constricted, and meandering path except in Chumathang and Leh where the valley becomes ~2–3 km wide and flows in a braided pattern. In a recent study based on several river parameters such as channel gradient, channel pattern, valley width, and knick points, Nag and Phartiyal (2015) divided the Indus River into four segments where the valley width varies from 100 to 6000 m and the channel gradient range from 0.75 to 7.5 m/km. Further, these authors also calculated the stream length gradient index (SL) and the steepness index (Ks) suggesting that tectonics is the major governing factor in deciding the valley width and channel flow.

Three tributary rivers namely the Tangtse, Zanskar and Nubra-Shyok and their watershed basins are the major contributor to the Indus (Phartiyal et al. 2013, 2018). The slope and drainage distribution (Fig. 10.2) indicate that the Indus attains 7th stream order, whereas its major tributary rivers (Tangtse, Zanskar, Shyok) are either of 5th or 6th stream order (Fig. 10.2) (Nag et al. 2016; Phartiyal et al. 2018; Sharma et al. 2017).

The Shyok River, the biggest tributary of Indus, largely flows in a wide valley, which becomes even wider at the confluence with the Nubra River. Shyok and Nubra have an unusual course, originating from the Rimo and the Siachen glaciers respectively: initially they flow in an SE direction but subsequently take an NW trend and join the Indus near Skardu in Pakistan (Sharma et al. 2017; Phartiyal et al. 2018).

10.2.2   Mountain Ranges

As mentioned above that the higher Himalaya limits the monsoon winds to enter into Ladakh, there are several other mountain ranges in Trans-Himalaya making the entire topography very rugged and in parts inaccessible too. As such the mountain ranges in the Ladakh Range are outcome of collision of Indian plate with the Eurasian plate ~50–55 Ma. The Himalayas were primarily formed by the basement rocks and overlying (Neo) Proterozoic meta-sedimentaries of the Indian plate, whereas the Zanskar range largely consists of Cambro-Ordovician sediments deposited in the Tethys sea (Fig. 10.1c). Another important mountain range is the Ladakh range, mostly comprised of magmatic rocks (granites with substantial mafic enclaves). The Indus Suture Zone comprised of Indus molasses and Ladakh

Batholith basically marks the collision that runs almost parallel to little south of the Ladakh range and also hosts the Indus River. The average height of the Ladakh range is ~6000 m with no major peaks, though few passes are ~5000 m or little less in height.

In the eastern sector near Pangong Lake, the Pangong range (highest peak ~6700 m) runs parallel to the Ladakh range. Further north of the Tangtse River lies the mighty Karakoram Range. The Karakoram Range has several very high massifs of over 7200 m such as the Apsarasas group, the Rimo group, the Teram Kangri group, etc. The Karakoram Range in Ladakh is not as mighty as in the Baltistan, besides the Kun Lun Mountains also lie in the north of the Karakoram Range (Nakata 1972; Molnar et al. 1987; Nakata et al. 1990; Chevalier et al. 2005; Dortch et al. 2009; Hintersberger et al. 2011). Interestingly, all these mountain ranges are the storehouse of numerous large and small glaciers, which not only ensure the fresh water supply to the entire Ladakh region but also the source of perennial Indus river system.

10.2.3 Glaciers

In the entire Ladakh region, numerous small or big glaciers and snow/ice fields are distributed atop every mountain hill and even high-altitude plains with Siachin being the most important and the largest (Fig. 10.1b). The meltwater from these glaciers/snowfields gives birth to streams (called Tokpo in the local dialect) that cater the need of agriculture as well as domestic usages of the villages, invariably situated along these streams. All these streams directly or through tributary rivers ultimately join the Indus River, which is considered as the backbone of socio-economic fabric of Ladakh.

The unique landscape of the Ladakh region points towards the important role of glaciers in defining the physiographic and the geomorphic architecture. The erosional and depositional glacial landforms distributed in the entire Ladakh region testify it further when we see that these glacial signatures are providing the history of glacier advance and retreat occurred during the Quaternary period. Quaternary Period, as we know is known for Ice Ages has witnessed several glacial and interglacial phases. During glacial phase glaciers usually, advance and in the process bring enormous amounts of glacial debris to relatively lower reaches of valleys and usually deposit them as moraines. Many other landforms form during the glacial advance. However, moraines are commonly used to establish the chronology of different stages of glaciation in a valley.

Glaciers as such are also very sensitive to climate (Kaab et al. 2007; Schaefer et al. 2008) because they effectively respond to both precipitation and temperature (Pratt-Sitaula et al. 2011). Several workers have used glaciers as effective probes to understand local as well as the past climate (Owen et al. 1997, 2008; Owen 2009; Owen and Dortch 2014; Benn et al. 2005; Ali and Juyal 2013). Compared to high latitude temperate regions, the glacial history of tropical/monsoonal regions of Himalaya (Schafer et al. 2002) is rather complex because of two major reasons (1)

The number of governing factors controlling the glacier dynamics is relatively more, and (2) The interplay/behavior of these factors is not well understood. Therefore, palaeoclimatic inferences based on past glaciations timing and patterns from mountainous regions, like the one in Ladakh, pose serious challenge, however, could be utilized as an opportunity not only to understand the glacial/climatic history but also to predict the role of glaciers in determining the future climate (Watson 1997; Dyurgerov and Meier 2000; Hughton et al. 2001; Owen 2009). It has become increasingly important in the context of contemporary global warming scenario wherein the anthropogenic factors also driving and modifying the climate responsible for glacier retreat (Hughton et al. 2001; Arendt 2002; Corell 2004).

10.2.4   Geology

The collision of the Indian and Asian plate was a major geological event of global significance. The collision also resulted into obliteration of Neotethys Ocean and formation of Himalaya. Both these activities (collision and obliteration) occurred during 50–60 Ma all along the Indus-Tsangpo Suture zone (ITSZ; extending ~2500 km in east-west direction), however, for Ladakh the use of Indian Suture Zine (ISZ) is more precise (Molnar and Tapponnier 1975; Gansser 1977; Klootwijk et al. 1992; Rowley 1996). It is interesting to note that prior to these activities, another suture zone referred as Shyok Suture Zone (SSZ; 75–85 Ma) is present, where the island arc system of Kohistan-Ladakh collided with the southern margin of Asian plate comprised of the Karakoram Mountains and Lhasa block (Rai 1982; Searle 1991; Robertson and Collins 2002; Jain 2014; Upadhyay 2014). The subduction of Indian plate continued, which resulted into the formation of Ladakh Batholith comprised of Jurassic to early Cenozoic calc-alkaline granitoids and associated volcanic rocks (Sinclair and Jaffey 2001; Steck 2003; Henderson et al. 2011; Jain and Singh 2008; Jain 2014).

The lithology of ISZ is highly complex where rocks of Cretaceous to Miocene age or younger formed under different conditions such as forearc, island arc, ocean basin setting. The major rocks present are continental slope forearc sediments, post-collision terrigenous clastic sedimentaries and calc-alkaline volcanic (Wadia 1937; Srikantia and Razdan 1980; Thakur and Misra 1984; Searle et al. 1987; Garzanti and Vanhaver 1988; Ahmad et al. 1998; Steck 2003; Henderson et al. 2011; Singh et al. 2015). Towards the south of ISZ lies Lamayuru-Karamba Complex representing slope to deep marine passive margin sedimentary rocks of the Indian plate including the remnant of a carbonate platform (Frank et al. 1977; Searle et al. 1987; Robertson 2000; Singh et al. 2015). In the western part of the ISZ, well-developed rock successions are observed and represented by Nindam Formation, Ophiolitic melange, Dras Formation, and Lamayuru Formation. Further, in the north the Indus Basin Sedimentary Rocks (ISBR) divided into Tar and Indus Group largely comprised of terrigenous clastic sediments derived both from Indian and Asian plates are exposed (Henderson et al. 2011; Singh et al. 2015). In the eastern part of the

Ladakh, some of the rock units belonging to ISZ are absent. However, the Nidar ophiolite along with the Indus Group sediments locally forming the Liyan Formation and correlated with Kargil formation (Indus Group) are present (Shanker et al. 1976; Thakur and Virdi 1979).

10.2.5 Geomorphology

Geomorphologically, the Ladakh region has the highest altitude region in India. The geomorphological evolution is governed by two sets of geological processes (1) Continental-scale geological processes, which includes the collision of Indian plate with Eurasian plate. The collision provided the basic framework for the landscape evolution of the Ladakh and, (2) Regional/local scale geological processes in which the role of climate and neo-tectonics is significant in defining the earth surface processes such as weathering and erosion through glacial, fluvial and aeolian agencies.

The present landscape of Ladakh has various landforms linked to the glacial, fluvial, lacustrine and aeolian origin (Fig. 10.3) of the Quaternary period. Throughout the Quaternary period, the earth surface processes that are sensitive to tectonoclimatic setup have modified the landscape and currently in operation as well. In a recent study Kumar and Srivastava (2018) reviewed the earlier works of several authors (Fort 1983; Burbank and Fort 1985; Brown et al. 2002; Damm 2006; Owen et al. 2006; Dortch et al. 2010, 2013) and summarized that moraines are the most extensively studied glacial landform in the Ladakh Himalaya suggesting ten glacial stages starting from the Indus Valley (~430 ka) to Pangong Cirque (0.4 ka). The sediment derived from these glacial events along with the sediments generated by other agencies aggraded in the river valleys of this region. Several workers have reported late Quaternary river valley infilling and incision in the Indus and Zanskar rivers (Burbank et al. 1996; Leland et al. 1998; Phartiyal et al. 2013, 2015; Blöthe et al. 2014). The strath terraces studied in the Indus and Zanskar rivers indicate that the valley infilling initiated ~83 and 50–20 ka (Blöthe et al. 2014; Kumar and Srivastava 2017) and also till recently (Nag and Phartiyal 2015), however, the incision occurred at differential rate depending on the spatiotemporal conditions experienced in different valleys (Burbank et al. 1996; Leland et al. 1998; Kumar and Srivastava 2018).

Palaeolake deposits can be seen distributed in all river valleys of Ladakh indicating an increase in temperature resulting into excessive melting of glaciers or increased runoff because of stronger monsoon in Ladakh Himalaya (Kotlia et al. 1997; Phartiyal et al. 2005; Nag and Phartiyal 2015; Phartiyal et al. 2015). The elaborated account of palaeolake deposits is dealt in the discussion section. Sand ramps are also studied recently by Kumar et al. (2017) and the authors suggested that the valley width and availability of material are two important factors in determining their distribution in Ladakh. Interestingly, the aeolian activity was pronounced between 25 and 17 ka, however, the studied ramps are >44 ka to ~8 ka old (Kumar et al. 2017) (Fig. 10.4a–c).

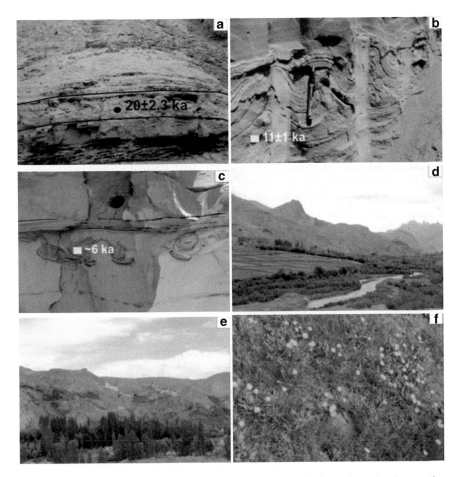

Fig. 10.4 Field photographs showing evidences of tectonic activities and restricted vegetation observed in the Ladakh region (**a–c**). Soft-sediment deformation structures observed at Tangtse, Spituk and Hanuthang palaeolake deposits respectively. (**d–f**) cropping pattern followed in the immediate vicinity of streams in Ladakh, also showing trees and herbs growing in this valley

10.3 Contemporary Climatic and Tectonic Scenario

10.3.1 Present Day Climate (Westerlies, ISM) and Extreme Hydrological Events)

Ladakh is characterized by arid to semi-arid continental climate. The region experiences prolonged winters extending from October to May and remains under snow cover for 3–4 months (November to February). The climate in this region is mainly governed by the Indian summer monsoon (from mid-June to mid-September). Besides, the region receives more rain/snow during Abnormal Monsoon Years

(AMYs) due to a northward shift of the Inter-Tropical Convergence Zone (ITCZ) (Bookhagen et al. 2005). The major sources of moisture in form of snow/ice are received from the Indian Ocean and the central Asian regions during the months of November to March (Benn and Owen 1998; Bookhagen and Burbank 2010; Bolch et al. 2012, Dimri et al. 2016, Sharma et al. 2017). In general, the average annual rainfall varies from ~50 mm to ~150 mm as one move from west to east direction in the Ladakh Himalaya. It is to be noted that the region seldom experiences extreme hydrological events, e.g. in the year 2010 the Leh city and adjoining areas were devastated by the cloudburst wherein 210 mm precipitation occurred only in 3 h period causing a destructive flash flood (Juyal 2010; Rasmussen and Houze 2012). Due to its high altitude, the intensity of solar radiation is significant resulting into large diurnal temperature variation. The large difference in diurnal temperature facilitates intense frost action (physical/mechanical weathering) responsible for the abundant supply of unconsolidated sediments, which are blanketing the mountain ranges and forming debris cones and piles in the valleys (Fig. 10.4). Overall, the cold and arid environment, >3000 m relief, large variation in seasonal temperature (varying from −20 °C in winter to 35 °C in summer), and extremely low rainfall, all these attributes qualifies to make Ladakh a high altitude cold desert.

10.3.2   Present Day Seismicity (Neotectonics)

Signatures of plate movement and associated intra-continental deformation features can be noticed all along the major faults and thrusts belts (Gansser 1964; Nakata 1972; Nakata et al. 1990; Aitchison et al. 2007). Therefore, these thrust and fault planes and their subsidiary faults are very prone to seismic activities. In a recent study, Hazarika et al. (2017) reconstructed the seismotectonic scenario of the northwestern part of the India-Asia collision zone by studying the local earthquake data (M ~ 1.4–4.3). According to this study, the most pronounced cluster of seismic activity was observed in the Karakoram Fault (KF) zone at a depth of ~65 km, whereas two microseismicity clusters were found at the northwestern and southeastern fringes of the Tso Morari gneiss dome at a depth range 5–20 km, which can be correlated to the Zildat and Karzok fault activities, respectively.

Similarly, the palaeoseismic records preserved in the fluvial and lacustrine Quaternary deposits indicate that the region must have experienced seismic activity. It is observed that in the Indus as well as all tributary river valleys fluvio-lacustrine facies sediments contain a series of seismically triggered soft-sediment deformation structures (SSDS). These SSDS's often have a continuous lateral extent and separated by undeformed beds commonly formed by an allogenic trigger mechanism. The SSDS is varied in nature such as sand dykes, pseudonodules, complex wavy laminations, flame structures, sand sills, clay diapirs, folding and micro-faulting. (Fig. 10.4a–c) (Phartiyal and Sharma 2009). Few of these structures are chronologically constrained in our earlier studies indicating that there were periods/time intervals when the seismic activity was pronounced, e.g. around 27 ka, 23 ka, 17–19 ka, 11–10 ka and 6 ka along the Karakorum Fault and Indus Suture Zone.

10.4   Discussion

The Ladakh region of NW India is relatively unexplored both by the geologists and the climatologist as compared to its counterpart Tibetan Plateau. However, the region provides ample opportunity to scientists to understand the dynamic relationship of the Indian and Eurasian plates, role in governing the global climate and controlling the earth surface temperature by carbon sequestration through silicate weathering of rocks (Walker et al. 1981; France-Lanord and Derry 1997; Bookhagen et al. 2005; Clift et al. 2008; Dosseto et al. 2015). The studies conducted earlier by the geologists were primarily inclined to resolve problems related to geodynamic evolution and their implications in understanding the structural setup, exhumation history and so on. It is indeed very important that these processes have provided the basic geomorphic setup over which the landscape of the Ladakh is evolved. In the contemporary geomorphic evolution of Ladakh region, the role of Quaternary glaciations is paramount.

The landscape evolution in the Indus River along with its all major tributary rivers, it is very well evident that in the entire Ladakh region glacial processes of the Quaternary period have played a major role. A variety of landforms such as U-shaped valleys, extensive outwash plains, lateral and terminal moraines, cirque valleys indicate that the region has experienced multiple glacial stages (Hewitt 1999; Pant et al. 2005; Dortch et al. 2010, 2011; Demske et al. 2009; Hedrick et al. 2011). Several workers based on field and remote sensing studies, carbon and optical chronology and other geomorphic parameters have shown three major events of glaciation (Sharma et al. 2016 and references therein). The oldest event predates Last Glacial Maxima (LGM) episode and corresponds to a relatively cold and wet Marine Isotopic Stage-IV (MIS-4), and it is corroborated with other studies as well (Benn and Owen 1998; Phillips et al. 2000). The second and third glacial advances were recorded ~20 and ~9 ka corresponding with LGM and 8.2 ka cooling events respectively. These later glacial phases were linked with enhanced winter precipitation coming from the Mediterranean region and accord well with the modern meteorological data (Dortch et al. 2013; Nagar et al. 2013). Further, some of the recent studies indicate that the Ladakh glaciers respond directly with mid-latitude westerlies and therefore the relatively recent glacial phases are in accordance with the northern latitude glaciations (Dortch et al. 2013; Nagar et al. 2013).

The glacial phases are punctuated by interglacial phases, and during these time intervals, fluvial and lacustrine regime modify and sometimes completely subdued the features developed by other environments. The earth surface processes, mainly the weathering and erosion, greatly help in obscuring the signatures of earlier activities. Lakes were formed by blockading of the channel by terminal moraines, or a river path is closed due to the debris avalanche triggered by higher precipitation or seismic activity. Compared to glacial deposits, the fluvial and particularly the lacustrine deposits are considered ideal archives for recoding the climatic signals of terrestrial origin (Gasse and Van Campo 1994; Gasse et al. 1996; Owen et al. 2001; Bookhagen et al. 2006; Herzschuh et al. 2006; Trivedi and Chauhan 2009; Demske et al. 2009; Wünnemann et al. 2010; Mischke and Zhang 2010; Dixit and Bera 2012; Nag and Phartiyal 2015; Phartiyal et al. 2015). In India as well as across the globe, multiproxy studies conducted over lake sediments have enhanced our understanding of palaeoclimate (Telford et al. 1999; Smith et al. 2002; Cohen 2003; Digerfeldt et al. 1993; Stone and Fritz 2004; Stevens et al. 2006; Fritz 2008; Nag et al. 2016 and many more). Keeping this in mind, we have studied the fluvial and fluvio-lacustrine sediments of different river valleys in greater detail to understand the role of climate and tectonics in the geomorphic evolution of the Ladakh sector as discussed in the following sections.

10.4.1 Fluvio-Lacustrine Deposits of Tangtse River Valley in Eastern Ladakh

Records from the Tangtse Valley in the Trans-Himalaya reveal depositional history since 48 ka, with fluvial aggradation followed by incision, lacustrine sediment fill, and later incision (Phartiyal et al. 2015). Varied sedimentary architecture with fluvial episodes intervened by lacustrine pulses, flood events, colluvial and glacial activity are well preserved. The valley that is located west of the Pangong Tso/ Bangong Co which is one of the largest lakes in Tibet, which has served as a spillway, flooding and damming the entire Tangtse Valley, resulting in the formation of a lake. Documentation based on ^{14}C and OSL chronologies of the sediment sections throughout the valley reveals evidence of a sixth basin of Pangong Tso towards west, occupying the present day Tangtse Valley between 9.6 and 5.1 ka (Phartiyal et al. 2015). This event coincides with periods of high lake levels in Tibet, China as well as intensified monsoon periods over the Indian subcontinent. A fluvial regime around 48 ka and 30–21 ka with comparatively arid conditions and dry phases interspersed by flooding at ~3.5 ka are also recorded by these researchers (Fig. 10.5). The valley has been incised to depths of 40–50 m in the upper part and 130 m in the lower part. The incision rate ranges from 0.3 to 1.2 mm year^{-1} in the upper part and reaches as high as 10.8 mm year^{-1} in the lower valley. Much of the incision took place between 22 and 9.6 ka although repeated sediment fill-incision cycles in the valley from 30 to 22 ka, 22–9.6 ka, 9.6–5.1 ka, and even to present time are observed by Phartiyal et al. (2015).

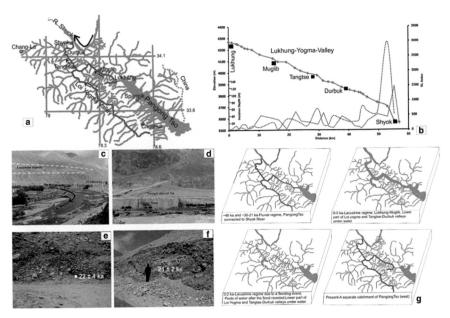

Fig. 10.5 (**a**) Drainage of the Tangtse River Valley; (**b**) Longitudinal profile of the River with Stream Length Index (very high at the confluence with Shyok River); (**c**). T1 and T2 River terraces along Tangtse; (**d**) Alluvial fans; (**e–f**) Fluvial terraces with chronology; (**g**) Schematic sketch of the sedimentation and fluvial scenario since past 35 ka in the Tangtse valley (Phartiyal et al. 2015)

10.4.2   Fluvio-Lacustrine Deposits of Shyok-Nubra River Valleys in the North and Northwest Ladakh

The Nubra-Shyok valley lies in the vicinity of the KF which is a right lateral, strike-slip fault with a bifurcation in two strands viz., the SW Tangtse Strand (Loi-Yogma Valley) and NE Pangong Strand (Lukhung-Muglib Valley). The neotectonic features are reflected in the form of deep incision of rivers, stream offsets, strath terraces, fossilized river valleys and truncation of alluvial fans; all indicate active nature of the KF in recent times. The U-shaped river valleys are filled with colluvial debris, alluvial fans, lacustrine, glacial and fluvial sediments. The width of the river at the confluence of the Nubra and Shyok rivers is around 10 km. Five main sediment facies types identified. Out of these five facies, lacustrine facies (buff and grey colored beds ranging from thin laminations to cm scale of massive clay, silty-clay and intermittent fine to medium sand) are the most important. To trace the lake towards the west from Tangtse into the Shyok river valley, either overlying the older glacial or fluvial deposit or the country rock (which are metamorphic rocks of the Pangong Tso Range, volcanics of Khardung formation, the Shyok group rocks or the Ladakh and Karakorum batholiths) (Fig. 10.6). The lacustrine sediments are exposed between 3986 and 4100 m altitudes along the entire stretch. A remnant lake between this range is reconstructed, which occupied the Shyok and the Nubra

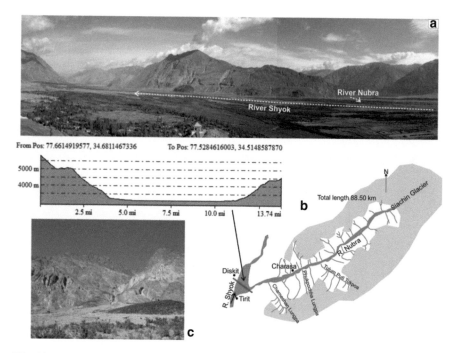

Fig. 10.6 (**a**) Panoramic view of the Shyok-Nubra valley; (**b**) Drainage of Nubra valley with a cross section near the confluence; (**c**) Moraine deposit observed along the right bank of the Nubra River

valleys during the Holocene times (Fig. 10.7). The lake existed downstream till Biagdangdo, occupying the Shyok, Nubra, and Tangtse river valleys.

10.4.2.1 Fluvio-Lacustrine Deposits of Zanskar River Valley in Southern Ladakh

The east-west trending Zanskar valley (32° 52′ 30″ N to 33° 52′ 30″ N latitudes and 76° 14′ 5″ E to 77° 32′ 4″ E longitude), a high altitude semi-desert, covering an area of ~7000 km² and situated between an altitudinal range of ~3500 to 7000 m asl is a sub-district of Kargil, Ladakh, Jammu and Kashmir (Raina and Koul 2011; Kumar et al. 2014; Quamar et al. 2016) (Fig. 10.8). The valley experiences a severe climate and remains cut off from rest of the region during November to June. The climate of Zanskar valley is extremely dry and cold with annual precipitation of around 239 mm/year (1987–2006), and humidity is very low (Raina and Koul 2011). The average January temperature of the valley is ~−20 °C, which drops to ~−40 °C (Kumar et al. 2014). Being a harsh terrain, the valley has limited vegetation because the diversity and distribution of flora is strongly influenced by environmental factors like altitude, topography, humidity, temperature, precipitation, exposure to radiation and nutrient supply (John and Dale 1990; Eldridge and Tozer 1997; Belnap and

Fig. 10.7 (**a**) River Shyok is showing braiding due to very low gradient even in the upper reaches; (**b**) View of Nubra River originating from Siachen Glacier and meeting the Shyok River near the Tirit Village; (**c**) Lacustrine sediments exposed in the left bank of Shyok River upstream Khalsar village; (**d**) Palaeo-strand lines showing gully erosion on the lacustrine sediments near Hunder village

Fig. 10.8 (**a**) DEM of the Zanskar region; (**b**) Confluence of the Zanskar and Indus Rivers; (**c**) Drum-Drum Glacier; (**d**) Padum fluvio-lacustrine section; (**e**) Fluvial terrace along Zanskar River

Gillette 1998; Ponzetti and McCune 2001; Körner 2003). Among various environmental variables, elevation gradient is among the most influencing factor, which effects atmospheric temperature, humidity, pressure, and precipitation and thereby determine the distribution dynamics of plant and animals (Vetaas and Grytnes 2002; Bhattarai et al. 2004; Baniya et al. 2012). Similarly, the area has a limited settlement, and that too is restricted to the banks of two main tributaries of Zanskar River. i.e., the Stot originating near the Pensi-La (4400 m asl) and the second tributary originating near the Baralacha-La. The Zanskar River passes through a deep and narrow gorge where the river freezes during winter months and provide a route to Leh district called the Chader route/track (Tandup 2014).

Geologically, the Zanskar valley has two different units, i.e., the Great Himalayan Range to the south-west and the Zanskar range to the northeast. The Zanskar lies on the leeward side of the Great Himalayan Range. The Central Crystalline rocks consisting of crystalline schist, stratified migmatites, porphyritic granites, gneisses, and feldspathic quartz-muscovite-biotite schist often garnetiferous are very commonly found in the southern parts (Koul 2017). The most prominent features present in the valley are glaciers and glaciated valleys. Survey of India Topographical maps based glacier inventory revealed ~268 glaciers in Zanskar valley basin confined between altitudes 3600 m to 6478 m (Koul 2017) and out of them, 131 glaciers are housed in Higher Himalaya and 137 glaciers in Zanskar Mountains.

Glacial and para-glacial landforms manifest multiple events of glaciation in the valley. The most prominent geomorphological landforms present in the area are (1) moraines, (2) glacio-fluvial terraces, (3) scree and alluvial fans, and (4) pro-glacial lake deposits. Besides these, occurrences of thin sheets of loess are found over moraines and alluvial fans (Pye 1995; Pant et al. 2005). Stratigraphy of the moraine using the conventional criteria such as morphology, sedimentary texture, relative elevation, the degree of weathering, and extent of vegetation cover helps us to understand the different stages of glaciation experienced in the river valley.

Stratigraphically, the oldest moraines are preserved along the flanks of the main Zanskar valley, while the younger stage moraines are restricted to the tributary valleys. The lateral moraines stratigraphically representing the oldest stage of glaciation are well preserved on the right flank of the Zanskar valley near Padum. However, their counterparts are faintly visible on the opposite flank of the valley. The oldest stage of glaciation in the valley is manifested by these moraines and from lateral ridges that can be traced up to a distance of ~4.5 km and terminates into a latero frontal moraine near the lower limit of Padum village. We have sampled this phase of moraines at three different locations for establishing the chronology. It is expected that the dates will help us in bracketing this phase of glaciation and understanding the forcing factors that might have led to such a huge event of glaciation. The second major phase of glaciation in the region is represented by unstable morainic ridges that are present at the mouth of the hanging valley that might have earlier been tributaries to the main trunk glacier.

Similarly, outwash gravel terraces are present, both in the main as well as the trunk valley. These gravel terraces present along the main valley. At places, lake deposits are exposed as discrete patches along the banks of the Zanskar River indicating that the lake must have existed at some point in time. These deposits are buff coloured, laminated with silty-clay layers. Alluvial fans and scree deposits are present all along the valley including the tributaries. Alluvial fans are differentiated from scree fans being associated with distinct catchments with streams and having finer grain size at the distal end (Blair and McPherson 1994; Hales and Roering 2005).

10.4.2.2   Fluvio-Lacustrine Deposits of Indus River Valley Across the Ladakh Region

The Indus river valley is tectonically unstable, exhibiting a complex topography, landscape relief and varied Quaternary sedimentation (Nag and Phartiyal 2015). The major geomorphic landforms are alluvial fans, debris cones, fluvio-lacustrine deposits, scree, and talus cone are present throughout the study area. During the Quaternary, a ubiquitous mass movements and catastrophic land sliding transported material from steep slopes to valley bottoms is seen, which was responsible for forming lakes (preserved as thick piles of fine sediment), while the outburst floods redistributed sediment down the valley. Three phases of lake formation are recorded- at ~35–26 ka (Lamayuru palaeolake) 17–13 ka (Rizong); 14–5 ka (Khalsi and Saspol) and 10–~1.5 (Spituk-Leh) (Fig. 10.9a–d). The seismic activity in the valley

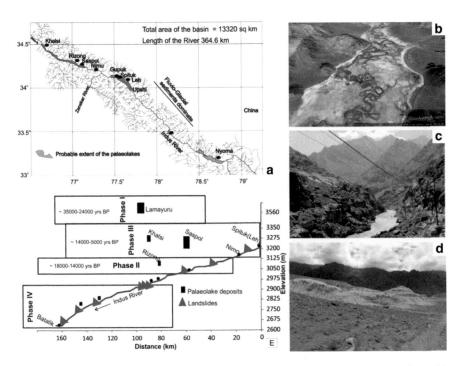

Fig. 10.9 (**a**) Drainage of the Indus river (**b**) Wide river valley in the upper course near Loma (**c**) Narrow Indus river valley downstream Khalsi (**d**) Khalsi palaeolake sequence (**e**) Four phases of depositional and climatic history after Phartiyal et al. (2013) and Nag and Phartiyal (2015)

is concentrated at 10 ka and 6 ka. The lake formation can be attributed jointly to the tectonic activity as well as the deglaciation after the Last Glacial Maximum (LGM) and Holocene warming and previous records show that Indian Summer Monsoon (ISM) intensity was more during the lake phases. Nag and Phartiyal (2015), described the following four prominent climatic phases from the 136 km transect of the Indus valley between Nimo and Batalik (Fig. 10.9e):

Phase I – 35,000–25,000 BP, Cold and Humid phase followed by the LGM.

Phase II – 17000–14,000 BP, a phase with deglaciation and rising lake levels forming the Rizong palaeolake (~35 km in length), due to rising temperatures and monsoonal activity.

Phase III – ~14,000–5000 BP with highest lake levels in the valley and the expansion of the lake to ~55 km occupying the valley length, due to high temperatures and monsoon contribution.

Phase IV – outburst and redistribution of the sediment downstream and change to the fluvial regime.

10.5   Conclusions

- The differential uplift along several thrust planes and associated climatic conditions are primarily governing the geomorphic evolution of the Ladakh.
- Among several agents of material movement, glaciers are the most effective factor and study of moraines suggests ten glacial stages (~430–0.4 ka) in Ladakh Himalaya.
- Ladakh glaciers respond directly with mid-latitude westerly winds and therefore the relatively recent glacial phases are in accordance with the northern latitude glaciations.
- The major source of moisture to Ladakh is supplied by monsoon and Central Asian westerlies but in subequal amounts (~76% and ~24% respectively).
- The region receives more rain/snow during Abnormal Monsoon Years (AMYs) due to a northward shift of the Inter-Tropical Convergence Zone (ITCZ).
- Several river parameters of the Indus River suggest that tectonics is governing the valley width and channel flow.
- The evidence of the most pronounced seismic activity recorded at 27 ka, 23 ka, 17–19 ka, 11–10 ka and 6 ka along the Karakorum Fault and ISZ.
- The fluviolacustrine deposits observed along the Indus and its tributary rivers are ideal archives for recoding the climatic signals of terrestrial origin. Our study suggests that ~35–25 ka the climate was cold and humid followed by the relatively dry LGM. Subsequently, a rise in lakes occurred during ~17–5 ka with major fluctuations in lake levels during the Older Dryas, Younger Dryas and 8.2 ka.
- The chronology is rather disturbing and detailed work is needed in this direction.

Acknoledgement We thank the Director BSIP, Lucknow for providing all necessary support in conducting studies in the Ladakh region of J&K, India. We are also thankful to field parties associated with us during these years. The DC, Leh and Wild Life Department, Jammu (J&K) is thanked for permission to carry out studies in the protected sanctuary area.

References

Ahmad T, Thakur VC, Islam R, Khanna PP, Mukherjee P (1998) Geochemistry and geodynamic implications of magamatic rocks from the Trans-Himalayan arc. Geochem J 32:303–404

Aitchison JC, Ali JR, Davis AM (2007) When and where did India and Asia collide? J Geophys Res 112:B05423

Ali SN, Juyal N (2013) Chronology of late quaternary glaciations in Indian Himalaya: a critical review. J Geol Soc India 82:628–638

Arendt AA, Echelmeyer KA, Harrison WD, Lingle CS, Valentine VB (2002) Rapid wastage of Alaska glaciers and their contribution to rising sea level. Science 297:82–386

Baniya C, Solhøy T, Gauslaa Y, Palmer MW (2012) Richness and composition of vascular plants and cryptogams along a high elevational gradient on Buddha Mountain, Central Tibet. Folia Geobot 47:135–151

Belnap J, Gillette DA (1998) Vulnerability of desert biological soil crusts to wind erosion: the influence of crust development, soil texture and disturbance. J Arid Environ 39:133–142

Benn DI, Owen LA (1998) The role of the Indian summer monsoon and the mid-latitude westerlies in Himalayan glaciation: review and speculative discussion. J Geol Soc 155:353–363

Benn DI, Owen LA, Osmaston HA, Seltzer GO, Porter SC, Mark B (2005) Reconstruction of equilibrium-line altitudes for tropical and sub-tropical glaciers. Quat Int 138-139:8–21

Besse J, Courtillot V, Pozzi JP, Westphal M, Zhou YX (1984) Palaeomagnetic estimates of crustal shortening in the Himalayan thrusts and Zangbo suture. Nature 311:621–626

Bhattarai KR, Vetaas OR, Grytnes JA (2004) Fern species richness along a central Himalayan elevational gradient, Nepal. J Biogeogr 31:389–400

Blair TC, McPherson JG (1994) Alluvial fan processes and forms. In: Geomorphology of desert environments. Springer, Dordrecht, pp 354–402

Blöthe JH, Munack H, Korup O, Fülling A, Garzanti E, Resentini A, Kubik PW (2014) Late quaternary valley infill and dissection in the Indus River, western Tibetan plateau margin. Quat Sci Rev 94:102–119

Bolch T, Kulkarni A, Koab A (2012) The state and fate of Himalayan glaciers. Science 336:310–314

Bookhagen B, Burbank DW (2006) Topography, relief, and TRMM-derived rainfall variations along the Himalaya. Geophys Res Lett 33:1–5

Bookhagen B, Burbank DW (2010) Toward a complete Himalayan hydrological budget: spatio-temporal distribution of snowmelt and rainfall and their impact on river discharge. J Geophys Res 115. https://doi.org/10.1029/2009JF001426

Bookhagen B, Thiede R, Strecker MR (2005) Abnormal monsoon years and their control on erosion and sediment flux in the high, arid northwestern Himalaya. Earth Planet Sci Lett 231:131–146

Bookhagen B, Fleitmann D, Nishiizumi K, Strecker MR, Thiede RC (2006) Holocene Monsoonal dynamics and fluvial terrace formation in the northwest Himalaya, India. Geology 34:601–604

Brown ET, Bendick R, Bourles DL, Gaur V, Molnar P, Raisbeck GM, Yiou F (2002) Slip rates of the Karakorum fault, Ladakh, India, determined using cosmic ray exposure dating of debris flows and moraines. J Geophy Res Solid Earth 107:ESE 7-1–ESE 7-13

Burbank DW, Fort MB (1985) Bedrock control on glacial limits: examples from the Ladakh and Zanskar ranges, north-western Himalaya, India. J Glaciol 31:143–149

Burbank DW, Leland J, Fielding E, Anderson RS, Brozovic N, Reid MR, Duncan C (1996) Bedrock incision, rock uplift and threshold hillslopes in the northwestern Himalayas. Nature 379(6565):505–510

Chevalier ML, Ryerson FJ, Tapponnier P, Finkel RC, Van Der Woerd J, Li H, Liu Q (2005) Slip-rate measurements on the Karakorum fault may imply secular variations in fault motion. Science 307:411–414

Clift PD, Giosan L, Biusztajn J, Campbell IH, Allen C, Pringle M, Tabrez AR, Danish M, Rabbani MM, Alizai A, Carter A, Lueckge A (2008) Holocene erosion of the Lesser Himalaya triggered by intensified summer monsoon. Geology 36:79–82

Cohen AS (2003) Paleolimnology: the history and evolution of lake systems. Oxford University Press, New York

Corell R, 30 Others (2004) Impacts of a warming Arctic, Arctic Climate Impact Assessment (ACIA report). Cambridge University Press, Cambridge

Dalai TK, Bhattacharya SK, Krishnaswamy S (2002a) Stable isotopes in the source waters of the Yamuna and its tributaries: seasonal and altitudinal variations and relation to major cations. Hydrol Process 16:3345–3364

Tandup C (2014) Natural environment of Cold Desert Region Zanskar (Ladakh). Int J Sci Res Publ 4(8):1–18

Dalai TK, Bhattacharya SK, Krishnaswamy S (2002b) Stable isotopes in the source waters of the Yamuna and its tributaries: seasonal and altitudinal variations and relation to major cations. Hydrol Process 16:3345–3364

Damm B (2006) Late quaternary glacier advances in the upper catchment area of the Indus River (Ladakh and Western Tibet). Quat Int 154–155:87–99

Demske D, Tarasov PE, Wünnemann B, Riedel F (2009) Late glacial and Holocene vegetation, Indian monsoon and westerly circulation in the Trans-Himalaya recorded in the lacustrine pollen sequence from Tso Kar, Ladakh, NW India. Palaeogeogr Palaeoclimatol Palaeoecol 279:172–185

Digerfeldt G, Almendinger JE, Bjorck S (1993) Reconstruction of past lake levels and their relation to groundwater hydrology in the Parkers Prairie sandplain, west-central Minnesota. Palaeogeogr Palaeoclimatol Palaeoecol 94:99–118

Dimri AP, Yasunari T, Kotlia BS, Mohanty UC, Sikka DR (2016) Indian winter monsoon: present and past. Earth Sci Rev 163:297–322

Dixit S, Bera SK (2012) Holocene climatic fluctuations from Lower Brahmaputra flood plain of Assam, northeast India. J Earth Syst Sci 121:135–147

Dortch JM, Owen LA, Haneberg WC, Caffee MW, Dietsch C, Kamp U (2009) Nature and timing of large landslides in the Himalaya and Transhimalaya of northern India. Quat Sci Rev 28:1037–1054

Dortch JM, Owen LA, Caffee MW (2010) Quaternary glaciation in the Nubra and Shyok valley confluence, northernmost Ladakh, India. Quat Res 74:132–144

Dortch JM, Owen LA, Schoenbohm LM, Caffee MW (2011) Asymmetrical erosion and morphological development of the central Ladakh Range, northern India. Geomorphology 135:167–180

Dortch JM, Owen LA, Caffee MW (2013) Timing and climatic drivers for glaciation across semi-arid western Himalayan-Tibetan orogen. Quat Sci Rev 78:188–208

Dosseto A, Vigier N, Joannes-Boyau R, Moffat I, Singh T, Srivastava P (2015) Rapid response of silicate weathering rates to climate change in the Himalaya. Geochem Perspect Lett 1:10–19

Dyurgerov MB, Meier MF (2000) Twentieth century climate change: evidence from small glaciers. Proc Natl Acad Sci U S A 97:1406–1411

Eldridge DJ, Tozer ME (1997) Environmental factors relating to the distribution of terricolous bryophytes and lichens in semi-arid eastern Australia. Bryologist 100:28–39

Fort M (1983) Geomorphological observations in the Ladakh area (Himalayas): quaternary evolution and present dynamics. In: Gupta VJ (ed) Stratigraphy and structure of Kashmir and Ladakh, Himalaya. Hindustan Publishing, New Delhi, pp 39–58

France-Lanord C, Derry LA (1997) Organic carbon burial forcing of the carbon cycle from Himalayan erosion. Nature 390:65–67

Frank W, Gansser A, Trommsdorff V (1977) Geological observations in the Ladakh area (Himalayas). A preliminary report. Schweiz Mineral Petrogr Mitt 57:89–113

Fritz SC (2008) Deciphering climatic history from lake sediments. J Paleolimnol 39:5–16

Gaina C, Müller RD, Brown B, Ishihara T (2007) Breakup and early seafloor spreading between India and Antarctica. Geophys J Int 170:151–169

Gansser A (1977) The great suture zone between Himalaya and Tibet, a preliminary account, in Himalaya. Sciences de la Terre, Colloque International Du CNRS Paris 268:181–191

Gansser A (1964) Geology of the Himalayas. Interscience, London, p 289

Garzanti E, Van Haver T (1988) The Indus clastics-forearc basin sedimentation in the Ladakh Himalaya (India). Sediment Geol 59:237–249

Gasse F, Van Campo E (1994) Abrupt post-glacial climate events in West Asia and North Africa monsoon domains. Earth Planet Sci Lett 126:435–456

Gasse F, Fontes JC, Van Campo E, Wei K (1996) Holocene environmental changes in Bangong Co basin (western Tibet). Part 4: discussions and conclusions. Palaeogeogr Palaeoclimatol Palaeoecol 120:79–82

Hales TC, Roering JJ (2005) Climate-controlled variations in scree production, Southern Alps, New Zealand. Geology 33(9):701–704

Hazarika D, Paul A, Wadhawan M, Kumar N, Sen K, Pant CC (2017) Seismotectonics of the trans-Himalaya, Eastern Ladakh, India: constraints from moment tensor solutions of local earthquake data. Tectonophysics 698:38–46

Hedrick KA, Seong YB, Owen LA, Caffee MW, Dietsch C (2011) Towards defining the transition in style and timing of Quaternary glaciation between the monsoon-influenced Greater Himalaya and the semi-arid Transhimalaya of Northern Indi. Quat Int 236:21–33

Heim A, Gansser A (1939) Central Himalaya; geological observations of the Swiss expedition 1936. Schweizer Naturf Ges Denksch 73(1):245

Henderson AL, Najman Y, Parrish R, Mark DF, Foster GL (2011) Constraints to the timing of India-Eurasia collision; a re-evaluation of evidence from the Indus Basin sedimentary rocks of the Indus-Tsangpo Suture Zone, Ladakh, India. Earth Sci Rev 106:265–292

Herzschuh U, Winter K, Wünnemann B, Li S (2006) A general cooling trend on the central Tibetan Plateau throughout the Holocene recorded by the Lake Zigetang pollen spectra. Quat Int 154–155:113–121

Hewitt K (1999) Quaternary moraines vs catastrophic rock avalanches in the Karakoram Himalaya, northern Pakistan. Quat Res 51:220–237. https://pubs.usgs.gov/publications/text/himalaya.html

Hintersberger E, Thiede RC, Strecker MR (2011) The role of extension during brittle deformation within the NW Indian Himalaya. Tectonics 30:TC3012

Hughton JT, Ding Y, Griggs DJ, Noguer M, Van Der Linden PJ, Dai X, Maskell K, Johnson CA (2001) Climate change 2001:the scientific basis. Cambridge University Press, Cambridge

Jain AK (2014) When did India- Asia collide and make the Himalaya? Curr Sci 106:254–266

Jain AK, Singh S (2008) Tectonics of the southern Asian Plate margin along the Karakoram Shear Zone: constraints from field observations and U-Pb SHRIMP ages. Tectonophysics 451:186–205

John E, Dale MRT (1990) Environmental correlates of species distributions in a saxicolous lichen community. J Veg Sci 1:385–392

Juyal N (2014) Ladakh: the high-altitude Indian Cold Desert. In: Kale VS (ed) Landscapes and landforms of India. Springer, Dordrecht, p 271

Juyal N, Sundriyal Y, Rana N (2010) Late Quaternary fluvial aggradation and incision in the monsoon-dominated Alaknanda valley, central Himalaya, Uttrakhand, India. J Quat Sci 25:1293–1304

Kääb A, Frauenfelder R, Roer I (2007) On the response of rock– glacier creep to surface temperature increase. Glob Planet Chang 56:172–187

Karim A, Veizer J (2002) Water balance of Indus river basin and moisture source in the Karakoram and western Himalayas: implications from hydrogen and oxygen isotopes in river water. J Geophys Res 107:ACH 9-1–ACH 9-12

Klootwijk CT, Conaghan PJ, Powell CMA (1985) The Himalayan Arc: large-scale continental subduction, oroclinal bending and back-arc spreading. Earth Planet Sci Lett 75:167–183

Klootwijk CT, Gee JS, Peirce JW, Smith GM (1992) An early India–Asia contact: Paleomagnetic constraints from Ninetyeast Ridge, ODP Leg 121. Geology 20:395–398

Körner C (2003) Alpine plant life – functional plant ecology of high mountain ecosystems, 2nd edn. Springer, Heidelberg

Kotlia BS, Shukla UK, Bhalla MS, Mathur PD, Pant CC (1997) Quaternary fluvio-lacustrine deposits of the Lamayuru Basin, Ladakh Himalaya: preliminary multidisciplinary investigations. Geol Mag 134:807–812

Koul MN (2017) Impact of climate changes on cryosphere in Suru-Zanskar Valley, Kargil: observed trends, and socio-economic relevance. Transactions 39(1):1–24

Kumar J, Rai H, Khare R, Upreti DK, Dhar P, Tayade AB, Chaurasia OP, Srivastava RB (2014) Elevational controls of lichen communities in Zanskar valley, Ladakh, a Trans Himalayan cold desert. Trop Plant Res 1(2):48–54

Kumar S, Wesnousky SG, Jayangondaperumal R, Nakata T, Kumahara Y, Singh V (2010) Paleoseismological evidence of surface faulting along the northeastern Himalayan front, India: timing, size, and spatial extent of great earthquakes. J Geophys Res 115:B12422. https://doi.org/10.1029/2009JB006789

Kumar A, Srivastava P (2018) Landscape of the Indus River. In: The Indian Rivers: scientific and socio-economic aspects, Springer Hydrogeology. Springer, Singapore, pp 47–60

Kumar A, Srivastava P (2017) The role of climate and tectonics in aggradation and incision of the Indus River in the Ladakh Himalaya during the late Quaternary. Quat Res 87:363–385

Kumar A, Srivastava P, Meena NK (2017) Late Pleistocene aeolian activity in the cold desert of Ladakh: a record from sand ramps. Quat Int 443:13–28

Leland J, Reid MR, Burbank DW, Finkel R, Cafee M (1998) Incision and differential bedrock uplift along the Indus River near Nanga Parbat, Pakistan Himalaya, from 10 Be and 26 Al exposure age dating of bedrock straths. Earth Planet Sci Lett 154:93–107

Li ZX, Bogdanova SV, Collins AS, Davidson A, De Waele B, Ernst RE, Fitzsimons ICW, Fuck RA, Gladkochub DP, Jacobs J, Karlstrom KE, Lu S, Natapov LM, Pease V, Pisarevsky SA, Thrane K, Vernikovsky V (2008) Assembly, configuration, and break-up history of Rodinia: a synthesis. Precambrian Res 160:179–210

Miashita Y, Yamamoto T (1996) Gondwanaland: its formation, evolution and dispersion. J Afr Earth Sci 23(2)

Mischke S, Zhang C (2010) Holocene cold events on the Tibetan Plateau. Glob Planet Chang 72:155–163

Molnar P, Tapponnier P (1975) Cenozoic tectonics of Asia: effects of a continental collision. Science 189:419–426

Molnar P, Burchfiel BC, Ziyun Z, Kuangyi L, Shuji W, Minmin H (1987) The geologic evolution of northern Tibet; results of an expedition to Ulugh Muztagh. Science 235:299–305

Nag D, Phartiyal B (2015) Climatic variations and geomorphology of the Indus River Valley, between Spituk and Batalik, Ladakh (NW trans Himalayas) in late quaternary. Quat Int 371:87–101

Nag D, Phartiyal B, Singh DS (2016) Sedimentary characteristics of palaeolake deposits along the Indus River valley, Ladakh, Trans-Himalaya: implications for the depositional environment. Sedimentology 63:1765–1785

Nagar YC, Ganju A, Satyawali PK (2013) Preliminary optical chronology suggests significant advance in Nubra valley glaciers during the Last Glacial Maximum. Curr Sci 105:96–101

Nakata T (1972) Geomorphic history and crustal movements of foothills of the Himalaya. Institute of Geography, Tohoku University, Sendai, p 77

Nakata T, Otsuki K, Khan SH (1990) Active faults, stress field, and plate motion along the IndoEurasian plate boundary. Tectonophysics 181:83–95

Owen LA (2009) Latest Pleistocene and Holocene glacier fluctuations in the Himalaya and Tibet. Quat Sci Rev 28:2150–2164

Owen LA, Gualtieri L, Finkel RC, Caffee MW, Benn DI, Sharma MC (2001) Cosmogenic radionuclide dating of glacial landforms in the Lahul Himalaya, northern India: defining the timing of Late Quaternary glaciation. J Quat Sci 16:555–563

Owen LA, Caffee MW, Bovard KR, Finkel RC, Sharma MC (2006) Terrestrial cosmogenic nuclide surface exposure dating of the oldest glacial successions in the Himalayan orogen. Ladakh Range, northern India. Geol Soc Am Bull 118:383–392

Owen LA, Bailey RM, Rhodes EJ, Mitchell WA, Coxon P (1997) Style and timing of glaciation in the Lahul Himalaya, northern India: a framework for reconstructing late Quaternary palaeoclimatic change in the western Himalayas. J Quat Sci 12:83–109

Owen LA, Caffee MW, Finkel RC, Seong BY (2008) Quaternary glaciations of the Himalayan Tibetan orogen. J Quat Sci 23:513–532

Owen LA, Dortch JM (2014) Nature and timing of Quaternary glaciation in the Himalayan Tibetan orogen. Quat Sci Rev 88:14–54

Pande K, Padia JT, Ramesh R, Sharma KK (2000) Stable isotope systematics of surface water bodies in the Himalayan and Trans-Himalayan (Kashmir) region. J Earth Syst Sci 109(1):109–115

Pant RK, Basavaiah N, Juyal N, Saini NK, Yadava MG, Appel E, Singhvi AK (2005) A 20-ka climate record from Central Himalayan loess deposits. J Quat Sci 20(5):485–492

Phartiyal B, Sharma A, Upadhyay R, Ram-Awatar, Sinha AK (2005) Quaternary geology, tectonics and distribution of palaeo- and present fluvio/glacio lacustrine deposits in Ladakh, NW Indian Himalaya – a study based on field observations. Geomorphology 65(3–4):241–256

Phartiyal B, Sharma A (2009) Soft-sediment deformation structures in the Late Quaternary sediments of Ladakh: evidence for multiple phases of seismic tremors in the North western Himalayan Region. J Asian Earth Sci 34:761–770

Phartiyal B, Sharma A, Kothyari GC (2013) Existence of Late Quaternary and Holocene lakes along the River Indus in Ladakh Region of Trans Himalaya, NW India: implications to lake formation-climate and tectonics. Chin Sci Bull 58(I):142–155

Phartiyal B, Singh R, Kothyari GC (2015) Late-quaternary geomorphic scenario due to changing depositional regimes in the Tangtse valley, Trans-Himalaya, NW India. Palaeogeogr Palaeoclimatol Palaeoecol 422:11–24

Phartiyal B, Singh R, Nag D (2018) Trans and Tethyan-Himalayan Rivers-in reference to Ladakh and Lahaul Spiti, NW Himalaya, India. In: Singh DS (ed) The Indian Rivers: scientific and socio-economic aspects, Springer hydrogeology book series (SPRINGERHYDRO). Springer, Singapore, pp 367–382

Phillips WM, Sloan VF, Shroder JF (2000) Asynchronous glaciation at Nanga Parbat, northwestern Himalaya Mountains, Pakistan. Geology 28:431–434

Ponzetti JM, McCune BP (2001) Biotic soil crusts of Oregon's shrub steppe: community composition in relation to soil chemistry, climate, and livestock activity. Bryologist 104:212–225

Pratt-Sitaula B, Burbank DW, Heimsath AM, Humphrey NF, Oskin M, Putkonen J (2011) Topographic control of asynchronous glacial advances: a case study from Annapurna, Nepal. Geophys Res Lett 38:L24502

Pye K (1995) The nature, origin and accumulation of loess. Quat Sci Rev 14(7):653–667

Quamar MF, Nawaz AS, Phartiyal B, Morthekai P, Sharma A (2016) Recovery of palynomorphs from the high-altitude cold desert of Ladakh, NW India: an aerobiological perspective. Geophytology 46(1):67–73

Rai H (1982) Geological evidence against the Shyok palaeo-suture, Ladakh Himalaya. Nature 297:142–144

Raina RK, Koul MN (2011) Impact of climatic change on agro-ecological zones of the Suru-Zanskar valley, Ladakh (Jammu and Kashmir), India. J Ecol Nat Environ 3(13):424–440

Rasmussen KL, Houze RA Jr (2012) A flash flooding storm at the steep edge of high terrain: disaster in the Himalayas. Bull Am Meteorol Soc 93:1713–1724. https://doi.org/10.1175/BAMS-D-11-00236.1

Robertson AHF (2000) Formation of melanges in the Indus Suture Zone, Ladakh Himalaya by successive subduction-related, collisional and post collisional processes during Late Mesozoic-Late Tertiary time. Geol Soc Lond Spec Publ 170:333–374

Robertson AHF, Collins AS (2002) Shyok suture zone, N. Pakistan: late Mesozoic- Tertiary evolution of a critical suture separating the oceanic Ladakh arc from the Asian continental margin. J Asian Earth Sci 20:309–351

Rowley DB (1996) Age of initiation of collision between India and Asia: a review of stratigraphic data. Earth Planet Sci Lett 145:1–13

Schaefer JM, Oberholzer P, Zhao ZZ, Ivy-Ochs S, Wieler R, Baur H, Kubik PW, Schlüchter C (2008) Cosmogenic beryllium-10 and neon-21 dating of late Pleistocene glaciations in Nyalam, monsoonal Himalayas. Quat Sci Rev 27:295–311

Schäfer JM, Tschudi S, Zhao ZZ, Wu XH, Ivy-Ochs S, Wieler R, Baur H, Kubik PW, Schlüchter C (2002) The limited influence of glaciations in Tibet on global climate over the past 170000 yr. Earth Planet Sci Lett 194:287–297

Searle MP (1991) Geology and tectonics of the Karakoram mountains. Wiley, Chichester

Searle MP, Windley BF, Coward MP, Copper DJW, Rex AJ, Rex D, Tindong L, Xuchang X, Jan MQ, Thakur VC, Kumar S (1987) The closing of Tethys and the tectonics of the Himalaya. Bull Geol Soc Am 98:678–701

Seton M, Müller RD, Zahirovic S, Gaina C, Torsvik T, Shephard G, Talsma A, Gurnis M, Maus S, Chandler M (2012) Global continental and ocean basin reconstructions since 200Ma. Earth Sci Rev 113:212–270

Shanker R, Pahdi N, Prakash G, Thussu JL, Wangdus C (1976) Recent geological studies in upper Indus valley and the plate tectonics. Miscellaneous Publications of the Geological Survey of India 34: 42–56

Sharma S, Chand P, Bisht P, Shukla AD, Bartarya SK, Sundriyal YP, Juyal N (2016) Factors responsible for driving the glaciation in the Sarchu Plain, eastern Zanskar Himalaya, during the late Quaternary. J Quat Sci 31:495–511

Sharma A, Kumar K, Laskar A, Singh SK, Mehta P (2017) Oxygen, deuterium, and strontium isotope characteristics of the Indus River water system. Geomorphology 284:5–16

Sinclair HD, Jaffey N (2001) Sedimentology of the Indus Group, palaeo-Indus River. J Geol Soc Lond 158:151–162

Singh IB, Sahni A, Jain AK, Upadhay R, Parcha SK, Parmar V, Agarwal KK, Shukla S, Kumar S, Singh MP, Ahmad S, Jigyasu DK, Arya R, Pandey S (2015) Post-collision sedimentation in the Indus basin (Ladakh, India): implications for the evolution of the Northern margin of the Indian plate. J Palaeontol Soc India 60(2):97–146

Smith AJ, Donovan JJ, Ito E, Engstrom DR, Panek VA (2002) Climate-driven hydrologic transients in lake sediment records: multiproxy record of mid-Holocene drought. Quat Sci Rev 21:625–646

Srikantia SV, Razdan ML (1980) Geology of parts of Central Ladakh Himalaya with particular references to Indus Tectonic Zone. J Geol Soc India 21:523–545

Steck A (2003) Geology of NW Indian Himalaya. Eclogae Geol Helv 96:147–196

Stevens LR, Stone JR, Campbell J, Fritz SC (2006) A 2200-yr record of hydrologic variability from Foy Lake, Montana, USA, inferred from diatom and geochemical data. Quat Res 65:264–274

Stone JR, Fritz SC (2004) Three-dimensional modeling of lacustrine diatom habitat areas: improving paleolimnological interpretation of planktic: benthic ratios. Limnol Oceanogr 49:1540–1548

Thakur VC, Misra DK (1984) Tectonic framework of Indus and Shyok suture zones in eastern Ladakh, NW Himalaya. Tectonophysics 101:207–220

Thakur VC, Virdi NS (1979) Lithostratigraphy, structural framework, deformation and metamorphism of the SE region of Ladakh Kashmir Himalaya India. Himal Geol 9:63–79

Telford RJ, Lamb HF, Mohammed MU (1999) Diatom-derived palaeoconductivity estimates for Lake Awassa, Ethiopia: evidence for pulsed inflows of saline groundwater. J Paleolimnol 21:409–421

Thompson LG, Yao T, Mosley-Thompson E, Davis ME, Henderson KA, Lin PN (2000) A high-resolution millennial record of the South Asian monsoon from Himalayan ice cores. Science 289:1916–1919

Torsvik TH, Cocks LRM (2013) Gondwana from top to base in space and time. Gondwana Res 24:999–1030

Trivedi A, Chauhan MS (2009) Holocene vegetation and climate fluctuations in northwest Himalaya, based on pollen evidence from Surinsar Lake, Jammu region, India. J Geol Soc India 74:402–412

Upadhyay R (2014) Palaogeographic significance of 'Yasin-type' rudist and orbitolinid fauna of the Shyok suture zone, Saltoro Hills, northern Ladakh, India. Curr Sci 106:223–228

Vetaas OR, Grytnes JA (2002) Distribution of vascular plant species richness and endemic richness along the Himalayan elevation gradient in Nepal. Glob Ecol Biogeogr 11:291–301

Wadia DN (1937) The cretaceous volcanic series of Astor-Deosai Kashmir and its intrusions. Record Geol Surv India 72:151–161

Walker JCG, Hays PB, Kasting JF (1981) A negative feedback mechanism for the long term stabilization of Earth's surface temperature. J Geophys Res 86:9776–9782

Watson RT, Zinyower MC, Moss RH (1997) The regional impacts of climate change: an assessment of vulnerability. A special report of IPCC Working Group II. Cambridge University Press, Cambridge, p 517

Wünnemann B, Demske D, Tarasov P, Kotlia BS, Reinhardt C, Bloemendal J, Diekmann B, Hartmann K, Krois J, Riedel F, Arya N (2010) Hydrological evolution during the last 15 kyr in the TsoKar lake basin (Ladakh, India), derived from geomorphological, sedimentological and palynological records. Quat Sci Rev 29:1138–1155

Yin A (2006) Cenozoic tectonic evolution of the Himalayan orogen as constrained by along-strike variation of structural geometry, exhumation history, and foreland sedimentation. Earth Sci Rev 76:1–131

Chapter 11
Climate, C/N Ratio and Organic Matter Accumulation: An Overview of Examples from Kashmir Himalayan Lakes

Hema Achyuthan, Aasif Mohmad Lone, Rayees Ahmad Shah, and A. A. Fousiya

Abstract This present study provides an overview of the C/N ratio and organic matter (OM) contents in surface sediments of four lakes that occur in the Kashmir Valley. Valleys are depressed areas of land that influence climate and weather at both global and local scales. The oval-shaped intermontane valley is dotted with numerous freshwater lakes of various shapes and depths. These lakes represent the remnants of a substantial Karewa Lake body that extended across the valley floor currently occupied by the modern lakes. The study presented is based on analysis of the organic material deposited on the lake basin floors, transported via both natural and anthropogenic sources. The present deposition of terrestrial organic material in the lake sediments has apparently altered the trophic status of these lakes from oligo- to eutrophic conditions. Analysis of the C/N ratio in the Holocene lake sediments suggests that these lakes are dynamic ecosystems the accumulation of organic material being sources and depositional processes occurring within these lake basins. The report reviews the accumulation of organic material and the C/N ratio of the surface sediments, and discusses their potential value in palaeoclimatic interpretation of fossil sediments in these lake basins to serve as an analogue with older materials. The organic material deposited in these lakes is predominantly a function of climatic conditions in the valley. The organic and isotopic composition of the lake sediments can potentially provide important information of past lake nutrient dynamics, palaeoenvironmental changes and depositional sources influencing the basins through time.

H. Achyuthan (✉) · A. M. Lone · R. A. Shah · A. A. Fousiya
Department of Geology, Anna University, Chennai, India

© Springer Nature Switzerland AG 2020
A. P. Dimri et al. (eds.), *Himalayan Weather and Climate and their Impact on the Environment*, https://doi.org/10.1007/978-3-030-29684-1_11

11.1 Introduction

Fresh water bodies constitute only 3% of the total volume of water on earth. These water bodies have a bearing on the economy as they provide potable water, fish and fodder for the people. The fresh water lakes are however, the most vulnerable habitats as they act as sinks for sewage and waste disposal. Limnological study has now firmly entrenched itself as an important branch of geo-environmental study. Physical and chemical characteristics of lake sediments are potential proxies, which offer important information on the long-term climate variability of the past. The lakes being relatively small bodies of water in comparison with the ocean, can respond relatively quickly to an external forcing variable. Changes in air temperature or regional landuse can be reflected within months to decades in the inputs to the lake's sedimentary record. The rapidity of this response varies both within a lake, for example from near shore to offshore environments, and between the lakes.

The organic content of the sediments reflects the amount of living organisms in and around the lake, i.e. the level of productivity and leaching from humus-rich catchment soils (Meyers and Teranes 2001). The lakes also have properties that allow us to use them as sentinels of stress in the terrestrial and wetland ecosystems that are found in their catchments. Changes to watersheds caused by climate warming can in turn affect the properties of lakes to which they drain. Examples include changes to nutrient inputs, the balance between base cations and strong acid anions, carbon cycles, and mercury, in some cases associated with insect outbreaks and forest fires.

Sedimentary organic matter (OM) is one of the most important components of terrestrial and marine lake sediments generally derived from in situ (autochthonous) and/or terrestrial (allochthonous) sources (Meyers and Ishiwatari 1993). Lake sediments play an important role as an efficient natural trap for organic rich sediments and hence act as a natural regulator of the various processes that occur inside the lake ecosystems (Burone et al. 2003). The particulate detritus of plants that grow in and around the lake catchment areas comprises the primary source of OM to the lake deposits. The relative contribution of OM from these two sources is strongly controlled by lake morphology, watershed topography and the relative abundances of lake and watershed plants (Meyers and Ishiwatari 1993). However, lake sediments are strongly influenced by the natural and human settlement characteristics around the catchments. While human induced changes in nutrient loads have had a marked effect on Kashmir Himalayan lakes, changes in climate also play a vital role. The key processes of climate variability are solar radiation and water balance (water level, retention time, stratification) and related factors (snow, wind). Thus, natural proxy signals preserved in lake sediments can be used for reconstruction of high-resolution Late-Quaternary-Holocene climatic variabilities.

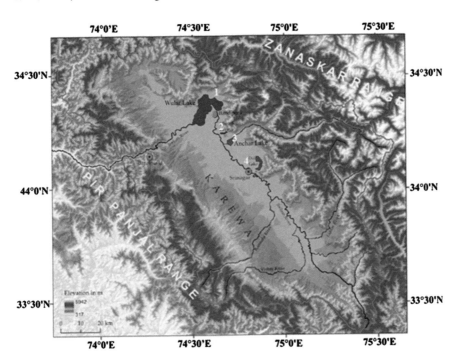

Fig. 11.1 DEM map of the Kashmir Valley along with location of four lakes (1) Wular, (2) Manasbal, (3) Anchar and (4) Dal Lake. Different colours indicate elevation in meters (Ganjoo 2014)

11.2 Climate and Geology of Kashmir Himalayan Region

The Kashmir Himalayas (Fig. 11.1) are extremely important because of their position and for the role; they play in the climate and weather phenomena and economy of the state. It is believed that the valley originated from the draining of the Karewa Lake, which was formed as a result of tectonic upheaval and subsequent tectonic uplift of the Pir Panjal and Zanskar Ranges. It is an intermountain valley fill, comprising of unconsolidated gravel, pebble, boulder and mud. A succession of plateaux is present above the plains of Jhelum and its tributaries. These plateaux-like terraces are called 'Karewas' or 'Vudr' in the local language. Due to the rise of the Pir-Panjal, the drainage was impounded, a lake of about 5000 km^2 area was developed, and thus a basin was formed.

The climate of the study area is characterized by warm summers and cold winters and varies between −5 and > 30 °C. According to Bagnolus and Meher-Homji (1959), the climate of Kashmir falls under Sub-Mediterranean type with four seasons based on mean temperature and precipitation. The precipitation is highly variable round the year with more rains occurring during the summer and snowfall

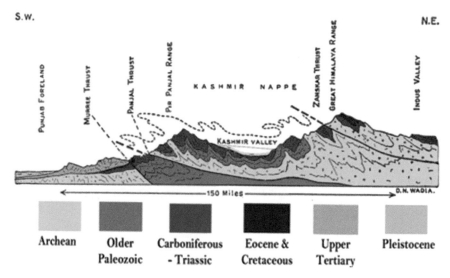

Fig. 11.2 Diagrammatic cross sections across the Kashmir Himalaya, showing broad tectonic features along with the Kashmir Nappe zone. (After Wadia 1976)

during winter. Western disturbances are mainly responsible for causing rainfall and snowfall from January to May. Both the daily minimum and maximum temperatures start rising and falling by April and September respectively, and are at their peak in July and January.

Kashmir Valley encompasses a very important place in the geo-tectonic framework of Kashmir Himalaya (Fig. 11.2). The geology of the Kashmir region displays a chronological record of the Great Alpine Orogeny, including sedimentation, tectonics and volcanic activity, that followed the Himalayan Orogeny. The valley is bounded on the southwest by the Pir Panjal and by the Greater Himalayas on the northeast. The bowl-shaped basin is represented by the Precambrian basement overlain by a thick succession of fossiliferous Palaeozoic and Triassic rocks with thrusted contacts in the north with the Great Himalayan Range and in the south with Pir Panjal Range (Ganjoo 2014). Panjal Volcanic Complex and the Triassic Limestone form the two main geological formations, underlain by the Archean metasedimentary rocks (Salkhala Formation). Salkhala formation constitutes carbonaceous slate, graphitic phyllite and schist associated with carbonaceous grey or white limestone, marble, calcareous slate and calcareous schist (Alam et al. 2015). Plio-Pleistocene glacio-fluvio-lacustrine sediments (approximately 1300 m thick) in turn overlie the Precambrian to Mesozoic basement rocks and constitute the Karewa Group. The Karewa Group of sediments is mostly composed of sand, silt, clay, shale, mud, lignite, gravel and loessic sediments (Alam et al. 2015; Babeesh et al. 2017). The landuse and land cover of the area may be divided into agricultural land with various types of crops and trees, forests, Karewa deposits, built-up, barren land and horticulture areas. The landuse and land cover classes along the lake catchment areas has been substantially modified by human population over the past few decades (Lone et al. 2017; Lone et al. 2018a, b, c).

11.3   The Karewa Lakes

One of the unique geomorphic features present in the Kashmir valley are the thick Quaternary Karewa deposits. These deposits are dominantly lacustrine, fluvial, glacial and aeolian in origin. The past existence of Karewa Lake, is thought to be an intermittent one attaining maximum extent during warmer interglacial periods and succeeded by periods of minimum extent. Nearly half of the present Kashmir Valley is covered by deposits of Karewa sediments and contain detailed systematic records of ice ages over western Himalayas (Wadia 1941). The important freshwater lakes (Anchar, Dal, Manasbal and Wular Lakes) found in the Kashmir Valley (Fig. 11.3) are the remnants of the great Karewa Lake and thus represent the present day geomorphic landscape of the region.

   In this chapter we present an overview of OM accumulation and C/N ratio from four Kashmir Himalayan lakes. This study was carried out to understand the accumulation and source of OM in surface sediments and the controlling factors that modulate the deposition of OM in recent surface sediments. The four lakes although located in the same climatic conditions prevailing in the valley display changes in the sources and depositional pattern of allochthonous and terrestrial OM contents.

Fig. 11.3   Google earth map (2017) showing location of important lakes in Kashmir Valley

This study was also aimed to understand the factors such as type of the lake (open or closed), catchment area changes and further, the paleoclimatic potential of the above two proxy records were also discussed for future references. The water in these lakes is added from in situ springs, drainage inlets from the upland catchment such as Jhelum river, Sindh river besides other small tributaries which bring in diverse type of sediments rich in OM (Sarkar et al. 2016). This study investigates the distribution and sources of organic matter within four Kashmir Himalayan lake basins and illustrates the usefulness of C/N ratio as a biomarker approach in discriminating organic matter sources within lacustrine environments.

11.3.1 *Wular Lake*

The Wular Lake (34°22′10.56″N and 74°33′28.8″E) located in the Bandipora district of north Kashmir is an oxbow type mono-basined lake (Fig. 11.4). The lake is situated nearly 34 km to the northwest of Srinagar city at an altitude of 1530 m a.s.l. The average surface area of the Wular lake is 189 km^2 and the water remains alkaline throughout the year and pH changes from 7.5 in summer to 8.8 in winter season (Shah et al. 2017). The average depth of the lake is 5 m. Based on its high biological, hydrological and socioeconomic value, the lake was declared as a wetland of National importance under the wetlands programme of the Ministry of Environment and Forests, Government of India, in 1986 and has been subsequently declared as a Ramsar Site in 1990 to give it the status of wetland of International importance.

Fig. 11.4 A view of the Wular lake, Bandipora Kashmir

Wular Lake, the largest freshwater lake in Asia within the Jhelum river basin, plays an important role in the hydrographic system of the Kashmir valley by acting as predominant absorption basin during the flood events besides playing a vital role in sustaining agriculture. The shrinkage in lake area is mainly because of a continuous siltation brought by the various tributaries besides River Jhelum. The lake is an open drainage type and hence experiences moderate eutrophication (Pandit 2002).

11.3.2   Manasbal Lake

The Manasbal Lake 34°14′47.66″ N and 74°41′20″ E is one of the deepest, meso-trophic freshwater lake, located in the Safapora area (Ganderbal district), at an altitude of about 1585 m a.s.l (Fig. 11.5). The lake is semi-drainage type, oblong in outline and covers an area of 2.80 km^2 with a maximum depth of 12.5 m (Maqbool and Khan 2013; Rashid et al. 2013; Lone et al. 2017; Lone et al. 2018a, b, c). The lake is surrounded by the Karewa deposits of Plio-Pleistocene age on the northeast and northwestern side. On the eastern side are the Panjal traps of Permian age overlain by the Triassic limestone extending to the southeast. The lake inflow is chiefly derived from the underground springs, local precipitation in addition to which a small irrigation stream (Laar Kul) on the eastern side also drains water into the lake during the rainy season from the higher catchment areas. The Laar-kul stream brings in diverse types of terrestrial organic and inorganic material into the lake, including major and minor nutrients and other materials. An outflow channel called Nunnyar Nalla drains excess water from the western side of the lake into the Jhelum River.

Fig. 11.5 Panoramic view of the Manasbal Lake with the Ahtung hills in the background

Fig. 11.6 A view of the Anchar Lake, Srinagar Kashmir Valley

11.3.3 Anchar Lake

Anchar Lake (34°08′37.58″N and 74°47′10.89″ E) located in the urban area, is a fresh water lake, situated about 25 km northwest of Srinagar city at an average elevation of ~1583 m a.s.l. with an average depth of about 4 m (Fig. 11.6). Presently, the lake occupies an area of 6.80 km² of which about 1.69 km² correspond to openwater and the remaining portion has been transformed into a marshland because of increased human impact on the lake ecosystem. The lake catchment covers a surface area of about 5.8 km². The lake surface is a single basined, open drainage type water body fed by a network of channels including local catchment watersheds, open springs and Sindh River (Ganaie et al. 2015). The Sindh River constituting one of the largest basins in the Kashmir Himalayas drains major part of its water into the Anchar Lake. Further, a number of channels from agricultural fields, effluents from the settlements around the Anchar Lake and surface runoff from the catchment area directly drain into the lake throughout the year (Lone et al. 2017; Lone et al. 2018a, b, c). Much of the excess water from the Anchar Lake drains through the Shalabugh wetland into the Jhelum River (Khanday et al. 2016).

Fig. 11.7 A view of the Dal Lake Srinagar, Kashmir Valley

11.3.4 Dal Lake

Dal Lake (Fig. 11.7), situated in Kashmir Himalaya, is a multi-basin drainage lake covering an area of 24 km^2 with the open water spread of about 10.5 km^2 and a water holding capacity of 15.45 × 10^6 m^3. The lake lies between 34°5′–34°9′N and 74°49′–74°53′E at a mean altitude of 1585 m a.s.l. The average depth of the Dal Lake is 1.42 m. Being a major tourist attraction in Kashmir, the lake is socio-economically important as the livelihood of a large section of population of Srinagar city is dependent on the services and products provided by the lake. The lake is an important source of vegetables, fisheries, recreation and drinking water to the people of Srinagar city. The catchment area, spread over 337 km^2, is geologically composed of Panjal Traps, agglomeratic slates, alluvium and Karewa deposits (Wadia 1976). As per Bagnolus and Meher-Homji (1959), climate of the area is sub-Mediterranean with four seasons. The total annual precipitation and average temperature recorded at the nearest metrological station (Srinagar) is 870 mm and 11 °C respectively.

11.4 Organic Matter Accumulation, C/N Ratio and Paleoclimate Inferences

Understanding the OM accumulation and diagenesis is one of the important parameters to distinguish between various depositional environments of the past as well as recent lake sediments. Terrestrial run off and organisms present within the water column generally contribute OM and calcium carbonate to the lake basins. The

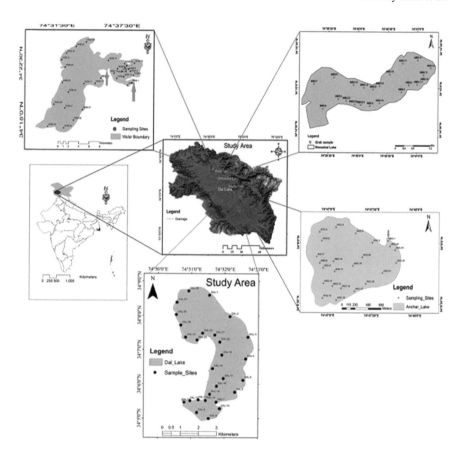

Fig. 11.8 Location map of the study area and the site location of the four freshwater lakes with sampling stations

concentration of OM and CaCO$_3$ varies within the lake basins (Dean 2006). This variation is owing to the changes in water chemistry, catchment geology, climate, lake-flora and fauna, anthropogenic influences along the catchments etc.

The sediment grain size variations, its sorting under hydraulic conditions and its distribution both spatially and temporally largely controls the distribution pattern and behaviour of organic matter in the lakes (Xiao et al. 2013). The modern surface sediment samples were collected from all the four lakes (Fig. 11.8). The sampling stations were carefully selected and these sampling sites represent the modern sediments that accumulate within these lake basins. Sediment texture of these lakes is dominantly fine fractions such as clay and silt reflecting that the source area catchment is contributing finer fractions into these lake basins (Lone et al. 2017). This also reflects the weathering conditions prevailing in the catchment area. Generally, coarser components such as sand fractions are higher along the lake margins supporting shallow water column and along the sites influenced by inlet streams. Further, it is noted that the small and large watershed drainage inlets also carry large volumes of coarser sediment from the upper catchment areas and deposit them along the lake margins. Higher clay contents reflect the deeper bathymetry of the lake-basin and the relatively stagnant depositional environment, which allows finer fractions to settle down.

The lakes in Kashmir Valley are the potable sources of freshwater for the region but unfortunately, due to their exploitation for various purposes like drinking, domestic, agriculture, hydropower, these freshwater ecosystems are getting eutrophicated and subsequently shrinking in area. These lakes are located within the intermountain settings and receive large amounts of weathered sediments from the nearby surrounding catchment areas, which consist of diverse lithological formations suggesting a variety of provenance with variable geological settings. Panjal traps and the Agglomeratic slates, granites and metamorphic schists, Quartzites, Triassic limestone and recent Alluvium represent the exposed lithology. These rock types form the important source of weathered sediments that are redeposited into these lakes. The unique climate of the valley (cold winters) and the steep slopes enhance the physical weathering of the exposed surfaces resulting in increased erosion and subsequent deposition of sediments into these lake basins. The sedimentation rate and pattern in these lakes vary spatially and temporally. The sedimentation rate in each Himalayan lake differs due to catchment lithology, slope, vegetation cover and energy of transporting media (inlet stream, rainfall and wind) causing differential erosion. The sedimentation rate reported from few of the Himalayan lakes reveals contribution of sediments due to both natural and human interference. Sarkar et al. (2016) reported sediment rates from the Kashmir Himalayan lakes and revealed that the Manasbal Lake shows a sedimentation rate of 0.44 cm/year while the average sedimentation rates of the Dal and Mansar lakes were measured and observed to be 0.93 and 0.23 cm/year, respectively.

The lake basins in the Kashmir Valley lie in the temperate zone, characterized by wet and cold winters driven by westerlies and relatively dry and moderate hot summers (subdued SW monsoon). Most of the precipitation falling in the area is owing to the western disturbances (westerlies) during the winter and early spring seasons. Additionally water from the local catchment streams, melting glaciers and insitu springs also add water to these lakes. The water and the sediment chemistry of these lake basins are comparable to catchment rock lithology as such the geochemical characteristics of the lake water is mostly influenced by the weathering of the catchment (carbonate and silicates) (Jeelani and Shah 2006; Rashid et al. 2013).

The C/N ratio and OM contents displayed a distinct spatial variation in all the four lakes (Fig. 11.9; Table 11.1). The C/N ratio of the Manasbal Lake sediments (Fig. 11.9a) showed a wide variation in the C/N ratio with an average value of 15, minimum of 7.73 and maximum value of 33.83. The OM in the lake sediments ranges between 3.2 and 29.6% with an average value of 16.89%. Wular Lake sediments (Fig. 11.9b) exhibited an average C/N value of 11.86 with a minimum and maximum value of 9.04 and 22.03. The OM percentage of the lake bottom samples showed a minimum value of 1.42% and maximum value of 14.89%. A lowest C/N value of 5.49 and a highest value of 27.27 were observed for Anchar Lake sediments (Fig. 11.9c). The overall concentration of OM varies along different sampling

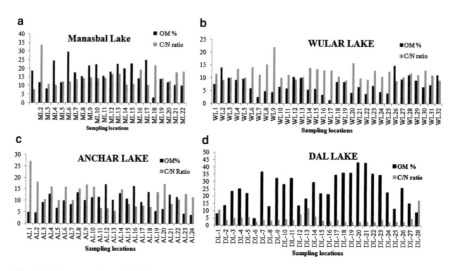

Fig. 11.9 An overview of C/N ratio and OM concentration in lake surface sediments from the four lakes

stations. The OM content ranges between 3.88 and 17.22% with an average value of 9.95%. Further, Dal Lake a hotspot centre for tourism reveals low C/N ratio indicating high N input due to anthropogenic contributions into the lake (Fig. 11.9d). The OM was determined by titration method and it was observed that 19 samples yielded more than 20% of OM contents. The OM in Dal lake sediments exhibited an average concentration of 24.70%, which is highest among all the four lakes. The overall the higher content of OM is concentrated in the lake towards western sampling sites, compared to the eastern and it could be the result of high household output due to large urbanization towards western fringe and addition of untreated municipal wastes directly drained into the lake. The highest value of C/N ratio was recorded to be 17.44 in the station DL-28 and the lowest value was recorded as 2.37 in station DL-6. Most of the sampling station shows C/N value of less than 4 pointing towards the secondary deposition of nitrogen to the lake sediments in addition by the lake vegetation. It is often noted that higher OM content and lower C/N ratio indicate higher productivity of algae and intense wetter conditions associated with high lake bathymetry.

The present-day landscape of the Kashmir Valley came into existence through various tectonic and climatic events. These changing climate conditions and tectonic pulsations have been instrumental in changing sediment transportation and accumulation rates in the Kashmir Himalayan lakes. Thus, the sediment and OM dynamics in Kashmir Himalayan lakes are governed by the presence of riverine and other terrestrial sources and later are accumulated towards the shallow margins of the lake. The amount of OM found in lake sediments is a function of the input of various sources reaching the sediment-water interface and the rate at which OM is degraded by microbial processes during burial (Hedges and Keil 1995). Aquatic flora and fauna, weeds growing within the lake, terrestrial vegetation growing along

Table 11.1 Concentration of OM (%) and C/N ratio in surface sediments of the (a) Manasbal, (b) Wular, (c) Anchar and (d) Dal Lakes

(a) Manasbal Lake			(b) Wular Lake			(c) Anchar Lake			(d) Dal Lake		
S. no.	OM %	C/N ratio	S. no.	OM %	C/N ratio	S. no.	OM%	C/N ratio	S. no.	OM %	C/N ratio
ML-1	18.6	7.73	WL-1	7.54	11.5	AN-1	4.98	27.27	DL-1	7.73	10.68
ML-2	11.8	33.83	WL-2	14	9.18	AN-2	4.77	18.2	DL-2	13.64	3.51
ML-3	8.4	11.03	WL-3	9.98	10.18	AN-3	9.16	10.48	DL-3	23.30	5.01
ML-4	24.6	10.49	WL-4	9.09	13.53	AN-4	12.9	15.99	DL-4	25.04	5.40
ML-5	11.8	12.31	WL-5	9.69	10.17	AN-5	6.68	10.11	DL-5	21.95	5.65
ML-6	29.6	12.32	WL-6	5.83	14.15	AN-6	9.92	16.15	DL-6	4.67	2.37
ML-7	17.6	13.75	WL-7	2.55	11.26	AN-7	8.34	10.15	DL-7	36.65	3.60
ML-8	15.6	14.39	WL-8	4.76	15.31	AN-8	13.49	15.11	DL-8	12.92	3.45
ML-9	21.8	14.81	WL-9	4.38	22.03	AN-9	10.12	16.9	DL-9	32.31	3.99
ML-10	22.4	14.23	WL-10	6.54	10.15	AN-10	11.31	16.15	DL-10	28.12	4.45
ML-11	15.8	14.43	WL-11	5.91	11.28	AN-11	11.63	6.67	DL-11	32.55	3.51
ML-12	18.2	16.79	WL-12	10.4	9.42	AN-12	17.22	6.56	DL-12	13.43	7.68
ML-13	22.8	16.9	WL-13	10.1	10.41	AN-13	10.35	5.49	DL-13	18.21	11.72
ML-14	20	10.61	WL-14	5.47	14.01	AN-14	13.12	14.92	DL-14	29.49	6.11
ML-15	23	10.93	WL-15	5.78	13.49	AN-15	10.91	8.55	DL-15	21.94	3.43
ML-16	14.4	19.23	WL-16	3.47	13.04	AN-16	16.45	7.36	DL-16	21.35	2.84
ML-17	25.2	10.51	WL-17	1.42	13.09	AN-17	9.46	7.69	DL-17	34.76	3.99
ML-18	3.2	22.02	WL-18	8.48	10.63	AN-18	13.82	7.25	DL-18	36.18	2.55
ML-19	14	14.37	WL-19	8.43	9.2	AN-19	5.47	13.68	DL-19	36.14	2.67
ML-20	12.2	12.83	WL-20	4.29	15.96	AN-20	6.27	17.4	DL-20	43.20	2.56
ML-21	10.4	17.99	WL-21	6.56	9.88	AN-21	12.64	8.45	DL-21	42.97	2.78
ML-22	10.2	18.47	WL-22	3.83	9.47	AN-22	11.61	10.49	DL-22	35.68	2.40
			WL-23	7.06	13.01	AN-23	4.38	13.06	DL-23	34.66	4.30
			WL-24	4.65	10.59	AN-24	3.88	11.59	DL-24	22.77	2.39
			WL-25	4.17	12.7				DL-25	11.64	2.89

(continued)

Table 11.1 (continued)

(a) Manasbal Lake			(b) Wular Lake			(c) Anchar Lake			(d) Dal Lake		
S. no.	OM %	C/N ratio	S. no.	OM %	C/N ratio	S. no.	OM%	C/N ratio	S. no.	OM %	C/N ratio
			WL-26	14.89	9.04				DL-26	25.98	2.57
			WL-27	9.57	10.5				DL-27	15.18	4.73
			WL-28	11.31	11.89				DL-28	9.00	17.44
			WL-29	9.22	11.17						
			WL-30	6.59	13.05						
			WL-31	7.37	11.04						
			WL-32	11.22	9.13						

the lake catchments and anthropogenic influences are the primary sources that control the quantity and purity of organic material delivered to the lakes. The amount and type of OM deposited within the lake reflect the ecology and environmental conditions of that ecosystem (Meyers and Lallier-Vergès 1999). The low to moderate content of OM in the lake sediments is attributed to the mixed contributions from both in situ lake flora and fauna and terrigenous detrital sources available along the lake catchment.

The C/N ratio is primarily controlled by the availability of organic material and fine fractions accumulating with a sedimentary basin. However, early digenesis can lead to both, a decline of C/N ratio inland derived material and an increase of C/N ratios in the planktonic OM. Nevertheless, the C/N ratio provides an important idea about the relative source contribution of terrestrial plants vs. planktonic OM in the sedimentary basins. Algae usually has a C/N ratio ranging from 4 to 10 because algae is rich in proteins and devoid of cellulose (Meyers and Teranes 2001; Meyers 1994). C/N ratios between 10 and 18 suggest that the source of OM was a combination of algae and vascular land-plants.

The C/N ratio of lake sediments reflects intermediate to moderately value of the C/N ratio for all the lake systems which suggest that the primary source of OM is governed by the mixing of both autochthonous algal matter and terrestrial OM. However, the high C/N ratio along certain sampling stations is attributed to the increased particulate matter loads and wastewater discharges from the upper catchments through the ephemeral channels and direct inflow of sewage from the households near the lake margins. This suggests that the increased utilization of the lake catchment for urbanization, tourism, water discharge and agricultural activities in recent years is the primary source contributing OM to the lake sediments.

The small and large watershed drainage inlets carry large volumes of organic rich sediment from the upper catchment areas and deposit them along the lake margins. Further, sediment texture within the lakes also controls the transport and deposition of OM accumulation and that the sediment OM acts as a source of recycled nutrients for lake productivity upon degradation and thereby decreasing the biological oxygen demand (BOD) and support large algal blooms and other aquatic plants in lake waters. Increased catchment soil erosion during the rainy season and the steep slopes of the catchment areas support the high influx of catchment derived organic rich materials into the lake basins. Further, in situ nutrient sources around the lake margins and macrophytes growing in the shallow parts contribute to local OM productivity thus augmenting OM sedimentation.

Paleoclimatic inferences based on OM accumulation and C/N can be deduced using sediment cores retrieved especially from the lake margins, as these margins are dynamic areas and are often subjected to the high and low lake level variations. Any change in the paleo-hydrology of the lake is reflected and preserved in sediments, and thus careful and high-resolution analyses of these sediments using OM and C/N ratio can help in reconstructing the paleo-environmental and vegetation changes around the lake catchments. Further, any anthropogenic impact such as deforestation, over exploitation of lake waters due to unplanned agricultural practices, application of pesticides around the lake catchment areas can be easily detected using these proxy records. A compilation of the C/N ratios in surficial sediments (Table 11.2) for different lakes over the world suggests that these bulk parameters retain source information despite large decreases in the total amount of organic matter during sinking. Considerable variation is evident, in the C/N values in the surface sediments. These changes in the C/N ratios reflect variations in the sources and accumulation of terrestrial and autochthonous organic inputs and prevailing climate. These variations also apparently reflect different diagenetic processes occurring at different water depths. The changes in C/N ratios is sensitive to collective input of OM from different sources along with its selective diagenesis within the lake basins. The two significant factors governing the amount of organic matter in surface sediments of these lake basins are (a) input of OM into the lake basin and (b) detrital sedimentation rates. As such, high sedimentation rates tend to dilute the OM whereas low sedimentation rates allow the OM to be oxidised (Hyne 1978).

From the above discussion, it is reiterated that the distribution of sediments, OM and C/N ratio in lake sediments from Kashmir Valley indicates that the catchment processes in context with present climate provides an important preliminary record of the factors governing the deposition and distribution of organic and inorganic sediments in lakes. Thus, the study suggests that the depositional processes in these lakes are controlled by OM influx, catchment geomorphology coupled with anthropogenic impacts. Further, recent eutrophication of these lakes is essentially simultaneous with large-scale human settlements and the application of agricultural and horticulture fertilizers in the catchment areas.

Table 11.2 A compilation of the present C/N ratio and OM content with other national and international studies

S. no	Lake	Location	Sediment type	C/N ratio	References
1	Bohai Bay	China	Surface sediments	21.3	Gao and Chen (2012)
2	Daihai Lake	China	Surface sediments	8.2–12.1	Hou et al. (2013)
3	Lake Michigan	N America	Surface sediments	9	Meyers and Ishiwatari (1993)
4	Lake Baikal	Russia	Surface sediments	11	Meyers and Ishiwatari (1993)
5	Lake Ohrid	Macedonia	Surface sediments	3–14.5	Vogel et al. (2010)
6	Mangrove Lake	Bermuda	Surface sediments	13	Meyers and Ishiwatari (1993)
7	Naples Harbour	Italy	Surface sediments	0.39–68.8	Rumolo et al. (2011)
8	Taapei Basin	Taiwan	Surface sediments	6.1	Ku et al. (2007)
9	Varthur Lake	India	Surface sediments	23–33	Mahapatra et al. (2011)
10	Walker Lake	Nevada	Surface sediments	8	Meyers and Ishiwatari (1993)
11	Welwich Marsh	England	Surface sediments	14.7	Lamb et al. (2007)
12	Lake Biwa	Japan	Surface sediments	6	Meyers and Ishiwatari (1993)
13	Karlad Lake	India	Surface sediments	9	Babeesh et al. (2016)
14	Berijam Lake	India	Surface sediments	10.95	Vijayaraj and Achyuthan (2016)
15	Kukkal Lake	India	Surface sediments	6	Vijayaraj and Achyuthan (2016)
16	Manasbal Lake	India	Surface sediments	15.0	Present study
17	Wular Lake	India	Surface sediments	11.86	Present study
18	Anchar Lake	India	Surface sediments	12.32	Present study
19	Dal Lake	India	Surface sediments	4.8	Present study
20	Lake Algae	–	–	4–10	Meyers and Ishiwatari (1993)
21	Terrestrial plants	–	–	>20	Meyers and Ishiwatari (1993)

11.5   Conclusions

In this chapter, a detailed overview was carried out on the changes in C/N ratio and OM content with paleoclimatic inferences of the modern lake sediments deposited in four freshwater lakes. The OM and C/N ratio content reveal that the sediments are organically moderate to highly productive (mesotrophic to eutrophic). The variation in C/N ratio and OM concentration of the lake sediments from the Manasbal, Wular and Anchar and Dal Lakes, revealed that sediment distribution, OM and, the C/N ratio in the lake is a function of both in situ and external sources. The C/N ratio reveals a mixed source of OM for all the four lakes suggesting both in situ production of OM by the lake biota and the allochthonous materials deposited by external sources. Anthropogenic impact, aided by landuse pattern changes and prevailing environment has resulted in the significant increase in the amount of OM accumulation in these lakes. More detailed and simultaneous analysis C/N and OM results from lake core sediments will help to better comprehend the past nutrient dynamics and palaeolimnological processes.

References

Alam A, Ahmad S, Bhat MS, Ahmad B (2015) Tectonic evolution of Kashmir basin in Northwest Himalayas. Geomorphology 239:114–126

Babeesh C, Achyuthan H, Sajeesh TP, Ravichandran R (2016) Spatial distribution of diatoms and organic matter of surface lake sediments, Karlad, North Kerala. J Palaeontol Soc India 61:239–247

Babeesh C, Achyuthan H, Jaiswal MK, Lone A (2017) Late Quaternary loess-like paleosols and pedocomplexes, geochemistry, provenance and source area weathering, Manasbal, Kashmir Valley. India Geomorphol 284:191–205. https://doi.org/10.1016/j.geomorph.2017.01.004

Bagnolus F, Meher-Homji VM (1959) Bioclimatic types of South East Asia. Travaux de la Section Scientific at Technique Institut Franscis de Pondicherry, Pondicherry, p 227

Burone L, Muniz P, Pires-Vanin A, Maria S, Rodrigues M (2003) Spatial distribution of organic matter in the surface sediments of Ubatuba Bay (Southeastern-Brazil). Ann Braz Acad Sci 75:77–80

Dean WE (2006) Characterization of organic matter in lake sediments from Minnesota and Yellowstone National Park. U.S. Geological Survey. Open-File Report 2006-1053, pp 1–39

Ganaie MA, Parveen M, Balkhi MH (2015) Physico-chemical profile of three freshwater flood plain lakes of River Jhelum, Kashmir (India). Int J Multidiscip Res Dev 2:527–532

Ganjoo RK (2014) The Vale of Kashmir: landform evolution and processes. In: Landscapes and landforms of India, World geomorphological landscapes. Springer, Dordrecht. https://doi.org/10.1007/978-94-017-8029-2_11

Gao XL, Chen CTA (2012) Heavy metal pollution status in surface sediments of the coastal Bohai Bay. Water Res 46:1901–1911

Hedges JI, Keil RG (1995) Sedimentary organic matter preservation: an assessment and speculative synthesis. Marine Chem 49:81–115

Hou D, He J, Lü C, Sun Y, Zhang F, Otgonbayar K (2013) Effects of environmental factors on nutrients release at sediment-water interface and assessment of trophic status for a typical Shallow Lake, Northwest China. Sci World J 2013:1–16

Hyne NJ (1978) The distribution and source of organic matter in reservoir sediments. Environ Geol 2:279–287

Jeelani G, Shah AQ (2006) Geochemical characteristics of water and sediment from the Dal Lake, Kashmir Himalaya: constraints on weathering and anthropogenic activity. Environ Geol 50:112–123

Khanday SA, Yousuf AR, Reshi ZA, Rashid I, Jehangir A, Romshoo SA (2016) Management of Nymphoides peltatum using water level fluctuations in freshwater lakes of Kashmir Himalaya. Limnology. https://doi.org/10.1007/s10201-016-0503-x

Ku HW, Chen YG, Chan PS, Liu HC, Lin CC (2007) Paleoenvironmental evolution as revealed by analysis of organic carbon and nitrogen: a case of coastal Taipei Basin in Northern Taiwan. Geochem J 41:111–120

Lamb LA, Vane HC, Wilson PG, Rees GJ, Moss-Hayes VL (2007) Assessing $\delta^{13}C$ and C:N ratios from organic material in archived cores as Holocene sea level and palaeoenvironmental indicators in the Humber Estuary, UK. Mar Geol 244:109–128

Lone A, Babeesh C, Achyuthan H, Chandra R (2017) Evaluation of environmental status and geochemical assessment of sediments, Manasbal Lake, Kashmir, India. Arab J Geosci 10:1–18

Lone AM, Achyuthan H, Shah RA, Sangode SJ (2018a) Environmental magnetism and heavy metal assemblages in lake bottom sediments, Anchar Lake, Srinagar, NW Himalaya, India. Int J Environ Res. https://doi.org/10.1007/s41742-018-0108-9

Lone AM, Shah RA, Achyuthan H, Fousiya AA (2018b) Geochemistry, spatial distribution and environmental risk assessment of the surface sediments: Anchar Lake, Kashmir Valley, India. Environ Earth Sci 77(65). https://doi.org/10.1007/s12665-018-7242-8

Lone AM, Shah RA, Achyuthan H, Rafiq M (2018c) Source identification of organic matter using C/N ratio in freshwater lakes of Kashmir Valley, Western Himalaya, India. Himal Geol 39(1):101–114

Mahapatra DM, Chanakya HN, Ramachandra TV (2011) C:N ratio of sediments in a sewage fed Urban Lake. Int J Geol 3:86–92

Maqbool C, Khan AB (2013) Biomass and carbon content of emergent macrophytes in Lake Manasbal, Kashmir: implications for carbon capture and sequestration. Internat J Sci Res Publ 3:1–7

Meyers PA (1994) Preservation of elemental and isotopic source identification of sedimentary organic matter. Chem Geol 114:289–302

Meyers PA, Ishiwatari R (1993) Lacustrine organic geochemistry – an overview of indicators of organic matter sources and diagenesis in lake sediments. Org Geochem 20(7):867–900

Meyers PA, Lallier-Vergès E (1999) Lacustrine sedimentary organic matter records of Late Quaternary paleoclimates. J Paleolimnol 21:345–372

Meyers PA, Teranes JL (2001) Sediment organic matter. In: Last WM, Smol JP (eds) Tracking environmental changes using lake sediments, Vol. 2: Physical and geochemical methods. Kluwer Academic, Dordrecht

Pandit AK (2002) Trophic evolution of lakes in Kashmir Himalaya. In: Pandit AK (ed) Natural resources of Western Himalaya. Valley Book House, Srinagar

Rashid SA, Masoodi A, Khan FA (2013) Sediment-water interaction at higher altitudes: example from the geochemistry of Wular Lake sediments, Kashmir Valley, northern India. Proc Earth Planet Sci 7:786–789

Rumolo P, Barra M, Gherardi S, Marsella E, Sprovieri M (2011) Stable isotopes and C/N ratios in marine sediments as a tool for discriminating anthropogenic impact. J Environ Monit 13:3399–3408

Sarkar S, Prakasam M, Banerji US, Bhushan R, Gaury PK, Meena NK (2016) Rapid sedimentation history of Rewalsar Lake, Lesser Himalaya, India during the last fifty years – estimated using ^{137}Cs and ^{210}Pb dating techniques: a comparative study with other north-western Himalayan Lakes. Himal Geol 37:1):1–1):7

Shah RA, Achyuthan H, Lone A, Ravichandran R (2017) Diatoms, spatial distribution and physi-cochemical characteristics of the Wular Lake Sediments, Kashmir Valley, Jammu and Kashmir. J Geol Soc India 90:159–168

Vijayaraj R, Achyuthan H (2016) Organic matter source in the freshwater tropical lakes of south-ern India. Curr Sci 111:168–176

Vogel H, Wessels M, Albrecht C, Stich HB, Wagner B (2010) Spatial variability of recent sedimen-tation in Lake Ohrid (Albania/Macedonia) – a complex interplay of natural and anthropogenic factors and their possible impact on biodiversity patterns. Biogeosci Discuss 7:3911–3930

Wadia DN (1941) Pleistocene Ice Age deposits of Kashmir. Proc Natl Inst Sci India 7(1):49–59

Wadia DN (1976) Geology of India, 4th edn. Reprint 1976. Macmillan, London, p 508

Xiao J, Fan J, Zhou L, Zhai D, Wen R, Qin X (2013) A model for linking grain-size component to lake level status of a modern clastic lake. J Asian Earth Sci 69:149–158

Chapter 12
Deciphering Climate Variability over Western Himalaya Using Instrumental and Tree-Ring Records

H. P. Borgaonkar, T. P. Sabin, and R. Krishnan

Abstract In this chapter, we discuss climate variability over the Western Himalaya based on instrumental records as well as information derived from tree-ring data. We focus here on two important climatological elements, namely rainfall and temperature of Western Himalayan region. Trend analyses of rainfall based on 14 stations covering entire western Himalayan region from Kashmir to Uttarakhanda indicate different patterns of rainfall variability. The database does not show any coherent patterns among the stations and does not indicate any significant trend during the twentieth century. Data on temperature show overall warming mainly contributed by maximum temperature. Most of the stations indicate significant increasing trends in maximum temperature for all the seasons. Cooling trend is observed mostly in minimum temperature of some stations for different seasons. Annual maximum, minimum and mean temperature series of all the stations indicate significant warming except slight cooling in minimum temperature of Deheradun, Mukteswar and Mussoorie. Climate projections for twenty-first century also indicate warming over the entire Himalayan region with significant warming in Tibetan plateau, and increasing trend in summer precipitation over the central Himalayan region including Nepal and Tibetan Plateau.

Dendroclimatic reconstructions give some information about summer climate conditions since past several centuries. They indicate some cool epochs associated with Little Ice Age (LIA). It is also seen that high altitude near glacier tree-ring records would be the potential source of information on long-term temperature variability and glacier fluctuations. Overall warming trends noted in different parts of the western Himalaya may be linked partially to global warming trends and rapid urbanization of the hill stations were the observatories are located.

H. P. Borgaonkar (✉) · T. P. Sabin · R. Krishnan
Indian Institute of Tropical Meteorology, Pune, India
e-mail: hemant@tropmet.res.in

© Springer Nature Switzerland AG 2020
A. P. Dimri et al. (eds.), *Himalayan Weather and Climate and their Impact on the Environment*, https://doi.org/10.1007/978-3-030-29684-1_12

205

12.1   Introduction

The Himalaya constitutes one of the youngest mountain systems in the world. The role of the Himalaya in maintaining and controlling the monsoon system over the Asian continent is so vital that it draws the foremost attention of climatologists as well as geologists and biologists. It consists of climatologically significant highly reflecting snow covered mountains, a vast biomass reserve and network of rivers with perennial water source. Vast differentiation in topography, climate, soil structure and rock along the Himalayan ranges from west to east has resulted in numerous types of vegetation. Himalayan environments are classic areas for the study of climate change because they contain a wide range of natural phenomena (e.g. glaciers and tree-line) that are strongly influenced by climate and provide easily detectable evidence of climate change. Mountain glaciers, particularly over the Himalayan region are much sensitive to climate change.

Over the Himalayan region, most of the climate studies so far dealt with the mean precipitation conditions over specific locales like Mount Everest (Dhar and Narayanan 1965), Cherrapunji (Ramaswamy 1972; Dhar and Farooqui 1973), GarhwalKumaon Himalaya (Dhar et al. 1984a) Uttar Pradesh Himalaya (Dhar et al. 1987) and Ladakh region (Dhar and Mulye 1987), extreme rainfall events (Dhar et al. 1975) and break – monsoon rains over the foothills of north India (Dhar et al. 1984b) and variation of rainfall with elevation (Dhar and Bhattacharya 1976). There are very few studies available on the climate change during the instrumental record of the past century, for the Himalayan region. Pant and Borgaonkar (1984) studied climate of the hill region of Uttar Pradesh on the basis of annual rainfall data of six stations and temperature data at one station, for a period of about a century. They did not find any long-term trend in the rainfall series. A marked increase in winter maximum and winter mean temperature is also observed after the year 1935. Bhutiyani et al. (2007) observed significant warming over northwest Himalaya during the last century at a rate, which is higher than the global average. Pant et al. (2003) also indicated overall warming trend since last few decades as observed in many other parts of the globe (Böhm et al. 2001; Jones and Moberg 2003; PAGES 2k Consortium 2013). Basistha et al. (2009) pointed out that rainfall has decreased in the Indian Himalayas lying in Uttarakhand State during last century as a sudden shift around 1964 CE, rather than gradual trend, whereas, over northwestern Himalaya, winter rainfall showed increasing trend while, monsoon and annual rainfall has decreased since last 100 years (Bhutiyani et al. 2010). It is likely that some of the climate changes that have taken place are the result of global scale events. However, it is obvious that the net change in the climate is due to a number of interactive physical mechanisms at global, regional and local scales.

Preliminary attempts on dendroclimatic analysis for the Indian region were made by Pant (1979, 1983). He suggested that the trees of the Himalayan zone are most appropriate for identifying well defined growth rings, which generally display a very prominent response to temperature. As a preliminary survey, few species of Alpine, temperature, subtropical and tropical regions of Himalayan zones were identified by Pant (1979). Bhattacharyya and Yadav (1989) also discussed the dendroclimatic potential and dating problems associated with the tropical and subtropical tree-rings form the Indian subcontinent. Pant (1979) also suggested two climatic zones in India for extensive sampling and tree-ring analysis, which are the forests of the mountains adjacent to Himalayan snow-line including the sub-Himalayan forest belts, and the semi-arid regions of the country.

The tree-ring based summer temperature reconstructions indicate cooling effect in recent decades over Western Himalaya (Yadav 2004), Kashmir (Hughes 2001) and Nepal (Cook et al. 2003). While tree-ring reconstructions of mean temperature over Karakoram Himalaya (Esper et al. 2002) and winter temperature of Nepal (Cook et al. 2003) follow warming trend in recent decade. However, few regional tree-ring summer temperature estimates did not find any warming or cooling trend during past 250 years (Borgaonkar et al. 1994, 1996). Recent study (Yadav et al. 2017) indicates increasing precipitation pattern and glacier expansion associated with cooling over northwest Himalaya. In general, all these reconstructions show some cool epochs related to Little Ice Age associated with intermittent warm episodes.

The geological, palaeobotanical and archaeological studies from this region point towards its palaeological history of glacial and interglacial phases in a broader sense; however, corroborative signals of climatic significance need careful examination (Borgaonkar 1996). In the recent past, with about a century of recorded data, the region shows the most varied climate throughout the year. Large variations in topography, elevation and location result in great contrasting climates within short distances. Besides these local and regional variations in climate, the entire Himalaya experience a general weather and climate pattern dictated by the monsoon systems of Asia and the chain of mid latitude extra-tropical systems. In addition to an overview of the overall climatological setting of the region this article summarizes the changes and fluctuations in temporal and spatial scales over the region for the period of observed meteorological data. Special mention is made of the recent studies in the detection of proxy signals of changing climate recorded in the annual growth rings of trees from the vast coniferous forests of the region.

12.2   Climatic Change over the Western Himalaya: The Twentieth Century Setting

Though the climatic effects of man-made changes in the Himalayan environment are not clearly known, it is causing serious concern because of the implications of possible chain reactions towards a deteriorated climatic regime. In view of this, we made an attempt to look at the available climate records on rainfall and temperature at some observatories in an around the Western Himalaya, for possible evidence of long-term changes. The data have been taken from the monthly weather Reports of the India Meteorological Department and other published reports. The data of most

Table 12.1 Details of Meteorological station data used in the analysis

Sr. no.	Station	State	Latitude	Longitude	Altitude (M)	Temperature	Rainfall
1	Srinagar	J&K	34° 05′	74° 50′	1587	1901–2016	1892–2016
2	Quazigund	J&K	33° 30′	75° 10′	1630	1962–2016	1962–2016
3	Banihal	J&K	33° 35′	75° 05′	1690	1962–2016	1962–2016
4	Leh	J&K	34° 09′	77° 34′	3506	1901–1990	1876–1990
5	Shimla	H.P.	31° 06′	77° 10′	2202	1901–2007	1863–2007
6	Bhunter	H.P.	31° 50′	77° 10′	1067	1964–2014	1964–2014
7	Dehra Dun	U.K.	30° 19′	78° 02′	682	1901–2015	1861–2015
8	Mukteswar	U.K	29° 28′	79° 39′	2311	1901–2014	1897–2016
9	Mussoorie	U.K.	30° 27′	78° 05′	2042	1901–1990	1869–1986
10	Joshimath	U.K.	30° 33′	79° 34′	2045	–	1871–1987
11	Almora	U.K.	29° 30′	79° 35′	1642	–	1856–1978
12	Nainital	U.K.	29° 25′	79° 27′	2084	–	1849–1978
13	Pauri	U.K.	30° 07′	78° 44′	1824	–	1871–1978
14	Pithoragarh	U.K.	30° 01′	80° 14′	1514	–	1864–1978

of the stations cover the long period of the twentieth century. The stations identified by their geographical locations and the periods of data for different climatic parameters are listed in Table 12.1 and locations of the stations are presented in Fig. 12.1. Total 14 stations covering entire western Himalaya including Jammu & Kashmir, Himachal Pradesh and Uttarakhand have been considered for the analysis.

These stations have long and continuous data. Out of 14 stations, 9 have both temperature and rainfall monthly records available. Rests of the stations have only rainfall data. The rainfall includes snowfall also, converted to rainwater terms, wherever applicable. The results have been discussed below. It may however, be noted that the study does not intend to be comprehensive for the whole Himalaya, as the data network available is meager. The results can only be an indicator of the broad signal of climatic change over the Himalaya. Of course, the complex terrain of the Himalaya makes generalizations extremely difficult, if not impossible. The study is mainly done for the western Himalaya.

12.2.1   Commonality in Climatological Patterns

The average monthly variations in rainfall and temperature based on the available data period for 14 stations over the western Himalaya are presented in Fig. 12.2. Lowest rainfall is recorded at the station in Leh. Rainfall values of Leh range from minimum 0.3 cm in November to maximum 1.3 cm in July and 1.4 cm in August. Average monthly temperature range of Leh is also lowest compare to all other stations in virtue of its higher elevation (3506 M amsl). Amount of rainfall received during any month of the year is very less. Therefore, Leh (Ladakh) region is known as cold desert. Climatological patterns of four northern latitude stations of Jammu

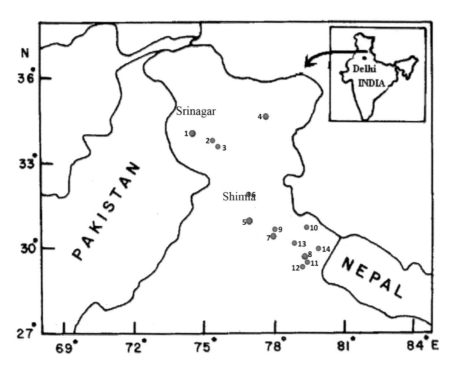

Fig. 12.1 Location map of meteorological station network in western Himalaya considered for the analysis.(*1* Srinagar, *2* Quazigund, *3* Banihal, *4* Leh, *5* Shimla, *6* Bhunter, *7* Dehra Dun, *8* Mukteswar, *9* Mussoorie, *10* Joshimath, *11* Almora, *12* Nainital, *13* Pauri, *14* Pithoragarh)

and Kashmir (Srinagar, Banihal, Quazugund and Leh) are different than the nine lower latitude stations of Himachal Pradesh and Uttarakhand (Shimla, Mukteswar, Mussoorie, Dehradun, Almora, Joshimath, Nainital, Pauri, Pithoragarh). Temperature data is not available for five stations Almora, Joshimath, Nainital, Pauri, Pithoragarh. Amount of rainfall received during winter (DJF) and pre-monsoon (spring; MAM) is more than the rainfall amount received during monsoon (JJAS) and post-monsoon (ON) in the Kashmir region (Srinagar, Banihal, Quazugund). Temperature reaches its maximum during July and August. Unlike Kashmir region, lower latitude stations of Himachal Pradesh and Uttarakhand receive more than 70% of annual total rainfall during monsoon season (Fig. 12.2). Temperature attains its highest value in May and June. These rainfall and tempera-ture patterns are similar to what is observed in stations located in the plains of India. The Kashmir region is out of the influence of monsoon currents. More rainfall (snowfall) is received in winter and spring due to western disturbances. Two stations Leh and Bhunter show very similar pattern of temperature and rainfall variations. Because of their more eastward longitude compared to the other Kashmir stations and more northward latitude compare to the stations of Himachal Pradesh and Uttarakhand, both the stations receive some rainfall in winter and monsoon seasons.

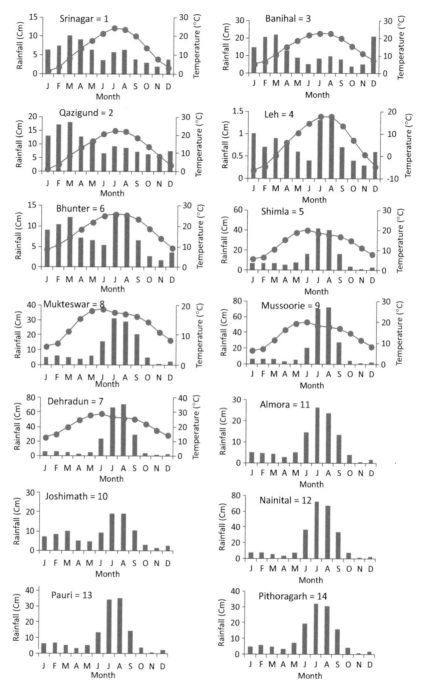

Fig. 12.2 Monthly variations of rainfall and temperature of western Himalayan stations based on long-term averages. Station numbers as per the Table 12.1

12.2.2 Trends in Rainfall

Mean standard deviation (s.d.) and linear trend during the available period for all the stations are presented in Table 12.2. The linear trends are expressed in cm/100 years to have an idea of the long-term trend. These statistics are given for the four seasonal totals as well as the annual total of rainfall. Among the stations, Leh located NE of Kashmir valley receives lowest annual rainfall (9 cm), whereas, Nainital in Uttarakhand gets highest annual rainfall (251.8 cm). Kashmir stations Srinagar, Banihal and Quazigund are well beyond the monsoon reach, show conspicuously different rainfall regime both in terms of mean and s.d. The means and s.d's of seasonal rainfall show the gradual transition of the rainfall regime from north to south and east to west. Shimla and Mussoorie located at nearly same elevation show significant decreasing trend in monsoon and annual rainfall since last 150 years. However; only Pithoragarh monsoon rainfall shows significant increasing trend. During post-monsoon season, most of the southward stations show increasing rainfall pattern with significant increasing trend for Shimla, Mussoorie, Almora, Joshimath and Pithoragarh, while, significant negative trend has been observed for Quazigund station since past five decades (Table 12.2). Significant negative trend has also been noted in pre-monsoon season for Quazigund. Table 12.2 reveals that during twentieth century western Himalaya did not experience significant increase or decrease in winter rainfall. Monsoon and annual rainfall patterns of Shimla and Mussoorie show significant decreasing trend. Deheradun station experienced increasing rainfall pattern in all the seasons with significant increase in pre-monsoon rainfall. Out of 14, three stations of Kashmir viz. Srinagar, Banihal, Leh and four stations viz. Bhunter, Mukteswar, Nainital and Pauri of Himachal and Uttarakhand do not register any significant trend throughout the year. Figure 12.3 shows annual rainfall time series of all the stations. Mussoorie shows decreasing tendency in most of the seasons, but the neighboring Dehradun which is located foot hills of Mussoorie mountain indicates opposite trend, particularly in monsoon and annual rainfall data. Thus, it can be concluded that there is no long-term trend in the annual rainfall over the Western Himalayan stations.

We also present mean precipitation patterns for different seasons over the Himalaya (Fig. 12.4a) based on APHRODITE 0.25 × 0.25 gridded data for the period 1951–2007 C.E. (Yatagai et al. 2012). It clearly depicts the same pattern as shown in observed data (Fig. 12.2). Winter (DJF) and pre-monsoon (MAM) seasons receive more rainfall over the Kashmir region than the lower latitude regions of Himachal Pradesh and Uttarakhand, whereas, Kashmir region gets much lower rainfall than Himachal Pradesh and Uttarakhand in Monsoon season. Lowest rainfall is received in post-monsoon (ON) season over the entire western Himalayan region. Figure 12.4b gives spatial pattern of trends in precipitation for different seasons. It does not indicate any significant coherent patterns of dry or wet conditions, however, more decreasing (dry) tendency is observed in monsoon season over the Himachal and Uttarakhand than Kashmir region. Figure 12.5 shows EOF1 of seasonal mean anomaly of precipitation for five seasons and their respective time series of PC1. The patterns simply demonstrate the dominant mode of variability in inter-annual time scale.

Table 12.2 Mean, standard deviation and long-term trends of rainfall over the Western Himalaya

Station	Data period	Parameter	Winter	Pre-monsoon	Monsoon	Post-monsoon	Annual
Srinagar	1893–2016	Mean	17.9	25.9	19.7	5.0	68.4
		s.d.	7.5	9.3	7.3	4.3	15.4
		Trend	−0.6	2.9	1.1	1.5	4.4
Banihal	1962–2016	Mean	57.2	43.6	31.6	8.8	129.1
		s.d.	90.8	18.3	16.7	6.3	97.3
		Trend	−115.8	−2.1	8.7	−5.0	−125.0
Quazigund	1962–2016	Mean	38.4	42.1	31.9	11.1	123.1
		s.d.	13.9	14.5	15.1	21.6	36.0
		Trend	8.0	−32.9**	10.3	−36.6*	−59.4
Leh	1876–1990	Mean	2.3	2.2	3.8	0.7	9.0
		s.d.	1.4	2.2	2.5	1.2	4.2
		Trend	0.1	0.0	−1.4	0.7	−0.8
Shimla	1863–2007	Mean	16.6	19.1	114.7	4.8	155.3
		s.d.	9.5	10.4	27.7	5.6	30.5
		Trend	−2.7	−1.0	−21.9**	1.6*	−24.0**
Bhunter	1964–2014	Mean	23.1	25.9	37.6	4.4	91.1
		s.d.	9.4	10.1	12.5	4.6	18.0
		Trend	3.4	−3.2	12.5	−1.9	8.9
Dehradun	1861–2015	Mean	14.0	10.9	189.9	4.9	219.6
		s.d.	8.4	8.2	47.8	6.2	48.7
		Trend	0.8	4.2*	12.7	2.3	19.7
Mukteswar	1897–2015	Mean	13.4	14.7	96.2	5.7	129.9
		s.d.	7.1	7.2	27.2	8.5	29.9
		Trend	−1.1	2.1	−9.9	−1.2	−10.6
Massoorie	1869–1986	Mean	15.2	15.0	190.2	5.3	225.8
		s.d.	8.1	8.1	44.5	7.3	46.6
		Trend	−4.1	0.6	−35.2**	4.2*	−34.4*
Almora	1856–1980	Mean	11.3	12.1	78.3	4.4	106.1
		s.d.	6.0	6.3	20.6	6.5	23.4
		Trend	−1.1	−1.9	6.3	2.9*	6.2
Joshimath	1871–1980	Mean	18.2	19.6	58.2	4.4	100.4
		s.d.	10.9	10.7	23.1	5.0	31.1
		Trend	−3.8	1.0	−9.9	1.8*	−11.0
Nainital	1849–1980	Mean	17.1	16.5	210.2	8.0	251.8
		s.d.	10.9	10..5	56.4	13.3	61.1
		Trend	−3.0	−0.4	12.7	5.1	14.3
Pauri	1871–1980	Mean	15.0	13.6	97.2	4.3	130.0
		s.d.	7.9	7.6	23.5	6.5	26.9
		Trend	−1.5	−1.9	5.2	2.2	3.8
Pithoragarh	1864–1980	Mean	12.4	15.0	97.8	4.9	130.2
		s.d.	7.0	7.8	25.0	6.9	28.2
		Trend	−1.0	−0.6	13.0*	3.4*	14.7

Trend values indicate Trend/100 years; *p < 0.05; **p < 0.01

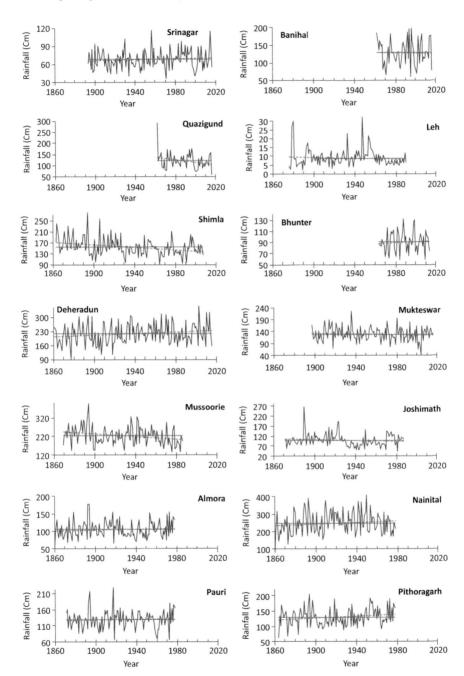

Fig. 12.3 Annual rainfall variations of 14 western Himalayan stations with linear trend in red dotted lines. Mean line is in green colour

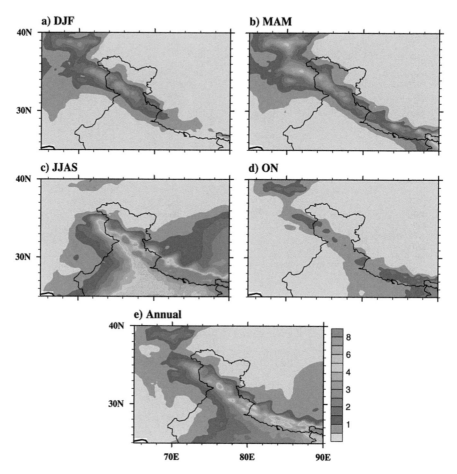

Fig. 12.4a Mean precipitation patterns for different seasons over the Himalaya based on APHRODITE data for the period 1951–2007 CE

12.2.3 Trends in Temperature

Mean, standard deviation and trend value per 100 years of maximum, minimum and mean temperature series of nine stations for four different seasons and annual series are presented in Table 12.3. The annual mean temperature series during the available data period for the stations along with trend line are presented in Fig. 12.6. Main feature of the trend analysis is, most of the stations show significant increasing trend in maximum and mean temperature series. Within the maximum temperature series of all the station only Quazigund maximum temperature of monsoon season indicates non-significant negative trend. Analysis also indicates warming trend of maximum, minimum and mean temperatures of Srinagar, Banihal, Shimla, Bhunter. All the stations except Leh indicates significant increasing trend in maximum and

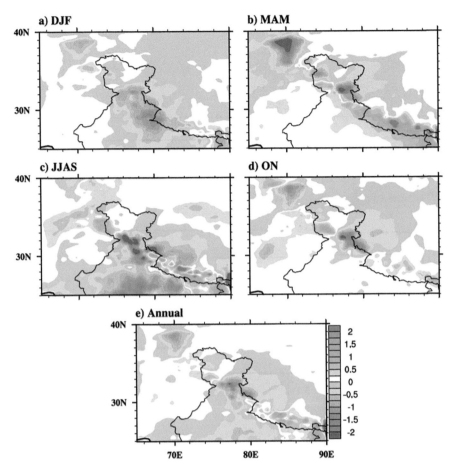

Fig. 12.4b The observed trend in precipitation over the Himalaya for different seasons based on APHRODITE data for the period 1951–2007 CE

mean winter temperature. Similar pattern is also noted in annual series of maximum and mean temperature. Pre-monsoon and monsoon minimum temperatures of Dehradun, Mukteswar and Mussoorie indicate cooling trend while winter and annual minimum temperature of Deheradun also shows cooling tendency. Most of the stations (Fig. 12.6) show a conspicuous increase in the temperature during a major part of the data period. However, the temperature appears to be on a slight downward trend after about 1960 till around 1980 followed by moderately sharp increase till recent. Borgaonkar et al. (2011) also noted significant warming trend in regional mean temperature of western Himalaya for different seasons. They showed prominent increasing trend in the time series of annual highest values of daily maximum and minimum temperatures for the period 1970–2003 for three stations viz. Srinagar, Shimla and Mukteswar.

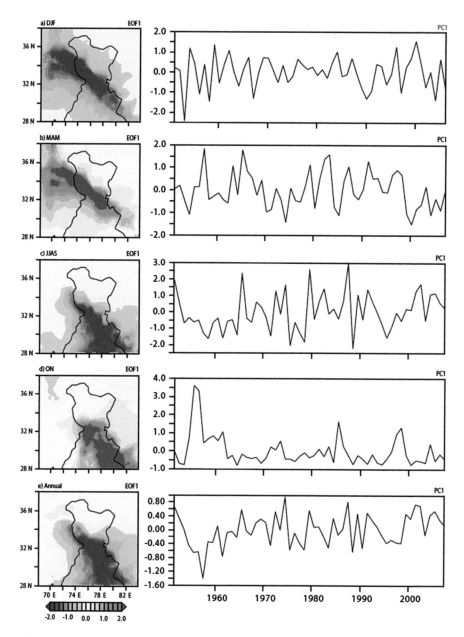

Fig. 12.5 EOF 1 and respective PC of seasonal mean anomaly of precipitation from APHRODITE data for the period 1951–2007 CE

Table 12.3 Mean, standard deviation and long-term trends (trend/100 years) of mean temperature over the Western Himalaya

Station	Data period	Parameter	Winter			Pre-monsoon			Monsoon			Post-monsoon			Annual		
			Max	Min	Mean	Max	Min	Mean	Max	Min	Mean	Max	Min	Mean	Max	Min	Mean
Shrinagar	1901–2016	Mean	7.3	−1.6	2.8	19.1	7.3	13.2	29.0	15.7	22.4	18.9	3.0	10.9	19.5	7.2	13.3
		s.d.	2.0	1.1	1.3	1.5	0.7	1.0	0.8	0.8	0.6	1.4	1.2	0.8	0.9	0.6	0.6
		Trend	2.2**	1.1**	1.7**	1.5**	0.7**	1.1**	0.0	0.3*	0.2*	0.6	1.5**	1.0**	1.0**	0.8**	0.9**
Banihal	1962–2016	Mean	11.7	0.7	6.2	21.2	8.1	14.7	28.1	15.5	21.8	21.5	5.3	13.4	21.1	8.3	14.8
		s.d.	1.7	0.9	1.2	1.5	0.9	1.1	0.6	0.7	0.5	1.3	0.9	0.9	0.9	0.6	0.7
		Trend	5.2**	2.4**	3.8**	3.9**	1.1	2.4**	0.7	0.5	0.6	4.0**	1.2	2.6**	3.1**	1.3*	2.2**
Quazigund	1962–2016	Mean	7.8	−2.1	2.8	19.1	6.4	12.7	27.3	14.4	20.9	18.8	3.3	11.1	19.0	6.4	12.7
		s.d.	2.1	1.1	1.4	1.4	0.6	0.9	0.7	0.7	0.6	1.5	0.8	0.8	1.0	0.4	0.6
		Trend	7.6**	2.9**	5.3**	3.4**	−0.8*	1.3*	−0.3	−1.1	−0.7	2.9*	−1.3*	0.8	3.1**	0.0	1.5**
Leh	1901–1990	Mean	1.8	−12.2	−5.2	12.3	−1.1	5.6	23.2	8.6	15.9	11.5	−3.8	3.8	13.2	−1.1	6.0
		s.d.	1.6	1.5	1.1	1.5	1.1	1.2	1.4	1.6	1.4	1.5	1.0	1.1	0.9	1.0	0.9
		Trend	1.4	−0.3	0.6	0.7	2.2**	0.6	3.4**	3.7**	3.6**	0.6	−0.2	0.2	1.8**	1.8**	1.8**
Shimla	1901–2007	Mean	10.1	3.1	6.6	18.8	10.7	14.8	21.3	15.1	18.2	16.6	9.0	12.8	17.1	10.0	13.5
		s.d.	1.7	1.1	1.3	1.6	1.2	1.4	1.1	0.7	0.8	1.5	0.8	1.1	1.2	0.7	0.9
		Trend	3.8**	1.2*	2.5**	2.9**	0.8*	1.9**	2.7**	0.2	1.4**	3.7**	0.6*	2.2**	3.2**	0.7*	1.9**
Bhunter	1964–2014	Mean	16.9	2.2	9.6	26.8	9.7	18.3	31.3	18.2	24.8	25.4	7.3	16.4	25.6	10.3	18.0
		s.d.	1.5	0.6	0.8	1.5	0.6	1.0	0.7	0.6	0.5	1.1	0.8	0.8	0.6	0.4	0.5
		Trend	5.0**	1.0	3.0**	2.2	1.1	1.6	0.3	1.0	0.6	1.2	1.6	1.4	1.7**	1.1**	1.4**

(continued)

Table 12.3 (continued)

Station	Data period	Parameter	Winter			Pre-monsoon			Monsoon			Post-monsoon			Annual		
			Max	Min	Mean	Max	Min	Mean	Mean	Max	Min	Max	Min	Mean	Max	Min	Mean
Dehradun	1901–2015	Mean	20.6	7.1	13.9	31.3	16.7	24.0	26.8	31.1	22.4	26.7	13.2	20.0	27.9	15.6	21.7
		s.d.	1.3	1.0	0.9	1.4	1.2	0.9	0.7	0.9	0.8	1.0	1.0	0.7	0.6	0.8	0.6
		Trend	1.5**	-0.1	0.7*	0.3	-0.5	-0.1	-.01	0.1	-0.5	1.0**	0.0	0.6**	0.6**	-0.3	0.2
Mukteswar	1901–2014	Mean	11.8	2.6	7.2	20.2	9.6	14.9	17.5	21.2	13.8	17.5	7.8	12.6	18.0	8.9	13.5
		s.d.	1.5	1.0	1.1	1.5	1.3	1.1	0.5	0.8	0.8	1.3	1.0	0.9	0.9	0.8	0.6
		Trend	3.2**	0.1	1.7**	2.0**	-0.6*	0.7	0.2	0.9**	-0.6*	2.2**	-0.3	1.0**	1.9**	-0.4*	0.8**
Massoorie	1901–1990	Mean	11.2	3.4	7.3	20.1	11.1	15.6	18.2	21.1	15.3	16.9	9.2	13.0	17.7	10.3	14.0
		s.d.	1.4	0.8	1.0	1.3	1.1	1.1	0.4	0.5	0.5	0.9	0.7	0.7	0.6	0.5	0.5
		Trend	2.6**	0.2	1.4**	0.8	-0.1	0.4	-0.1	0.2	-0.4*	1.9**	0.2	1.1**	1.2**	-0.1	0.5**

*p < 0.05; **p < 0.01

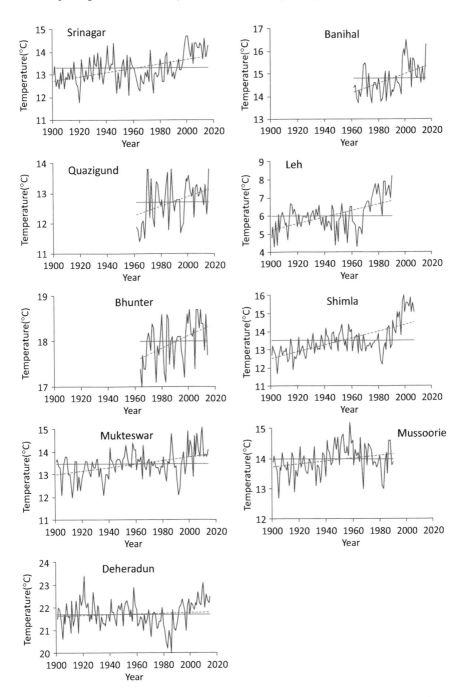

Fig. 12.6 Annual mean surface temperature variations of western Himalayan stations with linear trend in red dotted line. Mean line is in green colour

Figure 12.7a shows special patterns of surface mean temperature derived from CRU TS3.10 data (Harris et al. 2014) for different seasons. The data length used is from 1951–2017 C.E. Monsoon (JJAS) season is the warmest period; however, higher latitudes of western Himalaya including Kashmir and Tibetan plateau are comparatively less warm than the low latitude regions. Lower latitude regions are also warmer in pre-monsoon (MAM) season. Special maps of trend analysis of surface temperature (Fig. 12.7b) indicate overall warming trend in all the seasons. Warming is lowest in monsoon season with negative trend in some part of northwest India. Warming during post-monsoon and winter is comparatively higher.

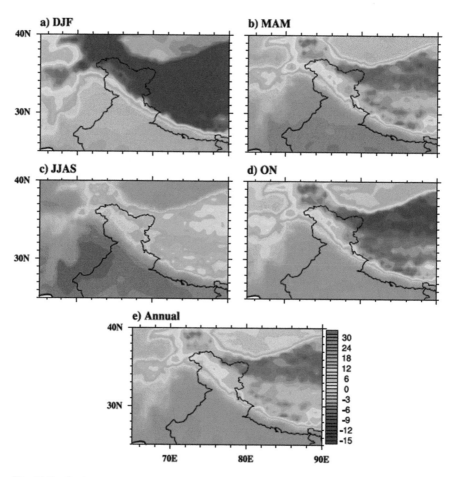

Fig. 12.7a Surface mean temperature patterns for different seasons over the Himalaya based on CRU data for the period 1951–2017 CE

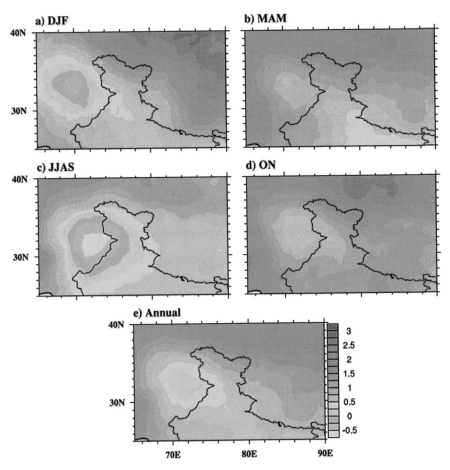

Fig. 12.7b The observed trend in temperature over the Himalaya for different seasons based on CRU data for the period 1951–2007 CE

12.2.4   Future Projections of Precipitation and Temperature

Climate models participated in the IPCC-CMIP5 (Coupled Model Inter-comparison Project phase-5) are the primary tools to understand the historical and projected climate change information. Here, we used the 28 models (Table 12.4) participated in the IPCC-CMIP5 assessment to understand the projected climate change signals over the Himalaya. The projected trends in precipitation for different seasons from the RCP4.5 and RCP8.5 scenario are shown in Fig. 12.8a. There is a moderate decrease in precipitation over the western Himalaya in winter and spring, which is in line with the present day climate signal, but during the summer season, the central Himalayan precipitation seems to significantly increase during the twenty-first century. Wu et al. (2017) noted that the precipitation seems to be increasing by 16% in

Table 12.4 List of 28 CMIP5 models used in this study, their sponsor, country and name

Sponsor and country	Model name
Commonwealth Scientific and Industrial Research Organization (CSIRO) and Bureau of Meteorology (BOM), Australia	ACCESS1-3
Beijing Climate Centre Climate System Model, China	BCC-CSM1-1 BCC-CSM1
Beijing Normal University Earth System Model, China	BNU-ESM
Canadian Centre for Climate Modelling and Analysis, Canada	CAN-ESM
National Center for Atmospheric Research, USA	CESM-BGC CESM-CAM5 CCSM4
Meteo-France/Centre National de Recherches Meteorologiques, France	CNRM-CM5
Centro Euro-Mediterraneo sui Cambiamenti Climatici, Italy	CMCC-CM CMCC-CM5
Commonwealth Scientific and Industrial Research Organisation (CSIRO), Australia	CSIRO-Mk3-6-0
Geophysical Fluid Dynamics Laboratory, Oceanic and Atmospheric Administration (NOAA), USA	GFDL-CM3 GFDL-ESM2M GFDL-ESM2G
Met Office Hadley Centre, UK	HadGEM2-AO HadGEM2-ES HadGEM2-CC
Institute for Numerical Mathematics, Russia	INM-CM4
Institute Pierre Simon Laplace, France	IPSL-CM5A-LR IPSL-CM5A-MR
Centre for Climate System Research (University of Tokyo), National Institute for Environmental Studies and Frontier Research Center for Global Change (JAMSTEC), Japan	MIROC-ESM-CHEM MIROC5 MIROC-ESM
Max-Planck-Institut für Meteorologie (Max Planck Institute for Meteorology)	MPI-ESM-LR MPI-ESM-LR
Meteorological Research Institute, Japan	MRI-CGCM3
Norwegian Climate Centre, Norway	NorESM1-ME

RCP4.5 and 24% in RCP8.5 scenario over the eastern Himalaya, while the western Himalaya undergoing a moderate decline in precipitation. However, the wide inter-model spread in the simulated precipitation changes over the Himalaya (Fig. 12.8b) makes the assessments of the hydroclimatic response even complex in reality. This highlights the requirement of a new high-resolution modeling approach which may possibly provide a better regional assessment over such a complicated terrain.

The spatial distributions of the trend in surface air temperature for 2006–2099 C.E. are provided in Fig. 12.9a. Significant warming is projected over the Himalayan region in future. Comparing to the present day climate (1961 to 1990 C.E.) Himalaya is projected to be warmer by 2 °C in RCP4.5 and 5 °C in RCP8.5 scenarios by the end of the twenty-first century which is consistent with the studies of Wu et al. (2017). The largest values of temperature increase are found in the Tibetan Plateau in accordance with the elevation dependent warming, which is one of the significant

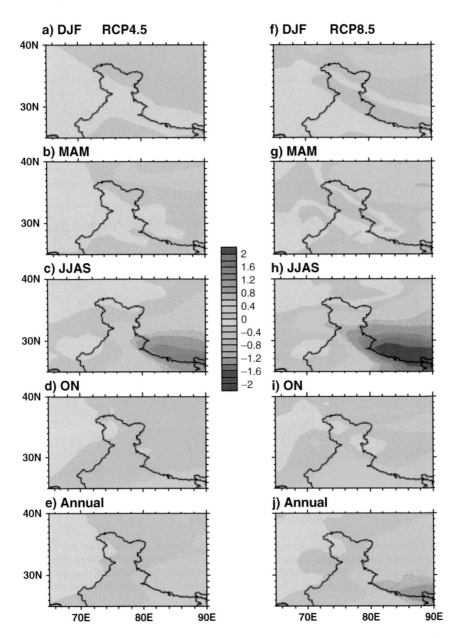

Fig. 12.8a Projected trends in precipitation for the period 2006–2099 CE over the Himalaya for different seasons from CMIP5 ensemble means from RCP4.5 and RCP8.5 simulations

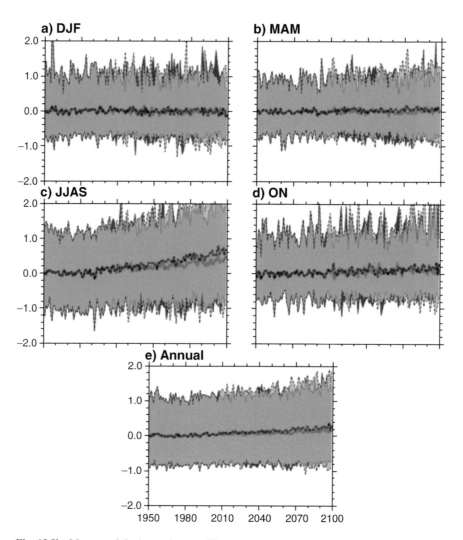

Fig. 12.8b Mean precipitation series over Himalaya (65°E-90°E, 25°N-40°N) for different seasons from the CMIP5 models. The anomalies are created by removing the Historical climatology for the period 1961–1990 CE. Future projection is shown with ensemble mean (solid/dash line) and ensemble spread (shaded) from RCP4.5 (dark gray) and RCP8.5 (light blue) simulations

global warming signals demonstrated by the significant warming at higher altitudes of the mountain environment relative to lower elevations (Beniston and Rebetez 1996; Diaz and Bradley 1997). This is clear from the spatial trend map, especially in winter months. The time series of projected changes with respect to its Historical mean (1961–1990 C.E.) for different seasons are given in the Fig. 12.9b. Even though there is huge inter-model spread, the projected temperature shows a clear rising tendency, both in RCP4.5 and RCP8.5 scenario.

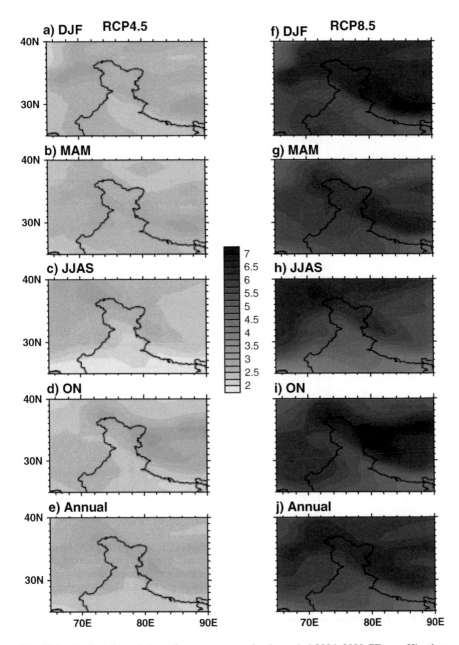

Fig. 12.9a Projected trends in surface temperature for the period 2006–2099 CE over Himalaya for different seasons from CMIP5ensemble means of from RCP4.5 and RCP8.5 simulations

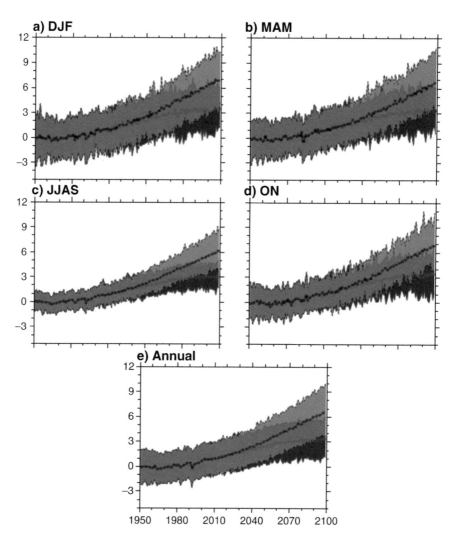

Fig. 12.9b Mean surface temperature series over Himalaya (65°E-90°E, 25°N-40°N) for different seasons from the CMIP5 models. The anomalies are created by removing the Historical climatology for the period 1961–1990 CE. Future projection is shown with ensemble mean (solid/dash line) and ensemble spread (shaded) from RCP4.5 (dark gray) and RCP8.5 (light red) simulations

12.2.5 The Palaeoclimatic Setting

The palaeoclimate of the Himalaya and north-west India on different time scales has been studied by various workers (Pilgrim 1932, 1944; de Terra and Paterson 1939; Morris 1938; Zeuner 1972; Agarwal 1985, Bhattacharyya 1989). The north-west Himalaya and Kashmir valley received particular attention and were studied using different techniques like lithostratigraphy, magnetostratigraphy, Palaeobotanical data, pollen evidences, diatom studies and stable isotopic ratios. Some clues from vertebrates from the sediments were also used to frame the Cenozoic changes of the region. The climatic records of Kashmir valley for the past four million years are well preserved in the Karewa sediments, having an estimated thickness of 1000 m. Most of the exposures of the Karewas in the valley were fully mapped and their mutual correlation studied. A reasonable chronological framework is now available, showing a satisfactory convergence between different climatic parameters within the limitations of the dating resolution (Agarwal et al. 1989). On the whole, the climatic pattern of Kashmir and western part of the Himalaya follows a global trend: the warming up to the Pliocene, the glacial and interglacial oscillations of the Pleistocene etc. But Pleistocene cooling was not abrupt but very gradual (Agarwal 1985).

Lake deposits in the Himalaya are potential source of palaeoclimate information and provide a continuous paleoclimate record since the Last Glacial Maximum (LGM). Juyal et al. (2009) studied the Goting Lake sediments in the Higher Central Himalaya to reconstruct the summer monsoon variability during the Last Glacial to early Holocene and presented Magnetic susceptibility and geochemical data which indicate a moderate to strong monsoon around 25 ka, 23.5–22.5 ka, 17 ka, 16.5 ka and after 14.5–13 ka, whereas around 22 ka the early LGM climate was mainly arid. First evidence of cooling during the younger Dryas was provided by mineral magnetic susceptibility data and elemental concentrations that reveal a high around 13 ± 2 ka to 11 ± 1 ka. The biochemical data of the Mansar Lake sediments, Lesser Himalaya indicated a hot and wet climate regime during the early Holocene and a dry and cold one during the late Holocene period (Das et al. 2010).

Several wet and dry periods were noticed in the lesser Himalaya since the beginning of the Holocene (Phadtare 2000; Kotlia et al. 2010). Monsoon intensity was weakened during the cold/dry events, i.e. Last Glacial Maximum (LGM), Older Dryas (OD), Younger Dryas (YD), 8.2 ka BP, 4.2 ka BP and Little Ice Age (LIA) (Sinha et al. 2005; Kotlia et al. 2010, Gupta et al. 2013). Few speleothem based $\delta^{18}O$ records also highlighted the role of the WDs along with the Indian summer monsoon variability over the northwestern Himalaya during the late Holocene (Kotlia et al. 2012, 2015; Duan et al. 2013; Sanwal et al. 2013; Sinha et al. 2015; Liang et al. 2015). Joshi et al. (2017) noted major drought event at ~ 3.4 ka BP from their speleothem $\delta^{18}O$ records from the Central Lesser Himalaya which could be correlated with the collapse of the Indus valley civilization in the NW India. Leipe et al. (2017) also presented mean annual precipitation (MAP) history during the last 12,000 years from the fossil pollen record analysed from Tso Moriri lake, northwest Himalaya and suggested that precipitation levels varied significantly during the Holocene. Moisture availability was higher at the end of the Younger Dryas, with hydrological optimum conditions in the study region occurring between ca. 11 and 9.6 cal ka BP and pronounced dry spells at ca. 4 and 3.2 cal ka BP, which possibly further causing the deurbanisation that occurred from ca. 3.9 cal ka BP and eventual collapse of the Harappan Civilisation between ca. 3.5 and 3 cal ka BP. Tso Kar lake sediments of northwest Himalaya dated from 15.2 ka to 14 ka reflect dry and cold conditions and the pollen studies indicate strengthening of the summer monsoon after 14 ka (Demske et al. 2009).

Though, the Little Ice Age (LIA) is believed to be a wide spread cooling episode in both the hemisphere during the last 1000 years, its intensity and duration was uneven in different parts of the globe (Grove 1988; Bradley 1992; Bradley and Jones 1993). In Asia, Himalayan foothills, the central Asia Indo-Pacific warm pool were wetter (Kotlia et al. 2015; Chen et al. 2006, 2010; Oppo et al. 2009), while monsoon regions were drier (Gupta et al. 2003; Zhang et al. 2008).

Continental wise temperature reconstruction during the past two millennia based on multi-proxy records indicates long-term cooling trend, which ended late in the nineteenth century (PAGES 2k Consortium 2013). They do not observe globally synchronous multi-decadal warm or cold intervals that define a worldwide Medieval Warm Period or Little Ice Age, but all reconstructions show generally cold conditions between 1580 and 1880 CE, punctuated in some regions by warm decades during the eighteenth century.

12.2.6 Dendroclimatic Studies

Climatic information over the Western Himalaya is available for about a century based on very sparse network of meteorological stations that too are located at low to middle elevations ranging from 500 to 2400 m above mean sea level (amsl). Within short distances contrasting features of climate are observed due to large variation in topography and vegetation cover. Information on glacier activities over high altitude Himalayan region is very limited and for a short period. Therefore, knowledge of climate change particularly at high altitude regions is very poor. In the context of long-term climate change, recent global warming and subsequent glacier activities, past information on climate and glacier activities is important. In view of this, tree-ring records from high elevation sites can provide valuable tool to understand the climate vis-à-vis glacier behavior under the present climatic scenario.

Tree-ring studies over the western Himalaya indicated high dendroclimatic potential of conifers (*Pinus, Abies, Picea, Cedrus*) to reconstruct summer and winter temperature and rainfall for a millennium period. Tree-ring based reconstructions of spring and summer climate (temperature and precipitation) over different parts of the Western Himalaya including Kashmir did not show any increasing or decreasing trend since last three to four centuries (Bhattacharyya et al. 1988; Borgaonkar et al. 1994, 1996, 2002; Hughes 1992; Yadav et al. 1999). It also revealed that Little Ice Age (LIA) effect was not prominent over this part of the Himalaya; however, some intermittent cool epochs were observed during LIA. Tree-ring based summer (May–September) precipitation reconstruction of Srinagar has been presented in Fig. 12.10 (Borgaonkar et al. 1994). The reconstructed rainfall series did not show any noticeable difference from present precipitation conditions; however, few wet and dry epochs are prominent. Bamzai (1962) and Koul (1978) reported famine condition (poor crop yield) in the valley due to heavy rainfall in 1814 and 1832 C.E. Same periods were observed as short wet epochs in the reconstructed rainfall series. Recent study (Yadav et al. 2017) indicated increasing pre-

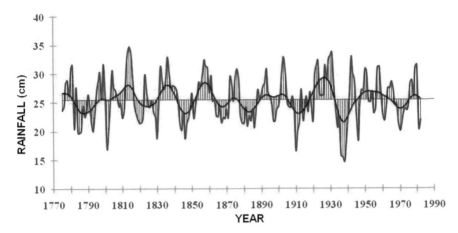

Fig. 12.10 Tree-ring reconstruction of summer (May–September) precipitation of Srinagar, J&K (Borgaonkar et al. 1994)

cipitation pattern and glacier expansion associated with cooling over northwest Himalaya. Similarly, reconstructed pre-monsoon (March–April–May) temperature and rainfall anomalies since 1747 C.E. over the western Himalaya using wide tree-ring chronology network did not show any increasing or decreasing trend since past three centuries (Fig. 12.11) (Borgaonkar et al. 2002). However; few tree-ring based regional temperature reconstructions indicated cooling trend in summer temperature (Hughes 2001; Cook et al. 2003; Yadav et al. 2004) and warming trend in annual and winter temperature during recent decades since last 4 centuries (Esper et al. 2002; Cook et al. 2003). Yadav et al. (2015) also demonstrated that standardised precipitation index developed from tree rings could be served as an important base line data to quantify the impact of droughts on forest as well as rabbi crop productivity in hilly terrains of the Kumaun Himalaya in long-term perspective.

High altitude tree-ring chronologies from western Himalaya provide signals of winter temperature variations and glacier fluctuations (Borgaonkar et al. 2009, 2011). Master tree-ring chronology prepared using the tree core samples collected from high elevation sites of Gangotri of Uttarcanchal and Kinnaur region of Himachal Pradesh of Western Himalaya goes back to fifteenth century (1452–2004 C.E.; 553 years; Fig. 12.12). Dendroclimatological investigation indicated significant positive relationship of tree-ring index series with winter (December–January–February) temperature and summer precipitation and inverse relationship with summer temperature (Borgaonkar et al. 2009, 2011). The chronology also indicated few decadal and longer epochs of Little Ice Age (LIA) cooling during 1453–1590 C.E. and 1780–1930 C.E. (Fig. 12.12). Many of these events have been observed to be well related to the other proxy records of glacial fluctuations of the region (Duan and Yao 2003; Mayewaski et al. 1980). Higher growth in recent few decades detected in the tree-ring chronology has been noticed coinciding with the warming trend and rapid retreat of the Himalayan glaciers. Suppressed and released growth

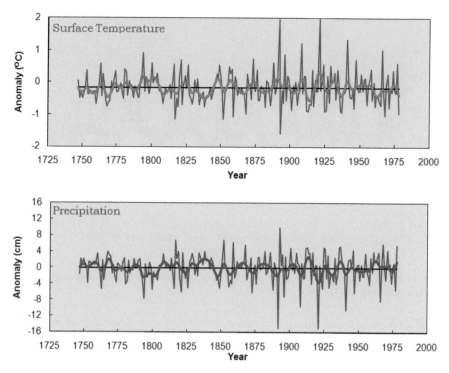

Fig. 12.11 Reconstructed pre-monsoon (March–April–May) temperature and rainfall anomalies since 1747 C.E. over the western Himalaya using tree-ring chronology network. Smooth lines indicate low-frequency variations (Borgaonkar et al. 2002)

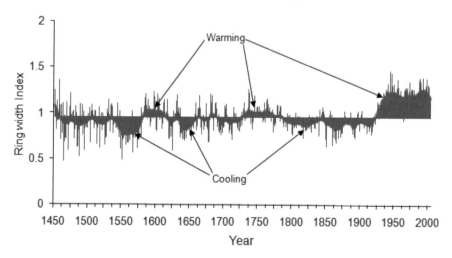

Fig. 12.12 553 years (1452–2004 C.E.) long tree-ring index chronology of high altitude Himalayan conifer from Western Himalaya. Smooth red line is 30 years cubic spline filter. Suppressed (cooling) and released (warming) growth patterns in tree-ring chronology have also been observed to be well related to the past glacial fluctuation records of the region. (Borgaonkar et al. 2011)

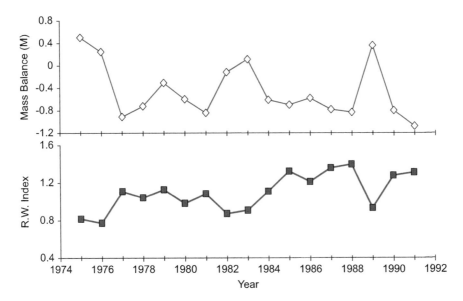

Fig. 12.13 Variations in average snow mass balance records of Western Himalaya and western Himalayan high altitude tree-ring chronology (R = −0.81) (Borgaonkar et al. 2009)

patterns in tree-rings have also been observed to be well related to the past glacial fluctuation records of the region. The higher tree growth in recent decades may be partially attributed to the warming trend over the region, particularly the increasing the winter warmth and thus to the regional manifestations of global warming.

The relationship between climate change and the Himalayan cryosphere is not understood sufficiently well. There have been few or no detailed investigations of snow and ice processes and their relevance to climate in the high mountain ranges. The interactions among different climate elements – processes related to freeze-thaw of glaciers, snowfall, wind systems, and seasonal or spatial balance between snow and rainfall is very complex. These changes will have significant impact on the trees growing in the vicinity. Borgaonkar et al. (2009) studied the relationship between annual variation in snow mass balance and tree growth (Fig. 12.13). They found highly significant inverse relationship (CC = −0.81; P < 0.001) between these two parameters during the period 1975–91 C.E. (17 years). The suppressed tree growth in 1976, 1982–83, 1989 C.E. are associated with positive mass balance and higher growth in 1977, 1984–85, 1987–88 C.E. as a negative mass balance. The mass balance data and winter (DJF) temperature has also significant negative relationship. Increasing temperature and decreasing precipitation particularly in winter and pre-monsoon months may result in less amount of snow accumulation during winter and more amounts of snow and ice removed by melting in the summer.

Dndroclimatic studies over the central Himalaya (Cook et al. 2003), eastern Himalaya including Sikkim, Arunachal Pradesh and Bhutan (Bhattacharyya and Chaudhary 2003; Yadava et al. 2015; Krusic et al. 2015; Borgaonkar et al. 2018),

Tibetan Plateau (Li et al. 2015; Wang et al. 2010; Bräuning and Mantwill 2004; Fan et al. 2009); east Asia (Cook et al. 2013) have indicated a warming trend in recent decades. Some of these reconstructions have provided strong signatures of few major volcanic eruptions (e.g. Tambora eruption in 1815 C.E.; Krakatau eruption in 1883). A noticeable cooling impact due to the Tambora eruption was seen in many temperature reconstructions. Post Tambora eruption cooling was sharp in Sikkim and Nepal as compared to that in Bhutan and East Asia. A temperature reversal started after 1817 C.E. at these two places. The temperature variability in the region might have been related to ENSO, Pacific Decadal Oscillation (PDO) and volcanic eruptions.

As discussed above, most of the dendroclimatic studies of Himalayan regions are mainly concentrated on long-term temperature changes. However, few tree-ring based hydroclimatic reconstructions (Treydte et al. 2006; Yadav et al. 2017) indicated wetting in recent decades over some regions of cold semi-arid to arid north-west (NW) Himalaya including Karakoram. Treydte et al. (2006) obtained millennial-scale precipitation reconstruction from oxygen isotope records of tree-rings from Karakoram mountain of northern Pakistan and suggested unprecedented wet condition during the twentieth century which could be the impact of industrialization and global warming. Similarly, based on the tree-ring records from semi-arid region of Kishtwar, Kashmir, India, Yadav et al. (2017) suggested wettest interval of recent decades (1984–2014 CE) in the past 576 years with prolonged drought condition during the fifteenth to early seventeenth centuries They attributed this as an impact of glacier expansion in the Kashmir and Karakoram region during the past few decades unlike central and eastern Himalaya where general receding trends have been observed.

12.3 Discussion and Conclusions

Climatological information based on century-long data of rainfall and temperature of 14 stations well spread over the western Himalaya including Kashmir, Himachal Pradesh and Uttarakand regions has been evaluated. Many stations' data are available up to recent and cover more than 100 years period. The stations are well spread over the entire western Himalayan region. The analysis indicates different patterns of rainfall variability within a short distance. Trend analyses do not show any coherent patterns among the stations and do not indicate any significant trend during the twentieth century except significant decreasing trend in monsoon and annual rainfall of Shimla and Mussoorie and significant increasing trend in Pithoragarh monsoon series. In a broader sense, long-term decreasing tendency of precipitation particularly in monsoon season is noticeable in Himachal and Uttarakand where as slight increasing trend in Kashmir and further northwest region.

Temperature analyses indicate increasing trends in maximum temperature for all the stations and in all the seasons. Most of these trends have significant values. Cooling trend is observed mostly in minimum temperature of some stations for different seasons. Annual maximum, minimum and mean temperature series of all the stations indicate significant warming except slight cooling in minimum temperature of Deheradun, Mukteswar and Mussoorie. The similar increasing trend in mean temperature was noted by Pant et al. (2003) over western Himalaya based on limited number of stations. The diurnal asymmetry of the temperature trends is quite similar to that observed over the large region of northwest India by Rupa Kumar et al. (1994).

Future projection scenarios by the end of twenty-first century based on the CMIP5 experiments indicated slight decrease in winter and spring precipitation in western Himalaya and significant increase in summer precipitation over the central Himalayan region including Nepal Himalaya and Tibetan Plateau. However, warming is projected over the entire Himalayan region with significant warming in Tibetan plateau.

Thus, the trend analyses and temperature curves indicate that the air temperature has undergone some long-term changes during the twentieth century over the Himalaya. The broad tendency is towards an increase, with some local differences. The rapid urbanization of the hill stations were the observatories are located may be one of the factors responsible for the observed increase. However, to gain a more comprehensive and conclusive evidences of the climate change, a better network of observatories in remote areas is required so that we can say with certainty whether the Himalayan climate has undergone perceptible change. The results can of course be used as a pointer towards possible changes if the natural ecosystems over the Himalaya are disturbed.

Dendroclimatic reconstructions from various Himalayan regions provide some clues and indications of long-term climate changes since last several centuries. Most of them give information on summer temperature conditions of the region. Few epochs of medieval warming, LIA cooling are the common pattern observed in these reconstructions. Significant warming trend since last few decades is also observed in most of the reconstructions. However, cooling trends are also observed in couple of pockets of western Himalaya. Few precipitation reconstructions indicate wetter conditions in recent years since last millennium particularly over the northwest Himalaya including Kashmir and Karakoram ranges. It is also seen that high altitude near glaciers tree-ring records would be the potential source of information on long-term temperature variability and glacier fluctuations. For better understanding of ecosystems and growing concern about the environmental impacts of climate change, it is necessary to have adequate knowledge of long-term climatic conditions prevailing over the region. Information on long-term climate variability based on tree-ring proxy records is important to understand the nature of different climate systems over the regions, particularly, when the observational data network is sparse.

Mountain systems are more sensitive to climate change. They are the most fragile ecosystem amongst all other ecosystem in the world (Diaz et al. 2003). Mountains are experiencing continued warming of more than the global average since late nineteenth century (Theurillat and Guisan 2001; Beniston 2003). Several studies (Shrestha et al. 1999; Liu and Chen 2000; Cook 2003; Pant et al. 2003; Bhutiyani et al. 2007) indicated warming trend in recent decades particularly over the western and central Himalaya.

Thus, the western Himalayan region has a more general climatological significance being the greatest mountain barrier on the earth where polar, tropical and Mediterranean influences interact. Climate variability of the region affects the large population directly or indirectly. The information on long-term climate change and variability based on instrumental records and tree-ring proxies would be useful to improve the understanding of climate variability in a much larger area and help to make robust policies of forest and water management.

Acknowledgement The authors are grateful to Director, Indian Institute of Tropical Meteorology, Pune for kindly providing the facilities to prepare this article. Meteorological data used in the analysis are kindly provided by the India Meteorological Department, Pune. The part of the work was supported by PACMEDY Project (No. MoES/16/06/2016-RDEAS) under the Belmont Forum, Paris, France.

References

Agarwal DP (1985) Cenozoic climate changes in Kashmir: the multidisciplinary data. In: Agarwal DP, Kusumagar S, Krishnamurthy RV (eds) Climate and geology of Kashmir, Current trends in Geology. VI. Today & Tomorrow's Printers and Publishers, New Delhi, pp 1–12

Agarwal DP, Dodia R, Kotlia BS, Razdan H, Sahni A (1989) The plioplestocene geologic and climatic record of the Kashmir valley, India: a review and new data. Palaeogeogr Palaeoclimatol Palaeocol 73:267–286

Bamzai PNK (1962) A history of Kashmir. Metropolitan Book Co., New Delhi

Basistha A, Arya DS, Goel NK (2009) Analysis of historical changes in rainfall in the Indian Himalayas. Int J Climatol 29:555–572

Beniston M (2003) Climatic change in mountain regions: a review of possible impacts. Clim Chang 59:5–31

Beniston M, Rebetez M (1996) Regional behavior of minimum temperatures in Switzerland for the period 1979–1993. Theor Appl Climatol 53:231–243

Bhattacharyya A (1989) Vegetation and climate during the last 30,000 years in Ladakh. Palaeogeogr Palaeoclimatol Palaeoecol 73:25–38

Bhattacharyya A, Chaudhary V (2003) Late-summer temperature reconstruction of the eastern Himalayan region based on tree-ring data of Abies densa. Arct Antarct Alp 35:1196–2002

Bhattacharyya A, Yadav RR (1989) Dendroglimatic research in India. Proc Indian Natl Sci Acad 55A:696–701

Bhattacharyya A, Lamarche VC Jr, Telewski FW (1988) Dendrochronological reconnaissance of the conifers of northwest India. Tree-Ring Bull 48:21–30

Bhutiyani MR, Kale VS, Pawar NJ (2007) Long-term trends in maximum, minimum and mean annual air temperatures across the northwestern Himalaya during the twentieth century. Clim Chang 85:159–177

Bhutiyani MR, Kale VS, Pawar NJ (2010) Climate change and the precipitation variations in the northwestern Himalaya: 1866–2006. Int J Climatol 30:535–548

Böhm R, Auer I, Brunetti M, Maugeri M, Nanni T, Schöner W (2001) Regional temperature variability in the European Alps: 1760–1998 from homogenized instrumental time series. Int J Climatol 21:1779–1801

Borgaonkar HP (1996) Tree growth – climate relationship and long-term climate change over western Himalaya: a dendroclimatic approach. PhD thesis, University of Pune, Pune

Borgaonkar HP, Pant GB, Rupa Kumar K (1994) Dendroclimatic reconstruction of summer precipitation at Srinagar, Kashmir, India since the late 18th century. The Holocene 4(3):299–306

Borgaonkar HP, Pant GB, Rupa Kumar K (1996) Ring-width variations in Cedrusdeodara and its climatic response over the western Himalaya. Int J Climatol 16:1409–1422

Borgaonkar HP, Rupa Kumar K, Sikder AB, Ram S, Pant GB (2002) Tree ring variations over the western Himalaya: little evidence of the Little Ice Age. PAGES News Lett 10:1

Borgaonkar HP, Somaru Ram, Sikder AB (2009) Assessment of tree-ring analysis of high elevation Cedrusdeodara D. Don from western Himalaya (India) in relation to climate and glacier fluctuations. Dendrochronologia 27(1):59–69

Borgaonkar HP, Sikder AB, Ram S (2011) High altitude forest sensitivity to the recent warming: a tree-ring analysis of conifers from western Himalaya, India. Quat Int 236:158–166

Borgaonkar HP, Gandhi N, Ram S, Krishnan R (2018) Tree-ring reconstruction of late summer temperatures in northern Sikkim (eastern Himalayas). Palaeogeogr Palaeoclimatol Palaeoecol 504:125–135. https://doi.org/10.1016/j.palaeo.2018.05.018

Bradley RS (1992) When was the "Little Ice Age"? In: Mikami T (ed) Proceedings of the international symposium on the Little Ice Age climate. Tokyo Metropolitan University, Tokyo, pp 1–4

Bradley RS, Jones PD (1993) Little Ice Age's summer temperature variations: their nature and relevance to recent global warming trends. The Holocene 3:367–376

Bräuning A, Mantwill B (2004) Summer temperature and summer monsoon history on the Tibetan plateau during the last 400 years recorded by tree rings. Geophys Res Lett 31:L24205. https://doi.org/10.1029/2004GL020793

Chen F, Xiaozhong H, Jiawu Z, Holmes JA, Jianhui C (2006) Humid Little Ice Age in arid central Asia documented by Bosten Lake, Xinjiang, China. Sci China 49(12):1280–1290

Chen F, Chen JH, Homes J, Boomer I, Austin P, Gates JB (2010) Moisture changes over the last millennium in arid central Asia: a review, synthesis and comparison with monsoon region. Quat Sci Rev 29:1055–1068

Cook ER, Krusic PJ, Jones PD (2003) Dendroclimatic signals in long tree-ring chronologies from the Himalayas of Nepal. Int J Climatol 23:707–732

Cook ER, Krusic PJ, Anchukaitis KJ, Buckley BM, Nakatsuka T, Sano M, PAGES Asia 2k members (2013) Tree-ring reconstructed summer temperature anomalies for temperate East Asia since 800 CE. Clim Dyn 41:2957–2972. https://doi.org/10.1007/s00382-012-1611-x

Das BK, Gaye B, Malik MA (2010) Biogeochemistry and paleoclimate variability during the Holocene: a record from Mansar Lake, Lesser Himalaya. Environ Earth Sci 61:565–574

De Terra H, Paterson TT (1939) Studies on the ice age in India and associated human cultures, vol 493. Carnegic Institute of Washington, Washington, DC, p 354

Demske D, Tarasov PE, Wünnemann B, Riedel F (2009) Late glacial and Holocene vegetation, Indian monsoon and westerly circulation in the Trans-Himalaya recorded in the lacustrine pollen sequence from Tso Kar, Ladakh, NW India. Palaeogeogr Palaeoclimatol Palaeoecol 279:172–185

Dhar ON, Bhattacharya BK (1976) Variation of rainfall with elevation in the Himalayas – a pilot study. Indian J Powr River Valley Dev 26:191–195

Dhar ON, Farooqui SMT (1973) A study of rainfall recorded at the Cherrapunji observatory. Hydrol Sci Bull 18:441–450

Dhar ON, Mulye SS (1987) Brief appraisal of precipitation climatology of the Ladakh region. In: Pangtey YPS, Joshi SC (eds) Western Himalaya, Vol. I – Environment. Gyanodaya Prakashan, Nainital, pp 87–98

Dhar ON, Narayanan J (1965) A study of precipitation distribution in the neighbourhood of Mount Everest. Indian J Meteorol Geophys 16:230–240

Dhar ON, Kulkarni AK, Sangam RB (1975) A study of extreme point rainfall over flash flood prone regions of the Himalayan foothills of north India. Hydrol Sci Bull 20:61–67

Dhar ON, Kulkarni AK, Sangam RB (1984a) Some aspects of winter and monsoon rainfall distribution over the Garhwal-Kumaon Himalayas – a brief appraisal. Himal Res Dev 2:10–19

Dhar ON, Soman MK, Mulye SS (1984b) Rainfall over the southern slopes of the Himalayas and the adjoining plains during breaks in monsoon. J Climatol 4:671–676

Dhar ON, Mandal BN, Kulkarni AK (1987) Some facts about precipitation distribution over the Himalayan region of Uttar Pradesh. In: Pangtey YPS, Joshi SC (eds) Western Himalaya, Environment, vol I. Gyanodaya Prakashan, Nainital, pp 72–86

Diaz HF, Bradley RS (1997) Temperature variations during the last century at high elevation sites. Clim Chang 36:253–279

Diaz HF, Grosjean M, Graumlich L (2003) Climate variability and change in high elevation regions: past, present and future. Clim Chang 59(1):1–4

Duan K, Yao T (2003) Monsoon variability in the Himalayas under the condition of global warming. J Meteorol Soc Jpn 81(2):251–257

Duan W, Kotlia BS, Tan M (2013) Mineral composition and structure of the stalagmite laminae from Chulerasim cave, Indian Himalaya and the significance for palaeoclimatic reconstruction. Quat Int 298:93–97

Esper J, Schweingruber FH, Winiger M (2002) 1300 years of climatic history for western Central Asia inferred from tree-rings. The Holocene 12:267–277

Fan ZX, Bräuning A, Yang B, Cao KF (2009) Tree ring density-based summer temperature reconstruction for the central Hengduan Mountains in southern China. Glob Planet Chang 65:1–11

Grove JM (1988) The Little Ice Age. Methuen, London

Gupta AK, Anderson DM, Overpeck JT (2003) Abrupt changes in the Asian southwest monsoon during the Holocene and their links to the North Atlantic Ocean. Nature 421:354–357

Gupta AK, Mohan K, Das M, Singh RK (2013) Solar forcing of the Indian summer monsoon variability during the Ållerød period. Sci Report 3:2753. https://doi.org/10.1038/srep02753

Harris I, Jones PD, Osbornaand TJ, Listera DH (2014) Updated high-resolution grids of monthly climatic observations – the CRU TS3.10 Dataset. Int J Climatol 34:623–642

Hughes MK (1992) Dendroclimatic evidence from the western Himalaya. In: Bradley RS, Jones Routledge PD (eds) Climate since A.D. 1500. Routledge, London

Hughes MK (2001) An improved reconstruction of summer temperature at Srinagar, Kashmir since 1660 AD based on tree-ring width and maximum latewood density of Abiespindrow [Royle] Spach. Palaeobotanist 50:13–19

Jones PD, Moberg A (2003) Hemispheric and large-scale surface air temperature variations: an extensive revision and an update to 2001. J Clim 16:206–223

Joshi LM, Kotlia BS, Ahmad SM, Sanwal J, Raza W, Singh AK, Shen C, Long T, Sharma AK (2017) Reconstruction of Indian monsoon precipitation variability between 4.0 and 1.6 ka BP using speleothem δ18O records from the Central Lesser Himalaya, India. Arab J Geosci 10:356

Juyal N, Pant RK, Basaviah N et al (2009) Reconstruction of last Glacial to early Holocene monsoon variability from relict lake sediments of the higher central Himalaya, Uttarakhand, India. J Asian Earth Sci 34:437–449

Kotlia BS, Sanwal J, Phartiyal B, Joshi LM, Trivedi A, Sharma C (2010) Late Quaternary climatic changes in the eastern Kumaun Himalaya, India, as deduced from multi-proxy studies. Quat Int 213:44–55

Kotlia BS, Ahmad SM, Zhao J-X, Raza W, Collerson KD, Joshi LM, Sanwal J (2012) Climatic fluctuations during the LIA and post-LIA in the Kumaun Lesser Himalaya, India: evidence from a 400y old stalagmite record. Quat Int 263:129–138

Kotlia BS, Singh AK, Joshi LM, Dhaila BS (2015) Precipitation variability in the Indian Central Himalaya during last ca. 4000 years inferred from a speleothem record: impact of Indian Summer Monsoon (ISM) and Westerlies. Quat Int 371:244–253

Koul A (1978) Geography of Jammu and Kashmir state (Revised by Bamzai PNK). Light and Life Publishers, New Delhi

Krusic PJ, Cook ER, Dukpa D, Putnam AE, Rupper S, Schaefer J (2015) Six hundred thirty-eight years of summer temperature variability over the Bhutanese Himalaya. Geophys Res Lett 42:2988–2994. https://doi.org/10.1002/2015GL063566

Leipe C, Demske D, Tarasov P, HIMPAC Project Members (2017) A Holocene pollen record from the northwestern Himalayan lake Tso Moriri: implications for palaeoclimatic and archaeological research. Quat Int 348:93–112

Li M-Y, Wang L, Ze-Xin F, Chen-Chen S (2015) Tree-ring density inferred late summer temperature variability over the past three centuries in the Gaoligong Mountains, southeastern Tibetan Plateau. Palaeogeogr Palaeoclimatol Palaeoecol 422:57–64

Liang F, Brook GA, Kotlia BS, Railsback LB, Hardt B, Cheng H, Edwards RL, Kandasamy S (2015) Panigarh cave stalagmite evidence of climate change in the Indian Central Himalaya since AD 1256: monsoon breaks and winter southern jet depressions. Quat Sci Rev 124:145–161

Liu X, Chen B (2000) Climatic warming in the Tibetan Plateau during recent decades. Int J Climatol 20(14):1729–1742

Mayewaski PA, Pregent GP, Jeschke PA, Ahmad N (1980) Himalayan and Trans-Himalayan glacier fluctuations and the South Asian monsoon record. Arct Alp Res 12:171–182

Morris TO (1938) The Bain Boulder bed, a glacial episode in the Siwalik series of the Marwet Kundi Range and Sheik Budin, North-west Frontier province, India. Q J Geol Soc Lond 94:385–421

Oppo DW, Rosentha Y, Linsley BK (2009) 2,000-year-long temperature and hydrology reconstructions from the Indo-Pacific warm pool. Nature 460:1113–1116

PAGES 2k Consortium (2013) Continental-scale temperature variability during the past two millennia. Nat Geosci 6:339–346

Pant GB (1979) Role of tree-ring analysis and related studies in palaeoclimatology: preliminary survey and scope for Indian region. Mausam 30:439

Pant GB (1983) Climatological signals from the annual growth rings of selected tree species of India. Mausam 34:251

Pant GB, Borgaonkar HP (1984) Climate of the hill regions of Uttar Pradesh. Himal Res Dev 3(I):13–20

Pant GB, Borgaonkar HP, Rupa Kumar K (2003) Climate variability over the western Himalaya since the little ice age: dendroclimatic implications. Jalvigyan Sameeksha (Hydrol Rev) 18(1–2):111–121

Phadtare NR (2000) Sharp decrease in summer monsoon strength 4000e3500 cal yr BP in the Central Higher Himalaya of India based on pollen evidence from alpine peat. Quat Res 53:122–129

Pilgrim GE (1932) The fossil Cornivora of India. Palaeol India Calcutta 18:232

Pilgrim GE (1944) The lower limit of the Pleistocene in Europe and Asia. Ged Mag Lond 81:28–38

Ramaswamy C (1972) Rainfall over Cherrapunji and Mawsynram. Vayu Mandal 2:119–124

Rupa Kumar K, Krishna Kumar K, Pant GB (1994) Diurnal asymmetry of surface temperature trends over India. Geophys Res Lett 21:677–680

Sanwal J, Kotlia BS, Rajendran C, Ahmad SM, Rajendran K, Sandiford M (2013) Climatic variability in central Indian Himalaya during the last 1800 years: evidence from a high resolution speleothem record. Quat Int 304:183–192

Shrestha AB, Wake CP, Mayewski PA, Dibb JE (1999) Maximum temperature trends in the Himalaya and its vicinity: an analysis based on temperature records from Nepal for the period 1971–94. J Clim 12(9):2775–2786

Sinha A, Cannariato KG, Stott LD, Li HC, You CF, Cheng H, Edwards RL, Singh IB (2005) Variability of Southwest Indian summer monsoon precipitation during the Bølling-Ållerød. Geology 33:813–816

Sinha A, Kathayat G, Cheng H, Breitenbach SFM, Berkelhammer M, Mudelsee M, Biswas J, Edwards RL (2015) Trends and oscillations in the Indian summer monsoon rainfall over the last two millennia. Nat Commun 6:6309. https://doi.org/10.1038/ncomms7309

Theurillat JP, Guisan A (2001) Potential impact of climate change on vegetation in the European Alps: a review. Clim Chang 50:77–109

Treydte KS, Schleser GH, Helle G, Frank DC, Winiger M, Haug GH, Esper J (2006) The twentieth century was the wettest period in northern Pakistan over the past millennium. Nature 440:1179–1182. https://doi.org/10.1038/nature04743

Wang LL, Duan JP, Chen J, Huang L, Shao XM (2010) Temperature reconstruction from tree-ring maximum density of Balfour spruce in eastern Tibet, China. Int J Climatol 30:972–979

Wu J, Xu Y, Gao X-J (2017) Projected changes in mean and extreme climates over Hindu Kush Himalayan region by 21 CMIP5 models. Adv Clim Chang Res 8(3):176–184

Yadav RR, Park WK, Bhattacharyya A (1999) Spring temperature fluctuations in the western Himalayan region as reconstructed from tree-rings; AD 1390–1987. The Holocene 9:85–90

Yadav RR, Park WK, Singh J, Dubey B (2004) Do the western Himalayas defy global warming? Geophys Res Lett 31:L17201

Yadav RR, Misra KG, Yadava AK, Kotlia B, Misra S (2015) Tree-ring footprints of drought variability in last ~300 years over Kumaun Himalaya, India and its relationship with crop productivity. Quat Sci Rev 117:113–123

Yadav RR, Gupta AK, Kotlia BS, Singh V, Misra KG, Yadava AK, Singh AK (2017) Recent wetting and glacier expansion in the northwest Himalaya and Karakoram. Sci Rep 7:6139. https://doi.org/10.1038/s41598-017-06388-5

Yadava AK, Yadav RR, Misra KG, Singh J, Singh D (2015) Tree ring evidence of late summer warming in Sikkim, northeast India. Quat Int 371:175–180. https://doi.org/10.1016/j.quaint.2014.12.067

Yatagai A et al (2012) PHRODITE: constructing a long-term daily gridded precipitation dataset for Asia based on a dense network of rain gauges. Bull Am Meteorol Soc 93:1401–1415

Zeuner FE (1972) Dating the past. Hafner Publishing Co, New York

Zhang P, Cheng H, Edwards RL et al (2008) A test of climate, sun and culture relationships from a 1810-year Chinese cave record. Science 322(5903):940–942

Chapter 13
Quaternary Glaciation of the Himalaya and Adjacent Mountains

Lewis A. Owen

Abstract The Himalaya and adjacent mountains are the most glaciated regions outside the polar realms. Abundant field studies, aided by remote sensing, and newly developing geochronological methods are aiding in reconstructing the timing and extent of Quaternary glaciation throughout the region. Abundant well-preserved glacial geologic evidence throughout the region shows that at least nine major regionally synchronous glacier advances occurred in the region over the past ~400 ka. The maximum extent of glaciation has been broadly defined and is characterized by extensive valley glaciers and expanded ice caps. The timing of maximum glacier extent is asynchronous throughout the region, with some areas experiencing maximum glaciation prior to the last glacial cycle (>100 ka), while in other regions it was during the early part of the last glacial (~30–70 ka), and in some regions, was possibly coincident with the global last glacial maximum (LGM) at ~18–24 ka. Glacier advances have been limited to a few kilometers beyond their present position in most regions since the LGM. The maximum glacier advance occurring during the Holocene at ~9–8 ka. The higher resolution of the Holocene glacial geologic record allows ~5 regional glacier advances to be resolved. All advances were somewhat restricted in extent. Glaciers have retreated in most areas during the last 100 years. However, in the Karakoram there has been little retreat and many glaciers have surged. Himalayan glaciation is forced by multiple drivers, and the timing and extent of glaciation is governed by a complex combination of different factors specific to each locality, including climate and microclimate regimes, topographic controls, and geomorphic and tectonic settings. Study of the Quaternary glacial history suggests that nature of future glacier fluctuations in response to human-induced climate change will be very complex and variable throughout the Himalaya and adjacent mountas.

L. A. Owen (✉)
Department of Marine, Earth, and Atmospheric Sciences, North Carolina State University, Raleigh, NC, USA
e-mail: lewis.owen@ncsu.edu

© Springer Nature Switzerland AG 2020
A. P. Dimri et al. (eds.), *Himalayan Weather and Climate and their Impact on the Environment*, https://doi.org/10.1007/978-3-030-29684-1_13

13.1   Introduction

The mountain of the Himalaya and the adjacent ranges in the Karakoram, Hindu Kush, Pamir and Tibet constitute the most glaciated region outside of the polar realms. Ranges such as Karakoram contain some of the world's longest extra-polar valley glaciers including Siachen (76 km long), Biafo (67 km long) and Baltoro (63 km long) glaciers (Fig. 13.1).

Glacier extent has varied greatly over throughout the Quaternary (last 2.58 Ma) with glaciers fluctuating in response to natural climate change on Milankovitch (10^{4-5} years) and sub-Milankovitch (10^{1-3} years) timescales (Owen and Dortch 2014). Over the past few decades, much evidence has emerged to show that glaciers throughout these regions are responding significantly to human-induced climate change; mostly the glaciers are retreating (National Academy 2012; Maurer et al. 2019). The glacial system in this region has a profound influence on the hydrology of the great rivers that drain into the forelands of the Himalaya and Tibet, and on the regional climate and biota. Understanding the nature of past, present and likely future glaciation, therefore, has important environmental, socio-economic and political implications (Sen and Kansal 2019). However, the complex and varied climatic, topographic and geomorphic settings throughout the Himalaya and its adjacent mountains makes the study of Himalayan glaciation challenging. Yet, over the past few decades a plethora of studies on the Quaternary glaciation is shedding much light on the nature and dynamics of the glacial systems, which in turn has important implications for understanding the varied and intricate nature of future

Fig. 13.1 View down Baltoro glacier from Concordia at 4600 m asl, some 36 km from its snout, in the Karakoram of Northern Pakistan

glaciation and climate in Central Asia. This chapter aims to describe the nature of Quaternary glaciation focusing on describing how glaciers have fluctuated throughout the Himalaya and adjacent mountain ranges.

13.2 Himalayan Glacial Systems

The glaciers of the Himalaya and adjacent regions reside in some of the world's greatest mountain ranges, including the Karakoram and Khumbu Himal, that are among the most impressive and contain all the world's 8000-m-high peaks. The Nyaingentanglha Shan, Tanggula Shan, Bayan Har Shan, Kunlun Shan, Altun Shan and Qilian Shan are significant mountain ranges that traverse the Tibetan Plateau and are also extensively glaciated. The mountain ranges of the Himalaya and adjacent regions trend approximately east-west trending stretching some 2000 km with the Hindu Kush at the western end and Namcha Barwa at the eastern end, and together form a broad belt ~1500 km in a north–south direction. The average elevation of the region is ~5000 m above sea level (asl; Fielding et al. 1994). Vast tracks of the interior of the Tibet, however, that were likely never glaciated. The region is still tectonically active as a consequence of the continued northward motion of the Indian continental lithospheric plate into the Eurasian continental lithospheric plate. As a consequence, earthquakes, some with magnitudes >7, are common and can trigger snow and ice avalanches onto glaciers which in turn contribute to the glacier mass balances (Van der Woerd et al. 2004; Kargel et al. 2016).

The glaciers throughout the Himalaya and adjacent mountains are influenced of the Asian summer monsoon and the mid-latitude westerlies (Benn and Owen 1998). Most of the southern and eastern regions experience a pronounced summer precipitation maximum due to moisture advected northwards from the Indian Ocean by the Asian summer monsoon. This summer precipitation declines northward across the Himalaya and little falls over western and central Tibet (Benn and Owen 1998). The mid-latitude westerlies produce a winter precipitation maximum at the western end of the Himalaya, Transhimalaya and western Tibet as a consequence of moisture advected from the Mediterranean, Black and Caspian seas (Benn and Owen 1998). Strong north-south and west-east precipitation gradients are the result of these two climate systems. However, there are also strong microclimatic variations within individual mountain ranges and valleys (Owen and Dortch 2014). The varied climatic settings and topographic extremes have a profound influence on the glacier systems and types of glaciers (Derbyshire 1981; Benn and Owen 2002; Owen and Dortch 2014). Derbyshire (1981) highlighted three main glacier types: (1) continental interior types in the central and western parts of the Tibet Plateau; (2) maritime monsoonal types in the Himalaya and in southeastern Tibet; and (3) continental monsoonal types in eastern and northeastern Tibet (Fig. 13.2).

The continental valley glaciers or small ice caps of central and western part of the Tibet and adjacent Transhimalaya are generally <10 km^2 in area and have high and cold accumulation areas, basal ice temperature much less than 0 °C. Their sur-

Fig. 13.2 Examples of different types of glaciers in the Himalaya and adjacent regions. (**a**) Relatively debris-free continental glacier in semi-arid southern Tibet advancing from the Gurla Mandhata at 7694 m asl. The glacier expands at its snout to become an almost piedmont-type glacier. (**b**) Sub-polar type continental glacier descending from the Lato massif in Zanskar, northern India. Minor amounts of transported debris contribute to small lateral moraines (seen on either side of the ice). (**c**) Khumbu glacier descending from Mount Everest (peak emerges in the distance most part of the photograph). This glacier becomes debris-mantled in its lower reaches. (**d**) Monsoon-influence debris mantled glacier in the Solang valley near Manali on the southern slopes of the Himalaya in northern India. This glacier has produced an impressive latero-frontal moraine complex

face velocities usually between 2 and 10 m/a, and may reach up to several 100 m/a (Derbyshire 1981; Benn and Owen 2002; Owen and Dortch 2014). In contrast, the maritime and continental monsoonal glaciers of the Himalaya and southeastern and eastern Tibet are warm-based and have summer accumulation and ablation. These cirque and valley glaciers are avalanche- and snowfall-fed, and many are very steep hanging glaciers; their surface velocities up to several hundred meters per year (Benn and Owen 2002; Owen and Dortch 2014).

Defining the exact extent of contemporary glaciers throughout the region is challenging because of the extremely large area to be covered, and because many of the glaciers are debris mantled making it difficult to define their limits from morainic debris along their margin left by recent retreat. Glaciers extend to the lowest elevations in the wettest regions with the lowest elevations at ~ 3000 m asl. The snout elevation of glaciers can be quite varied even within the same climatic region because numerous factors such as the shape and aspect of the valley, and the amount of snow blow and/or avalanche contribution to the glacier surface (Benn et al. 2005). In addition, glacier types within and across region may have varied throughout the Quaternary as a consequence of climate change.

Benn and Owen (2002) describe the range of landforms associated with Himalayan glaciers, including impressive latero-frontal moraines, hummocky moraines, lateral moraine valley complexes, glacially eroded bedrock surfaces and deeply entrenched valleys (Fig. 13.3). Derbyshire and Owen (1997), Hewitt (1999, 2011), Benn and Owen (2002) and Owen and Dortch (2014) highlight the importance of understanding the nature and formation of glacial landforms for accurate reconstructions of the former extent of glaciation.

Fig. 13.3 Examples of geomorphic evidence for former glaciation. (**a**) View of the western slope of the Lato Massif in Zanskar, Northern India. The moraine in foreground with numerous weathered boulders dates back to ~250 ka, while the hummocky moraines in the middle of the view represent the position of a glacier that advance downvalley at between 25–15 ka (see Orr et al. 2018 for more details). (**b**) Impressive lateral moraine enclosing the village of Periche in the Khumbu Himal of Nepal. This moraine represents the position of Khumbu Glacier when it advanced from Mount Everest during the global last glacial maximum at ~22–18 ka (see Richards et al. 2000 and Finkel et al. 2003 for more details). (**c**) Glacial eroded and polished valley sides in the Braldu valley in the Central Karakoram of Northern Pakistan. These slopes were eroded when Baltoro Glacier advanced from K2 at ~15 ka (see Seong et al. 2007 for more details). (**d**) Lateral moraine ridges radiating from the mouth of the Bara Shugri valley into the Chandra Valley of Lahul, northern India, and boulder moraine ridges in the foreground. These landforms mark the position of Bara Shugri Glacier when it advances across the Chandra valley at the end of the nineteenth century (see Owen et al. 1997 for more details)

13.3 Reconstructing the Extent and Timing of Quaternary Glaciation

Reconstructing Quaternary glaciation requires detailed geologic studies involving remote sensing, field mapping, analysis of landforms and sediments, and geochronology. Over the last few decades, reconstructing of the timing and extent of glaciation has been stimulated by the view that the Quaternary glacial record provides an important climate archive and that mountain glaciers are good climate proxies. Benn and Owen (1998) hypothesized that the two major climatic systems that dominate the region, mid-latitude westerlies and the Asian monsoon, would result in asychronouity of glaciation across the region. Study of the glacial geologic record would, therefore, provide important insights into the nature of Quaternary climate change, especially the relative importance the different climate systems throughout the region.

Reconstructing Quaternary glaciation in the region has proved challenging because of the logistical and political hurdles. Despite these hurdles, there have been numerous field studies that have concentrated on reconstructing the former extent of glaciation (most notably Frenzel 1960; Porter 1970; Derbyshire et al. 1984; Cronin et al. 1989; Holmes and Street-Perrott 1989; Watanabe et al. 1989; Li et al. 1991; Shiraiwa and Watanabe 1991; Shi 1992; Lehmkuhl and Lui 1994; Osmaston 1994; Lehmkuhl 1995, 1997, 1998; Owen et al. 1995, 1996, 1997, 2000; Taylor and Mitchell 2000; Lehmkuhl et al. 2004; Dortch et al. 2010; Heyman et al. 2010, 2011; Hedrick et al. 2017). Moreover, remote sensing studies have aided in regional reconstruction in recent years (Duncan et al. 1998; Heyman et al. 2008; Morén et al. 2011). These reconstructions rely on the accurate interpretation of landforms and sediments. Misinterpretations of landforms and sediments have in some instances led to erroneous reconstructions of former glacier extents (Hewitt 1999; Hewitt et al. 2011; Owen and Dortch 2014). Extreme misinterpretations led to the erroneous view that the Himalaya and Tibet was covered by an ice sheet during the last glacial (e.g., Kuhle 1985, 1987, 1988, 1991, 1995). Abundant evidence, however, shows that glaciation was limited to expanded ice caps and extended valley glaciers during the last ~500 ka (Derbyshire 1987; Burbank and Kang 1991; Derbyshire et al. 1991; Shi et al. 1992; Hövermann et al. 1993a, b; Lehmkuhl 1998; Rutter 1995; Lehmkuhl et al. 1998; Zheng and Rutter 1998; Schäfer et al. 2002; Owen et al. 2003a, 2008; Seong et al. 2008, 2009; Dortch et al. 2013; Orr et al. 2018). Shi et al. (1992) provides the most complete reconstruction of the maximum extent of glaciation across the Himalaya and Tibet (Fig. 13.4) However, the maximum extent of glaciation in their reconstruction did not occur at the same time across the region.

Once glacial landforms and sediments have been mapped and the former extent of glaciers have been reconstructed, the ages of the landforms and/or sediments need to be determined to define the timing of glaciation. The standard method to date landforms and sediments in most glaciated regions has been radiocarbon dating. However, this method relies on the presence of organic material, such as trees or charcoal, that has been incorporated into the moraine or associated sediments.

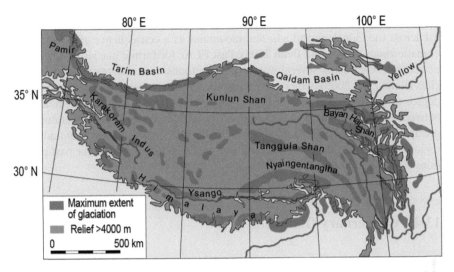

Fig. 13.4 Reconstruction for the maximum extent of glaciation across the Himalaya, Tibet and the bordering mountains. (After Shi et al. 1992 and adapted from Owen et al. 2008 and Owen 2010, 2017)

Unfortunately, organic material is not commonly found in glacial and associated landforms and sediments in the Himalaya and adjacent regions because the extreme mountain environments are not inductive to its preservation and abundant vegetation is rare. Fortunately, dating techniques such as terrestrial cosmogenic nuclide (TCN) surface exposure and optically stimulated luminescence (OSL) dating that have been developed over the past few decades can now be readily employed to date moraines and associated landforms devoid of organic material (Benn and Owen 1998; Owen et al. 2009; Owen and Dortch 2014). These dating methods also allow dating of glacial landforms and sediments older than the limit (~50 ka) of radiocarbon dating. Both these methods have been applied in a plethora of studies throughout the Himalaya and adjacent mountains (see Owen and Dortch 2014 and references therein). TCN studies, mainly using [10]Be, are by far the most common. TCN dating is normally used to date glacial boulders on moraines and glacially rock surfaces, and generally provide a minimum age for a glacier advance. OSL dating provides a means to date the deposition of sediment and can provide a minimum or maximum age dependent upon the sample's context. Both methods have inherent challenges and ages have large uncertainties (from a few to >50% of the age) associated with them. The various methods applied date different landform or sediments that might represent the advantage and/or retreat of a glacier. As such, the different age methods might give the impression of widely different ages for glaciation if the context of what is being dated is not fully described in detail. A comprehensive discussion of these dating methods, approaches and associated challenges is provided by Owen and Dortch (2014).

The standard method to quantify glaciation is to determine the past equilibrium-line attitude (ELA), which is the line on a glacier where accumulation and ablation of snow and ice are in balance over several years. ELAs vary from <4300 to >6200 m asl across the Himalaya and adjacent mountains as a consequence of the different climatic zones (Owen and Benn 2005). Past ELAs for former glacier advances can be reconstructed using several geomorphic methods on reconstructed former glaciers (Benn et al. 2005), which in turn provide a means to quantify and compare the degree of past glaciation. The ELA depression (ΔELA), the difference in altitude between the present and past ELA, can be used to help determine temperature and precipitation changes (Benn et al. 2005). ΔELAs for the Himalaya and adjacent region have been in the order of a few tens to few hundred meters for past glaciations (Owen and Benn 2005).

13.4   Quaternary Glaciation

Several reviews have summarized aspects of the nature and characteristics of Quaternary glaciation across the Himalaya, including Owen and Dortch (2014), and regional syntheses have been provided in Ehlers et al. (2004, 2011), Kamp and Owen (2011), Owen (2011), Zhou et al. (2011), Dortch et al. (2013), Murari et al. (2014), Solomina et al. (2015) and Owen (2017). The major characteristics of the timing and style of glaciation varies across the range are summarized in Fig. 13.5.

Dortch et al. (2013), Murari et al. (2014), and summarized in Owen and Dortch (2014), provide a framework for comparing glacial advances across the semi-arid western end of the Himalaya-Tibetan orogen and the monsoon-influenced regions of the Himalaya and Tibet. They base their timing of glacier advances on an extensive dataset of [10]Be ages, and define local and then regionally significant glacier

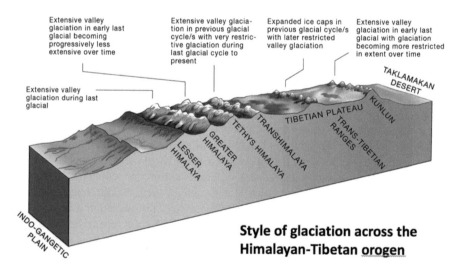

Fig. 13.5 Summary of the dominant style of Quaternary glaciation across the Himalaya and adjacent regions

advances, which for the semi-arid areas they term semi-arid western Himalaya-Tibetan glacial stages (SWHTS) and for monsoon-influences areas they term monsoon Himalayan-Tibetan glacial stages (MOHITS). These are summarized in Fig. 13.6 together with Quaternary marine isotopic stages (MIS) and insolation curves, and the Holocene climatostratigraphy (Owen and Dortch 2014). The ages presented by Dortch et al. (2013), Murari et al. (2014) and Owen and Dortch (2014) may subject to change in the coming years with the refinement of [10]Be production rates and scaling models.

The major characteristic of glaciation throughout the Himalaya are described chronologically in the sections below. Possible forcing factor that help drive glaciation are discussed, which include a combination of changes in Earth's orbital parameters (Milankovitch timescales of 10^{4-5} year), and sub-Milankovitch autocyclicity such as changes in ice-sheet movement and oceanic circulation (10^{1-4} years), solar variability (10^{1-3} years) and volcanic activity (10^{0-2} years), and geomorphic factors (10^{2-5} years).

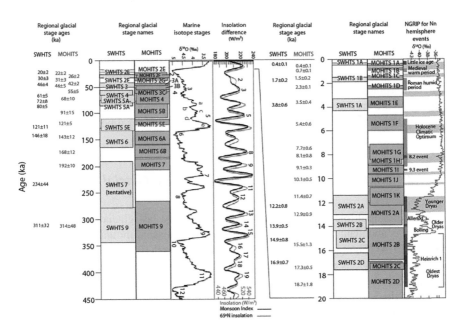

Fig. 13.6 Regional glacial stages for the semi-arid regions at the western end of the Himalaya and Tibet (*SWHTS* semi-arid western Himalaya-Tibetan Stage) defined by Dortch et al. (2013) and for the monsoon-influenced Himalayan-Tibetan orogen (*MOHITS* monsoon Himalayan-Tibetan stages) defined by Murari et al. (2014) (after Owen and Dortch 2014). The $\delta^{18}O$ curve NGRIP (2004) is provided for comparison and the duration of specific climatostratigraphy events are marked by black and grey bars in the far-right column. The Monsoon Index is the insolation difference between 30°N and 30°S for 1 August from Leuschner and Sirocko (2003)

13.4.1 Glacier Fluctuations Prior to the Last Glacial

There is abundant geologic evidence throughout the semi-arid regions of the Himalaya and adjacent mountains for glaciation prior to the last glacial cycle (>100 ka; Fig. 13.3a). Amongst the oldest moraines are those of the Indus glacial stage in Ladakh that were dated by Owen et al. (2006a) to >430 ka. Other extremely old moraines are present on the west side of Gurla Mandata in southernmost central Tibet dated by Owen et al. (2010) to >300 ka. Other moraines of great antiquity have been dated in the Kunlun Shan, Mustag Ata-Kongur Shan, Tashjurgan valley in the Pamir, Ayilari Range, Nyalam, Xainza graben in south central Tibet, and Zaskar (Owen et al. 2006b, 2012; Chevalier et al. 2005, 2011; Schäfer et al. 2008; Seong et al. 2009; Hedrick et al. 2011) Erratics on slopes along the Rongbuk valley on the northern side of Mt Everest date to >330 ka (Owen et al. 2009). In addition Orr et al. (2018) dated erratics around the Lato Massif in Ladakh that are to ~250 ka.

In these regions, glaciation have changed in style from expanded ice caps, to large piedmont valley glaciers, to cirque and small valley glaciers over the past few glacial cycles (Fig. 13.7). Owen et al. (2005, 2008, 2010), Seong et al. (2009) and Orr et al. (2018) argue that these shifts in glacial style may be climatically controlled by means of a reduction of precipitation over the interior of the Himalaya and Tibet over the last few glacial cycles, which effects the moisture flux necessary to maintain positive glacier mass balances. Alternatively, over several glacial cycles, these glaciated massifs may have geomorphically evolved from isolated alpine plateaus to a dissected and steep relief mountain range, where deep valleys were calved by geomorphic processes including glacial and fluvial erosion. These changes to the landscape are argued to be associated with local climatic and environmental conditions that were sufficient to influence the glaciation style over time.

13.4.2 Glacier Fluctuations During the Last Glacial

Evidence of multiple glacier advances throughout the last glacial cycle (~100 to 11.6 ka) is abundant throughout the Himalaya and adjacent mountains (Owen and Dortch 2014) (Figs. 13.3b, c and 13.8). Owen and Dortch (2014) show that glaciers in semi-arid regions of the Himalaya and adjacent mountains reached their maximum extent during the early part of the last glacial, as compared to the maximum extent of the northern hemisphere ice sheets that reached their maximum late in the last glacial cycle at global last glacial maximum (LGM), the time of maximum global ice volume at which is defined by Mix et al. (2001) to between ~24–18 ka, Clark et al. (2009) as ~26–19 ka in MIS 2 and Hughes and Gibbard (2015) as ~28-23 ka. Researchers in other mountain regions of the world, e.g., Gillespie and Molnar (1995) and Thackray et al. (2008), have also suggested that mountain glaciers reached their maximum extent earlier in the last glacial, which is often referred to a local last glacial maximum (lLGM).

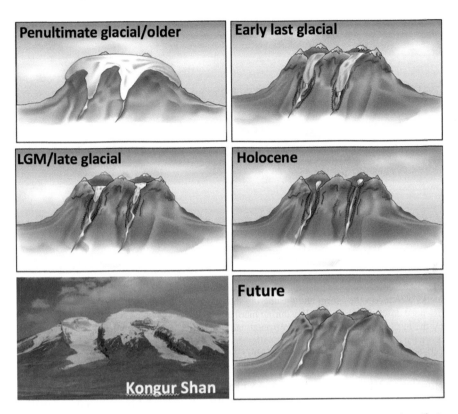

Fig. 13.7 Changes of style of glaciation on massifs such as Kongur Shan, Mustag Ata, Gurla Mandata and Lato over the last few glacial cycles. Kongur Shan illustrates the typical geomorphology of the massifs. In the future, glaciers are likely to retreat and/or melt away (*LGM* global last glacial maximum)

Owen and Dortch (2014) points out that in the Himalaya and adjacent mountains, the pattern of glaciation during the last glacial cycle is more complex. In the more monsoon-influenced Himalaya, e.g., there is increasing evidence to suggest that glaciation was more extensive later in the last glacial cycle as compared to most other regions. Moreover, Owen and Dortch (2014) illustrates this complexity further by highlighting the contrasting timing and extent of the lLGM at the western end of the Himalaya and Tibet. At the western end of the Himalaya and Tibet, the local last glacial maximum occurring at very different time and climate modeling predicting vast differences in snowfall across the regions for different times during the last glacial and through the Holocene highlights this complexity (Bishop et al. 2010; Owen and Dortch 2014). The climate modeling shows that the pattern of change is not similar in any of the modelled time slices, and it should not be surprising that the style and timing of glaciation across the mountains is complex.

Using ice sheet modeling, Yan et al. (2018) showed that glaciers in western and southern Himalayan-Tibetan region exhibit high region variability in Glaciation across the Himalaya and Tibet. Glaciers in the western and southern regions of the

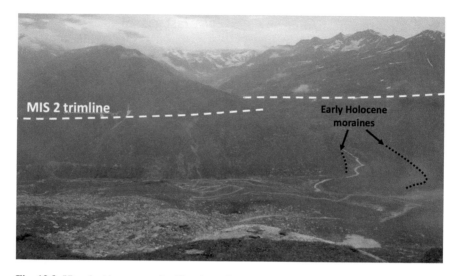

Fig. 13.8 View looking across the Chandra valley toward the Kulti valley. The MIS 2 trimline marks a change valley form and weathering that indicated the top of a large valley glacier that extended down the Chandra valley during marine isotope stage (MIS) 2. The small ridges marked by the black dashed lines are latero-frontal end moraines that formed when a glacier advanced down the Batal valley into the Chandra valley during the early Holocene. The glacial history of this valley is described in Owen et al. (1995, 1997, 2001) and Eugster et al. (2016)

Himalayan-Tibetan region are the most sensitive to climate change and those in the interior are the least sensitive. Their modelling constrained by climatic parameters based on geologic proxy data broadly reproduced the restricted glaciation across the Himalaya and Tibet during the LGM that geologic data shows. In particular, decreased precipitation during the LGM would have hampered glacier growth over northern Himalayan-Tibetan region, while insufficient cooling would have hampered glacier advance over eastern regions. In addition, both reduced precipitation and insufficient cooling during the LGM would have inhibited large-scale glaciation over inner Tibet.

Dortch et al. (2013) recognized 15 SWHTS during the last glacial and suggested that SWHTS older than 21 ka are broadly correlated with a greater monsoonal influence and that the six SWHTS that are 21 ka or younger broadly correlate with global ice volume and Northern Hemisphere climatic events (Oldest Dryas, Older Dryas, Younger Dryas, Roman Humid Period, and Little Ice Age). Murari et al. (2014) defined 13 MOHITS for the last glacial and Holocene, and suggested strong correlations with both periods of strong monsoons and northern hemisphere climate events.

13.4.3   Holocene Glacier Fluctuations

Summaries of Holocene glacier fluctuations in the Himalaya and adjacent mountains are provided by Yi et al. (2008), Owen (2011), Owen and Dortch (2014) and Solomina et al. (2015, 2016). In addition, Saha et al. (2018, b) provides summaries for the northwestern end of the Himalaya. The radiocarbon dating of moraines in Pakistan, India and Nepal by Röthlisberger and Geyh (1985a, b) is noteworthy and defined glacier advances to ~ 8.3, 5.4–5.1, 4.2–3.3, and 2.7–2.2 ka with relatively small extensions at 2.6–2.4, 1.7–1.4, 1.3–0.9, 0.8–0.55 and 0.5–0.1 ka; however this work has yet to be validated/duplicated. Compilations of radiocarbon, TCN and OSL ages for the Himalaya and adjacent regions suggested that glaciers were responding to periods of Holocene rapid climate change in the North Atlantic, with climate teleconnected via the mid-latitude westerlies to Central Asia (Yi et al. 2008; Owen and Dortch 2014). Owen and Dortch (2014) highlight at least 4 SWHTS and 11 MOHITS during the Holocene in the semi-arid and monsoonal areas of the Himalaya and Tibet, respectively (Fig. 13.6). Saha et al. (2018) recognizes ~7 regional glacier advances during the Holocene at the western end of the Himalaya, albeit somewhat restricted in extent. While Saha et al. (2019) extend their study across the Himalaya and Tibet to show that 5 regional Himalayan-Tibetan Holocene glacial stages (HTHS) at ~11.5–9.5, ~8.8–7.7, ~7.0–3.2, ~2.3–1.0, and <1 ka.

The most extensive glacier advances during the early Holocene occurred between ~11.5 and ~8.0 ka (Fig. 13.6; Sharma and Owen 1996; Phillips et al. 2000; Richards et al. 2000; Owen et al. 2001, 2002a, b, 2003a, b, 2005, 2006a, b, 2009, 2010, 2012; Finkel et al. 2003; Zech et al., 2003, 2005; Barnard et al. 2004a, b; Spencer and Owen 2004; Abramowski et al. 2006; Gayer et al. 2006; Jiao and Shen 2006; Seong et al. 2007, 2009; Meyer et al. 2009; Chevalier et al. 2011; Murari et al. 2014; Orr et al. 2018; Saha et al. 2018, 2019). But, there is considerable regional variability in the extent of glaciation and ΔELA between regions (Owen and Benn 2005; Owen and Dortch 2014; Saha et al. 2018, 2019).

Glaciers advanced to within a kilometer of the present glacier margins throughout the region during the mid-Holocene (~8.0–3.0 ka) and late Holocene (Neoglacial, <3.0 ka) (Owen 2009; Owen and Dortch 2014; Solomina et al. 2015, 2016; Saha et al. 2018, 2019). Five MOHITS and 3 SWHTS are evident for this time, and Owen and Dortch (2014) suggested these relate to changes in Northern Hemisphere climate. Little Ice Age (LIA) glacier advances, although not particularly well defined, occurred during the last few hundred years probably responding to Northern Hemisphere cooling (Owen and Dortch 2014; Rowan 2017). Glaciers began to retreat after the LIA at the beginning of the twentieth Century in monsoon-influenced areas of Himalaya (Owen and Dortch 2014).

13.4.4   Recent Glacier Fluctuations

There is much geological and historic evidence for recent glacier retreat (Figs. 13.3d and 13.9). Numerous studies are showing that glaciers in the Himalaya and adjacent mountains are generally retreating at rates similar to glaciers elsewhere in the world due to human-induced climate change, although glaciers in the Himalaya west of the Sutlej River have experienced little or no retreat (Bagla 2009; Bolch et al. 2012; National Academy 2012). Cogley (2011) suggests that up to about one-fifth of the

Fig. 13.9 View of Ngozumpa Glacier in the Khumbu Himal of Nepal. (**a**) View east looking toward Everest from Goyko Ri at 5357 m asl. Note how the glacier is inset into impressive lateral moraines, providing unequivocal evident of glacier wasting during the past few hundred years. (**b**) Major ice wasting at near snout of the glacier and development of supraglacial lakes during the last few years. (Photographs taken November 2017)

glaciers present in 1985 may have disappeared by ~2008 and some 3000–13,000 more glaciers might disappear by 2035. However, Copland et al. (2011) suggests that the rate of glacier retreat over the past 30 years in some regions has been less than during the earlier part of the twentieth century, and certain glaciers have begun to stabilize and/or advance during the past few years. Some glaciers such as Baltoro Glacier in northern Pakistan have fluctuated only a few hundred meters from their present positions since the 1850s (Mayer et al. 2006). In the Karakoram, some glaciers have advanced during the last century, sometimes surging, which has become known as the Karakoram anomaly (Hewitt 2005; Copland et al. 2011; Bolch et al. 2012). This is likely a consequence of the Karakoram summer vortex proposed by Forsythe et al. (2017), which is forced by the interaction of the mid-latitude westerlies and South Asian monsoon. Recently Maurer et al. (2019) have observed consistent glacier loss along the Himalaya and find a doubling of the average loss rate during 2000–2016 compared to 1975–2000. They suggest that the similar magnitude and acceleration of glacier loss across the Himalayas suggests a regionally coherent climate forcing, consistent with atmospheric warming and associated energy fluxes as the dominant drivers of glacier change.

Predicting how human-induced global climate change may force glaciers to melt, retreat and eventually disappear is challenging because of the inherent uncertainties of climate modeling and prediction, and the complex nature of Himalayan glacial systems as shown from the Quaternary evidence, especially given the large altitudinal ranges, and the complex topographic and geomorphic settings, and the microclimatic diversity (Owen and Dortch 2014; Owen 2017). Given the complex response of glaciers to climate change in the past, it is highly likely that future glacier changes under human-induced climate change will be equally as complex, with contrasting degrees of retreat between one glacier and the next, across regions and between regions.

13.5   Conclusion

The glaciers of the Himalaya and adjacent regions range from sub-polar continental to maritime types and are influenced by a diverse and complex range of factors including climate, topography and geology. Until relatively recently, defining the extent and timing of Quaternary glaciation throughout the Himalaya and adjacent mountains has been challenging. However, new satellite remote sensing and geochronological methods such as TCN and OSL dating has increased the attention being paid to defining the extent and timing of Quaternary glaciation throughout the region. Much focus has been placed on the desire to examine the relative roles of the Asian summer monsoon and mid-latitude westerlies in driving glaciation for paleoclimatic/paleoenvironmental studies.

These studies show significant glacier advances over the past several glacial cycles, with many valleys preserving glacial geologic evidence for numerous glacier advances during the Quaternary. At least nine major regionally synchronous glacier advances are evident throughout the semi-arid and monsoon influenced regions of the Himalaya and Tibet over the past ~400 ka. The maximum extent of glaciation has been broadly defined, characterized by extensive valley glaciers and expanded ice caps. The timing of maximum glacier extent is asynchronous throughout the region, with some areas experiencing maximum glaciation prior to the last glacial (>100 ka), while in other regions it was during the early part of the last glacial (~30–70 ka) and in some regions, it was possibly coincident with the LGM at ~18–24 ka. Glacier advances since the LGM have been limited to a few kilometers beyond their present position in most regions, with the maximum advance occurring during the early Holocene at ~9–8 ka. The higher resolution of the Holocene glacial geologic record allows ~5 regional glacier advances to be resolved, albeit somewhat restricted in extent. However, complex variations in the timing and extent of late Quaternary glaciation are apparent over relatively short distances (10–100 km). This is well illustrated at the western end of the Himalaya, where the lLGM occurred at different times with vastly different glacier extents. Within semi-arid regions, the style of glaciation has changed significantly over the last several glacial cycles, from expanded ice caps to entrenched valley glaciation. Throughout the Holocene there have been numerous minor glacier advances with the most recent, the Little Ice Age, starting several hundred years ago and ending at approximately the beginning of the twentieth Century. During the last 100 years, glaciers have retreated in most areas, but in regions such as the Karakoram there has been little retreat and numerous glaciers have surged.

These patterns of glaciation suggest future glacier fluctuations in response to human-induced climate change will be complex. Rather than a single driver of Himalayan glaciation, the timing and extent of glaciation, and the preservation of glacial evidence, is likely governed by a complex combination of different factors specific to each locality, including climate and microclimate regimes, tectonic setting, topographic controls and geologic setting.

Acknowledgments Many thanks to two anonymous reviewers for their very useful and constructive comments and Dr. A.P. Dimri for encouraging me to write this chapter.

References

Abramowski U, Bergau A, Seebach D, Zech R, Glaser B, Sosin P, Kubik PW, Zech W (2006) Pleistocene glaciations of Central Asia: results from [10]Be surface exposure ages of erratic boulders from the Pamir (Tajikistan) and the Alay-Turkestan range (Kyrgyzstan). Quat Sci Rev 25:1080–1096

Bagla P (2009) No sign yet of Himalayan meltdown, Indian report finds. Science 326:924–925

Barnard PL, Owen LA, Finkel RC (2004a) Style and timing of glacial and paraglacial sedimentation in a monsoonal influenced high Himalayan environment, the upper Bhagirathi Valley, Garhwal Himalaya. Sediment Geol 165:199–221

Barnard PL, Owen LA, Sharma MC, Finkel RC (2004b) Late Quaternary (Holocene) landscape evolution of a monsoon-influenced high Himalayan valley, Gori Ganga, Nanda Devi, NE Garhwal. Geomorphology 61:91–110

Benn DI, Owen LA (1998) The role of the Indian summer monsoon and the mid-latitude westerlies in Himalayan glaciation: review and speculative discussion. J Geol Soc 155:353–363

Benn DI, Owen LA (2002) Himalayan glacial sedimentary environments: a framework for reconstructing and dating former glacial extents in high mountain regions. Quat Int 97(98):3–26

Benn DI, Owen LA, Osmaston HA, Seltzer GO, Porter SC, Mark B (2005) Reconstruction of equilibrium-line altitudes for tropical and sub-tropical glaciers. Quat Int 138(139):8–21

Bishop MP, Bush A, Copland L, Kamp U, Owen LA, Seong YB, Shroder JF (2010) Climate change and mountain topographic evolution in the Central Karakoram, Pakistan. Ann Geogr 100:1–22

Bolch T, Kulkarni A, Kääb A, Huggel C, Paul F, Cogley JG, Frey H, Kargel JS, Fujita K, Scheel M, Bajracharya S, Stoffel M (2012) The state and fate of Himalayan Glaciers. Science 336:310–314

Burbank DW, Kang JC (1991) Relative dating of Quaternary moraines, Rongbuk Valley, Mount Everest, Tibet: implications for an ice sheet on the Tibetan Plateau. Quat Res 36:1–18

Chevalier M-L, Ryerson FJ, Tapponnier P, Finkel RC, Van Der Woerd J, Haibing L, Qing L (2005) Slip-rate measurements on the Karakoram Fault may imply secular variations in fault motion. Science 307:411–414

Chevalier M-L, Hilley G, Tapponnier P, Van Der Woerd J, Liu-Zeng J, Finkel RC, Ryerson FJ, Li Haibing L, Liu X (2011) Constraints on the late Quaternary glaciations in Tibet from cosmogenic exposure ages of moraine surfaces. Quat Sci Rev 30:528–554

Clark PU, Dyke AS, Shakun JD, Carlson AE, Clark J, Wohlfarth B, Mitrovica JX, Hostetler SW, McCabe AM (2009) The last glacial maximum. Science 325:710–714

Cogley JG (2011) Present and future states of Himalaya and Karakoram glaciers. Ann Glaciol 52:69–73

Copland L, Sylvestre T, Bishop MP, Shroder JF, Seong YB, Owen LA, Bush A, Kamp U (2011) Expanded and recently increased glacier surging the Karakoram. Arct Alp Antarc Res 43:503–516

Cronin VS, Johnson WP, Johnson NM, Johnson GD (1989) Chronostratigraphy of the upper Cenozoic Bunthang sequence and possible mechanisms controlling base level in Skardu intermontane basin, Karakoram Himalaya, Pakistan. Geol Soc Am Spec Pap 232:295–309

Derbyshire E (1981) Glacier regime and glacial sediment facies: a hypothetical framework for the Qinghai-Xizang Plateau. In: Proceedings of symposium on Qinghai-Xizang (Tibet) Plateau, Beijing, China, Geological and ecological studies of Qinghai-Xizang Plateau, vol 2. Science Press, Beijing, pp 1649–1656

Derbyshire E (1987) A history of the glacial stratigraphy in China. Quat Sci Rev 6:301–314

Derbyshire E, Owen LA (1997) Quaternary glacial history of the Karakoram Mountains and Northwest Himalayas: a review. Quat Int 38(39):85–102

Derbyshire E, Li J, Perrott FA, Xu S, Waters RS (1984) Quaternary glacial history of the Hunza valley Karakoram Mountains, Pakistan. In: Miller K (ed) International Karakoram project. Cambridge University Press, Cambridge, pp 456–495

Derbyshire E, Shi Y, Li J, Zheng B, Li S, Wang J (1991) Quaternary glaciation of Tibet: the geological evidence. Quat Sci Rev 10:485–510

Dortch JM, Owen LA, Caffee MW (2010) Quaternary glaciation in the Nubra and Shyok valley confluence, northernmost Ladakh, India. Quat Res 74:132–144

Dortch JM, Owen LA, Caffee MW (2013) Timing and climatic drivers for glaciation across semiarid western Himalayan-Tibetan orogen. Quat Sci Rev 78:188–208

Duncan CC, Klein AJ, Masek JG, Isacks BL (1998) Late Pleistocene and modern glaciations in Central Nepal from digital elevation data and satellite imagery. Quat Res 49:241–254

Ehlers J, Gibbard P (eds) (2004) Quaternary glaciations – extent and chronologies. Part III: South America, Asia, Africa, Australia, Antarctica. Dev Quat Sci 2:380pp

Ehlers J, Gibbard P, Hughes PD (eds) (2011) Quaternary glaciations – extent and chronology: a closer look, Dev Quat Sci, vol 15, 2nd edn. Elsevier, Amsterdam, pp 929–942

Eugster P, Scherler D, Thiede RC, Codilean AT, Strecker MR (2016) Rapid last glacial maximum deglaciation in the Indian Himalaya coeval with midlatitude glaciers: new insights from 10Be-dating of ice-olished bedrock surfaces in the Chandra Valley, NW Himalaya. Geophys Res Lett 43:1589–1597

Fielding E, Isacks B, Barazangi M, Duncan C (1994) How flat is Tibet? Geology 22:163–167

Finkel RC, Owen LA, Barnard PL, Caffee MW (2003) Beryllium-10 dating of Mount Everest moraines indicates a strong monsoonal influence and glacial synchroniety throughout the Himalaya. Geology 31:561–564

Frenzel B (1960) Die Vegetations- und Landschaftszonen Nordeurasiens während der letzten Eiszeit und während der Postglazialen Warmezeit. Akademie der Wissenschaften und der Literatur in Mainz, Abhandlungen der Mathematisch-Naturwissenschaftlichen Klasse 13:937–1099

Gayer E, Lavé J, Pik R, France-Lanord C (2006) Monsoonal forcing of Holocene glacier fluctuations in Ganesh Himal (central Nepal) constrained by cosmogenic ^{3}He exposure ages of garnets. Earth Planet Sci Lett 252:275–288

Gillespie A, Molnar P (1995) Asynchronous maximum advances of mountain and continental glaciers. Rev Geophys 33:311–364

Hedrick KA, Seong YB, Owen LA, Caffee MC, Dietsch C (2011) Towards defining the transition in style and timing of Quaternary glaciation between the monsoon-influenced Greater Himalaya and the semi-arid Transhimalaya of Northern India. Quat Int 236:21–33

Hedrick K, Owen LA, Chen J, Robinson A, Yuan Z, Yang X, Imrecke DB, Li W, Caffee MW, Schoenbohm LM, Zhang B (2017) Quaternary history and landscape evolution of a high-altitude intermountain basin, Waqia Valley, Chinese Pamir. Geomorphology 284:156–174

Hewitt K (1999) Quaternary moraines vs catastrophic avalanches in the Karakoram Himalaya, northern Pakistan. Quat Res 51:220–237

Hewitt K (2005) The Karakoram anomaly? Glacier expansion and the 'elevation effect,' Karakoram Himalaya. Mt Res Dev 25:332–340

Hewitt K, Gosse J, Clague JJ (2011) Rock avalanches and the pace of late Quaternary development of river valleys in the Karakoram Himalaya. Geol Soc Am Bull 123:1836–1850

Heyman J, Hattestrand C, Stroeven AP (2008) Glacial geomorphology of the Bayan Har sector of the NE Tibetan plateau. J Maps 2008:42–62

Heyman J, Stroeven AP, Caffee MW, Hättestrand C, Harbor J, Li YK, Alexanderson H, Zhou LP, Hubbard A (2010) Palaeoglaciology of Bayan Har Shan, NE Tibetan Plateau: the case of a missing LGM expansion, In: Heyman J (ed) Palaeoglaciology of the northeastern Tibetan Plateau. PhD thesis, Stockholm University, Stockholm

Heyman J, Stroeven A, Harbor J, Caffee MW (2011) Too young or too old: evaluating 884 cosmogenic exposure dating based on an analysis of compiled boulder exposure 885 ages. Earth Planet Sci Lett 302:71–80

Holmes JA, Street-Perrott FA (1989) The Quaternary glacial history of Kashmir, North-West Himalaya: a Revision of de Terra and Paterson's sequence. Z Geomorphol 76:195–212

Hövermann J, Lehmkuhl F, Pörtge K-H (1993a) Pleistocene glaciations in Eastern and Central Tibet-preliminary results of the Chinese-German joint expeditions. Z Geomorphol 92:85–96

Hövermann J, Lehmkuhl F, Süssenberger H (1993b) Neue Befunde zur Paläoklimatologie Nordafrikas und Zentralasiens. Abhandlungen der Braunschweigischen Wissenschaftlichen Gesellschaft 43:127–150

Hughes PD, Gibbard PL (2015) A stratigraphical basis for the last glacial maximum (LGM). Quat Int 383:174–185

Jiao KQ, Shen YP (2006) Quaternary glaciations in Tanggulha Mountains. In: Shi YF, Su Z, Cui ZJ (eds) The Quaternary glaciations and environmental variations in China. Science and Technology Press of Hebei Province, Shijiazhuang, pp 326–356

Kamp U, Owen LA (2011) Late Quaternary glaciation of Northern Pakistan. In: Elhers J, Gibbard P, Hughes PD (eds) Quaternary glaciations – extent and chronology: a closer look, Developments in Quaternary science, vol 15. Elsevier, Amsterdam, pp 909–927

Kargel JS, Leonard GJ et al (2016) Geomorphic and geologic controls of geohazards induced by Nepal's 2015 Gorkha earthquake. Science 351:8353-1-10

Kuhle M (1985) Ein subtropisches Inlandeis als Eiszeitauslöser, Südtibet un Mt. Everest expedition 1984. Georgia Augusta, Nachrichten aus der Universität Gottingen, May, pp 1–17

Kuhle M (1987). The problem of a Pleistocene Inland Glaciation of the Northeastern Qinghai-Xizang Plateau. In: Reports of the Qinghai-Xizang (Tibet) Plateau, Hövermann J, Wang W. (eds). Beijing, pp 250–315

Kuhle M (1988) Geomorphological findings on the built-up of Pleistocene glaciation in Southern Tibet and on the problem of inland ice. GeoJournal 17:457–512

Kuhle M (1991) Observations supporting the Pleistocene inland glaciation of High Asia. GeoJournal 25:131–231

Kuhle M (1995) Glacial isostatic uplift of Tibet as a consequence of a former ice sheet. GeoJournal 37:431–449

Lehmkuhl F (1995) Geomorphologische Untersuchungen zum Klima des Holoza"ns und Jungpleistoza"ns Osttibets. Göttinger Geographische Abhandlungen 102:1–184

Lehmkuhl F (1997) Late Pleistocene, Late-glacial and Holocene glacier advances on the Tibetan Plateau. Quat Int 38(39):77–83

Lehmkuhl F (1998) Extent and spatial distribution of Pleistocene glaciations in Eastern Tibet. Quat Int 45(46):123–134

Lehmkuhl F, Lui S (1994) An outline of physical geography including Pleistocene glacial landforms of Eastern Tibet (Provinces Sichuan and Qinghai). GeoJournal 34:7–30

Lehmkuhl F, Owen LA, Derbyshire E (1998) Late Quaternary glacial history of northeastern Tibet. Quat Proc 6:121–142

Lehmkuhl F, Klinge M, Stauch G (2004) The extent of late Pleistocene Glaciations in the Altai and Khangai Mountains. In: Ehlers J, Gibbard PL (eds) Quaternary glaciations – extent and chronologies, Part III: South America, Asia, Africa, Australia, Antarctica. Elsevier, Oxford, pp 243–254

Leuschner DC, Sirocko F (2003) Orbital insolation forcing of the Indian Monsoon – a motor for global climate change. Palaeogeogr Paleoclimatol Palaeoecol 197:83–95

Li B, Li J, Cui Z (eds) (1991) Quaternary glacial distribution map of Qinghai-Xizang (Tibet) Plateau 1:3,000,000. Shi Y. (Scientific Advisor), Quaternary Glacier, and Environment Research Center, Lanzhou University. Science Press, Beijing

Maurer JM, Schaefer JM, Rupper S, Corley A (2019) Acceleration of ice loss across the Himalayas over the past 40 years. Sci Adv 5:eaav7266

Mayer C, Lambrecht A, Belò M, Smiraglia C, Diolaiuti G (2006) Glaciological characteristics of the ablation zone of Baltoro glacier, Karakoram, Pakistan. Ann Glaciol 43:123–131

Meyer MC, Hofmann C-C, Gemmell AMD, Haslinger E, Häusler H, Wangda D (2009) Holocene glacier fluctuations and migration of Neolithic yak pastoralists into the high valleys of northwest Bhutan. Quat Sci Rev 28:1217–1237

Mix AC, Bard E, Schneider R (2001) Environmental processes of the ice age: land, ocean, glaciers (EPILOG). Quat Sci Rev 20:627–657

Morén B, Heyman J, Stroeven AP (2011) Glacial geomorphology of the central Tibetan Plateau. J Maps 2011:115–125

Murari MK, Owen LA, Dortch JM, Caffee MW, Dietsch C, Fuchs M, Haneberg WC, Sharma MC, Townsend-Small A (2014) Timing and climatic drivers for glaciation across monsoon-influenced regions of the Himalayan-Tibetan orogen. Quat Sci Rev 88C:159–182

National Academy (Committee on Himalayan Glaciers, Hydrology, Climate Change, and Implications for Water Security) (2012) Himalayan glaciers: climate change, water resources, and water security. The National Academies Press, Washington, DC, 156 pages

NGRIP Members (2004) High-resolution record of Northern Hemisphere climate extending into the last interglacial period. Nature 431:147–151

Orr EN, Owen LA, Saha S, Caffee MW, Murari MK (2018) Quaternary glaciation of the Lato Massif, Zanskar Range of NW Himalaya. Quat Sci Rev 183:140–156

Osmaston H (1994) The geology, geomorphology and Quaternary history of Zangskar. In: Crook J, Osmaston H (eds) Himalayan Buddhist villages. University of Bristol Press, Bristol, pp 1–36

Owen LA (2011) Quaternary glaciation of Northern India. In: Elhers J, Gibbard P, Hughes PD (eds) Quaternary glaciations – extent and chronology: a closer look, Dev Quat Sci, vol 15, 2nd edn. Elsevier, Amsterdam, pp 929–942

Owen LA (2017) Late Quaternary glacier fluctuations in the Himalaya and adjacent mountains. In: Prins HT, Namgail T (eds) Bird migration across the Himalayas: wetland functioning amidst mountains and glaciers. Cambridge University Press, Cambridge, pp 155–176

Owen LA, Benn DI (2005) Equilibrium-line altitudes of the last glacial maximum for the Himalaya and Tibet: an assessment and evaluation of results. Quat Int 138(139):55–78

Owen LA, Dortch JM (2014) Quaternary glaciation of the Himalayan-Tibetan orogen. Quat Sci Rev 88:14–54

Owen LA, Benn DI, Derbyshire E, Evans DJA, Mitchell WA, Thompson D, Richardson S, Lloyd M, Holden C (1995) The geomorphology and landscape evolution of the Lahul Himalaya, Northern India. Z Geomorphol 39:145–174

Owen LA, Derbyshire E, Richardson S, Benn DI, Evans DJA, Mitchell WA (1996) The Quaternary glacial history of the Lahul Himalaya, Northern India. J Quat Sci 11:25–42

Owen LA, Mitchell W, Bailey RM, Coxon P, Rhodes E (1997) Style and timing of Glaciation in the Lahul Himalaya, northern India: a framework for reconstructing late Quaternary palaeoclimatic change in the western Himalayas. J Quat Sci 12:83–109

Owen LA, Scott CH, Derbyshire E (2000) The Quaternary glacial history of Nanga Parbat. Quat Int 65(66):63–79

Owen LA, Gualtieri L, Finkel RC, Caffee MW, Benn DI, Sharma MC (2001) Cosmogenic radionuclide dating of glacial landforms in the Lahul Himalaya, Northern India: defining the timing of late Quaternary glaciation. J Quat Sci 16:555–563

Owen LA, Finkel RC, Caffee MW, Gualtieri L (2002a) Timing of multiple glaciations during the late Quaternary in the Hunza Valley, Karakoram Mountains, Northern Pakistan: defined by cosmogenic radionuclide dating of moraines. Geol Soc Am Bull 114:593–604

Owen LA, Kamp U, Spencer JQ, Haserodt K (2002b) Timing and style of late Quaternary glaciation in the eastern Hindu Kush, Chitral, northern Pakistan: a review and revision of the glacial chronology based on new optically stimulated luminescence dating. Quat Int 97-98:41–56

Owen LA, Finkel RC, Ma H, Spencer JQ, Derbyshire E, Barnard PL, Caffee MW (2003a) Timing and style of late Quaternary glaciations in NE Tibet. Geol Soc Am Bull 11:1356–1364

Owen LA, Ma H, Derbyshire E, Spencer JQ, Barnard PL, Nian ZY, Finkel RC, Caffee MW (2003b) The timing and style of late Quaternary glaciation in the La Ji Mountains, NE Tibet: evidence for restricted glaciation during the latter part of the last glacial. Z Geomorphol 130:263–276

Owen LA, Finkel RC, Barnard PL, Ma H, Asahi K, Caffee MW, Derbyshire E (2005) Climatic and topographic controls on the style and timing of Late Quaternary glaciation throughout Tibet and the Himalaya defined by [10]Be cosmogenic radionuclide surface exposure dating. Quat Sci Rev 24:1391–1411

Owen LA, Caffee M, Bovard K, Finkel RC, Sharma M (2006a) Terrestrial cosmogenic surface exposure dating of the oldest glacial successions in the Himalayan orogen. Geol Soc Am Bull 118:383–392

Owen LA, Finkel RC, Ma H, Barnard PL (2006b) Late Quaternary landscape evolution in the Kunlun Mountains and Qaidam Basin, Northern Tibet: a framework for examining the links between glaciation, lake level changes and alluvial fan formation. Quat Int 154-155:73–86

Owen LA, Caffee MW, Finkel RC, Seong YB (2008) Quaternary glaciation of the Himalayan–Tibetan orogen. J Quat Sci 23:513–532

Owen LA, Robinson R, Benn DI, Finkel RC, Davis NK, Yi C, Putkonen J, Li D, Murray AS (2009) Quaternary glaciation of Mount Everest. Quat Sci Rev 28:1412–1433

Owen LA, Yi C, Finkel RC, Davis N (2010) Quaternary glaciation of Gurla Mandata (Naimon'anyi). Quat Sci Rev 29:1817–1830

Owen LA, Chen J, Hedrick KA, Caffee MW, Robinson A, Schoenbohm LM, Zhaode Y, Li W, Imrecke D, Liu J (2012) Quaternary glaciation of the Tashkurgan Valley, Southeast Pamir. Quat Sci Rev 47:56–72

Phillips WM, Sloan VF, Shroder JF Jr, Sharma P, Clarke ML, Rendell HM (2000) Asynchronous glaciation at Nanga Parbat, northwestern Himalaya Mountains, Pakistan. Geology 28:431–434

Porter SC (1970) Quaternary glacial record in the Swat Kohistan, West Pakistan. Geol Soc Am Bull 81:1421–1446

Richards BWM, Owen LA, Rhodes EJ (2000) Timing of late Quaternary glaciations in the Himalayas of northern Pakistan. J Quat Sci 15:283–297

Röthlisberger F, Geyh MA (1985a) Gletscherschwankungen der letzten 10.000 Jahre – Ein Verleich zwischen Nord- und Südhemisphare (Alpen, Himalaya, Alaska, Südamerika, Neuseeland). Verlag Sauerländer, Aarau

Röthlisberger F, Geyh M (1985b) Glacier variations in Himalayas and Karakoram. Z Gletscherk Glazialgeol 21:237–249

Rowan AV (2017) The 'Little Ice Age' in the Himalaya: A review of glacier advance driven by Northern Hemisphere temperature change. The Holocene 27:292–308

Rutter NW (1995) Problematic ice sheets. Quat Int 28:19–37

Saha S, Owen LA, Orr EN, Caffee MW (2018) Timing and nature of Holocene glacier advances at the northwestern end of the Himalayan-Tibetan orogen. Quat Sci Rev 187:177–202

Saha S, Owen LA, Orr EN, Caffee MW (2019) High-frequency Holocene glacier fluctuations in the Himalayan-Tibetan orogen. Quat Sci Revs. https://doi.org/10.1016/j.quascirev.2019.07.021. (in press)

Schäfer JM, Tschudi S, Zhao Z, Wu X, Ivy-Ochs S, Wieler R, Baur H, Kubik PW, Schluchter C (2002) The limited influence of glaciations in Tibet on global climate over the past 170000 yr. Earth Planet Sci Lett 194:287–297

Schäfer JM, Oberholzer P, Zhao ZZ, Ivy-Ochs S, Wieler R, Baur H, Kubik PW, Schluchter C (2008) Cosmogenic beryllium-10 and neon-21 dating of late Pleistocene glaciations in Nyalam, monsoonal Himalayas. Quat Sci Rev 27:295–311

Sen SM, Kansal A (2019) Achieving water security in rural Indian Himalayas. J Environ Manag 245:398–408

Seong YB, Owen LA, Bishop MP, Bush A, Clendon P, Copland P, Finkel RC, Kamp U, Shroder JF (2007) Quaternary glacial history of the Central Karakoram. Quat Sci Rev 26:3384–3405

Seong YB, Owen LA, Bishop MP, Bush A, Clendon P, Copland P, Finkel RC, Kamp U, Shroder JF (2008) Reply to comments by Matthias Kuhle on Seong YB, Owen LA, Bishop MP, Bush A, Clendon P, Copland P, Finkel RC, Kamp U, Shroder JF, (2007) Quaternary Glacial History of the Central Karakoram. Quat Sci Rev 27:1656–1658

Seong YB, Owen LA, Yi C, Finkel RC (2009) Quaternary glaciation of Muztag Ata and Kongur Shan: evidence for glacier response to rapid climate changes throughout the Late Glacial and Holocene in westernmost Tibet. Geol Soc Am Bull 121:348–365

Saha S, Owen LA, Orr EN, Caffee MW (2018b) Timing and nature of Holocene glacier advances at the northwestern end of the Himalayan-Tibetan orogen. Quat Sci Rev 187:177–202

Sharma MC, Owen LA (1996) Quaternary glacial history of NW Garhwal Himalayas. Quat Sci Rev 15:335–365

Shi Y, Zheng B, Li S (1992) Last glaciation and maximum glaciation in the Qinghai-Xizang (Tibet) Plateau: a controversy to M. Kuhle's ice sheet hypothesis. Z Geomorphol 84:19–35

Shiraiwa T, Watanabe T (1991) Late Quaternary glacial fluctuations in the Langtang Valley, Nepal Himalaya, reconstructed by relative dating methods. Arct Alp Res 23:404–416

Solomina O, Bradley RS, Hodgson DA, Ivy-Ochs S, Jomelli V, Mackintosh AN, Nesje A, Owen LA, Wanner H, Wiles GC, Young NE (2015) Holocene glacier fluctuations. Invited review. Quat Sci Rev 111:9–34

Solomina O, Bradley R, Jomelli V, Geirsdottir A, Kaufman D, Koch J, McKay NP, Masiokas M, Miller G, Nesje A, Nicolussi K, Owen LA, Putnam AE, Wanner H, Wiles G, Yan B (2016) Glacier fluctuations during the past 2000 years. Quat Sci Rev 149:61–90

Spencer JQ, Owen LA (2004) Optically stimulated luminescence dating of late Quaternary glaciogenic sediments in the upper Hunza valley: validating the timing of glaciation and assessing dating methods. Quat Sci Rev 23:175–191

Taylor PJ, Mitchell WA (2000) Late Quaternary glacial history of the Zanskar Range, North-west Indian Himalaya. Quat Int 65(66):81–100

Thackray GD, Owen LA, Yi C (2008) Timing and nature of late Quaternary mountain glaciation. J Quat Sci 23:503–508

Van der Woerd J, Owen LA, Tapponnier P, Xiwei X, Kervyn F, Finkel RC, Barnard PL (2004) Giant, M~8 earthquake-triggered, ice avalanches in the eastern Kunlun Shan (Northern Tibet): characteristics, nature and dynamics. Geol Soc Am Bull 116:394–406

Watanabe T, Shiraiwa T, Ono Y (1989) Distribution of periglacial landforms in the Langtang Valley, Nepal Himalaya. Bull Glacier Res 7:209–220

Yan Q, Owen LA, Wang H, Zhang Z, Sun J (2018) Climate constrains glacier advance over high Mountain Asia during the last glacial maximum. Earth Planet Sci Lett 45:9024–9033

Yi C, Chen H, Yang J, Liu B, Fu P, Liu K, Li S (2008) Review of Holocene glacial chronologies based on radiocarbon dating in Tibet and its surrounding mountains. J Quat Sci 23:533–558

Zech W, Glaser B, Sosin P, Kubik PW, Zech W (2003) Evidence for long-lasting landform surface instability on hummocky moraines in the Pamir Mountains (Tajikistan) from ^{10}Be surface exposure dating. Earth Planet Sci Lett 237:453–461

Zech R, Abramowski U, Glaser B, Sosin P, Kubik PW, Zech W (2005) Late Quaternary glacier and climate history of the Pamir Mountains derived from cosmogenic [10]Be exposure ages. Quat Res 64:212–220

Zheng B, Rutter N (1998) On the problem of Quaternary glaciations, and the extent and patterns of Pleistocene ice cover in the Qinghai-Xizang (Tibet) plateau. Quat Int 45(46):109–122

Zhou S, Li J, Zhao J, Wang J, Zheng J (2011) Quaternary glaciations: extent and chronology in China. In: Elhers J, Gibbard P, Hughes PD (eds) Quaternary glaciations – extent and chronology: a closer look, Dev Quat Sci, vol 15, 2nd edn. Elsevier, Amsterdam, pp 981–1002

Chapter 14
Summer Monsoon Variability in the Himalaya Over Recent Centuries

Masaki Sano, Chenxi Xu, A. P. Dimri, and R. Ramesh

Abstract Hydroclimatic variations during the summer monsoon season (June–September) across the Himalaya are examined over the past several hundred years using tree-ring oxygen isotope records. Owing to their strong associations with hydroclimatic variables including precipitation, relative humidity, and the Palmer Drought Severity Index, tree-ring $\delta^{18}O$ chronologies from the Himalaya can be used to reconstruct summer monsoon intensity precisely. A regional chronology derived from five local chronologies across the Himalaya shows a significant correlation with Indian summer rainfall data. One of the most noteworthy features of the regional chronology is a drying trend over the past 180 years, indicating that summer monsoon intensity in the Himalayan region has weakened. A declining land–ocean thermal gradient over South Asia seems to be responsible for the weakened summer monsoon. By analyzing spatio-temporal correlations between zonally distributed tree-ring data over the Himalaya, we also explore possible changes in the relative contributions of source water originating in the Arabian Sea and the Bay of Bengal.

M. Sano (✉)
Faculty of Human Sciences, Waseda University, Tokorozawa, Japan
e-mail: fokienia@gmail.com

C. Xu
Institute of Geology and Geophysics, Chinese Academy of Sciences, Beijing, China

A. P. Dimri
School of Environmental Sciences, Jawaharlal Nehru University, New Delhi, India

R. Ramesh
School of Earth and Planetary Sciences, National Institute of Science Education and Research, Odisha, India

© Springer Nature Switzerland AG 2020
A. P. Dimri et al. (eds.), *Himalayan Weather and Climate and their Impact on the Environment*, https://doi.org/10.1007/978-3-030-29684-1_14

261

14.1 Introduction

Rain, snow, and glacial meltwater generated in mountainous regions such as the Himalaya play a critical role in sustaining the water supply to populations in adjacent lowlands (Immerzeel et al. 2010; Viviroli et al. 2007). Climatic changes are expected to seriously affect hydrological processes (Barnett et al. 2005). Future changes over the Himalaya and the Tibetan Plateau in terms of flood/drought amplitude and frequency, the hydrological cycle, and seasonality should therefore be accurately predicted to ensure the welfare of people in South Asia. As instrumental weather data available in the Himalaya and on the Tibetan Plateau are fairly limited both temporally and spatially, annually or decadally resolved paleoclimatic records derived from tree rings, speleothems, lake sediments, and ice cores are essential to improve our understanding of the dynamics and causes of the Indian monsoon.

Climate model experiments conducted by Meehl and Washington (1993) and May (2002) predicted that the Indian summer monsoon rainfall would increase under enhanced greenhouse gas levels. In addition, sediment records derived from the Arabian Sea reveal a strengthening of monsoon winds over recent centuries (Anderson et al. 2002). However, a significant decrease in summer monsoon rainfall over the Himalayan foothills was detected from instrumental data over the past 56 years (Wang and Ding 2006). This finding contrasts with model experiments, indicating the complex dynamics of the Indian monsoon. The spatial coverage of annually resolved paleoclimatic records must therefore be improved to allow a better understanding of the long-term changes in atmospheric circulation induced by the Indian monsoon system.

Tree rings have been widely used to reconstruct past temperatures and precipitation for the Himalayan region (e.g., Borgaonkar et al. 1994; Cook et al. 2003; Hughes 1992; Sano et al. 2005; Singh et al. 2006; Yadav et al. 2015). Previous studies have generally focused on ring widths; however, the widths are regulated by moisture availability in the hot and dry pre-monsoon season (e.g., Sano et al. 2005; Singh et al. 2006). Reconstructions of temperature and precipitation in the monsoon season (June–September) are therefore rather limited, owing partly to the reduced climatic sensitivity of tree growth associated with abundant summer precipitation. In spite of the limitations, Krusic et al. (2015) produced a 638-year summer temperature reconstruction for the Bhutan Himalaya and found a significant correlation with major volcanic eruptions, and a possible link between solar variability and decadal-scale temperature variability.

Recent progress in isotope dendroclimatology has revealed that oxygen isotope ratios ($\delta^{18}O$) recorded in tree-ring cellulose can be utilized to reconstruct hydroclimatic variations precisely (precipitation, relative humidity, cloud cover, and drought index) during the monsoon season in the Himalaya and on the Tibetan Plateau (e.g., Liu et al. 2013; Sano et al. 2012, 2013, 2017; Shi et al. 2012; Wernicke et al. 2015; Xu et al. 2012). In this chapter, we review monsoon variability over the past few centuries in the Himalaya and on the Tibetan Plateau, focusing on summer monsoon dynamics based on tree-ring oxygen isotope records (Sano et al. 2012, 2013, 2017; Xu et al. 2018).

14.2 Past Changes in Summer Monsoon Rainfall

A regional tree-ring $\delta^{18}O$ chronology based on five local chronologies distributed across the Himalaya (Fig. 14.1) was developed to reconstruct summer monsoon rainfall in the northern Indian subcontinent (Xu et al. 2018). The chronologies originate from five study sites: Wache in Bhutan (Sano et al. 2013), Ganesh (Xu et al. 2018) and Humla (Sano et al. 2012) in Nepal, and Jageshwar (Xu et al. 2018) and Manali (Sano et al. 2017) in India. For each study site, 3–5 trees were selected for isotopic analysis. The oxygen isotope ratio ($\delta^{18}O$) of each growth ring in each tree sample was individually measured using an isotope ratio mass spectrometer. Thus, 3–5 $\delta^{18}O$ values were obtained for each year at each site. The local $\delta^{18}O$ chronology was then produced by averaging the 3–5 separate $\delta^{18}O$ time series at each site (Fig. 14.2a–e). The strength of common variations in tree-ring $\delta^{18}O$ between

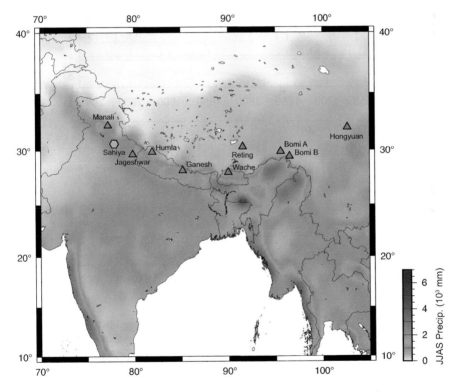

Fig. 14.1 Map of the study region showing the locations of sites with tree-ring $\delta^{18}O$ chronologies (red triangles) and speleothem $\delta^{18}O$ data (orange hexagon). The tree-ring data are from the Manali (Sano et al. 2017), Jageshwar (Xu et al. 2018), Humla (Sano et al. 2012), Ganesh (Xu et al. 2018), Wache (Sano et al. 2013), Reting (Grießinger et al. 2011), Bomi A (Shi et al. 2012), Bomi B (Liu et al. 2013), and Hongyuan sites (Xu et al. 2012). The speleothem $\delta^{18}O$ data are from Sahiya Cave (Sinha et al. 2015). Background colors represent the long-term (1951–2010 CE) mean June–September precipitation, calculated using the GPCC ver.6 datasets (Schneider et al. 2011)

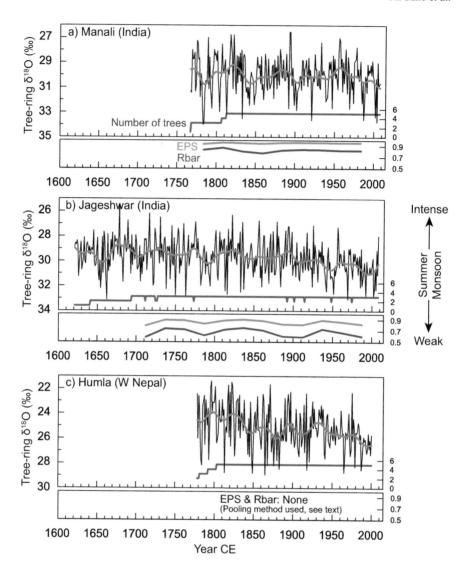

Fig. 14.2 Local tree-ring δ¹⁸O chronologies with sample sizes, EPS, and RBar, for: (**a**) Manali (Sano et al. 2017), (**b**) Jageshwar (Xu et al. 2018), (**c**) Humla (Sano et al. 2012), (**d**) Ganesh (Xu et al. 2018), and (**e**) Wache (Sano et al. 2013). (**f**) Regional tree-ring δ¹⁸O chronology (vertical axis inverted). Running EPS and Rbar values are calculated over 50 years, lagged by 25 years

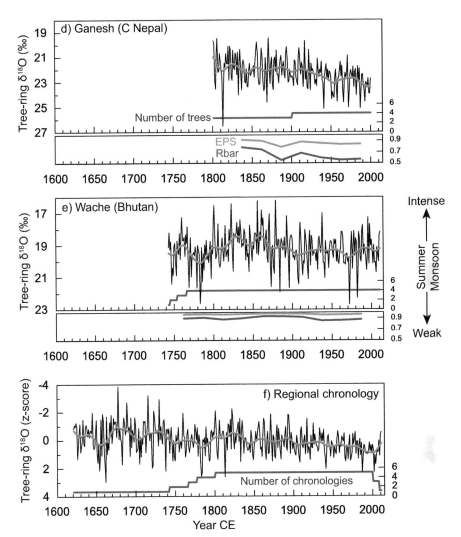

Fig. 14.2 (continued)

individual trees was evaluated by calculating the inter-series correlation (Rbar) and the expressed population signal (EPS) (Wigley et al. 1984). As the Rbar and EPS statistics were significant at all study sites (Fig. 14.2), we were able to develop robust local chronologies. For the five trees sampled at the Humla site, the five individual growth rings for each calendar year were pooled and homogenized before isotopic measurement. We therefore produced only a single $\delta^{18}O$ series, which was used as the local $\delta^{18}O$ chronology for the Humla site (Sano et al. 2012).

We next conducted linear correlation analyses between each local $\delta^{18}O$ chronology and the monthly climate records (i.e., precipitation, temperature, relative humidity, and the drought index) to identify the climatic response of tree-ring $\delta^{18}O$. For the drought index, we used the self-calibrating Palmer Drought Severity Index or scPDSI (van der Schrier et al. 2013), a drought metric based on soil moisture availability. As we described the climatic responses in previous work (Sano et al. 2012, 2013, 2017; Xu et al. 2018), only the responses of local $\delta^{18}O$ chronologies to area-averaged precipitation (CRU TS4.01 dataset, Harris et al. 2014) and scPDSI are presented in Fig. 14.3. Overall, all the local chronologies show significant negative correlations with precipitation, relative humidity and/or the scPDSI in the monsoon season (June–September), indicating that tree-ring $\delta^{18}O$ is controlled primarily by moisture conditions during the growth season (Sano et al. 2012, 2013, 2017; Xu et al. 2018). These climatic responses are consistent with the theory of tree-ring $\delta^{18}O$, as follows.

The oxygen isotope composition in tree-ring cellulose is controlled mainly by two climatic factors: $\delta^{18}O$ of the source water and relative humidity, although there are several isotopic fractionations before cellulose synthesis (McCarroll and Loader 2004; Roden et al. 2000; Sternberg 2009). The source water for tree growth is soil moisture, supplied mainly by precipitation. The $\delta^{18}O$ signal in precipitation is therefore preserved in tree-ring $\delta^{18}O$. Negative correlations between precipitation $\delta^{18}O$ and precipitation amount have been observed in low latitudes, known as the 'amount effect' (Araguás-Araguás et al. 1998; Dansgaard 1964). The precipitation amount signal is therefore preserved in tree-ring $\delta^{18}O$. Soil water, with its preserved precipitation $\delta^{18}O$ signal, is taken up by tree roots without isotopic fractionation (White et al. 1985) and is transported to the leaves through xylem. The oxygen isotope composition in leaf water is modulated by transpiration through the stomata, leading to the preferential loss of lighter isotopes (^{16}O) and consequent enrichment of heavier isotopes (^{18}O) in the leaf water (Roden et al. 2000; Sternberg 2009). During transpiration, lower (higher) relative humidity increases (decreases) the $\delta^{18}O$ in the leaf water. Consequently, there is an inverse correlation between tree-ring $\delta^{18}O$ and relative humidity. Further details of climatic responses in the five local $\delta^{18}O$ chronologies are given by Sano et al. (2012, 2013, 2017) and Xu et al. (2018).

As listed in Table 14.1, the five $\delta^{18}O$ chronologies across the Himalaya are significantly correlated, even though the locations are distant from one other (Xu et al. 2018). This indicates that regional hydroclimate signals related to the Indian monsoon are well recorded in the five chronologies. We therefore integrated the five $\delta^{18}O$ chronologies into a single regional chronology (Fig. 14.2f). A spatial correlation analysis by Xu et al. (2018) revealed that the regional chronology is inversely

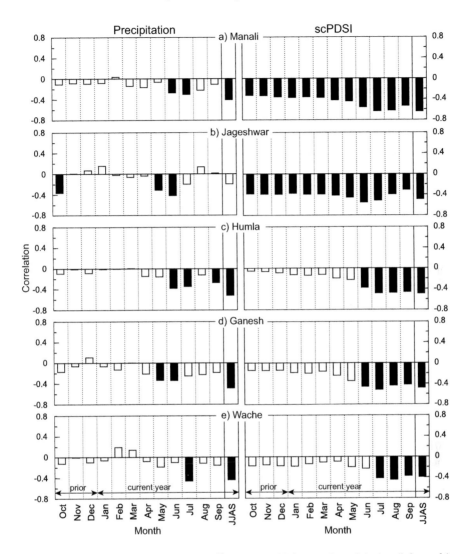

Fig. 14.3 Correlations between tree-ring $\delta^{18}O$ and monthly/seasonal precipitation (left panels) and scPDSI (right panels) for the period 1901–2000 CE, at (**a**) Manali (Sano et al. 2017), (**b**) Jageshwar (Xu et al. 2018), (**c**) Humla (Sano et al. 2012), (**d**) Ganesh (Xu et al. 2018), and (**e**) Wache (Sano et al. 2013). Area-averaged precipitation (for $5° \times 5°$ grids covering each study site) and scPDSI derived from gridded datasets were correlated with tree-ring $\delta^{18}O$. A 12-month window (preceding-year October–September) was used to calculate correlation coefficients. Black bars represent significance at the 1% level

Table 14.1 Correlation matrix calculated using five tree-ring $\delta^{18}O$ chronologies for the common period 1801–2000 CE

R (1801–2000 CE)	Manali (WH, India)	Jageshwar (WH, India)	Humla (W Nepal)	Ganesh (C Nepal)
Jageshwar (Western Himalaya, India)	0.49			
Humla (Western Nepal)	0.50	0.53		
Ganesh (Central Nepal)	0.47	0.66	0.60	
Wache (Bhutan)	0.21	0.33	0.39	0.52

correlated with gridded June–September precipitation in northern India and Nepal. The regional chronology also shows a significant correlation ($r = -0.50$, $p < 0.001$) with the All India Summer (June–September) Monsoon Rainfall (Kothawale and Rajeevan 2017; Parthasarathy et al. 1994) for the period 1871–2011 (Fig. 14.4), clearly demonstrating that the synthesized chronology reflects large-scale variations in summer monsoon intensity. We further calculated split-period calibration and verification statistics, commonly used in dendroclimatology, to scrutinize the reliability of the transfer function used to reconstruct the All India Summer Monsoon Rainfall. As listed in Table 14.2, the linear regressions and sign tests are significant at the 1% level. In addition, the reduction of error (RE) and the coefficient of efficiency (CE) are positive, passing rigorous tests for verification. Our regional tree-ring $\delta^{18}O$ chronology is therefore considered a robust proxy of the All India Summer Monsoon Rainfall.

Perhaps one of the most noteworthy features in the regional chronology is a significant increasing $\delta^{18}O$ trend since 1820 CE (Mann–Kendall trend test, $p < 0.001$), indicating a weakening tendency of the summer monsoon in the Himalaya (Xu et al. 2018). Similar strong trends are also seen in the local $\delta^{18}O$ chronologies from Jageshwar in India, and Humla and Ganesh in Nepal, whereas the other local chronologies show rather weak trends. Tree-ring $\delta^{18}O$ chronologies from the Tibetan Plateau (Grießinger et al. 2011; Liu et al. 2013; Shi et al. 2012; Xu et al. 2012) also show weakening trends in summer monsoon intensity (Fig. 14.5; see Fig. 14.1 for the locations). Furthermore, Liu et al. (2014) identified a shift in the hydroclimatic regime from wet/cold to dry/warm conditions since 1860 CE in their regional tree-ring $\delta^{18}O$ chronology and temperature-sensitive tree-ring data from the Tibetan Plateau. These findings indicate that summer monsoon incursion into the Tibetan Plateau has weakened over the past 150 years.

Sinha et al. (2015) reported a drying trend since the 1950s in speleothem $\delta^{18}O$ data from Sahiya Cave, located 200 km southwest of the Manali site (see Fig. 14.1) in northern India. Interestingly, the summer rainfall reconstructed from the Sahiya Cave data is positively correlated with Northern Hemisphere (NH) temperatures on multi-centennial scales over the past two millennia. In contrast, the drying trend since the 1950s is opposite to the NH temperature rise, suggesting that anthropogenic aerosol release over South Asia has counterbalanced an increase in rainfall

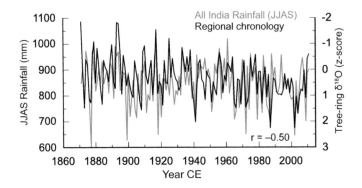

Fig. 14.4 Comparison of the regional tree-ring δ¹⁸O chronology and the All India Summer (June–September) Monsoon Rainfall (Kothawale and Rajeevan 2017; Parthasarathy et al. 1994) for the common period 1871–2011 CE

Table 14.2 Calibration and verification statistics for the June–September All-India Rainfall reconstruction

Calibration period	R^2	Verification period	r	RE	CE	sign test
Full (1871–2011)	0.251[a]		–	–	–	–
Early half (1871–1941)	0.248[a]	Late half (1942–2011)	0.565[a]	0.215[b]	0.215[b]	50/20[a]
Late half (1942–2011)	0.319[a]	Early half (1871–1941)	0.498[a]	0.059[b]	0.058[b]	49/22[a]

[a]$p < 0.01$
[b]RE, CE > 0

over the Indian Subcontinent. Sinha et al. (2015) also noted that the current drying trend detected in the speleothem δ¹⁸O data is not unprecedented in the past 2000 years.

What mechanisms have contributed to the weakening of the summer monsoon in the Himalaya and on the Tibetan Plateau? A significant decreasing trend in summer precipitation over the Himalayan foothills during the period 1901–2012 was reported by Roxy et al. (2015). Their analyses, based on observed records and model experiments, indicate that a declining land–ocean thermal gradient over South Asia was responsible for the weakening summer monsoon. Specifically, a rapid warming in the Indian Ocean and a relatively subdued warming over the Indian continent have weakened the land–ocean thermal gradient, thereby dampening the summer monsoon Hadley circulation (Roxy et al. 2015). In this context, Xu et al. (2018) explored the dynamics of the land–ocean thermal gradient in the past using available proxy records (Cook et al. 2013; Shi et al. 2015; Tierney et al. 2015; Wang et al. 2015). As shown in Fig. 14.6, variations in the thermal contrast derived from land and ocean temperature records are consistent with those in the regional tree-ring δ¹⁸O chronology over the past 250 years, supporting the interpretation of a link between the weakened thermal contrast and intensified aridity since 1820 CE in the Himalaya.

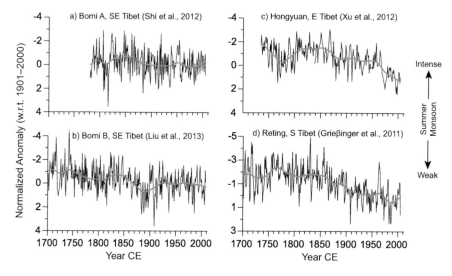

Fig. 14.5 Tree-ring δ¹⁸O chronologies from the following study sites: (**a**) Bomi A (Shi et al. 2012) and (**b**) Bomi B (Liu et al. 2013) on the southeastern Tibetan Plateau, (**c**) Hongyuan on the eastern Tibetan Plateau (Xu et al. 2012), and (**d**) Reting on the southern Tibetan Plateau (Grießinger et al. 2011), modified from Sano et al. (2013). The red lines represent 50-year low-pass spline filtered values. The locations of sampling sites are shown in Fig. 14.1. Note that the original chronologies for (**b**) and (**d**) cover the past 409 and 842 years, respectively (data prior to 1700 are not shown)

Whereas climate simulation experiments predict a strengthening of the Indian summer monsoon under global warming scenarios (May 2002; Meehl and Washington 1993), tree-ring records provide compelling evidence of a weakened summer monsoon in the Himalaya (Xu et al. 2018) and on the Tibetan Plateau (Liu et al. 2014). Our results also contrast with a tree-ring δ¹⁸O chronology from northern Pakistan, in which the twentieth century was the wettest period over the past millennium (Treydte et al. 2006). Furthermore, a network of tree-ring-width chronologies from the semi-arid region of the northwest Himalaya reveals a general wetting trend since the mid-seventeenth century, with the wettest interval detected in recent decades (Yadav et al. 2017). The regions examined in those studies are mainly dominated by westerly synoptic fronts throughout the year, with the maximum precipitation in winter and spring. It is therefore not surprising that their tree-ring records show the opposite trend to our monsoon reconstruction. The regional wetting trends during winter and spring in Pakistan and the northwest Himalaya indicate a strengthening of western disturbances, contrasting with the weakened summer monsoon precipitation found in our tree-ring data.

It is widely recognized that the El Niño–Southern Oscillation (ENSO) plays an important role in modulating the summer monsoon in South Asia (e.g., Rasmusson and Carpenter 1983; Ropelewski and Halpert 1987; Shukla and Paolino 1983). Spectral analyses based on the regional δ¹⁸O chronology from the Himalaya reveal

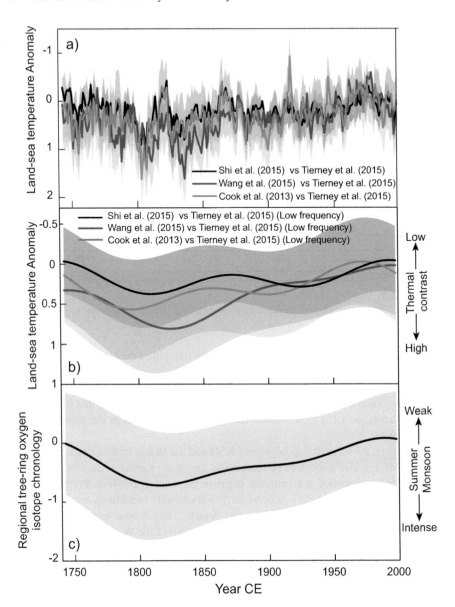

Fig. 14.6 (a) Land–ocean temperature anomalies (thermal gradients) calculated using three sum-mer temperature reconstructions for the Tibetan Plateau (Cook et al. 2013; Shi et al. 2015; Wang et al. 2015) and one Indian Ocean SST reconstruction (Tierney et al. 2015). (b) Same as (a) but for low-frequency anomalies. (c) Low-frequency variations in the regional tree-ring $\delta^{18}O$ chronology, modified from Xu et al. (2018). Shading indicates the uncertainty for each time series

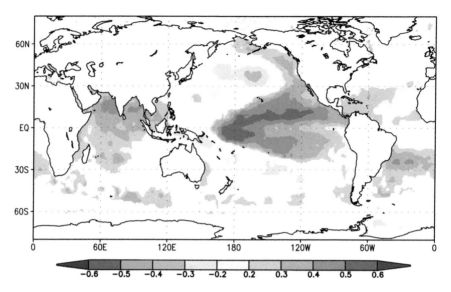

Fig. 14.7 Spatial correlations between the regional tree-ring δ^{18}O chronology and global SSTs for the period 1871–2011 CE

significant peaks over 2–5 year intervals, which fall within the range of ENSO variability (Xu et al. 2018). In addition, our regional chronology is significantly correlated with sea surface temperatures (SSTs) for central and eastern parts of the equatorial Pacific over the period 1871–2011 (Fig. 14.7), indicating that El Niño (La Niña) phases have led to decreased (increased) precipitation in the summer monsoon season.

However, the relationship between ENSO and the Indian summer monsoon rainfall (ISMR) is also known to be temporally unstable (Kumar et al. 1999). Kumar et al. (1999) identified a significant negative correlation between the ISMR and NINO3 SSTs for the century prior to the 1980s (drought conditions arising from El Niño phases), but no correlation after that decade. They demonstrated that a southeastward shift in the location of the descending limb of the Walker circulation contributed to this time-dependent result. Kumar et al. (2006) also pointed out that the central Pacific El Niño, characterized by the warmest SSTs prevailing in the central equatorial Pacific, is more effective in generating the descending limb over India (thereby resulting in drought) than the conventional eastern Pacific El Niño.

Thirty-one-year running correlations between the regional δ^{18}O chronology and two reconstructed ENSO indices (Emile-Geay et al. 2013; McGregor et al. 2010) also show an unstable relationship between ENSO and summer monsoon rainfall over the past 250 years (Xu et al. 2018). As pointed out by Kumar et al. (2006), the unstable association might be attributed in part to the two different types of ENSO (developed mainly in the eastern or central Pacific). Here, we cannot evaluate the influence of different ENSO types on this time dependency, because the two types of ENSO have not been reconstructed separately. Continued effort towards a spatial reconstruction of SSTs, divided into two distinct chronologies corresponding to the two ENSO types, will shed more light on the ENSO–monsoon linkage during the pre-instrumental era.

It should also be noted that the ISMR is known to be linked to the equatorial Indian Ocean. For example, Ashok et al. (2001) found that whenever the ISMR–ENSO correlation is high (low), the ISMR–IOD (Indian Ocean Dipole) correlation is low (high), indicating that the IOD plays an important role in modulating the ISMR. Our regional $\delta^{18}O$ data show significant correlations with SSTs in the Indian Ocean (Fig. 14.7), although the correlations are rather weak compared with those in the tropical Pacific. Thirty-one-year running correlations between the reconstructed Indian Ocean SSTs (Tierney et al. 2015) and the two reconstructed ENSO indices (Emile-Geay et al. 2013; McGregor et al. 2010) were calculated to evaluate the relationship between the Indian Ocean and the ENSO over the past 250 years. The results indicate that variations in the Indian Ocean SSTs are mostly consistent with those of the ENSO indices, but that the association collapsed or was reversed during the period 1820–1860 CE (Fig. 14.8b) (Xu et al. 2018). Interestingly, the association between Indian Ocean SSTs and ENSO indices (Fig. 14.8b) decoupled when the ISMR–ENSO association was weak (Fig. 14.8a). These results suggest that the weak ISMR–ENSO correlation seen in Fig. 14.8a may be related to the decoupled association between Indian Ocean SSTs and ENSO indices.

14.3 Past Changes in Moisture Flux from the Oceans

Tree-ring oxygen isotopes contain not only local hydroclimate signals, but also moisture-source signals originating far from the sampling site. This is because the oxygen isotope compositions of precipitation at a given site are governed by vapor sources and transport processes (Araguás-Araguás et al. 1998; Midhun and Ramesh 2016). Therefore, the integrated history of upstream processes, such as evaporation of water from the ocean and subsequent condensation of water vapor through atmospheric circulation, is recorded in the precipitation $\delta^{18}O$ at the sampling site (Araguás-Araguás et al. 1998; Kurita et al. 2009). This is, in turn, preserved in the tree-ring $\delta^{18}O$. Summer precipitation over the Himalayan foothills has two moisture sources: the Arabian Sea and the Bay of Bengal (Sengupta and Sarkar 2006; Sinha et al. 2015). Water vapor supply to the Himalaya originates predominantly in the Bay of Bengal. Water vapor originating in the Arabian Sea is another important source, especially for the western Himalaya.

Figure 14.9 shows spatial correlations for the period 1951–2000 CE between each local chronology and the gridded CRU TS4.01 precipitation dataset (Harris et al. 2014). The highest correlations are not observed between tree-ring $\delta^{18}O$ and precipitation at the sampling site, but with precipitation in the upstream region. More specifically, water vapor originating in the Arabian Sea seems to be transported to the western Himalaya (India) (Fig. 14.9a). On the other hand, moisture-source signals for the eastern Himalaya (Bhutan) appear in the Bay of Bengal,

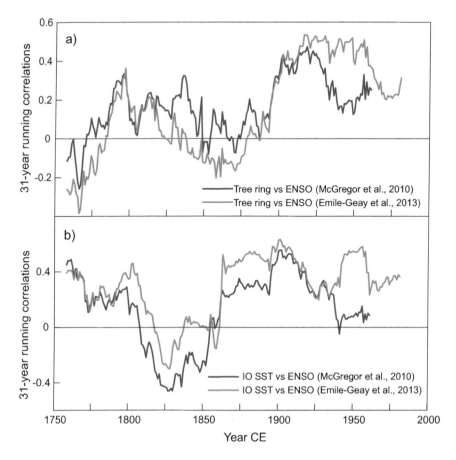

Fig. 14.8 Thirty-one-year running correlations between two ENSO reconstructions (Emile-Geay et al. 2013; McGregor et al. 2010) and (**a**) the regional tree-ring $\delta^{18}O$ chronology and (**b**) Indian Ocean SST reconstruction (Tierney et al. 2015), modified from Xu et al. (2018)

although the upstream signal is rather weak (Fig. 14.9c). With its location between western and eastern sites, the local $\delta^{18}O$ chronology from the Humla site (Nepal) indicates moisture supply from both the Arabian Sea and the Bay of Bengal (Fig. 14.9b). Although these analyses were limited to the past 50 years, they do reveal the mean states of moisture sources at the three sites.

Sano et al. (2017) investigated possible changes in the relative contributions of source water originating in the Arabian Sea and the Bay of Bengal at three sampling sites across the Himalaya. To examine the patterns over time, they calculated 31-year running correlations between the local $\delta^{18}O$ chronologies (Fig. 14.10d). As the sites in India and Nepal are located close to each other, the two $\delta^{18}O$ chronologies are significantly correlated over the entire period, except ca. 1930–1960 CE. In contrast, $\delta^{18}O$ values of the India chronology are less commonly correlated with the Bhutan chronology. Interestingly, the periods showing weak correlations between

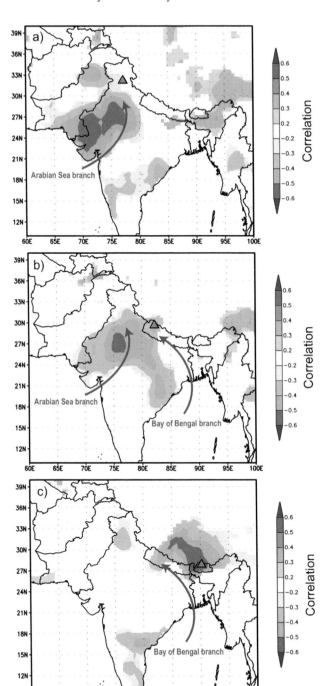

Fig. 14.9 Spatial correlations between local tree-ring δ^18O chronologies generated for (**a**) Manali in India (Sano et al. 2017), (**b**) Humla in Nepal (Sano et al. 2012), and (**c**) Wache in Bhutan (Sano et al. 2013), and June–September precipitation (CRU TS4.01 dataset) for the period 1951–2000 CE. Sites are represented by red triangles

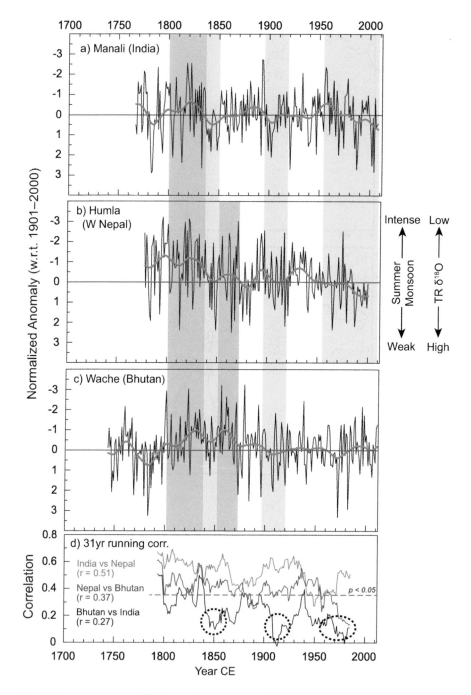

Fig. 14.10 Local tree-ring $\delta^{18}O$ chronologies across the Himalaya from (**a**) Manali in India (Sano et al. 2017), (**b**) Humla in Nepal (Sano et al. 2012), and (**c**) Wache in Bhutan (Sano et al. 2013). (**d**) Thirty-one-year running correlations between the chronologies, modified from Sano et al. (2017). The red lines in (**a**)–(**c**) represent 30-year low-pass spline filtered values

the India and Bhutan $\delta^{18}O$ data correspond to relatively dry conditions for the entire Himalaya, as highlighted by the yellow bands in Fig. 14.10a–c. These weakened correlations between the India and Bhutan $\delta^{18}O$ chronologies probably originate from changes in relative contributions of the source water transported from the Arabian Sea and the Bay of Bengal to these sampling sites.

A backward moisture trajectory analysis performed by Sinha et al. (2015) revealed that a weak (strong) monsoon circulation enhanced (reduced) the flux of ^{18}O-enriched Arabian Sea moisture and reduced (enhanced) the flux of ^{18}O-depleted Bay of Bengal moisture to their speleothem site in the western Himalaya. Their results indicate that the moisture flux from the Arabian Sea (the Bay of Bengal) to our India site increased in dry (wet) years. In contrast, the flux of moisture originating in the Arabian Sea seems to be negligible for the Bhutan site, even in dry years, because the moisture at this site originates mainly from the Bay of Bengal (Fig. 14.9c).

In general, in dry years with an enhanced flux of Arabian Sea moisture to the India site, but with no such flux to the Bhutan site, the correlation between the tree-ring $\delta^{18}O$ records of India and Bhutan collapses (Fig. 14.10). In wet years, on the other hand, an enhanced flux of Bay of Bengal moisture to the India site, with a similar flux prevailing at the Bhutan site, can result in significant correlation between the two sites (Fig. 14.10).

14.4 Conclusions

We developed a regional tree-ring $\delta^{18}O$ chronology based on five local chronologies across the Himalaya. The annually resolved chronology precisely records summer monsoon rainfall over the past several hundred years. The chronology shows an increasing $\delta^{18}O$ trend since 1820 CE, clearly indicating a weakening tendency in the summer monsoon over the Himalaya. Similarly, a regional tree-ring $\delta^{18}O$ chronology and temperature-sensitive tree-ring data from the Tibetan Plateau show a shift in the hydroclimatic regime from wet/cold to dry/warm conditions since 1860 CE.

These findings indicate that summer monsoon incursion into the Himalaya and the Tibetan Plateau has weakened over the past 150 years. A declining land–ocean thermal gradient over South Asia is likely to be responsible for the weakening of the summer monsoon. In addition to the reconstruction of summer precipitation, tree-ring $\delta^{18}O$ chronologies across the Himalaya can be used to understand changes in the relative contributions of source water originating in the Arabian Sea and the Bay of Bengal to the sampling sites. Continued effort towards the development of a dense tree-ring $\delta^{18}O$ network, combined with other proxies, will improve our understanding of the dynamics and causes of monsoon variability over the Himalaya and the Tibetan Plateau.

References

Anderson DM, Overpeck JT, Gupta AK (2002) Increase in the Asian southwest monsoon during the past four centuries. Science 297:596–599

Araguás-Araguás L, Froehlich K, Rozanski K (1998) Stable isotope composition of precipitation over southeast Asia. J Geophys Res 103(D22):28721–28742

Ashok K, Guan Z, Yamagata T (2001) Impact of the Indian Ocean dipole on the relationship between the Indian monsoon rainfall and ENSO. Geophys Res Lett 28:4499–4502

Barnett TP, Adam JC, Lettenmaier DP (2005) Potential impacts of a warming climate on water availability in snow-dominated regions. Nature 438:303–309

Borgaonkar HP, Pant GB, Rupa Kumar K (1994) Dendroclimatic reconstruction of summer precipitation at Srinagar, Kashmir, India, since the late-eighteenth century. The Holocene 4(3):299–306

Cook ER, Krusic PJ, Jones PD (2003) Dendroclimatic signals in long tree-ring chronologies from the Himalayas of Nepal. Int J Climatol 23:707–732

Cook ER et al (2013) Tree-ring reconstructed summer temperature anomalies for temperate East Asia since 800 C.E. Clim Dyn 41(11):2957–2972

Dansgaard W (1964) Stable isotopes in precipitation. Tellus 16:436–468

Emile-Geay J, Cobb KM, Mann ME, Wittenberg AT (2013) Estimating central equatorial Pacific SST variability over the past millennium. Part II: reconstructions and implications. J Clim 26(7):2329–2352

Grießinger J, Bräuning A, Helle G, Thomas A, Schleser G (2011) Late Holocene Asian summer monsoon variability reflected by $\delta^{18}O$ in tree-rings from Tibetan junipers. Geophys Res Lett 38(3):L03701. https://doi.org/10.1029/2010GL045988

Harris I, Jones PD, Osborn TJ, Lister DH (2014) Updated high-resolution grids of monthly climatic observations – the CRU TS3.10 dataset. Int J Climatol 34(3):623–642

Hughes MK (1992) Dendroclimatic evidence from the western Himalaya. In: Bradley RS, Jones PD (eds) Climate since A.D. 1500. Routledge, London, pp 415–431

Immerzeel WW, van Beek LPH, Bierkens MFP (2010) Climate change will affect the Asian water towers. Science 328(5984):1382–1385

Kothawale DR, Rajeevan M (2017) Monthly, seasonal and annual rainfall time series for all-India, homogeneous regions and meteorological subdivisions: 1871–2016. IITM research report no. RR-138

Krusic PJ et al (2015) Six hundred thirty-eight years of summer temperature variability over the Bhutanese Himalaya. Geophys Res Lett 42(8):2988–2994

Kumar KK, Rajagopalan B, Cane MA (1999) On the weakening relationship between the Indian monsoon and ENSO. Science 284(5423):2156–2159

Kumar KK, Rajagopalan B, Hoerling M, Bates G, Cane M (2006) Unraveling the mystery of Indian monsoon failure during El Niño. Science 314(5796):115–119

Kurita N, Ichiyanagi K, Matsumoto J, Yamanaka MD, Ohata T (2009) The relationship between the isotopic content of precipitation and the precipitation amount in tropical regions. J Geochem Explor 102(3):113–122

Liu X et al (2013) A 400-year tree-ring $\delta^{18}O$ chronology for the southeastern Tibetan Plateau: implications for inferring variations of the regional hydroclimate. Glob Planet Chang 104:23–33

Liu X et al (2014) A shift in cloud cover over the southeastern Tibetan Plateau since 1600: evidence from regional tree-ring $\delta^{18}O$ and its linkages to tropical oceans. Quat Sci Rev 88(0):55–68

May W (2002) Simulated changes of the Indian summer monsoon under enhanced greenhouse gas conditions in a global time-slice experiment. Geophys Res Lett 29(7). https://doi.org/10.1029/2001GL013808

McCarroll D, Loader NJ (2004) Stable isotopes in tree rings. Quat Sci Rev 23:771–801

McGregor S, Timmermann A, Timm O (2010) A unified proxy for ENSO and PDO variability since 1650. Clim Past 6(1):1–17

Meehl GA, Washington WM (1993) South Asian summer monsoon variability in a model with doubled atmospheric carbon dioxide concentration. Science 260(5111):1101–1104

Midhun M, Ramesh R (2016) Validation of $\delta^{18}O$ as a proxy for past monsoon rain by multi-GCM simulations. Clim Dyn 46(5):1371–1385

Parthasarathy B, Monot AA, Kothawale DR (1994) All-India monthly and seasonal raifall series: 1871–1993. Theor Appl Climatol 49:217–224

Rasmusson EM, Carpenter TH (1983) The relationship between eastern equatorial Pacific sea surface temperatures and rainfall over India and Sri Lanka. Mon Weather Rev 111:517–528

Roden JS, Lin G, Ehleringer JR (2000) A mechanistic model for interpretation of hydrogen and oxygen isotope ratios in tree-ring cellulose. Geochim Cosmochim Acta 64(1):21–35

Ropelewski CF, Halpert MS (1987) Global and regional scale precipitation patterns associated with the El Niño/southern oscillation. Mon Weather Rev 115:1606–1626

Roxy MK et al (2015) Drying of Indian subcontinent by rapid Indian Ocean warming and a weakening land-sea thermal gradient. Nat Commun 6:7423

Sano M, Furuta F, Kobayashi O, Sweda T (2005) Temperature variations since the mid-18th century for western Nepal, as reconstructed from tree-ring width and density of *Abies spectabilis*. Dendrochronologia 23:83–92

Sano M, Ramesh R, Sheshshayee M, Sukumar R (2012) Increasing aridity over the past 223 years in the Nepal Himalaya inferred from a tree-ring $\delta^{18}O$ chronology. The Holocene 22(7):809–817

Sano M et al (2013) May–September precipitation in the Bhutan Himalaya since 1743 as reconstructed from tree ring cellulose $\delta^{18}O$. J Geophys Res Atmos 118(15):8399–8410

Sano M et al (2017) Moisture source signals preserved in a 242-year tree-ring $\delta^{18}O$ chronology in the western Himalaya. Glob Planet Chang 157:73–82

Schneider U et al (2011) GPCC full data reanalysis version 6.0 at 0.5°: monthly land-surface precipitation from rain-gauges built on GTS-based and historic data. https://doi.org/10.5676/DWD_GPCC/FD_M_V6_050

Sengupta S, Sarkar A (2006) Stable isotope evidence of dual (Arabian Sea and Bay of Bengal) vapour sources in monsoonal precipitation over north India. Earth Planet Sci Lett 250(3–4):511–521

Shi C et al (2012) Reconstruction of southeast Tibetan Plateau summer climate using tree ring $\delta^{18}O$: moisture variability over the past two centuries. Clim Past 8(1):205–213

Shi F et al (2015) A multi-proxy reconstruction of spatial and temporal variations in Asian summer temperatures over the last millennium. Clim Chang 131(4):663–676

Shukla J, Paolino DA (1983) The Southern Oscillation and long-range forecasting of the summer monsoon rainfall over India. Mon Weather Rev 111:1830–1837

Singh J, Park W-K, Yadav RR (2006) Tree-ring-based hydrological records for western Himalaya, India, since A.D. 1560. Clim Dyn 26:295–303

Sinha A et al (2015) Trends and oscillations in the Indian summer monsoon rainfall over the last two millennia. Nat Commun 6:6309

Sternberg LSLOR (2009) Oxygen stable isotope ratios of tree-ring cellulose: the next phase of understanding. New Phytol 181(3):553–562

Tierney JE et al (2015) Tropical sea surface temperatures for the past four centuries reconstructed from coral archives. Paleoceanography 30(3):226–252

Treydte KS et al (2006) The twentieth century was the wettest period in northern Pakistan over the past millennium. Nature 440:1179–1182

van der Schrier G, Barichivich J, Briffa KR, Jones PD (2013) A scPDSI-based global data set of dry and wet spells for 1901–2009. J Geophys Res Atmos 118(10):4025–4048

Viviroli D, Dürr HH, Messerli B, Meybeck M, Weingartner R (2007) Mountains of the world, water towers for humanity: typology, mapping, and global significance. Water Resour Res 43(7):W07447

Wang B, Ding Q (2006) Changes in global precipitation over the past 56 years. Geophys Res Lett 33:L06711. https://doi.org/10.1029/2005GL025347

Wang J, Yang B, Ljungqvist FC (2015) A millennial summer temperature reconstruction for the eastern Tibetan Plateau from tree-ring width. J Clim 28(13):5289–5304

Wernicke J, Grießinger J, Hochreuther P, Bräuning A (2015) Variability of summer humidity during the past 800 years on the eastern Tibetan Plateau inferred from $\delta^{18}O$ of tree-ring cellulose. Clim Past 11(2):327–337

White JWC, Cook ER, Lawrence JR, Broecker WS (1985) The D/H ratios of sap in trees: implications for water sources and tree ring D/H ratios. Geochim Cosmochim Acta 49:237–246

Wigley TML, Briffa KR, Jones PD (1984) On the average value of correlated time series, with applications in dendroclimatology and hydrometeorology. J Clim Appl Meteorol 23:201–213

Xu H, Hong Y, Hong B (2012) Decreasing Asian summer monsoon intensity after 1860 AD in the global warming epoch. Clim Dyn 39(7–8):2079–2088

Xu C et al (2018) Decreasing Indian summer monsoon on the northern Indian sub-continent during the last 180 years: evidence from five tree-ring cellulose oxygen isotope chronologies. Clim Past 14(5):653–664

Yadav RR, Misra KG, Yadava AK, Kotlia BS, Misra S (2015) Tree-ring footprints of drought variability in last ~300 years over Kumaun Himalaya, India and its relationship with crop productivity. Quat Sci Rev 117:113–123

Yadav RR et al (2017) Recent wetting and glacier expansion in the northwest Himalaya and Karakoram. Sci Rep 7(1):6139

Chapter 15
The Uplift of the Himalaya-Tibetan Plateau and Human Evolution: An Overview on the Connection Among the Tectonics, Eco-Climate System and Human Evolution During the Neogene Through the Quaternary Period

Tetsuzo Yasunari

Abstract This paper reviews the recent studies on the uplift of the Tibet-Himalaya mountains (TH) and its association with the human origin and evolution through the climate and ecosystem changes in Afro-Eurasian continents.

The uplift of TH since the late Tertiary Era gradually formed the Asian monsoon system and dry climate in southwest Asia through North Africa. Meanwhile, during 5–10 Ma the formation of the Rift valley of east Africa brought about drier climate and grassland in the equatorial east Africa, which has an important implication to the early hominid evolution. The uplift of TH also caused and/or enhanced decrease of atmospheric CO_2 content through chemical weathering of mountain slopes, which has induced colder climate through the late Tertiary to the Quaternary Era. The lowering of CO_2 content caused expansion of grassland of the C4-plant and associated evolution of Ungulata (e.g., antelope), which may have also affected the early hominid evolution.

The Quaternary period was characterized with glacial cycles of 40–100k year periods. The ice/snow albedo feedback of Tibetan Plateau may have played as an amplifier of the climate change of this Era. Large temporal and spatial variability of wet/dry zones in east Africa affected by the glacial cycles is very likely to induce further evolution and diffusion, including the migration to Eurasia. During this period, the cold climate and weakened Asian monsoon formed a broad zone of steppe and grassland in central Asia through Europe, and enabled large variety of herbivorous mammals there. The codependent relation between these mammals and the hominid species was essential for the evolution of the later hominid species (Homo erectus) to the modern hominid (Homo sapiens).

T. Yasunari (✉)
Research Institute for Humanity and Nature, Kyoto, Japan
e-mail: yasunari@chikyu.ac.jp

© Springer Nature Switzerland AG 2020
A. P. Dimri et al. (eds.), *Himalayan Weather and Climate and their Impact on the Environment*, https://doi.org/10.1007/978-3-030-29684-1_15

Under the warm and stable climate of the Holocene since 10 ka the modern hominid heuristically started agriculture and civilization, which has, however, been a new epoch called "the Anthropocene" when the human beings are changing the earth system itself.

15.1 Introduction

Recent research into the origins and evolution of humans is making great progress with the addition of new methods such as the molecular clock and DNA analysis. It is thought that seven million years ago during the latter half of the Neogene period of the present Cenozoic Era Hominia, the direct ancestors of humankind, branched off from subhuman primates in equatorial Africa. The Quaternary period (since approximately 2.6 million years ago up to present) is defined as the period when humans started to become active. Due to recent progress in geosciences including stable isotope analysis and numerical experiments using climate models, we are coming to realize that, during the time from the latter half of the Neogene period to the Quaternary period when humans originated and evolved, an uplift of the Tibetan and Himalayan massif and tectonic changes in the African region started, leading to enormous changes in the climate and environment of the Afro-Eurasian continent.

This paper attempts a comprehensive view on the global-scale tectonic changes centering on the uplift of the Himalayan mountains and the Tibetan Plateau (hereafter called HTP) and how the resultant changes in climate and the biosphere regulated or encouraged the origins and evolution of humankind, by reviewing the recent results of research in geoscience, paleoclimatology and paleoecology, and anthropology.

15.2 The Origin and Evolution of Humans and Environmental Change – Why Africa?

It was Molnar, famous for his research in the tectonics of Tibet and the Himalayas, who first pointed out the possible connection between the uplift of the HTP from the Neogene period to the Quaternary period and the origin and evolution of humans (Molnar 1990). Figure 15.1 shows the average height evolution of the HTP from the start of the Eocene epoch (60Ma: 60 Million Years Ago) to the present and the changes in human brain size since the Miocene epoch (40 Ma) (from the ape-man Australopithecus to the modern man Homo sapiens). Molnar suggested that both these figures show a sudden increase toward modernity, particularly after the Late Neogene period, and that rather than mere coincidence, the global scale tectonic

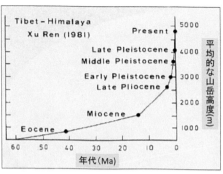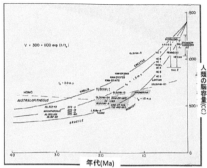

Fig. 15.1 Left: The average altitude of the Tibetan and Himalayan massif from the second half of the Quaternary (60 Ma) and (right) changes in the capacity of human brains since the end of the Late Tertiary (4 Ma). The continuous line shows changes based on various inferences, and the broken line divides the Australopithecus and Homo genus. (Modified the figure by Molnar 1990)

changes typified by the uplift of the HTP and human evolution may have a close relationship. Of course, the latest findings of both geoscience and anthropology indicate, as we shall see later, that the relationship between the Earth's tectonic changes and the associated climatic and environmental changes and human evolution is not that simple.

The ancestors of humans branched off from the apes several million years ago as ape-men (Australopithecus) (DeMenocal 2004). From this start, there emerged primitive man, paleanthropic man (Neanderthals), and over a hundred thousand years ago, our direct ancestor, modern man (Homo sapiens). Recent anthropology has reconsidered these distinctions, but this paper will avoid getting into these questions, using designations like primitive man and modern man as they are. Rather, what I want to consider here is why Africa was the stage where ape-man emerged and evolved through primitive man to modern man. If humans evolved from the relatives of apes inhabiting tropical rain forests, then humans should have originated from South America or Southeast Asia. Here, the distribution of continents and oceans in the Eocene (50 ma) and the Miocene (20 Ma) when the mountains started to rise (Fig. 15.2). At the start of the Eocene, there were already tropical rain forests on all the continents around the equator, where primates would have inhabited. However, the continent of South America was still separated from the North American continent, and the African continent was also separated from the Eurasian continent by the Tethys Sea (the ancient Mediterranean). The tropical and subtropical forests were extended over a wide area from the equator to the subtropical zone of Southeast Asia due to the monsoon climate that already existed along the Southeast and East Asian coast. During the Miocene era, the Tethys Sea closed, and the African continent were joined to the Eurasian continent, and a subtropical arid or semi-arid area were

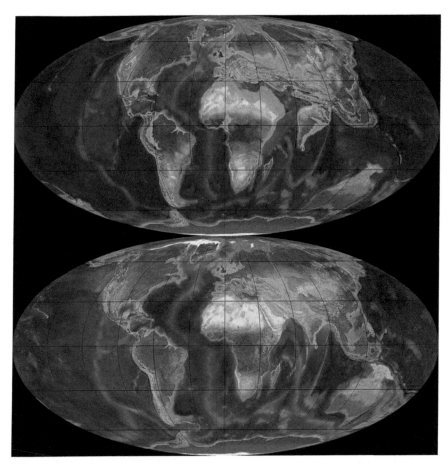

Fig. 15.2 Distribution of sea and land around the Paleogene Eocene (50 Ma) and the Neogene Miocene (20 Ma). http://en.wikipedia.org/wiki/File:Neogene-MioceneGlobal.jpg

gradually expanding between both continents. This geographical contrast of climate and ecosystem in the Southeast Asia through African tropics may be very important for the issue of human evolution. In the late Neogene era, the uplift of the HTP became more pronounced. As will be discussed in the next, the differences in climate and ecosystems between the African and Southeast Asian tropics was further reinforced. It is likely that these differences have a profound implication for the environmental conditions that enabled the origin and evolution of humans on the African continent.

15.3   Uplift of the HTP and Global Climate Change in the Neogene Period

15.3.1   When Did the Uplift of the HTP Occur?

After the Indian subcontinent (the Indian Plate) collided with the Eurasian continent at 50 Ma, the uplift of the HTP began, and by 20 Ma, mountains of a certain height had already formed (Fig. 15.2). At this time, the Tethys Sea had already disappeared, and the African continent and Eurasian continent were largely joined. Not only the HTP but also the world's main mountains existing today such as the Rockies, the Andes, and the Alps, started their upheaval at almost the same time.

In the past few decades there has been much discussion on the tectonics of the Himalayan uplift and the formation of the Tibetan plateau. Some recent research has suggested that uplift proceeded after 50 Ma when the Indian subcontinent (plate) collided with the Eurasian continent, and the mountains were already quite high in the Neogene period (from 23 Ma). For example, based on tectonic dynamics and structural geological research, Molnar et al. (1993) assume that by up to 8 Ma, they had reached heights of around 1000–2500 m on average. Other geological surveys estimated that a significant portion of the plateau had already reached its current height around 11–14 Ma (Sakai 2005; Sakai et al. 2006). From isotopic analysis of water in lakes on the Tibetan plateau, Rowley and Currie (2006) estimated that the area from the Himalayas to the southern part of the plateau had already reached its current height around 35 Ma, and gradually expanded northward. Wang et al. (2008) assumed that although the central part of the plateau was already at its current height in the period before 30 Ma, the uplift of the northern part of the plateau and the southernmost part of the Himalayas was more recent, from 10 Ma onwards. On the other hand, from analysis of the teeth of herbivorous animals that inhabited the plateau around 7 Ma, Wang et al. (2006),for example, assume that they were eating carbon fixing plants (described below) that prefer a warmer climate, so it was at most about 3000 m high, and after 7 Ma, it rose considerably to today's height. For this reason, much discussion continues today concerning the characteristics of the vast Himalayan region, such as when and how high it reached, and conclusive results have yet to emerge. However, when we consider the impact of this massif on the global climate, average elevation is an only one of the important indices. To evaluate moisture transportation to the plateau and the influence on the atmosphere, it is also important whether the Tibetan plateau was widely covered with vegetation, and the height of the Himalayas, which act as a barrier of moisture transport to the Tibetan plateau.

15.3.2   Reduction of CO_2 Concentration and Cooling of the Climate in the Cenozoic Era

Here, we discuss the temperature trend of the Earth as a whole in the whole Cenozoic Era. Figure 15.3 shows the change global deep ocean temperature estimated from changes in the oxygen isotope ratio ($\delta^{18}O$) (Zachos et al. 2001) from 60 Ma to the present. This figure clearly shows global cooling trend after a warm peak around 50 Ma. The change in temperature from around 40 Ma to 10 Ma appears to be associated with the formation of the Antarctic ice sheet, its melting and re-expansion, linked to the separation of Antarctica. Then from 10 Ma (Mid Miocene) through the Quaternary period, there was significant cooling up to the present, and we are now coming to realize that the uplift of the HTP made a significant contribution.

The remarkable uplift of the HTP between subtropical and tropical areas simultaneously caused intense weathering and erosion of its slopes by rain and river water. As we will explain next, the uplift of this massif strengthened the monsoon, and heavy rain intensified the weathering and erosion of its slopes. Silicate, a principal constituent of rock, captures atmospheric CO_2 in the process of chemical weathering, generating calcium carbonate and silicic acid, which are washed away by the water. Through this weathering and erosion, the uplift of the mountains works to reduce the CO_2 concentration in the global atmosphere. The overall global cooling from the second half of the Tertiary (around 10 Ma) to the Quaternary period seen in Fig. 15.3 is basically due to the reduction in atmospheric CO_2 concentration from the active weathering and erosion of the HTP, and associated weakening of the greenhouse effect (Molnar and England 1990; Raymo and Ruddiman 1992). What is notable here is that, as shown in Fig. 15.4, C_3 and C_4 plants that have different photosynthesis and physiological characteristics predominate, according to the atmospheric CO_2 concentration and growing season temperature. C_3 plants

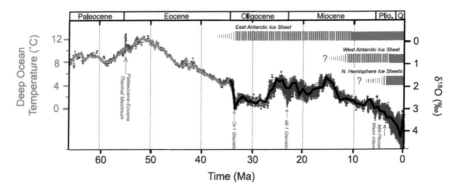

Fig. 15.3 Changes in the global sea temperature in the past 65 million years from the Tertiary to the Quaternary period (today) based on oxygen isotope ratios (Delta-O18) of marine sediment cores. Particularly rapid cooling from around the Late Tertiary (5 Ma) is noticeable. (Some changes made to the figure by Zachos et al. (2001) found in the report by the IPCC (2007))

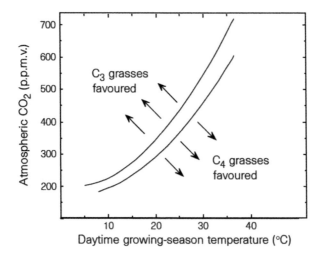

Fig. 15.4 Prevalence of herbaceous C_3 and C_4 plants with temperature (daytime temperature in the growth phase) and atmospheric CO_2 concentration as functions (Cerling et al. 1997)

are those with conventional photosynthetic characteristics. C_4 plants incorporate a cycle called the C_4 pathway for further concentrating CO_2, for more efficient photosynthesis (CO_2 absorption) than C_3 plants. In addition, C_4 plants photosynthesize more efficiently than C_3 plants in arid climates with harsh moisture conditions, and as Fig. 15.4 shows, they are highly adaptable to a low-CO_2 atmosphere. In particular, around 6–8 Ma, there was a change in grasslands in arid areas in all parts of the world from C_3 plants that prefer high concentrations of CO_2 to C_4 plants that are adapted to low CO_2 concentrations. It is inferred that at this time, concentrations of atmospheric CO_2 were declining on a global level due to weathering and erosion (Cerling et al. 1993, 1997). The cooling of the climate due to a decline in CO_2 levels through the weathering process as a result of the uplift of the HTP is an important condition for the emergence of glacial cycles as explained in Fig. 15.4. However, the fact that humans evolved at the same time as this cooling of the climate has interesting implications considering current global environmental issues.

15.3.3   The Emergence of the Asian Monsoon and an Arid Climate

Since the 1970s, numerical simulations by General Circulation Models (GCMs) have already demonstrated that the vast mountain massif of the HTP located in the relatively low latitude subtropics generates a vast monsoon climate over a broad area of the Eurasian continent from the south to the east of the HTP, and an arid climate on its west and northwest side, due to the thermal and dynamical effects induced by the orography of the HTP (Hahn and Manabe 1975; Manabe and Broccoli 1990; Ruddiman 1997). The author's group conducted more quantitative research through numerical simulation with a Coupled General Circulation Model (CGCM), to determine how different (mean) heights of the HTP affect climate formation, including the interactive processes how change of atmospheric circulation affect ocean circulation through atmosphere-ocean interaction, and its feed back to the atmosphere (Abe et al. 2003, 2004, 2005). As shown in Fig. 15.5, the results of these numerical simulations indicate that the Asian monsoon becomes pronounced when the average height of the HTP is about 60% of their current elevation, and the arid land extending to the west of the Tibetan plateau also appears in tandem with the occurrence of the Asian monsoon, at largely the same time. By combining these results by the climate model studies with those inferred from the paleoclimatology and paleoecology of the Asian and African regions (e.g., An et al. 2001), we can approximately estimate the chronology of tectonic evolution (uplift) of the HTP and relevant climate formation. These studies show that as the climate of this region, about 7–10 Ma, a monsoon close to today's Indian monsoon (South Asia) was formed, and almost simultaneously, the arid areas of North Africa, Southwest and Central Asia also appeared (Sakai 2005). Some paleo-ecological studies also show that flora and fauna indicating an arid climate emerged in significant numbers in this region from approximately 10 Ma (Cerling et al. 1993; Amer and Kumazawa 2005). Through these results we can estimate that at this time, the Tibetan plateau had reached about 60% of its average height.

On the other hand, in largely tropical East Africa, around the time when the Great Rift Valley stretching 7000 km north to south formed (5–10 Ma), a long plateau with a total height of 1000–2000 m was formed, with a 5000 m volcano at its western tip and deep gorges. Increased uplift of the East African plateau since ~15–10 Ma might be connected to climate change in East Africa and human evolution. The uplifting East African plateau intercepted those winds and contributed to the increased aridification of East Africa (Ring 2018). The results of the numerical simulation of the authors using the CGCM climate model (Abe et al. 2003) indicated that this East African plateau forms a barrier to the damp easterlies over the equatorial Indian Ocean, and causes the aridification of equatorial East Africa, playing a significant role in the formation of today's savanna climate as shown in Fig. 15.6.

It is noteworthy to state that the formation of this East African plateau (and the Great Rift Valley) and the remarkable uplift of the Himalayas and Tibetan plateau inferred from the speed of weathering and erosion (with the associated change in CO_2 concentration) are likely to have occurred nearly simultaneously. These huge tectonic event in Afro-Eurasian continents may have caused enhanced South Asian summer monsoon, desert climate in central to west Asia (An et al. 2001), and the appearance of expansive arid Savanna climate of East Africa (savanna) at almost the same time, around 5–10 Ma.

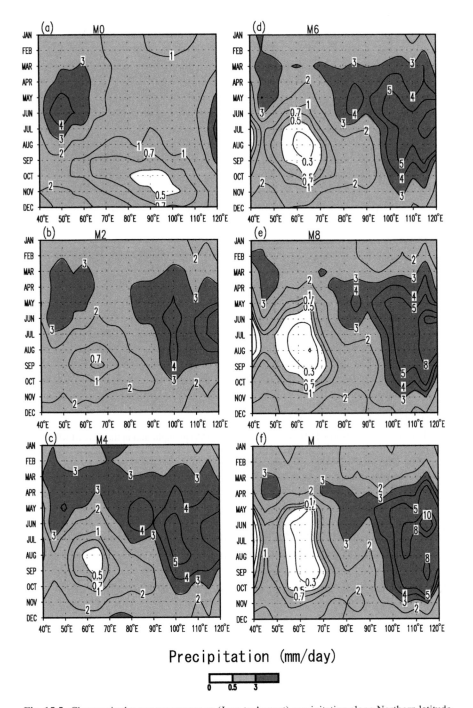

Precipitation (mm/day)

Fig. 15.5 Changes in the summer monsoon (June to August) precipitation along Northern latitude 30–45 N when the average height of the Himalayas and Tibetan plateau is changed. When average altitude of the massif is (**a**) 0% of today (M0 no mountains), (**b**) 20% (M2), (**c**) 40% (M4), (**d**) 60% (M6), (**e**) 80% (M8), (**f**) 100% (current M altitude) (Abe et al. 2005). The unit is mm/day. Dark hatching shows regions with 3 mm/day or more, and white space shows regions with 0.5 mm/day or less. (Abe et al. 2005)

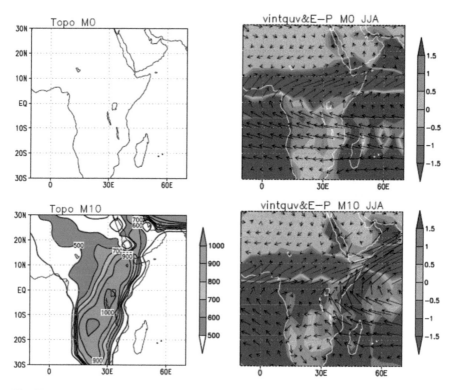

Fig. 15.6 Changes in Northern Hemisphere summer (JJA) precipitation in the CGCM when there was no East African plateau (top right) and today (M10) (bottom right). Blue is the real precipitation (P-E). The arrows show the lower atmosphere wind vectors. The left is the terrain elevation used for numerical modeling. (Drawn with the simulated data from Abe et al. (2003))

15.4   The Quaternary Period Glacial Cycle and the Role of the Himalayas and Tibetan plateau

15.4.1   The Start of the Ice Age and a 40,000-Year Glacial Cycle

The Quaternary period is defined as the newest epoch in the history of the Earth, including the present, when humans appeared and started their explosive evolution. Its start has recently been revised from 1.8 Ma to 2.6 Ma (Gibbard et al. 2009). At almost the same time, the climate of the entire globe cooled, and in the northern hemisphere, ice sheets and glaciers repeatedly expanded and contracted in an ice age. The timing of human evolution and the ice age correspond, but as we noted earlier, rather than mere coincidence, they have a necessary relationship.

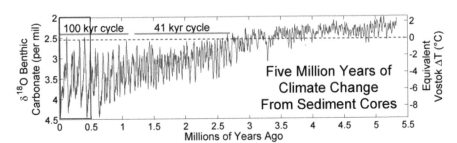

Fig. 15.7 Global temperature change for the past 5.5 million years. Dashed lines show the average temperature of today's Earth. (Lisiecki and Raymo 2005)

Figure 15.7 shows the global temperature changes since 5.5 Ma (Lisiecki and Raymo 2005). It shows that from around 2.5 to 3 Ma, the trend in declining temperature strengthened, and variations also became larger. In the first half of the period from 5.5 to 2.5 Ma, the cycles ranged from 20,000 to 30,000 years with small amplitude, but in the first half of the Quaternary period from 2.5 to 1 Ma, 40,000-year cycles, and from 1 Ma to the present, 100,000-year cycles become prominent, with very large amplitude.

The mechanism behind the modulation of these ice age climate cycles occured is still widely debated, but the basic mechanism for this modulation is thought to be attributed to long-term seasonal changes in the flux of solar radiation on the Earth's surface and the complex variation in latitude distribution resulting from a combination of the periodic movement of eccentricity, axial tilt, precession that make up the Earth's orbital elements (the elements that determine the characteristics of its revolutionary movement), caused by the nonlinear interaction of the attraction of the Earth, Sun and other planets in the solar system. This variation is called the Milankovitch cycle, after Milankovitch who first identified the mechanism. Orbital eccentricity has a cycle of about 100,000 years and about 400,000 years, axial tilt about 41,000 years, and precession about 20,000 years (Milankovitch 1941). Recent studies have suggested that the flux of solar radiation due to the Milankovitch cycle has a complex non-linear relationships between the climate, the continental ice sheets and the lithosphere–asthenosphere system, combined with changes in atmospheric greenhouse gas composition (e.g., CO_2), which are thought to cause these changes and modulations (Lisiecki and Raymo 2005; Abe-Ouchi 2013).

What, then, caused initiation of the cold glacial period in the Quaternary period? As we explained in Sect. 15.3.2, the decrease in atmospheric CO_2 concentration (weakening of the greenhouse effect) due to weathering and erosion accompanying the uplift of the Himalayas and Tibetan plateau and the cooling trend should have been important conditions, but no conclusive answers have been found for the emergence of glacial cycles. Currently the Earth's tilt is 23.5 °, but it moves in a range from 22.5 ° to 24.5 °. In the first half of the Quaternary period (2.6 to 1 Ma), a glacial cycle continued, corresponding to variations in axial tilt on an approximately

40,000-year cycle. In periods when the tilt is small, latitudes with enormous insolation come to the low latitude side, so there is a possibility that the south-north temperature gradient increases. However, a change in some sort of mechanism within the climate system that reduces south-north heat transport efficiency (the strength of heat transport) is necessary for the polar regions to become colder. Tectonic changes that occurred at this time (around 3 Ma) included the formation of the Isthmus of Panama (the separation of the Atlantic and Pacific oceans), and the northern edge of the Australia-New Guinea continent reached the equator. The former reinforced the upwelling in the equatorial eastern Pacific Ocean (and increased the east-west water temperature difference), establishing the east-west circulation system (Walker circulation) connecting the atmosphere and ocean in the tropics (Maslin and Christensen 2007). The latter changed the Indonesian throughflow flowing from the Pacific Ocean to the Indian Ocean from the warm seawater originating in the southern Pacific Ocean to the cold seawater originating in the northern Pacific Ocean, reducing the water temperature of the whole equatorial Indian Ocean (Cane and Molnar 2001). These changes are thought to be linked to the start of the glacial period with the 40,000 year period.

In the equatorial Africa, climate change was characterized by a decline in rainfall and aridification in eastern Africa caused by the lower water temperature of the equatorial Indian Ocean (particularly the western Indian Ocean) (Hastenrath et al. 1993; Goddard and Graham 1999). In the tropical Pacific Ocean, it is inferred that the Walker circulation was strengthened about 1.7 Ma, and heat transport to high latitudes was weakened, contributing to cooling and expanded glaciers in polar regions (Molnar and Cane 2002). In the current climate system, dominant fluctuations of the Walker circulation system are reprented as the ENSO (El Niño Southern Oscillation) phenomenon itself, which causes extreme climate events such as drought and heavy rain in the tropics including the African region. This implies that the ENSO cycle may have started in this period, suggesting that the range of interannual variability of the climate in the tropics increased.

On the other hand, it has been proved that the Walker circulation system spanning the Pacific and Indian oceans has a close connection with the Asian monsoon, tending to appear the La Niña (El Niño) phenomenon with strong (weak) Asian monsoons (Yasunari 1990, 1991). Through the simulation using CGCM conducted by the authors (Abe et al. 2003), this coupling of the ENSO and monsoon is shown to be strongly apparent at the time when the mean height of the HTP is about 80% or more of today's height. It can be inferred from recent tectonics research that the HTP had already reached this height level in this period (from 3 Ma). However, as we explain in the next Sect. (15.4.2), the Asian monsoon was generally weak in the glacial period, as indicated both in the paleoclimate data and in climate models. That being the case, how can we interpret the combination of a strong Walker circulation and a weak Asian monsoon in the glacial period, unlike today? As noted above, the east-west gradient of the ocean surface temperature along the equator was maintained, possibly by a mechanism that differs from that today. This issue remains for the future study.

15.4.2 Dynamics of the 100,000-Year Glacial Cycle and Possible Role of the HTP

Cooling progressed further in the Quaternary period, and about 1 Ma, a glacial cycle of about 100,000 years prevailed, with very significant expansion of ice sheets and glaciers in the cold periods. The fluctuations of this time scale are well understood in considerable detail based on ocean sediment cores and ice cores from Antarctica and Greenland. Figure 15.8 shows global temperature fluctuations over the past 800,000 years revealed in ice cores from Antarctica (Jouzel et al. 2007). It is known that not only temperature but also mass of snow and ice (ice sheets, glaciers and snowfall) worldwide, seawater temperature, the concentration of CO_2 and CH_4 greenhouse gases in the atmosphere, and so on fluctuated in ways that explain climate warming and cooling. In other words, in this 100,000-year cycle, all elements

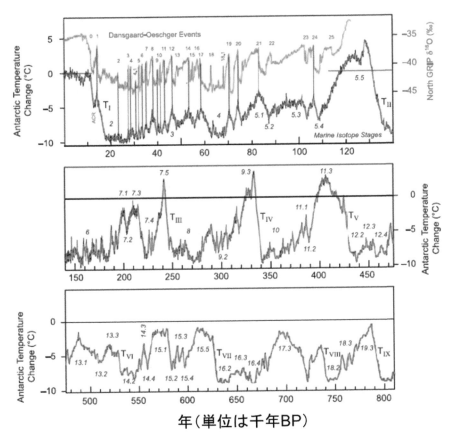

Fig. 15.8 Average temperature of the Earth in the past 810 Kyr reconstructed from Antarctic ice sheet Dome C ice cores. From 120 Kyr, the temperature deviation reconstructed from Greenland ice cores is shown (Jouzel.et al. 2007)

of the global climate including the deep-ocean thermohaline circulation have been involved in the climate system from the polar regions to the tropics. The fluctuations of this cycle have been thought to harmonize with the 100,000-year cycle of the Earth's orbital element (change in eccentricity).

The dynamics of this 100,000 year glacial cycles is still one of the big mysteries in the glaciation of the Quaternary Period. However, numerous studies have pointed to the importance of the modulatory timing by which insolation of summer in the Northern Hemisphere fluctuates significantly due to the nonlinear interaction of the 40,000-year cycle (axial tilt) and/or 20,000-year cycle (precession), which have significant amplitudes. A recent climate model study including ice-sheet dynamics (Abe-Ouchi et al. 2013) has proved that the viscoelastic response of the earth's crust to the continental ice-sheet is essential for producing the 100,000 year cycle. This study also shows that the fluctuations in summer insolation in the Northern Hemisphere plays a key role as a pacemaker of snow-ice sheet cover on the Eurasian and North American continents. The extent of the area covered by snow in summer on the Eurasian and North American continents, which constitute the greater part of the land, significantly affects global-scale surface heat balance, by controlling insolation through the albedo (reflection rate of solar radiation) effect of snow. In current summers, accumulated snow largely disappears from the continents. However, if it were to remain over a fairly wide area even in the summer, it would trigger climate cooling on a global scale, with the expansion of ice sheets and glaciers at high latitudes. This particular study (Abe-Ouchi et al. 2013) emphasizes the effect of stagnant snow cover in summer on the North American continent, where quasi-stationary cold trough can easily be formed due to the orographic effect of the Rockies.

However, our previous study by using a GCM (Yasunari et al. 1991) showed that a remarkable atmospheric teleconnection pattern over the north Pacific through the North American continent is likely to be induced by the anomalous snow cover over the Tibetan Plateau and East Asia during spring through late summer. These circulation patterns with anomalous cold trough over Canada are responsible for the considerable decrease of surface temperature particularly over the northeastern part of North America. These results are partly supported by the observed circulation anomalies associated with the weak Asian monsoon condition (Yasunari and Seki 1992). These evidences suggest the possible role of the anomalous snow cover over Eurasia on the initiation of the glacial period of the 100,000 year cycle. The authors also carried out a numerical simulation using the CGCM to investigate the possibility that the extent of snow on the Tibetan plateau 115 ka when insolation was at its lowest in the Northern Hemisphere summer after rapid cooling from the peak of the interglacial period about 120 ka shown in Fig. 15.8 may have contributed to global scale cooling. The results showed that from 120 ka to 115 ka, the change in solar radiation in summer reduced summer temperatures on the continents of the Northern Hemisphere by 5–8 °C, the presence of snow on the plateau reduced temperatures particularly at mid and low latitudes by a further several degrees, and the extent of snow around the North Pole expanded, which persisted from year to year (Yasunari et al. 2006). These studies suggest that the albedo effect of the sufficiently high

Tibetan plateau at relatively low latitudes effectively may effectively reduce the absorption of solar radiation at the Earth's surface, and further promoted a temperature drop on a hemispheric scale, representing an important element causing glacial-interglacial cycles.

Under the current climate, the area of snow cover over the Tibetan plateau in summer is very small, with the majority bare ground without snow. This absorbs large amounts of solar radiation, becoming a strong heat source for the atmosphere, as explained in Sect. 15.3.3. However, if snow remained over the entire plateau throughout the summer, it might weaken the Asian monsoon, as well as cooling the atmosphere of the whole Northern Hemisphere, changing atmospheric circulation significantly, even in the north America through the atmospheric teleconnection (Yasunari et al. 1991).

15.5 Changes in the Climatic and Environmental Conditions Related to the ORIGIN and evolution of Humans

15.5.1 Climatic and Ecological Changes Prompting the Origin of Humans (Ape-Men)

The definition of an ape-man, considered as the ancestor of humans, is upright bipedal walking, and degeneration (reduction) of the canine teeth (Mitsui 2005). According to this definition, ape-men are said to have emerged 5–7 Ma, in equatorial East Africa. Around that time, as described in Sect. 15.3.2, the formation of the Great Rift Valley in East Africa stretching for several thousand kilometers from Ethiopia to South Africa caused the aridification of the regional climate. The shift from forests to grasslands caused the primates of the area to start walking on two feet, becoming ape-men. This is the so-called "East Side Story". But when an older ape-man (Sahelanthropus tchadensis) was discovered near Lake Chad in the Sahel region of West Africa, this theory came into doubt. However, the East-West contrast between the humid Asian monsoon in the East, and the dry climate in the West was intensified at this time due to the rise of the HTP, and it is possible that the whole North African region including Lake Chad entered an arid period.

What needs to be clarified is what necessity for survival drove the first humans to walk upright due to the change in climate and ecosystem from forest to grassland. In both East Africa and around Lake Chad, there was probably a long period during which gradual aridification changed the ecosystem from forest to mixed grasslands and lakes. The aridification at this time (6–8 Ma) probably lead to the sudden spread of grasslands of C_4 plants suited to low CO_2 concentrations (Cerling et al. 1993), promoting the evolution of today's herbivorous fauna (and carnivorous animal groups as their predators). In this kind of ecological environment, if we assume that it became necessary to switch from simply collecting nuts and berries from trees in the forest to feeding on the grassland animals, then we can assume that the need to hold sticks required walking on two legs. Bipedalism was probably a strategy for survival among the herbivores living in the grasslands and the carnivores that fed on them.

15.5.2 The Evolution of Primitive Man and Climatic and Ecological Changes from the Late Neogene to the First Half of the Quaternary Period

The evolution from ape-man to primitive man who actively used tools such as stone tools began around 3 Ma with Australopithecus, followed by Homo habilis and Homo erectus. Evolution continued with the appearance of modern man, Homo sapiens, around 0.1–0.2 Ma. We will not review the study of human evolution any further here. Figure 15.9 shows the evolution from primitive man from 3 to 4 Ma to modern man, with climate change on a global scale and in Africa, and the change in vegetation in East Africa on the same timeline, prepared by DeMenocal (2004).

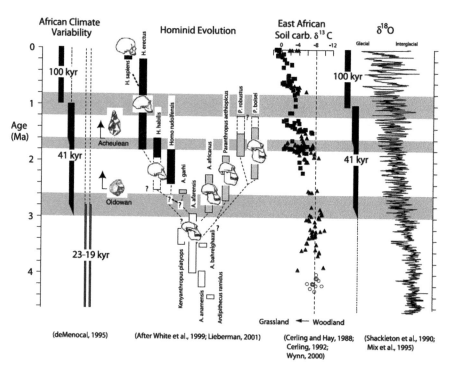

Fig. 15.9 The evolutionary process from hominins to modern man from the Late Tertiary (4.6 Ma) to the present. It also shows variations in the global climate (δ^{18}O) and the process of evolution of East African grassland vegetation (δ^{13}C) at the same time. (DeMenocal 2004)

As this figure makes clear at a glance, from the late Pliocene (5 to 3 Ma) to the first half of the Quaternary period (3 to 1 Ma), various species of Australopithecus emerged, one of which evolved into modern man via Homo habilis and Homo erectus. Paranthropus, with its strong jaw adapted for eating hard nuts and the like diverged at this time, becoming extinct before 1 Ma. This was a time of dramatic differentiation and selection for primitive man. As the figure shows, in this period global cooling occurred, with a 40,000-year glacial cycle becoming prominent (see Sect. 15.4.1). We know that aridification in East Africa progressed around 2.8 Ma, around 1.7 Ma and around 1.0 Ma, gradually changing from forest to grassland. The Quaternary period glacial period began 2.8 Ma when global cooling began. 1.0 Ma roughly corresponds to the period when the 100,000-year glacial cycle began. Around 1.7 Ma, cooling of the western Indian Ocean proceeded due to tropical East-West circulation (due to upwelling), and aridification in East Africa made further advances. This figure strongly suggests that differentiation proceeded in response to these gradual changes in climate and ecosystem, where primitive man evolved into modern man (Homo) including evolution of the use of stone tools, while hominins such as Paranthropus went extinct.

One of the keys to determining the relationship between the aridification of the climate and the evolution (differentiation) of primitive man is the evolution of herbivores such as antelope (the generic name of bovine ungulates including gazelles, impala etc.) in response to changes in the ecosystem (Vrba 1995; DeMenocal 2004). With the shrinking of the forests and expansion of grasslands, these herbivores speciated explosively in periods corresponding roughly to the three stages mentioned above. This probably encouraged the evolution of stone tools and primitive man who hunted them with such tools, while on the other hand, hominins who relied on the forests for food declined. The primitive man Homo erectus, who shifted his focus to hunting, left Africa at this time, and moved to Central Asia, Europe, and East Asia. The background to this was no doubt the deteriorating conditions for hunting and gathering due to changes in flora and fauna resulting from the increasing aridification of the East African climate. In a vast area from North Africa to the Levant region of the Middle East to Central Asia, a dry climate prevailed due to the presence of the HTP (see Sect. 15.3.2). It was already a cold glacial period, and the Asian monsoon was weak and the dry climate in the west was also weakened, resulting in an expansive steppe grassland bordering the desert regions, forming a vast corridor between Southwestern Asia and Central Asia. The northern half of Europe was already covered in ice sheets, while the southern regions were grassland. You can imagine primitive men gradually making their way north in pursuit of the herds on the steppe corridor, or pushing to the East and West. Some primitive men such as Peking Man, discovered at Zhoukoudian in the suburbs of Beijing, are thought to have reached East Asia. Peking Man is said to have used fire, but he was probably unable to withstand the cold climate of the glacial period and died out. Much further south, Java Man is believed to have reached the island of Java in Indonesia. In the glacial period, the level of the South China Sea decreased, and so-called Sundaland emerged. Java Man is thought to have moved south to Java at that time.

15.5.3   The Emergence of Homo sapiens and the Glacial Cycle of the Late Quaternary Period

As mentioned in Sect. 15.4.2, the global climate from about 1 Ma became very cold, and as far as it has been ascertained to date, from 0.8 Ma, a pronounced 100,000-year glacial-interglacial cycle has prevailed. The past 10,000 years until now is known as the postglacial age, but this cycle is still continuing, and we should note that we are living in a short interglacial period in the middle of a long, ongoing cold glacial period.

The direct ancestor of today's humans, Homo sapiens, is said to have appeared in Africa from 150 to 200 ka based on mitochondrial DNA analysis (Cann et al. 1987). As Fig. 15.8 shows, this period was cooling after two previous inter-glacials (around 200 ka) towards a glacial period on a 100,000-year cycle. From this time, our ancestors from Africa were moving or dispersing to the Eurasian continent, evolving into the present human species, Homo sapiens. During this time, basically a cold glacial climate prevailed in the whole globe, though it was interrupted by a short interglacial period. A question may be, then, why humans did evolve into modern man in this particular period?

Typical savanna grasslands where C_4 plants predominate, found in the current East African tropics (such as the Serengeti National Park), finally emerged around 1 Ma (Cerling 1992). At the same time, animals such as ungulates that are linked to today's fauna, which are adapted to grasslands, speciated actively (Vrba 1995). From tropical and subtropical East Africa to the Arabian Peninsula and Southwest and Central Asia, the glacial period was, if anything, paired with a weakening of the Asian monsoon. It is highly likely that the dry climate of this region weakened, becoming more humid than today's, and grassland formed an almost unbroken corridor. Alternatively, it is highly likely that repeated North-South displacements of dry and humid climate zones due to glacial cycles caused North-South displacement or expansion and contraction of the grasslands. For example, in numerical simulations reproducing the climate model combined with a vegetation model for the climate and vegetation of the last glacial periods (21 ka, 16 ka), as Fig. 15.10 shows, it is inferred that the Arabian Peninsula, which is a completely desert region today, was covered with grassland vegetation in the glacial periods (Kutzbach et al. 1998). Considering these climatic and ecological environmental conditions, as when primitive man left Africa, some modern men would have pursued the ungulates that proliferated on the grasslands, encouraged by the expansion and contraction of the grassland, gradually moving to the Eurasian continent. While pursuing a hunting lifestyle, these people would gradually have tamed the grassland ungulates such as wild cows and sheep, leading to domestication and nomadism (Imanishi 1995).

Although we have developed our thesis based on the currently prevailing African single origin hypothesis for Homo sapiens, there are naturally arguments against this theory, and the inferences made in this paper are not definitive. However, it can

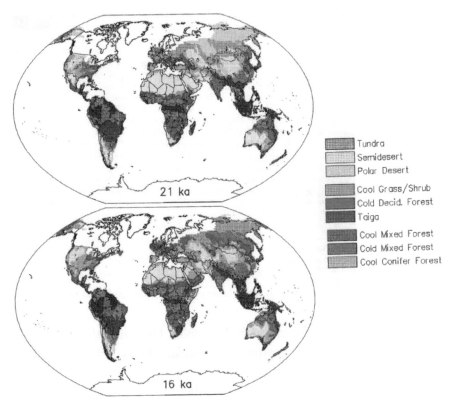

Fig. 15.10 The distribution of vegetation in the interglacial (21 Ka) and glacial period (16 Ka) estimated using a vegetation model and temperature, precipitation and insolation based on a climate model (Kutzbach et al. 1998)

be said that as with the origin and evolution of primitive man, our direct ancestor modern man evolved in a cold climate during a glacial period, in a region with a relatively dry climate and ecosystem formed due to the presence of the HTP. When considering the formation of the Mongoloids of East Asia, the geographical dispersion of modern man from Africa as shown in Fig. 15.11 is an interesting problem. This figure shows possible main routes by which modern man entered Southeast Asia via the Indian subcontinent south of the Himalayas. A question may how he could pass through the dense tropical and subtropical forests of the Yunnan and Assam regions with their monsoon climate. There is rather, a high possibility that his main route was via the steppe grassland between the southern rim of the giant Siberian lake (Mangerud et al. 2004) formed in the West Siberian lowland, and the Tibetan plateau. Alternatively, the distinctive Mongoloid features may have resulted from mixing of the human races that took both routes.

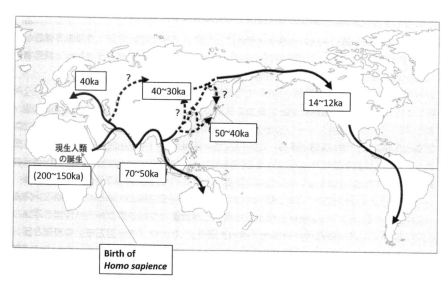

Fig. 15.11 Global dispersion of modern man (the period is estimated) (after from Mitsui 2005)

15.5.4 Human Development in Monsoon Asia and Discovery of Paddy Rice Farming on Alluvial Plains

Around 10 ka, the Earth's most recent glacial period ended, followed temporarily by the Younger Dryas period (13 to 12 ka), a short period of recurrence of cold climate. After the Younger Dryas period, the Earth entered a postglacial age (interglacial period), with a relatively warm and stable climate. This period has continued until now (or very recently), which is named as the Holocene period as a geological classification as shown in Fig. 15.12 (Rockstrom et al. 2009; Alley 2000). The peak of warming occurred 8000 to 6000 years ago. In the Holocene, humans, who evolved on the grasslands under a dry climate in the glacial period, along with the C_4 plants and ungulates unique to Quaternary period arid lands, entered the humid forest valleys and wetlands of the mountains under a monsoon climate, where malaria was rampant and which they had hitherto avoided as the haunt of evil spirits.

However, in this particular humid region called monsoon Asia, humans invented (or discovered) paddy-field rice farming. With the uplift of the HTP, the complex mountain folds have been created by the rivers that flowed east, southeast and south of the HTP. Under these tectonic and climatic situations, numerous deep valleys and alluvial plains have been formed due to active erosion and earth/sand deposition along the valleys and near the river mouths. These valleys and alluvial plains were the habitat of the grasses that were to become rice plants (Sato 1996). Many arguments on the origin place of rice have been made (e.g., Yunnan Province of China or delta area of Yangtze river). Whichever it was, the existence of complex valleys and alluvial plains formed in the periphery zone of the HTP in southern China and

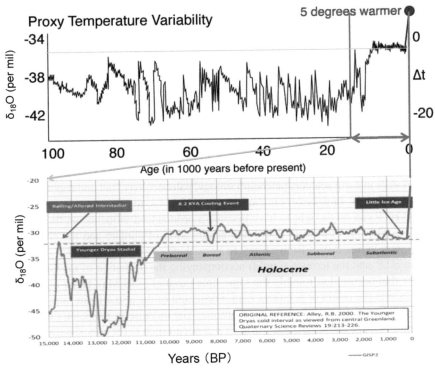

Fig. 15.12 (Top) Changes in the Earth's average temperature from 100,000 years ago to the present, estimated from Greenland ice cores, etc. It also shows temperatures rising rapidly in the late twentieth century. (Original figure: Rockstrom et al. 2009 original drawing) (Bottom) Expanded view of the changes in the period from the end of the glacial period from 15,000 years ago to the Holocene. (Original figure: Alley 2000) It also shows temperatures rising based on observation since 1950. Note that the vertical axis is shown as $\delta^{18}O$ ratio, and the standards for temperature conversion differ between top and bottom

Southeast Asia or India is thought to be a necessary geographical condition for the start of rice farming. Thereafter, for thousands of years rice farming in monsoon Asia made big developments so that today, 60% of the world's population is engaged in it. This humid monsoon Asia produced diverse traditional cultures based on rice paddy farming, and became one of the centers for world civilization.

Incidentally, the progenitor of wheat, another of the three major crops, was found under the dry climate in Southwest Asia. Here, the formation of a wetland ecosystem along the humid valleys arising from the extensive snow melt from the Zagros Mountains can be considered important for the formation of many agricultural progenitors. Wheat farming also started in this region about 12,000 years ago in the postglacial age, becoming the basis of the Mesopotamian civilization. However, the drying of the climate during the Holocene, presumably associated with enhanced Asian monsoon climate to the east (refer to Sect. 15.3.3), prompted the development of irrigation agriculture, which finally had caused the salinization of soils and loss of sustainability, ultimately leading to the demise of the Mesopotamian civilization.

15.6 Summary and Remarks

The association and implication of the uplift of the Himalaya and Tibetan Plateau (HTP) during the Neogene period through the Quaternary period to the evolution of humans are comprehensively overviewed and discussed, based on the recent scientific results in earth science, paleoclimatology, paleoecology, and anthropology. We have focused the earth history in the Neogene (20 Ma ~) to present, when the height of the HTP can be thought to have become a pronounced influence on the Earth's climate. There is almost no doubt that aridification of climate and the change of ecosystem in East Africa from a forest to a grassland ecosystem, are due to the uplift of the HTP, and the terrestrial uplift accompanying the formation of the Great Rift Valley that occurred at roughly the same time (around 5–10 Ma), had a major significance in the origin of primitive man. Formation of the arid climate was coupled with the establishment of the Asian monsoon. The increase in monsoon rainfall on the uplifted HTP caused a reduction in CO_2 concentration in the atmosphere through weathering and erosion along slopes of the HTP, which should have enhanced global colder climate from the Neogene to the Quaternary period. In addition, the lower CO_2 concentration caused expansion of grassland comprising C_4 plants, promoting the evolution of diverse ungulates. These factors were important for the evolution of hominins.

From the Quaternary period (since 2.6 Ma), the Earth entered a period of climate change with expanding and contracting of ice sheets on 20,000- or 40,000- to 100,000-year glacial cycles. A definitive answer to the role played by the HTP in these glacial cycles has yet to be fully understood. However, research using climate models suggests the possibility of the albedo effect of snow and ice as an amplifier of climate change. Hominid evolution advanced in this glacial cycle. During this period, along with these glacial cycles, the climate of East Africa repeatedly varied between humid and dry climate, associated with wax and vane in the Asian monsoon, which probably should have been important opportunities prompting the evolution of hominins.

In the region from East Africa to the Arabian Peninsula, a humid–dry–humid climate and ecosystem was distributed south to north. In the glacial and interglacial periods, this distribution was characterized by displacement and variation in the strength of the contrast. These variations represented an evolutionary pressure on primitive and modern man, and they also became the opportunity for moving to Eurasia. In addition, during the Quaternary period, the cold glacial periods of the climate change cycles were also long compared to the warmer interglacial periods, which in turn could be a demanding climate for the humans even after their migration to the Eurasian continent. Nevertheless, Central and Southwest Asia was, unlike today's desert, a vast grassland steppe, and was inhabited by diverse herbivorous animals, providing a main stage for the co-evolution of humans and herbivorous animals.

In the warm, relatively stable Holocene climate starting after the end of the last glacial period about 10 K, humans developed agriculture cultivating rice and wheat to the East and West of the HTP respectively. This new settled life-style for human-kind induced a new era of civilization. It should be noted that based upon the agri-culture and civilization including urbanization the humans achieved the industrial revolution started in Europe in the eighteenth century. Since the industrial revolu-tion, human activities have significantly been changing the Earth's surface includ-ing the atmosphere, the hydrosphere and the biosphere, and this human impact has become so enormous particularly since the middle of the twentieth century. The recent global warming of climate may be one of the tangible evidences (IPCC 2013). It is said that the Holocene is over and we are now entering a new era called the Anthropocene (Crutzen 2002).

The uplift of the HTP is still continuing due to tectonic mechanisms, and the basic mechanism of the glacial cycle can also be operated even today. However, there is a big question how the current global warming will change the glacial cycle of the Quaternary period. Considering the complex, nonlinear earth (climate) sys-tem, it is not easy for us to prognose the future state of the earth's climate.

On the other hand, the changing climate in the Anthropocene is causing serious rapid changes of the environment of the HTP region, including the overall retreat of glaciers, which are threatening human society of this region. We should not forget that traditional human-nature system in various regions of the world have been formed and maintained under the relatively stable Holocene climate and ecosystem. What should the relationship between the natural environment and humans be like in the Anthropocene? This question is also a pressing issue in the HTP region.

References

Abe M, Kitoh A, Yasunari T (2003) An evolution of the Asian summer monsoon associated with mountain uplift -simulation with the MRI atmosphere-ocean coupled GCM. J Meteorol Soc Jpn 81(5):909–933

Abe M, Yasunari T, Kitoh A (2004) Effects of large-scale orography on the coupled atmosphere-ocean system in the tropical Indian and Pacific Oceans in boreal summer. J Meteorol Soc Jpn 82(2):745–759

Abe M, Yasunari T, Kitoh A (2005) Sensitivity of the central Asia climate to uplift of the Tibetan Plateau in the coupled climate model (MRI-CGCM1). Island Arc 14(4):378–388

Abe-Ouchi A, Saito F, Kawamura K, Raymo ME, Okuno J, Takahashi K, Blatter H (2013) Insolation-driven 100,000-year glacial cycle and hysteresis of ice-sheet volume. Nature 500(7461):190–193

Alley RB (2000) The Younger Dryas cold interval as viewed from central Greenland. Quat Sci Rev 19:213–226

Amer SAM, Kumazawa Y (2005) Mitochondrial DNA sequences of the Afro-Arabian spiny-tailed lizards (genus Uromastyx; family Agamidae):phylogenetic analyses and evolution of genear-rangements. Biol J Linn Soc 85:247–260

An Z, Kutzbach JE, Prell WL, Porter SC (2001) Evolution of Asian monsoons and phased uplift of the Himalaya-Tibetan plateau since Late Miocene times. Nature 411:62–66

Cane MA, Molnar P (2001) Closing of the Indonesian seaway as the missing link between Pliocene East African aridication and the Pacic. Nature 6834:157–161

Cann RL, Stoneking M, Wilson AC (1987) Mitochondrial DNA and human evolution. Nature 325:32–36

Cerling TE (1992) Development of grasslands and savannas in East Africa during the Neogene. Palaeogeogr Palaeoclimatol Palaeoecol 97(1992):241–247

Cerling TE, Wang Y, Quade J (1993) Expansion of C4 ecosystems as an indicator of global ecological change in the late miocene. Nature 361:344–345

Cerling TE et al (1997) Global vegetation change through the Miocene/Pliocene boundary. Nature 389:153–158

Crutzen PJ (2002) Geology of mankind: the Anthropocene. Nature 415:23

DeMenocal PB (2004) African climate change and faunal evolution during the Pliocene-Pleistocene. Earth Planet Sci Lett 220:3–24

Gibbard et al (2009) Formal ratification of the quaternary system/period and the Pleistocene series/epoch with a base at 2.58 Ma. J Quat Sci 25:96–102

Goddard L, Graham NE (1999) Importance of the Indian Ocean for simulating rainfall anomalies over Eastern and Southern Africa. J Geophys Res 104:19099–19116

Hahn DG, Manabe S (1975) The role of mountains in the south Asian monsoon circulation. J Atmos Sci 32:1515–1541

Hastenrath S, Nicklis A, Greischar L (1993) Atmospheric-hydrospheric mechanisms of climate anomalies in the western equatorial Indian Ocean. J Geophys Res 98:20219–20235

Imanishi K (1995) On the theory of Nomadism (Yuboku-ron). (in Japanese) Heibonsha

IPCC (2007) In: Solomon S, Qin D, Manning M, Chen Z, Marquis M, Averyt KB, Tignor M, Miller HL (eds) Climate change 2007: the physical science basis. Contribution of working group I to the fourth assessment report of the intergovernmental panel on climate change. Cambridge University Press, Cambridge/New York, 996p

IPCC (2013) In: Stocker TF, Qin D, Plattner G-K, Tignor M, Allen SK, Boschung J, Nauels A, Xia Y, Bex V, Midgley PM (eds) Climate change 2013: the physical science basis. Contribution of Working Group I to the fifth assessment report of the Intergovernmental Panel on Climate Change. Cambridge University Press, Cambridge/New York. 1535 p

Jouzel J, Masson-Delmotte V, Cattani O, Dreyfus G, Falourd S, Hoffmann G, Minster B, Nouet J, Barnola JM, Chappellaz J, Fischer H, Gallet JC, Johnsen S, Leuenberger M, Loulergue L, Luethi D, Oerter H, Parrenin F, Raisbeck G, Raynaud D, Schilt A, Schwander J, Selmo E, Souchez R, Spahni R, Stauffer B, Steffensen JP, Stenni B, Stocker TF, Tison JL, Werner M, Wolff EW (2007) Orbital and Millennial Antarctic Climate Variability over the Past 800,000 Years. Science 317:793–796

Kutzbach et al (1998) Climate and biome simulations for the past 21,000 years. Quat Sci Rev 17:473–506

Lisiecki LE, Raymo ME (2005) A Pliocene-Pleistocene stack of 57 globally distributed benthic $\delta^{18}O$ records. Paleoceanography 20:1003

Manabe S, Broccoli AJ (1990) Mountains and arid climates of middle latitudes. Science 247:192–195

Maslin M, Christensen B (2007) Tectonics, orbital forcing, global climate change, and human evolution in Africa: introduction to the African paleoclimate special volume. J Hum Evol 53:443–464

Milankovitch M (1941) Kanon der Erdbestrahlung und seine Anwendung auf das Eiszeitproblem. Royal Serbian Academy, Belgrade

Mitsui M (2005) Seven million years of human evolution (Jinrui Shinka no 700 mannen) (in Japanese) Kodan-sha paperback 265 p

Molnar P (1990) The rise of mountain ranges and the evolution of humans: a causal relation? Irish J Earth Sci 10:199–207

Molnar P, Cane MA (2002) El Nin͂o's tropical climate and teleconnections as a blueprint for pre–Ice Age climates. Paleoceanography 17(2):1021

Molnar P, England P (1990) Late cenozoic uplift of mountain ranges and global climate change: chicken or egg? Nature 346:29–34

Molnar P, England P, Martinod J (1993) Mantle dynamics, uplift of the Tibetan Plateau and the Indian Monsoon. Rev Geophys 31:357–396

Raymo ME, Ruddiman WF (1992) Tectonic forcing of late Cenozoic climate. Nature 359:117–122

Ring U (2018) Tectonic dynamics in the African Rift Valley and climate change. Climate science. Oxford Research Encyclopedias. Future Climate Change Scenarios, Climate of Africa, Online Publication Date: January 2018. https://doi.org/10.1093/acrefore/9780190228620.013.524

Rockstrom J et al (2009) Planetary boundaries: exploring the safe operating space for humanity. Ecol Soc 14:32

Rowley DB, Currie BS (2006) Paleo-altimetry of late eocene to Miocene sediments from the Lunpola Basin, Central Tibet: implications for growth of the Tibetan plateau. Nature 439:677–681

Ruddiman WF (ed) (1997) Techtonic uplift and climate change. Pienum Press, New York

Sakai H (2005) Uplift of the Himalayan range and Tibetan Plateau-From a viewpoint of birth ofmonsoon system and its changes. J Geol Soc Jpn 111:701–716. (in Japanese with English Abstract)

Sakai H et al (2006) Pleistocene rapid uplift of the Himalayan frontal ranges recorded in the Kathmandu and Siwalik basins. Palaeogeogr Palaeoclimatol Palaeoecol 241:16–27

Sato Y (1996) The civilization of rice-production revealed from DNA—its origin and development (in Japanese) NHK book series

Vrba E (1995) The fossil record of African antelopes (Mammalia, Bovidae) in relation to human evolution and paleoclimate. In: Vrba E, Denton G, Burckle L, Partridge T (eds) Paleoclimate and evolution with emphasis on human origins. Yale University Press, New Haven, pp 385–424

Wang Y, Deng T, Biasatti D (2006) Ancient diets indicate significant uplift of southern Tibet after ca. 7 Ma. Geology 34(4):309–312. https://doi.org/10.1130/G22254.1

Wang C, Zhao X, Liu Z, Lippert PC, Graham SA, Coe RS, Yi H, Zhu L, Liu S, Li Y (2008) Constraints on the early uplift history of the Tibetan Plateau. Proc Natl Acad Sci 105:4987–4992

Yasunari T (1990) Impact of Indian Monsoon on the Coupled Atmosphere/Ocean System in the Tropical Pacific. Meteorog Atmos Phys 44:29–41

Yasunari Y (1991) The monsoon year – a new concept of the climatic year in the tropics. Bull Am Meteorol Soc Jpn 72(9):1331–1338

Yasunari T, Seki Y (1992) Role of the Asian monsoon on the interannual variability of the global climate system. J Meteor Soc Japan 70(1):177–189

Yasunari T, Kitoh A, Tokioka T (1991) Local and remote responses to excessive snow mass over Eurasia appearing in the northern spring and summer climate. A study with the MRI·GCM. J Meteor Soc Japan 69(4):473–487

Yasunari et al (2006) Abstract of the international conference on the Ice Age climate. Nagoya University, 21th Century COE Program

Zachos J, Pagani M, Sloan L, Thomas E, Billups K (2001) Trends, rhythms, and aberrations in global climate 65 Ma to present. Science 292(5517):686–693

Part III
Snow, Glaciers and Hydrology

Chapter 16
Climate Change and Cryospheric Response Over North-West and Central Himalaya, India

H. S. Negi, A. Ganju, Neha Kanda, and H. S. Gusain

Abstract This study summarizes the results of several climate studies conducted using field observed data of winter period over the North-West Himalaya (NWH) and Central Himalaya (CH). It also summarizes the latest conclusions about winter-time trends over NWH and its constitutive zones that have been drawn from the study conducted by Negi et al. (Curr Sci 114(4):760–770, 2018), which incorporates the results and inferences of all other studies as well. Wintertime climatic variability over CH has also been discussed for the first time in this study. The salient deductions are as under:

- Overall warming trends in mean and maximum temperature of NWH (1991–2015) and CH (2001–2012) have been observed. In contrast to the situation at the global scale, the data of both NWH and CH reflect higher rate of warming in maximum temperature than minimum temperature. Consequently, there has been an increase in Diurnal Temperature Range (DTR) over both NWH and CH.
- Regionally, long term (~30 years) warming trends have been observed in all zones of NWH except for the minimum temperature over the Lower Himalaya (LH) which shows cooling trends.
- The rate of warming (mean temperature) is found to be highest in the Greater Himalaya (GH) than the Karakoram Himalaya (KH) and LH, which partly explains the higher rate of glacier melt in regions of GH than KH. In addition, no conclusive trends in Elevation Dependent Warming (EDW) were observed in NWH.
- Short term trends (2000–2015) depict cooling in maximum temperature of LH and GH, which though unexplained, may have some links with rising concentration of aerosols in atmosphere in recent decades as reported in a study by Krishnan and Ramanathan (Geophys Res Lett 29(9):54-1–54-4, 2002).

H. S. Negi (✉) · A. Ganju · N. Kanda · H. S. Gusain
Snow and Avalanche Study Establishment, Chandigarh, India
e-mail: hs.negi@sase.drdo.in

© Springer Nature Switzerland AG 2020
A. P. Dimri et al. (eds.), *Himalayan Weather and Climate and their Impact on the Environment*, https://doi.org/10.1007/978-3-030-29684-1_16

- The cryosphere of NWH and CH show heterogeneous behaviour to climate change.
- Long term warming trends over LH, GH and CH have manifested in retreat of glaciers lying in these areas. Though KH also reports warming but this marginal increase in temperature field has not yet made a dent in KH where temperatures are still in subfreezing range even during ablation period. This obviously has resulted in less ablation indirectly implying marginal gain in mass, which has resulted in bringing more stability to the glaciated region of Karakoram Himalaya.

16.1 Introduction

The Hindu Kush Himalayan (HKH) region, which encompasses a major part of world's ice and snow cover is often termed as 'Third Pole of the World' and is spread over 08 countries namely Afghanistan, Pakistan, India, Nepal, Bhutan, China, Bangladesh and Myanmar. Being the source of 10 major river systems, it sustains almost one-fifth of total world's population (Schild 2008). As per fifth assessment report of IPCC, mean surface temperature has increased by about 0.84 °C globally since 1880 (Stocker et al. 2013) and warming is found to be predominant over high altitudes (Diaz and Bradley 1997). Since glaciers are known as sensitive indicators of climate change (Thompson 2000), many studies have been conducted in high altitude snow bound areas of Himalaya to ascertain the impact of changing climate over glaciated areas. Rise in temperature and consequent decline in snowfall amount over high altitudes has been reported by many studies (Dimri and Dash 2012; Bhutiyani et al. 2007, 2010; Gusain et al. 2014, 2015; Shekhar et al. 2010). This change has been attributed to rising concentration of greenhouse gases (Schneider 1990) and decreased cloud cover (Shekhar et al. 2010) etc. The cryosphere responds to rising temperature and declining snowfall mainly in the form of retreating glaciers (Bolch et al. 2012; Azam et al. 2014; Saurabh and Braun 2016). This response is quite spatially heterogeneous since stable/advancing glaciers have also been observed at certain locations despite significant retreat of glaciers reported at majority of locations (Gardelle et al. 2012; Kääb et al. 2012).

During recent decades, the advent of various remote sensing (satellite, aerial and terrestrial) and model generated gridded products has facilitated better analysis of climate variables of high altitude regions. Simultaneously, a decline in observatory setup across the globe has been observed (Strachan et al. 2016). This decline in observation network has hindered the reliability tests of available gridded modeled products especially for high mountain areas where paucity of observations already existed (Hasenauer et al. 2003; Pepin and Seidel 2005). In addition, some of these gridded products (especially interpolated datasets) utilize very less observations from upper Himalayan region e.g. APHRODITE which has no single observation from altitudes above 5000m. This makes it challenging to have a reliable gridded dataset with high accuracy for high altitude areas like Karakoram (Kumar et al. 2015). Further, many approximations are made while estimating snowfall (solid) contribution from total (solid + liquid) precipitation when using satellite based products (Rasmussen et al. 2012). Conclusively, although the use of gridded products is gaining popularity yet the importance of ground based observatories can never be overlooked.

Various studies using field observed data of high altitude have been conducted regularly to ascertain the existence as well as impact of global warming in various parts of north-west Himalayas (NWH) during different seasons. The major drawback, in such studies, is the lack of agreement on the inferences drawn from the studies, which could be attributed to many things like selection of different stations, period of study, disparity in selection of months for a season, missing data and lack of standardized approaches etc. This study summarizes the overall climatic changes that prevailed in the NWH and Central Himalaya (CH) along with references of various other studies conducted for analysis of climate trends over these regions. Our study focuses mainly on winter season since during winter months, the precipitation which is under the influence of westerly winds is critical for avalanche study as well as for the growth of glacier in terms of mass gain. The response of cryosphere to observed climate change is also discussed.

16.2 Study Area and Data

Snow and Avalanche Study Establishment (SASE), India has a scanty observational network of surface observatories and Automatic Weather Stations (AWS) over NWH and CH. The altitude of these stations ranges approx. Between 2000 m and 6000 m above msl as shown in Fig. 16.1 NWH by virtue of its high spatial variability has been divided into three climatic zones viz., lower, middle and upper Himalayas (Sharma and Ganju 2000). Pir Panjal (PP) and Shamashavari (SS) range constitute Lower Himalaya (LH) whereas the Great Himalayan (GH) range dwells in Middle Himalayas. Likewise, Karakoram Range (KH) lies in Upper Himalayan zone. Thus NWH comprises LH, GH and KH Himalayan zones. However, most of the studies discussed are from the data of Western Himalaya (WH), which comprises LH and GH. SASE observatories are present in all zones and the analysis of recorded data is conducted at zonal as well as range level. In addition to the above, one observatory named 'Bhojbasa' (S17), which represents snow met conditions of Gangotri glacier in CH has also been included in this study. The data at these observatories is recorded as per World Meteorological Organization (WMO) guidelines twice at 0830 and 1730 (IST) daily. The length of data period varies differently at different stations. As explained in preceding sections that divergent results of climate trends were reported by different studies despite all having used the same source of data for drawing the deductions. Authors feel that the foremost reason for it could be the disparity in selection of months for a season. For example, some researchers have considered 6 months (Nov–Apr) whereas others have chosen

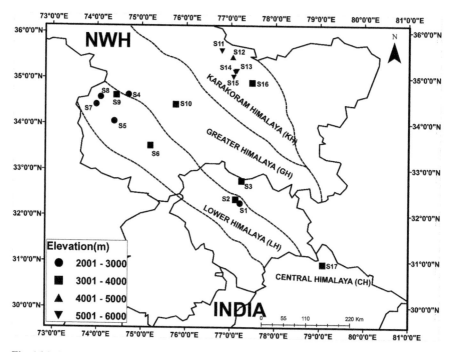

Fig. 16.1 Location of SASE observatories in different parts of Himalayas

3 month (DJF) period as winter season. Apart from the disparity in month selection, stations and period selected for study also vary from one study to another. Since topography (e.g. Elevation ranges between 2000 m and 6000 m above m.s.l.) and consequently land use is highly variable over NWH, the data is likely to have effects of location. Thus, it becomes imperative to standardize the data prior to any analysis considering the impact of different locations. Standardization is known to remove such biases since standardized anomalies are calculated by dividing the anomalies by respective standard deviation (Wilks 1995). But some studies in which standardization approach was not followed, the results may have reflected dominating influence of some locations than others on overall trends.

The conclusions over NWH have been drawn mainly from the latest study conducted by (Negi et al. 2018) since their study was based on standardized anomalies, which were calculated after filling of missing data. Moreover, their study is based on latest climatic records as well i.e. 1991–2015. The analysis of wintertime trends in CH brings novelty to this study since only annual trends at 'Bhojbasa' have been published so far by Gusain et al. 2015. Trends in various climatological parameters at LH, GH, KH, NWH and CH are shown as stacked figures for each variable under consideration.

16.3 Observed Changes in Temperature Over NWH

16.3.1 Average Temperature Values and Overall Trends at NWH

Average maximum, minimum, mean temperature and DTR values over NWH and CH along with its constituent Himalayan zones have been presented as box charts in Fig. 16.2a–d respectively. Table 16.1 summarizes the studies conducted for different durations with their respective reported trends.

Overall temperature trends at NWH depict warming since last century (1901–1999) by 1.7 °C during winter period (Bhutiyani et al. 2007). This long term warming trend corroborates well with warming trends reported by other studies (using SASE data) conducted for comparatively shorter durations e.g. (Shekhar et al. 2010) ~24 years, (Dimri and Dash 2012) ~31 years, (Gusain et al. 2014) ~35 years, (Singh et al. 2015) ~30 years and (Negi et al. 2018) ~25 years. However, the magnitude of warming is found to be higher e.g. ~2.2 °C (1991–2002) and ~2 °C (1984–1985 till 2007–2008) as reported by (Bhutiyani et al. 2007) and (Shekhar et al. 2010) respectively during recent decades than the century average (i.e. 1.7 °C) as suggested by (Bhutiyani et al. 2007). In addition, the warming in NWH is found to be more pronounced in case of maximum temperature than minimum temperature (Bhutiyani et al. 2007; Shekhar et al. 2010; Dimri and Dash 2012; Singh et al. 2015; Negi et al. 2018). Figures 16.3, 16.4, 16.5 and 16.6 depict standardized time series of maximum, minimum, mean temperature and DTR respectively for LH, GH, KH, NWH and CH during 1991–2015.

16.3.2 Regional Long Term Trends at NWH

16.3.2.1 Warming Trends in Various Pockets of NWH

Since Himalaya is a complex region, several cooling as well as warming domains could be observed within different mountain ranges. Apart from overall warming trends in NWH, warming in mean temperature is also evident over its various zones i.e. LH and GH along with their constitutive ranges viz. Pir Panjal and Shamashavari range (Shekhar et al. 2010). In addition, Dimri and Dash (2012) reported that besides the overall warming trend over Pir Panjal and Shamashavari range, some part of ranges lying in Himachal Pradesh (HP) have warmed more than that lying in Jammu and Kashmir (JK). Similarly, an elevation wise study conducted by Gusain et al. (2014) reveals that majority of stations having elevation below 4000 m show warming in maximum and minimum temperature. Similar study conducted by Singh et al. (2015) reveals warming trends at all ranges of WH during 1979–2009. Such regional warming is also depicted by a latest study conducted by Negi et al. (2018) which shows warming in maximum, minimum and mean temperature in all

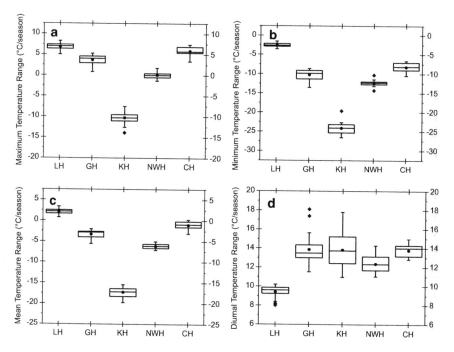

Fig. 16.2 Box-charts representing values of various parameters during winter (Nov-Apr) of NWH (25 years) and CH (12 years). (**a**) Average maximum temperature at LH, GH, KH, NWH and CH (6.8 °C, 3.6 °C, −10.3 °C, 0.05 °C and 6.07 °C respectively). (**b**) Average minimum temperature at LH, GH, KH, NWH and CH (−2.6 °C, −10.3 °C, −24.13 °C, −12.3 °C and − 8.10 °C respectively). (**c**) Mean temperature at LH, GH, KH, NWH and CH (2.1 °C, −3.3 °C, −17.3 °C, −6.2 °C and − 0.86 °C respectively) (**d**) DTR values at LH, GH, KH,NWH and CH (9.4 °C, 13.85 °C, 13.82 °C, 12.36 °C and 13.86 °C respectively)

zones of NWH as seen in Figs. 16.3, 16.4 and 16.5 respectively with an exceptional cooling in minimum temperature over LH. The exceptional cooling at LH is explained in succeeding paragraphs. In addition, Dimri and Dash (2012) found that number of days with maximum temperature above 90th percentile i.e. warm days have increased and number of days with maximum temperature below 10th percentiles i.e. cold days have decreased in Western Himalaya (WH). Similarly, percent number of nights with minimum temperature above 90th percentile i.e. warm nights have increased, and percent number of nights with minimum temperature below 10th percentile i.e. cold nights have decreased during last three decades over WH. Conclusively highest and lowest temperatures have increased over WH. All studies attributed these warming trends to rising concentration of greenhouse gases as a result of anthropogenic activities, within or outside the region. The probability attributing this warming trend to anthropogenic activities outside the region stands more than within the region.

Table 16.1 Summary of studies depicting change in maximum, minimum and mean temperature over different parts of NWH during different durations

Reference	Region	Time period	Parameter	Reported change (°C)
Bhutiyani et al. (2007)	Western Himalaya	1901–1999	Mean temp	1.7
		1991–2002	Max temp	3.2
			Min. temp	0.8
			Mean temp	2.2
Shekhar et al. (2010)	Western Himalaya	1984–1985 till 2007–2008	Max temp	2.8
			Min. temp	1.0
			Mean temp	2.0
	PP	1984–1985 till 2007–2008	Max temp	0.8
			Min. temp	0.6
	SS	1984–1985 till 2007–2008	Max temp	2.0
			Min. temp	1.0
	GH	1984–1985 till 2007–2008	Max temp	1.0
			Min. temp	3.4
	KH	1984–1985 till 2007–2008	Max temp	−1.6
			Min. temp	−3.0
Dimri and Dash (2012)	PP(HP)	1975–2006	Max temp	1.4
			Min. temp	0.5
			Mean temp	0.7
	PP(JK)	1975–2006	Max temp	1.1
			Min. temp	0.3
			Mean temp	0.6
	GH(HP)	1975–2006	Max temp	2.0
			Min. Temp	0.2
			Mean temp	1.3
	GH(JK)	1975–2006	Max temp	2.5
			Min. temp	0.3
			Mean temp	0.6
Singh et al. (2015)	PP range	1979–2009	Max temp	0.98
			Min. temp	0.28
	S range	1979–2009	Max temp	0.27

(continued)

Table 16.1 (continued)

Reference	Region	Time period	Parameter	Reported change (°C)
			Min. Temp	0.11
	GH range	1979–2009	Max temp	1.0
			Min. temp	0.75
Negi et al. (2018)	LH	1991–2015	Max temp	1.2
			Min. temp	−0.83
			Mean temp	0.34
	GH	1991–2015	Max temp	0.77
			Min. temp	0.98
			Mean temp	0.87
	KH	1991–2015	Max temp	0.72
			Min. temp	0.44
			Mean temp	0.56

16.3.2.2 Exceptional Long Term Cooling Behavior

A recent study conducted by Negi et al. (2018) reports cooling in minimum temperature over LH by −0.83 °C during 1991–2015 as is seen in Fig. 16.4. Yadav et al. (2004) also reported sharp rate of cooling in minimum temperature over Western Himalaya during latter part of twentieth century and they attributed it to local forcing factors i.e. large scale deforestation and land degradation which could be seen as a direct impact of increased population and industrialization in and nearby areas. Negi et al. (2018) also observed cooling in maximum temperature while studying short term trends i.e. during 2001–2015 over LH and GH. A similar cooling in maximum temperature during winter season over part of Indian subcontinent affected by haze was reported by Krishnan and Ramanathan (2002). They reported that due to increased industrialization post 1950s, concentration of aerosols has been increasing. These aerosols by virtue of their absorbing nature don't allow much of the incoming solar radiation to reach to the surface of earth leading to surface cooling. Since, industrialization has also made an impact in form of changed land use/land cover over the Himalayan foothills and in Indo-Gangetic plains, the observed anomalous cooling could be a resultant of rising aerosol concentrations from the foothills and Indo-Gangetic plains. The aerosol in Upper Atmosphere could also have been brought from outside India from the western regions as wind dominates from west to east in this region.

Fig. 16.3 Inter annual variability of average winter maximum temperature at different climatic zones of NWH (1991–2015) and CH (2001–2012)

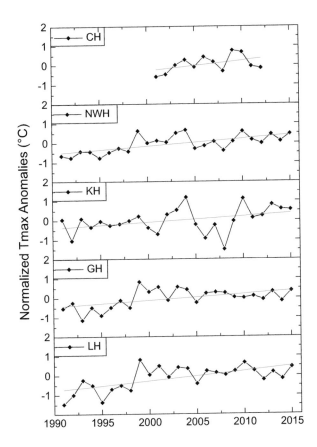

16.3.3   Elevation Dependent Warming (EDW) and NWH

The differential rate of warming along an altitude gradient is often termed as 'Elevation Dependent Warming' (EDW). This rate of warming is usually different from global warming rate since many factors like snow albedo feedback, aerosols, clouds, water vapor, soil moisture etc. affect the response of any surface towards climatological forcing (Tudoroiu et al. 2016). Many studies have shown positive correlation between altitude and warming (Beniston and Rebetez 1996; Liu et al. 2009; Diaz and Eischeid 2007) while many others have observed negative altitudinal dependence of warming (Beniston and Rebetez 1996; Rangwala et al. 2009) in different parts of the world. NWH shows inconclusive trends along the altitude gradient since no clear cut relationship between altitude and rate of warming could emerge from the picture so far.

Fig. 16.4 Inter annual variability of average winter minimum temperature at different climatic zones of NWH (1991–2015) and CH (2001–2012)

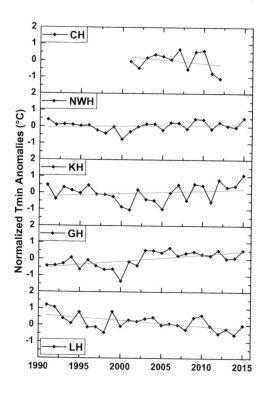

16.3.3.1 Positive Elevation Dependent Warming Trend

Shekhar et al. (2010) reports higher rate of warming in Shamashavari range (higher elevation) than the Pir Panjal range (lower elevation) along with higher warming in minimum temperature of GH than PP and SS. Similarly, Dimri and Dash (2012) reported higher rate of warming in GH than the PP range. Singh et al. (2015) also reports higher rate of warming in GH than LH (PP and SS ranges).

16.3.3.2 Negative Elevation Dependent Warming Trend

A study by Gusain et al. (2014) reports that most of the observatories located at elevations beyond 4000 m (which covers most of the observatories in KH) showed cooling trends though the observed cooling trends were significant in case of minimum temperature only. Shekhar et al. (2010) also reports cooling at KH, which is approx. −1.6 °C in maximum temperature and −3 °C in minimum temperature during 1984–1985 till 2007–2008. Whereas the latest study conducted by Negi et al. (2018) reports warming at KH (highest elevation amongst all zones) by 0.72 °C, 0.44 °C and 0.56 °C in maximum, minimum and mean temperature respectively during 1991–2015. Even short term trends (2000–2015) studied by Negi et al.

Fig. 16.5 Inter annual variability of average winter mean temperature at different climatic zones of NWH (1991–2015) and CH (2001–2012)

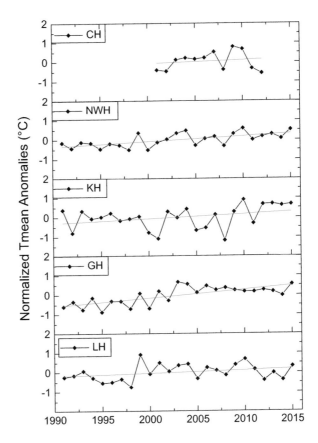

(2018) reveal that warming in maximum, minimum and mean temperature is consistent over KH. It is interesting to note that while the period from 1991 till 2008 in KH, as revealed in Fig. 16.7, shows steady cooling trend, the reverse trend is seen drastically happening 2008 onwards. As a result, the long and short term trend lines have shifted from declining to rising ones. This anomaly needs to be investigated in order to establish a robust relationship between rising temperature and other parameters.

16.3.3.3   Highest Warming Rates at Intermediate Altitudes

Negi et al. (2018) reported that the extent of warming in mean temperature was highest over GH (0.87 °C) followed by KH (0.56 °C) and LH (0.37 °C) during 1991–2015 as seen in Fig. 16.5. Similar observance was reported by McGuire et al. (2012) in Rocky Mountains, which as per them shows maximum warming along the front range at intermediate elevations.

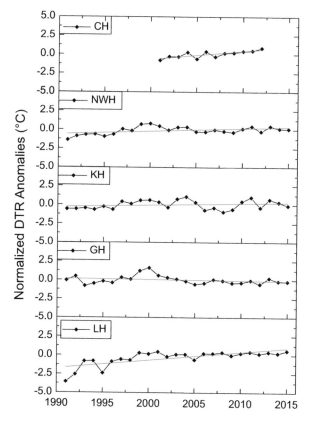

Fig. 16.6 Inter annual variability of average winter DTR at different climatic zones of NWH (1991–2015) and CH (2001–2012)

16.3.4 Trends in DTR Over NWH

Diurnal Temperature Range (DTR) is used as an index of climate change and variability globally (Braganza et al. 2004). Unlike, worldwide trends where DTR is decreasing due to more pronounced warming minimum temperature than maximum temperature, DTR over NWH has increased significantly by 0.94 °C ($\alpha = 0.05$) during 1991–2015 at the rate of +0.037 °C per year as reported by (Negi et al. 2018). However, same study reports a declining trend in DTR during short term (2001–2015) in GH. It is to be noted that in regional context, DTR has increased at all zones except GH. The rise in DTR over LH is significant ($\alpha = 0.001$) which is due to significant rise/fall in maximum/minimum temperature at the same time.

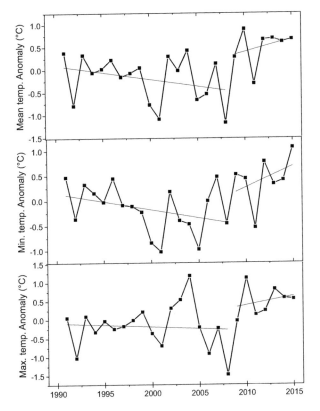

Fig. 16.7 Trends in Maximum, minimum and mean temperature over KH during 1991–2008 and 2009–2015

16.4   Observed Changes in Precipitation Over NWH

16.4.1   Average Values and Trends in Winter Precipitation at NWH

The average winter precipitation (Snowfall + Rainfall) values at NWH along with its constitutive zones and CH are shown in Fig. 16.8. Since WDs encounter the LH zone first during their eastward motion, LH receives maximum highest amount of winter precipitation primarily in form of wet snow (Dimri and Dash 2012) as can be seen in Fig. 16.8. It is clearly discernable that contribution of WDs in fetching wintertime precipitation is minimal in case of CH since this part of Himalayas receives winter precipitation after NWH had it, leaving little moisture to precipitate as the system moves eastwards. Alternatively, occasionally when Westerlies descend to lower latitudes moisture incursion from Arabian Sea may precipitate higher amount

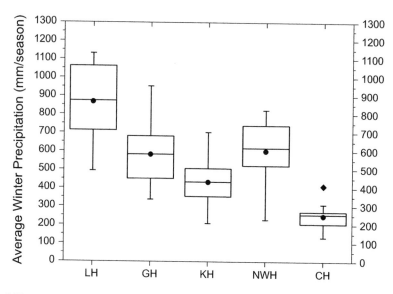

Fig. 16.8 Box-charts representing values of winter precipitation during winter (Nov-Apr) of NWH (25 years) and CH (12 years). Average precipitation at LH, GH, KH, NWH and CH is 804 mm, 549 mm, 431 mm, 595 mm and 249 mm respectively)

of snow in CH, but that situation happens very rarely. As such the CH region has its share of solid precipitation during summer when it is mainly under the influence of Indian Summer Monsoon Rainfall (ISMR).

16.4.2 Overall Precipitation Trends at NWH

The precipitation trends are highly variable owing to high spatial variability of WH (Dimri and Dash 2012). No significant trend during 1866–2006 was observed since alternate episodes of precipitation above and below normal prevailed over NWH (Bhutiyani et al. 2010). Gusain et al. (2014) studied trends of winter precipitation at various altitudes and reported a decrease in winter precipitation at majority of stations lying at altitudes below 4000 m whereas those lying above 4000 m showed erratic behavior with some stations showing increase in precipitation while others followed declining trends. Recent findings over NWH during 1991–2015 by Negi et al. (2018) reveal that an increasing trend in precipitation is observed at all zones of NWH i.e. LH, GH and KH, the trends being significant for LH only as can be seen in Fig. 16.9.

In addition, the extreme events study conducted by Dimri and Dash (2012) reported a decline in heavy precipitation days with increasing consecutive dry days and decreasing consecutive wet days during 1975–2006 over WH. Shekhar et al. (2017) also reported that low and heavy precipitation events have increased significantly but medium precipitation events decreased over Pir Panjal Range of Lower

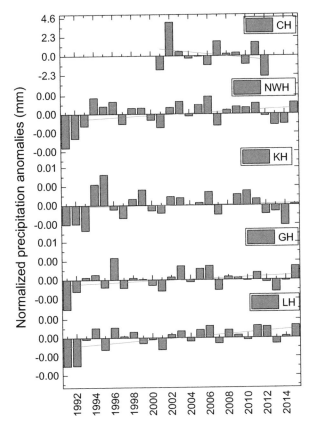

Fig. 16.9 Inter annual variability of winter precipitation expressed as standardized precipitation anomaly at NWH (1991–2015) and CH (2001–2012)

Himalayas. In addition, prolonged dry days (PDD) have increased in all the ranges and altitudes of Himalayas.

16.4.3 Snowfall vs. Rainfall Trends

During recent decades, total precipitation showed increasing trend but snowfall contribution towards total precipitation has declined significantly at many locations (Bhutiyani et al. 2010; Shekhar et al. 2010; Dimri and Dash 2012; Gusain et al. 2014; Singh et al. 2015; Negi et al. 2018). This observance of increased liquid precipitation and decreased solid precipitation has been attributed to rising temperature, weakening of teleconnections between temperature and precipitation due to, perhaps increased aerosol emissions (Bhutiyani et al. 2010). Since winter precipitation in the area is mainly due to WDs, the rising variations in amplitude of WDs over the area partly explains the increased contribution of rainfall to total precipitation in the area than snowfall (Madhura et al. 2015). Singh et al. (2015) further added that decrease in snowfall amount is more at high altitudes whereas increase in rainfall amount is predominant at lower altitudes owing to higher temperature at lower altitudes and overall warming trends over NWH.

Singh et al. (2015) also reported decreased snowfall days and increased rainfall days in last three decades over NWH. They also reported that seasonal snow cover days and seasonal snow cover depth has also declined over NWH in last three decades.

16.5   Observed Changes in Temperature and Precipitation Over CH

SASE observatory at 'Bhojbasa' is located approx. 5 km northwest to the snout of Gangotri glacier which is commonly known as 'Gaumukh'. Gusain et al. (2015) has already reported the mean annual and winter time conditions in the area. He reported that average annual maximum temperature over CH is ~11.7 ± 0.7 °C and average annual minimum temperature is ~ −2.3 ± 0.4 °C. Similarly, average winter maximum and minimum temperature is observed ~3.0 ± 1.0 °C and − 10.4 ± 1.3 °C respectively. They also found that annual maximum and minimum temperature have increased by 0.9 °C and 0.05 °C respectively during 2000–2012. As a consequence of warming temperature, perhaps snowfall amount has reduced by ~37 cm during same period.

This study presents inter-annual variability of temperature and precipitation over CH during 2001–2012 (winter period : Nov-Apr). It is to be noted that temperature and precipitation variability is expressed as Standardized Temperature/Precipitation anomaly in this study. As depicted by Figs. 16.3, 16.4, 16.5 and 16.6 maximum temperature, mean temperature and DTR have increased, whereas minimum temperature and snowfall amount has reduced over CH during 2001–2012. However, the rise in DTR only is significant at $\alpha = 0.01$. Maximum temperature has increased by ~0.6 °C. Conversely, minimum temperature has decreased by ~ −0.5 °C during same period. Consequently, mean temperature and DTR increased by ~ 0.2 °C and 1.2 °C respectively. However, snowfall amount has also reduced. In absolute terms, the per year reduction in wintertime precipitation was found to be approx. 8 mm.

16.6   Cryosphere and Climate Change

Western Disturbances (WDs) is the important source of moisture providing maximum precipitation in the western Himalaya and KH during winter. Hence, WDs are the major contributors for nourishment of snow cover and glaciers in NWH. CH on the other hand receives maximum precipitation during summers under the influence of Monsoon and thus receives less snowfall during winter compare to NWH (Fig. 16.8). The western Himalaya is a transition region receiving precipitation from both the summer monsoon and WDs. Therefore, the role of geographical locations and regional orography controls climate of the different Himalayan regions and thus the changes in cryosphere (Azam et al. 2018).

Snow Cover Area (SCA) SCA is generally considered as a proxy indicator of climate change owing to its negative relationship with temperature. Changes in temperature greatly affect the occurrence of snowfall patterns, which further manifests in changes in SCA and many other factors (Gurung et al. 2017). Thus assessment of SCA over different parts of NWH has been investigated by various researchers for different periods. But such studies often show disparity among each other owing to lack of selection of appropriate scale for study since SCA is highly sensitive to micro-climatic effects (Gurung et al. 2017). Also, since assessment of SCA is based solely on satellite remote sensing products which have come into picture quite lately, the availability of data for study is very short termed and hence wide fluctuations in results are inevitable. We believe that such fluctuations may even out and may be transient in nature when studied on long term basis. A recent study by Gurung et al. (2017) reported a declining trend in SCA (2003–2012) over all elevation belts and aspects of Himalayas along with a strong negative correlation (statistically significant) between temperature and SCA. Thus the impact of rising temperature cannot be overlooked especially over western Himalayan region, where strongest negative correlation between temperature and SCA during 2000–2007 was found (Gurung et al. 2017). A significant negative trend in SCA of Upper Indus basin (UIB) during 2000–2008 was reported by Immerzeel et al. (2009) and they attributed it to rising temperature. A similar decreasing tendency of wintertime SCA over UIB during 2001–2012 was reported by Hasson et al. (2014). Our study also indicates that mean temperature has increased over all zones of NWH during 1991–2015, as depicted by Fig. 16.5. However as explained earlier that when studies pertaining to snow-cover are focused on small areas, microclimatic effects might interfere, thereby producing contrary results when compared regional to sub-regional scales. For an instance, Singh et al. (2014) analyzed SCA variations over three basins of Himalayas i.e. Indus, Ganga and Brahmaputra during 2000–2011 and found a rising trend in SCA over Indus basin as compared to other basins. It is to be noted that Indus basin wholly covers all parts of NWH. Regional climate trends as reported by Negi et al. (2018) depict cooling trends at LH and GH during recent decades (2000–2015) which could partly explain the observed increasing trends in SCA over such huge area. On the contrary, warming trends during 2000–2015 over KH were reported by same study, which is in sharp contrast to the observation of increased SCA in the UIB as reported by Singh et al. (2014). Despite getting warmed in recent decades, KH remains below freezing temperature for most of the time even during ablation period (Singh et al. 2014) which could have sustained the snow cover for longer durations. Conclusively, climatic variability is manifesting itself in the form of varying snow cover amounts reaffirming the sensitivity of cryosphere to climate change.

The use of **albedo** values as a proxy to estimate winter mass balance was supported by Sirguey et al. 2016 when they found a high positive correlation between mass balance and albedo. Long term (1991–2010) albedo trends depict a declining trend over GH (Negi et al. 2017) which is in agreement to long term (1991–2015) warming trends as shown in Fig. 16.5. This reduced albedo could have resulted in higher rate of glacier retreat in GH as reported by many researchers (Kulkarni and Karyakarte 2014). However, an increased albedo over GH post year 2000 as reported by Negi et al. (2017) is in agreement to short- term temperature trends reported by Negi et al. (2018) wherein a cooling trend over GH during 2000–2015 was observed. This also explains that most of the glaciers are in a steady state compared to the rate of retreat prior to 2001 (Bahugana and Rathore 2014). Thus, we feel that winter snow albedo has also an important role in governing the rate of overall glacier retreat in NWH.

Apart from SCA and albedo, glaciers also serve as very sensitive indicators of climate change. Many glacier features like length, percent glacierized area and mass balance are being studied in Indian Himalaya to ascertain the impact of ongoing climatic changes. Any change in **glacier mass balance** affects the hydrological cycle of a region (Diaz and Bradley 1997). A positive mass balance may reflect reduced run off and negative mass balance depicts increased run off from glacierized catchments (Scherler et al. 2011). The response of Himalayan glaciers is quite heterogeneous owing to existent high spatial variability. Some studies report that the glaciers lying in zones, which are under the influence of summer monsoons (e.g. Eastern Himalaya and Central Himalaya) are retreating at faster pace while those lying under the influence of Westerlies (Karakoram Himalaya) are known to gain mass or have stable fronts (Scherler et al. 2011). Glaciological assessments of mass balance reveal more negative mass balance values for CH than western Himalaya, while minimum negative or say slight positive mass balance values are reported for Karakoram region (Table S4, Azam et al. 2018). Temperature and precipitation undoubtedly serve as powerful drivers for mass budget of a glacier. Fig. 16.2c reveals that lowest wintertime mean temperature is recorded at KH followed by GH and highest at CH. At the same time, Fig. 16.8 depicts that minimum wintertime precipitation is received at CH followed by KH, GH and LH. Thus due to prevalence of highest wintertime temperature along with minimum wintertime precipitation that too when temperature trends depict warming (Fig. 16.5), higher rate of mass loss of glaciers in CH is inevitable. Even GH and KH depict long term (1991–2015) warming with more warming at GH (Table 16.1), its impact is discernible by observed more negative mass balance values for GH than KH (Table S8, Azam et al. 2018). Consequently, glaciers in Western Himalayas are found to be receding at faster pace whereas those in KH are reported to have negligible mass loss or sustained stable condition (Scherler et al. 2011). Surprisingly, Gardelle et al. (2012, 2013) reported slight positive mass balance values for glaciers of KH which are in striking contrast to other studies. Despite the observation of warming trends (Figs. 16.3, 16.4 and 16.5), slight mass gain by few glaciers can be attributed to the fact that the observed mean temperature in KH is very low that is subfreezing temperature (Fig. 16.2c), due to which slight warming does not make a discernible impact in shorter durations. However, if such warming trends are likely to continue for few decades, then a striking difference in mass budget of KH glaciers would emerge. It is to be noted that glaciers with mass gain were reported in Central Karakoram only. Hence we believe that along with very low temperature in Karakoram range, effect of geographical position and local microclimates could have contributed towards mass gain phenomenon in Central Karakoram.

Apart from changes in glacier thickness, the impact of climatic change is revealed by the spatial variations in **glaciated area and length**. Kulkarni (2007) has reported that the response of a glacier to climate change is a function of glacier size, area-altitude distribution, moraine cover and orientation. Thus, the extent of area shrinkage for small glaciers can reach as high as -1.34% year^{-1} for CH (Chen et al. 2007) to as low as -0.002% year^{-1} for large glaciers like Siachen in KH (Agarwal et al. 2017). Kulkarni (2007) also added that differential rate of glaciers retreat is because of differences in glacier thickness, mass balance and rate of melting at the terminus. Thus, loss of glaciated area is highly dependent on areal extent of glaciers leading to more sensitivity of small sized glaciers to climatic changes. Hence, it can be inferred that the differential rates of retreat/area shrinkage/mass loss etc. reported by different studies (Tables 2, 3, 4 and 8, Azam et al. 2018) are a function of response time of a glacier and these rates are found higher in case of small size glaciers.

16.7 Conclusions

This study discussed the importance of ground based observations despite the emergence of modern data sources like gridded satellite/reanalysis products and modeled outputs etc. Various studies conducted over high altitude areas of NWH and CH using field observed data were analyzed. This study affirms the claim by many studies that high altitude areas are warming at faster pace than the lower counterparts. Alike many parts of the world, NWH and CH are no exception to ravaging impacts of 'Climate Change' with few exceptions at regional level. But interestingly, NWH and CH show more warming in maximum temperature than minimum temperature whereas other parts of the world report opposite trends. Warming is evident in mean and maximum temperature at NWH and CH with different magnitudes. Minimum temperature however anomalously shows a significant dip over LH. Also, short term trends i.e. after year 2000, maximum temperature over LH and GH are found to experience cooling, which can be attributed to rising concentration of absorbing aerosols, which due to their absorbing nature shield the underlying surfaces from incoming solar radiation thereby showing a decline in surface temperature. DTR trends over NWH and CH show rising trends unlike global trends which depict narrowing DTR due to differential warming rates of maximum and minimum temperature. Total precipitation (Snowfall+ Rainfall) has increased at all zones of NWH but contribution of snowfall towards total precipitation has decreased at NWH and its constitutive zones. Similar decline in snowfall amount has been observed over CH. However, rainfall contribution towards total precipitation has been increasing consistently at all zones of NWH which is seen as a direct impact of warming temperatures. Factually, the erratic behavior of extreme precipitation events is linked with increased variability in amplitudes of WDs and weakening of teleconnections over the study area. Climate change is manifesting itself in form of cryospheric changes like decreased snow cover area, retreating glaciers, negative mass balance of much of the glaciers and decreasing albedo values as well. However, few exceptions like mass gain in glaciers of Central Karakoram imply that due to high spatial variability of Himalayas, the response of glaciers to climatic forcing is heterogeneous. Moreover, comparatively higher rate of warming in mean temperature over GH than KH substantially explains the exceptional behavior of glaciers in KH.

Acknowledgements The authors are thankful to technical staff of SASE for data collection from rugged terrain in extreme harsh climatic conditions. This work is carried out under DRDO project 'Him-Parivartan'.

References

Agarwal V, Bolch T, Syed TH, Pieczonka T, Strozzi T, Nagaich R (2017) Area and mass changes of Siachen glacier (East Karakoram). J Glaciol 63:148–163

Azam MF, Wagnon P, Vincent C et al (2014) Reconstruction of the annual mass balance of Chhota Shigri glacier, Western Himalaya, India, since 1969. Ann Glaciol 55(66):69–80

Azam MF, Wagnon P, Berthier E et al (2018) Review of the status and mass changes of Himalayan-Karakoram glaciers. J Glaciol 64(243):61–74

Bahugana M, Rathore BP (2014) Are the Himalayan glaciers retreating? Curr Sci 106(7):1008–1013

Beniston M, Rebetez M (1996) Regional behavior of minimum temperatures in Switzerland for the period 1979–1993. Theor Appl Climatol 53:231–243

Beniston M, Diaz HF, Bradley RS (1997) Climatic change at high elevation sites: an overview. Clim Chang 36(3–4):233–251

Bhutiyani MR, Kale VS, Pawar NJ (2007) Long-term trends in maximum, minimum and mean annual air temperatures across the Northwestern Himalaya during the twentieth century. Clim Chang 85:159–177

Bhutiyani MR, Kale VS, Pawar NJ (2010) Climate change and precipitation variations in the northwestern Himalaya: 1866–2006. Int J Climatol 30:535–548

Bolch T, Kulkarni A, Kääb A et al (2012) The state and fate of Himalayan glaciers. Science 336:310–314

Braganza K, Karoly DJ, Arblaster JM (2004) Diurnal temperature range as an index of global climate change during the twentieth century. Geophys Res Lett 31:L13217. https://doi.org/10.1029/2004GL019998

Chen X, Cui P, Li Y, Yang Z and Qi Y (2007) Changes in glacial lakes and glaciers of post1986 in the Poiqu River basin, Nyalam, Xizang (Tibet). Geomorph 88(3): 298–311

Diaz HF, Bradley RS (1997) Temperature variations during the last century at high elevation sites. Clim Chang 36:253–279

Diaz H, Eischeid J (2007) Disappearing 'alpine tundra' Köppen climatic type in the western United States. Geophys Res Lett 34:L18707

Dimri A, Dash S (2012) Wintertime climatic trends in the western Himalayas. Clim Chang 111:775–800

Gardelle J, Berthier E, Arnaud Y (2012) Slight mass gain of Karakoram glaciers in the early twenty-first century. Nat Geosci 5:322–325. https://doi.org/10.1038/NGEO1450

Gardelle J, Berthier E, Arnaud Y, Kääb A (2013) Region-wide glacier mass balances over the Pamir-Karakoram-Himalaya during 1999–2011. Cryosphere 7:1263–1286

Gurung DR, Maharjan SB, Shrestha AB et al (2017) Climate and topographic controls on snow cover dynamics in the Hindu Kush Himalaya. Int J Climatol 37:3873–3882. https://doi.org/10.1002/joc.4961

Gusain HS, Mishra VD, Bhutiyani MR (2014) Winter temperature and snowfall trends in the cryospheric region of north-west Himalaya. Mausam 65(3):425–432

Gusain HS, Kala M, Ganju A et al (2015) Observations of snow-meteorological parameters in Gangotri glacier region. Curr Sci 109(11):2116–2120

Hasenauer H, Merganicova K, Petritsch R et al (2003) Validating daily climate interpolations over complex terrain in Austria. Agric For Meteorol 119:87–107

Hasson S, Böhner J, Lucarini V (2014) Early 21st century snow cover state over the western river basins of the Indus River system. Hydrol Earth Syst Sci 18:4077–4100

Immerzeel WW, Droogers P, de Jong SM, Bierkens MF (2009) Large scale monitoring of snow cover and runoff simulation in Himalayan river basins using remote sensing. Remote Sens Environ 113:40–49. https://doi.org/10.1016/j.rse.2008.08.010

Kääb A, Berthier E, Nuth C et al (2012) Contrasting patterns of early twenty-first-century glacier mass change in the Himalayas. Nature 488(7412):495–498

Krishnan R, Ramanathan V (2002) Evidence of surface cooling from absorbing aerosols. Geophys Res Lett 29(9):54-1–54-4

Kulkarni AV (2007) Effect of global warming on the Himalayan cryosphere. Jalvigyan Sameeksha 22:93–108

Kulkarni AV, Karyakarte Y (2014) Observed changes in Himalayan glaciers. Curr Sci 106(2):237–244

Kumar P, Kotlarski S, Moseley C et al (2015) Response of Karakoram-Himalayan glaciers to climate variability and climatic change: a regional climate model assessment. Geophys Res Lett 42:1818–1825. https://doi.org/10.1002/2015GL063392

Liu X, Cheng Z, Yan L, Yin Z (2009) Elevation dependency of recent and future minimum surface air temperature trends in the Tibetan Plateau and its surroundings. Glob Planet Chang 68:164–174

Madhura RK, Krishnan R, Revadekar JV, Mujumdar M, Goswami BN (2015) Changes in western disturbances over the Western Himalayas in a warming environment. Clim Dyn 44:1157–1168

McGuire CR, Nufio CR, Bowers MD, Guralnick RP (2012) Elevation-dependent temperature trends in the Rocky Mountain Front Range: changes over a 56- and 20-year record. PLoS One 7(9):12

Negi HS, Datt P, Thakur NK, Ganju A, Bhatia VK, Vinay Kumar G (2017) Observed spatio-temporal changes of winter snow albedo over the north-west Himalaya. Int J Climatol 37(5):2304–2317. https://doi.org/10.1002/joc.4846

Negi HS, Kanda N, Shekhar MS, Ganju A (2018) Recent wintertime climatic variability over North West Himalayan cryosphere. Curr Sci 114(4):760–770

Pepin NC, Seidel DJ (2005) A global comparison of surface and free-air temperatures at high elevations. J Geophys Res 110:D03104. https://doi.org/10.1029/2004JD005047

Rangwala I, Miller J, Xu M (2009) Warming in the Tibetan Plateau: possible influences of the changes in surface water vapor. Geophys Res Lett 36:L06703

Rasmussen R, Baker B, Kochendorfer J et al (2012) How well are we measuring snow? Bull Am Meteorol Soc 93:811–829. https://doi.org/10.1175/BAMS-D-11-00052.1

Saurabh V, Braun M (2016) Elevation change rates of glaciers in the Lahaul-Spiti (Western Himalaya, India) during 2000–2012 and 2012–2013. Remote Sens 8:1038

Scherler D, Bookhagen B, Strecker MR (2011) Spatially variable response of Himalayan glaciers to climate change affected by debris cover. Nat Geosci 4:156–159. https://doi.org/10.1038/NGEO1068

Schild A (2008) ICIMOD's position on climate change and mountain systems. Mt Res Dev 28:328–331

Schneider S (1990) The global warming debate heats up: an analysis and perspective. Bull Am Meteorol Soc 71:1292–1304

Sharma SS, Ganju A (2000) Complexities of avalanche forecasting in Western Himalaya – an overview. Cold Reg Sci Technol 31:95–102

Shekhar MS, Chand H, Kumar S, Srinivasan K, Ganju A (2010) Climate-change studies in the western Himalaya. Ann Glaciol 51(54):105–112

Shekhar MS, Devi U, Paul S et al (2017) Analysis of trends in extreme precipitation events over Western Himalaya region: intensity and duration wise study. J Indian Geophys Union 21(3):225–231

Singh SK, Rathore BP, Bahuguna IM, Ajai (2014) Snow cover variability in Himalayan Tibetan region. Int J Climatol 34:446–452

Singh D, Sharma V, Juyal V (2015) Observed linear trend in few surface weather elements over the Northwest Himalayas (NWH) during winter season. J Earth Syst Sci 124:553–565

Sirguey P, Still H, Cullen NJ et al (2016) Reconstructing the mass balance of Brewster glacier, New Zealand, using MODIS-derived glacier-wide albedo. Cryosphere 10:2465–2484

Stahl K, Moore RD, Floyer JA et al (2006) Comparison of approaches for spatial interpolation of daily air temperature in a large region with complex topography and highly variable station density. Agric For Meteorol 139:224–236

Stocker TF, Qin D, Plattner GK et al (2013) Summary for policymakers of climate change 2013: the physical science basis. Contribution of Working Group I to the Fifth Assessment Report of the Intergovernmental Panel on Climate Change. Cambridge University Press, Cambridge/New York

Strachan S, Kelsey EP, Brown RF et al (2016) Filling the data gaps in mountain climate observatories through advanced technology, refined instrument siting, and a focus on gradients. Mt Res Dev 36(4):518–527

Thompson LG (2000) Ice core evidence for climate changes in the tropics: implications for our future. Quat Sci Rev 19:19–35

Tudoroiu M, Eccel E, Gioli B et al (2016) Negative elevation-dependent warming trend in the Eastern Alps. Environ Res Lett 11:12

Yadav RR, Park WK, Singh J and Dubey B (2004) Do the western Himalayas defy global warming? Geophys Res Lett, 31, L17201; https://doi.org/10.1029/2004GL020201

Wilks DS (1995) Statistical methods in the atmospheric sciences, 2nd edn. Academic, San Diego, p 467

Chapter 17
Impacts of Climate Change on Himalayan Glaciers: Processes, Predictions and Uncertainties

L. Parry, S. Harrison, R. Betts, S. Shannon, D. B. Jones, and J. Knight

Abstract The glaciers of the Hindu Kush Himalaya region (HKH) produce the water for around 40% of the world's population. Over the past century these glaciers have lost mass in response to recent climate change and they are predicted to lose more in the future. The precise ways in which glaciers will respond to future climate change are still unknown; many will melt entirely, but some will undergo a transition to debris-covered glaciers which will retard melting, and others will undergo a further transition to form rock glaciers whose response to atmospheric warming and changes in precipitation is as yet unclear. As a result, this chapter stresses the paraglacial response of mountain systems to deglaciation to better understand future glacier recession in the region.

17.1 Introduction

In recent decades, climate change has impacted severely the mass balance of many of the glaciers in the Himalaya and the wider Hindu Kush-Himalaya region (HKH), and the status of the seasonal snow pack. Both these elements contribute meltwater to local hydrological systems. However, modelling studies of the glaciers, snow

L. Parry
Ove Arup & Partners, Leeds, UK

S. Harrison (✉) · D. B. Jones
College of Life and Environmental Services, Exeter University, Exeter, UK
e-mail: stephan.harrison@exeter.ac.uk

R. Betts
College of Life and Environmental Sciences, Exeter University and Met Office Hadley Centre, Exeter, Devon, UK

S. Shannon
Bristol Glaciology Centre, Department of Geographical Science, University of Bristol, Bristol, UK

J. Knight
Geography Department, University of Witwatersrand, Johannesburg, South Africa

© Springer Nature Switzerland AG 2020
A. P. Dimri et al. (eds.), *Himalayan Weather and Climate and their Impact on the Environment*, https://doi.org/10.1007/978-3-030-29684-1_17

331

packs and hydrology, to inform adaptation and mitigation, are severely limited by the extreme spatial and temporal variability of contemporary climate, extreme topographic variability in the region which affects microclimate regimes, and the sparsity of instrumental data on both glacier mass balance and variability of river discharge. This chapter discusses the current state of Himalayan glaciers in the context of ongoing climate change and the likely future evolution of these glaciers, and highlights the issue of paraglaciation, without which regional geomorphological responses to future deglaciation cannot be fully understood.

17.2 The Himalayas

The mountain ranges of the Greater Himalayas run in a north-west to south-east arc, some 2500 km in length, across the north of the Indian subcontinent, and form the location for the largest concentration of glacier ice outside of the poles (Fig. 17.1). The region contains glaciers covering an area of ~22,800 km^2 (Bolch et al. 2012), forming part of the "Asian Water Towers" (Immerzeel et al. 2013). These glaciers and associated snowpacks provide much of the runoff which contributes to the flow of the Brahmaputra, Yangtze, Indus and Yellow Rivers amongst others, and through

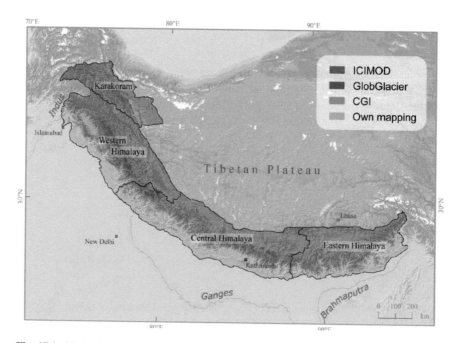

Fig. 17.1 Glacier inventory of the Himalayas from mapping carried out by ICIMOD, Chinese Glacier Inventory (CGI), GlobGlacier and R Bhambri. From Bolch et al. (2012)

these river systems to approximately 1.3 billion people downstream. In the upper catchments within the Himalayas and foothills, approximately 210 million people are dependent upon the hydrological regimes of Himalayan rivers for survival.

Recent estimates of the volume of water stored across the Himalayas range from 3600 to 6500 km³ (Bolch et al. 2012), although Himalayan glaciers are generally losing mass, with estimated glacial mass change rates of −26 ± 12 Gt year⁻¹ (2003–2009) across the wider High Mountain Asia region (Gardner et al. 2013). Substantial further long-term glacial mass losses are projected under future climate warming (e.g. Bolch et al. 2012; Huss et al. 2014; Kraaijenbrink et al. 2017). Recession and, in some locations, the complete loss of high-altitude frozen water stores has potentially significant consequences for downstream water supply (Immerzeel et al. 2010, 2012; Bolch et al. 2012; Lutz et al. 2014; see Fig. 17.2), particularly following peak non-renewable water (Gleick and Palaniappan 2010; Bliss et al. 2014), with long-

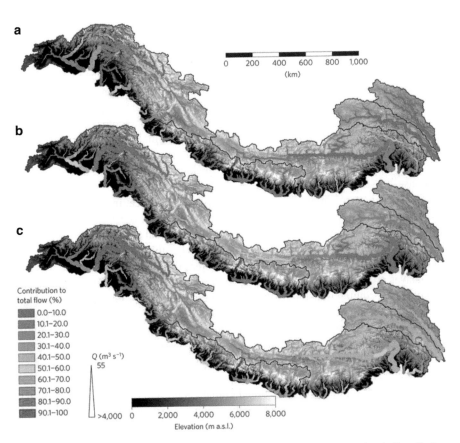

Fig. 17.2 Contribution to total flow by flow components in major streams. (a–c), Contribution to total flow by glacier melt (**a**), snow melt (**b**) and rainfall runoff (**c**) for major streams during the reference period (1998–2007). Line thickness indicates the average discharge (Q) during the reference period. From Lutz et al. 2014

term decreased summer runoff being now projected based upon glacier hydrological and climate modelling (e.g., Sorg et al. 2014).

The climate and partly therefore the behavior of the glacial systems of the region today is dominated by spatial and temporal variability of the Indian Summer Monsoon (ISM) which produces distinct wet (June to September) and dry seasons. This seasonal cycle is reflected in the strong seasonality of the associated hydrological regimes, and in the mass balance behaviour of Himalayan glaciers. Although the glacier and snow coverage and contribution to hydrological regimes vary significantly over space, it is during the pre-monsoon period of the annual cycle that the glacier and snowmelt components of the hydrological regimes are of particular importance in augmenting low river flows. This baseflow also helps smooth interannual variability in streamflow resulting from variations in the onset, strength and duration of the monsoon.

In addition to these inter-and intra-annual patterns of precipitation, strong precipitation gradients also exist over the Himalayan region. In general annual precipitation totals decreases away from the moisture sources of the Bay of Bengal, including during the monsoon, as does the length of the monsoon season. Precipitation also decreases markedly from south to north across the Himalayas, from the Ganges plains and Himalayan foothills to the Tibetan Plateau (TP) (Anders et al. 2006; Barry 2008; Bookhagen and Burbank 2006, 2010; Kansakar et al. 2004; Nandargi and Dhar 2011). As a result, precipitation totals range from 2000 to 5000 mm year^{-1} in the Arunachal Pradesh region of India (south east Himalayas) (Dhar and Nandargi 2004) to below 100 mm year^{-1} in the region of Ladakh (north west Himalayas) (Schmidt and Nüsser 2012). These precipitation gradients have significant potential impacts on snowpack thickness and duration, and thus river response.

Temporal patterns in precipitation also exist. For instance, in eastern Nepal, the monsoon accounts for 75–85% of annual precipitation (e.g. Barry 2008), whereas at Leh in Ladakh only approximately one third of the annual precipitation occurs during the monsoon season (June–August), with another third between December and February (Schmidt and Nüsser 2012). The winter precipitation peaks in the western Himalayas are due to the occurrence of westerly low pressure systems in the lower troposphere bringing moisture from the Mediterranean Sea or Persian Gulf, or even the mid-west Atlantic Ocean (Ridley et al. 2013). Overall, Bookhagen and Burbank (2010) estimated the spatial trends in annual precipitation to show a ~tenfold decrease from south to- north, and ~sixfold decrease from east-to-west. Since the mid-twentieth century a slight decreasing trend in monsoon precipitation is evident (Turner and Annamalai 2012).

The effect of topography on regional climate is profound, and forms part of the reason for the spatial and temporal climate variability experienced across the region, and the problems of using sparse data sets to characterise the climatological regimes for the purpose of climate modelling and prediction (see Pepin et al. 2015). The HKH and TP form a physical topographic barrier to air masses from north to south and west to east throughout the year and during monsoon times. The barrier also affects the path of the sub-tropical jet stream in the upper atmosphere and creates orographic enhancement of precipitation. As a result, the highest precipitation in the Himalayas is found within a few kilometers of the southern side of the highest mountains, where the orographic rise in air masses is most rapid, producing the steep precipitation gradients found in the region (Anders et al. 2006; Bookhagen and Burbank 2006).

Temperature trends over northern India, the HKH and TP show significant variability. For instance, from 1971 to 2007, the Indian Himalayas warmed more rapidly than the rest of the Indian sub-continent, at a rate of 0.46 °C per decade (Kothawale et al. 2010). This trend supports earlier research by Bhutiyani et al. (2007) who reported a lower rate of 0.16 °C per decade for mean annual and winter temperature changes over the last century for this region, but a higher rate of change of 1.1 °C per decade over the period 1981–2002. Overall temperature trends in north western India are driven by changes in annual maximum temperatures (T_{max}), estimated to be around four times higher than minimum temperature (for T_{min}) changes (Bhutiyani et al. 2007), but around double those reported by Kothawale and Rupa Kumar (2005) which were 0.53 °C per decade (for T_{max}) compared to 0.37 °C (for T_{min}). Most studies show that winter exhibits the greatest rate of warming (Kothawale and Rupa Kumar 2005; Fowler and Archer 2006; Bhutiyani et al. 2007; Kothawale et al. 2010; Forsythe et al. 2012).

Although there are considerable spatial variations in precipitation and temperature patterns, detailed analysis of these patterns is restricted by the absence of a well-developed instrumental data network. For example, in Nepal with an area of over 147,000 km² and with elevations ranging from approximately 60 m in lowlands to the summit of Mount Everest at 8850 m asl, the precipitation gauge network consists of around 440 stations, of which 160 have only been established since 1998. High altitude observations are especially sparse with few stations above 3000 m elevation (Kansakar et al. 2004). This restricts our understanding of contemporary temperature and precipitation trends and variability in high mountains and their impacts on glacier mass balance. As a result, satellite data have played a major role in understanding precipitation patterns in the region. TRMM (Tropical Rainfall Measurement Mission) data show that there are large variations in precipitation across even small spatial scales; for example between valley floors and surrounding ridges (Anders et al. 2006; Bookhagen and Burbank 2006, 2010), invalidating broad generalisations about spatial and temporal climate trends.

This spatial and temporal climate variability, and the topographic variability which partly accounts for these trends, also limits the applicability of low-resolution General Circulation Models (GCMs) for detailed climate projections in this region, unless suitable downscaling methods are applied. It also makes trends difficult to detect in the relatively short and spatially-sparse observed time series of meteorological, glacial and hydrological data which are available for the region. Despite these caveats, it is clear that recent climate change has driven glacier recession over much of the Himalayas (e.g. Kaab et al. 2012). This is combined with reduction in the strength of Indian summer monsoon rainfall (Kumar et al. 2006, 2011) which has reduced high altitude snow accumulation.

Climate projections by 2080s provided by CMIP5 GCMs are highly variable but ensemble averages for India range from around 2.0 °C of warming (using RCP2.5) to 4.8 °C using RCP8.5, with temperature increases amplified by up to 8 °C over higher elevation areas (Chaturvedi et al. 2012). Using PRECIS (a version of the Hadley Centre regional model, HadRM3), the projections for annual all-India mean surface air temperature increases by 2100 range from 3.5 to 4.3 °C under a medium emissions scenario (A1B), with precipitation projected to rise by 9–16% by 2100 (Krishna Kumar et al. 2011). The median of an ensemble of 22 GCMs from CMIP3 (downscaled using HadRM3) project a rise in annual temperatures of 2.5 °C by 2070–2099 over India and surrounding areas, based on the same emissions scenario. Projections of mean precipitation for the summer months JJAS show increases of 10% (Kumar et al. 2013).

Projections from Regional Climate Models (RCMs) show warming over northern India and the Himalayas (e.g. Rupa Kumar et al. 2006; Mathison et al. 2013, 2015; Kumar et al. 2013; see Fig. 17.3). Using two RCMs (and four simulations) at 25 km-resolution Mathison et al. (2013) reported average annual temperature changes between 2.5 and 3.0 °C by 2040–2070 from a 1970 to 2000 baseline for the Ganges and Brahmaputra region, but with net changes between 2 and 4 °C across seasonal cycles, as warming was greater in the winter than summer. Using three common RCMs (HadRM3, REMO and CCLM) Kumar et al. (2013) reported projections of warming between 2.5 and 5.5 °C for a medium emissions scenario, but with maxima over the Himalayas, north, central and west India. In addition, daily T_{min} were found to be rising more rapidly than daily T_{max}.

Projections for monsoon precipitation show maximum increases over the western peninsula of India, including the Western Ghats, and north east India. Summer JJAS precipitation increases by approximately 20–40% over peninsula India in an ensemble of three RCMs, and by 10–20% over the Western Ghats and north east India (Kumar et al. 2013). For both temperature and precipitation, there are marked increases in extreme values, and again for precipitation where the greatest increases are most pronounced over the Western Ghats and north western peninsula India (Rupa Kumar et al. 2006).

17.3 Himalayan Glaciers

Net mass accumulation of glaciers in much of the Himalayas is dominated by the monsoon season. Accumulation and ablation occurs in summer and this is typical for at least ~75% of glaciers in Nepal and adjacent regions (Barry 2008; Bolch et al. 2012). However, in the Karakoram sector of the Himalayas, accumulation occurs throughout the year associated with winter precipitation (Hewitt 2005, 2011). Accumulation is also partly driven by avalanches from surrounding mountain slopes, and this process brings rock debris as well as snow and ice to glacier surfaces, which serves to retard melting by changing the surface heat balance of glaciers. This process is particularly important in the Karakoram and south of the main

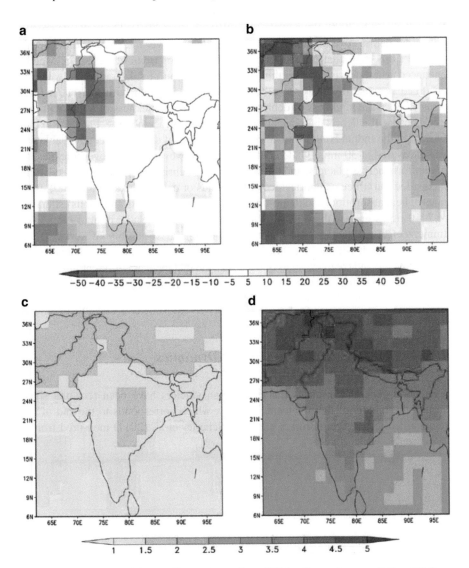

Fig. 17.3 Projections of future climate change from GCMs. Change in precipitation (%) from 1970–1999 to (i) 2020–2049 (ii) from 2070–2099, over south Asia. Lower panels show the temperature change (°C). From Kumar et al. 2013, who used CMIP3 models and downscaled HadRM3 driven by SRES A1B

Himalaya crest where glaciers have steep slopes in their accumulation areas, and greater debris cover than in other regions (Scherler et al. 2011) which affects the spatial patterns and rates of ablation, and dynamics of the glacier terminus.

Mass loss from a glacier occurs in four main areas: from melting of clean ice exposed on the surface, melting beneath a surface debris cover, melting of ice cliffs and around supraglacial ponds, and calving at an ice terminus located in water (e.g.

Sakai et al. 1998, 2000; Benn et al. 2012; Watson et al. 2017). The melt rate around and under debris covered areas is sensitive to debris cover thickness and the debris thermal properties, that have implications for albedo (Benn et al. 2012). Melt is enhanced when debris cover is thin due to conduction of solar radiation, which is absorbed because of the lower albedo of debris compared with ice, but melt is reduced once the thickness of debris increases above a thickness of a few centimetres, because the debris then insulates the ice from surface heating. For the Rakhiot Glacier in the Indian Himalaya, the critical thickness of debris for reducing ablation was measured at 30 mm (Mattson et al. 1993). However, debris cover thickness is not uniform across glaciers, and the surface of the debris-covered area can generally be divided into three categories: ice cliffs, ponds or debris (Sakai et al. 2000), with rates of heat absorption, and therefore melt, of areas of ice cliffs and ponds estimated at 7–10 times greater than the rest of the ablation zone (Sakai et al. 1998, 2000; Immerzeel et al. 2014). This means that the spatial variability of surface melt dynamics is high within the debris covered area of a glacier, but also between the debris covered and clean ice areas of a glacier; and that there are different sensitivities of climate variations of debris covered glaciers to clean ice glaciers or debris-free zones of the glacier (Figs. 17.4 and 17.5).

17.3.1   Recent Changes in Glacier Dynamics

There is a general consensus that Himalayan glaciers have been receding since around the middle of the nineteenth century, which corresponds to the end of the regional Little Ice Age. (Brun et al. 2017; Scherler et al. (2011) measured frontal

Fig. 17.4 The debris covered Khumu glacier, Nepal (right) and its lateral moraine (left)

Fig. 17.5 The Ama Dablam glacier with the north face of Ama Dablam behind it. The glacier is currently undergoing a transition from debris covered glacier to rock glacier. The front of the glacier has already made this transition

changes for 255 glaciers across the Himalayas for the period 2000–2008 and calculated varying rates of change between −80 and + 40 m year⁻¹. Estimates of recent mass balance trends vary, however. For the period 2003–2008, Kaab et al. (2012) used remote sensing data from ICESat, and satellite-derived topographical data (SRTM, from 2002) to estimate the specific mass balance for the HKH. This produced a mass loss lower than the global average at −0.21 ± 0.05 m² year⁻¹ water equivalent, which relates to a 2003–2008 mass balance of −12.8 ± 3.5 Gt year⁻¹. However, for a similar time period (2003–2010), Jacob et al. (2012) used GRACE derived satellite gravity fields and estimated a lower mass loss of around 4 ± 20 Gt year⁻¹. More recent studies (e.g. Bolch et al. 2017) based on remote sensing imagery from 1973 to 2009 suggest that glaciers in parts of the central Karakoram were in balance or showed small amounts of mass loss.

In addition, Bhambri and Bolch (2009) report that the rate of mass balance change has not been constant; they suggest that glaciers in the Indian Himalayas are retreating at a faster pace than those in Nepal although this has yet to be confirmed. They also show that about 60% of the glaciers in the Karakoram are either advancing or display stable termini. This is termed the 'Karakoram Anomaly' (Hewitt 2005; Scherler et al. 2011; Gardelle et al. 2012).

Terminus change is regularly measured by remote sensing. However, terminus position is not always a good metric for glacier behaviour; glaciers with negative mass balance may have stable termini particularly if they have extensive supraglacial debris cover (Bolch et al. 2011). Despite the importance of long-term mass balance measurements to assess glacier behaviour, in the Himalayas few glaciers have been monitored using field-based measurements, and this restricts our under-standing of glacier responses to climate trends and variability. In Nepal, Fujita and Nuimura (2011) analysed the mass balance of three benchmark glaciers, two located in the humid south-eastern part of Nepal and one in the more arid north-west region. They found that glacier mass wastage had initially been higher than the global aver-age until the last decade. At this point, the rate of mass loss accelerated for the two glaciers in the humid east, while the glacier in the arid west underwent a reduction in the rate of mass loss.

Variations in mass loss are probably caused by a range of climatic and site-specific topographic and geomorphological variables. Bhambri and Bolch (2009) summarise the latter as variations in valley topography, supraglacial debris cover, hypsometry, and contributions from tributary glaciers in the accumulation zone. Recently, Azam et al. (2018) have suggested that mass wastage in the region results in increasing debris cover, the growth of glacial lakes and probably decreased gla-cier velocities.

17.3.2 Projections of Changes in Glacier Extents

Wiltshire (2014) used RCM projections for a medium emissions scenario for gla-ciers across the HKH and concluded that glaciers in the east (Nepal and Bhutan) would be most vulnerable to climate change despite projected increases in precipi-tation. Glaciers in the western Himalayas were expected to melt at a lower rate given their higher mean elevations. This spatial pattern was also reported by Viste and Sorteberg (2015) using a subset of 16 GCMs from CMIP5, and they estimated a 20–40% reduction in annual snowfall over the Upper Indus (UI), with projected larger reductions of 50–60% in the Ganges Basin and 50–70% in the Brahmaputra basin by 2071–2100.

Using a glacial-hydrological model, Immerzeel et al. (2013) compared the response of two glacierized catchments in climatologically different settings: the Baltoro in the Upper Indus (UI) (Karakoram) and the Langtang catchment in central Nepal. For both watersheds, glacier area and volume projections showed decreases under both low and high emissions scenarios. For the Langtang catchment, glacier area projections showed shrinkage of 37% and 54% for these emission scenarios, respectively, by 2100, and ice volume reduction of 60% under the high emissions scenario by 2100. In the Baltoro catchment where glaciers are larger, glacier volume reduction was 50% under this scenario by 2100.

Climate model projections therefore suggest a continued reduction in glacier mass balance throughout the remainder of this century and beyond (Collins et al. 2013). However, the precise evolution of mountain glaciers in the Himalayas is cur-rently unclear. What is clear, however, is that glacier recession in a range of moun-tain settings produces a range of linked geomorphological processes, and it is important to understand the operation of these processes if we are to provide better projections of mountain glacier behavior in response to future climate warming.

17.4   Paraglacial Processes and Himalayan Glaciers

Current retreat patterns of Himalayan glaciers are relatively well known from remote sensing data, discussed above, but less is known about geomorphic responses to ice retreat, either now or during lateglacial times. Glaciated valleys contain typical glacial landforms including terminal and dead-ice moraines, kame terraces and limited proglacial outwash spreads (Owen et al. 1995; Barnard et al. 2004). These landforms mark successive stages of ice retreat. What is notable however is the presence throughout different sectors of the Himalayas of landforms that reflect slope instability under cold climate regimes following ice retreat, which would be mainly during the lateglacial and Holocene. These landforms include alluvial and colluvial fans, talus cones, rock glaciers, landslides, rock slope failures (RSFs), and landforms associated with the formation and drainage of glacial lakes. The descriptive term paraglacial can be used to describe these landform assemblages in this context. The term paraglacial is defined as "… nonglacial earth-surface processes, sediment accumulations, landforms, landsystems and landscapes that are directly conditioned by glaciation and deglaciation" (Ballantyne 2002, p. 1938, following the seminal work of Church and Ryder 1972). With continued climate-driven deglaciation, high mountain systems such as the Himalayas are in initial stages of transitioning from glacial- to paraglacial-dominated process regimes (Harrison 2009; Knight and Harrison 2012), and thus the landform assemblages found here provide a useful case study of the impacts on the land surface of climate warming and ice retreat. Several studies in different Himalayan regions focus on macroscale landform assemblages that correspond to this glacial–paraglacial transition (Owen et al. 1995; Owen and Sharma 1998). For example, Iturrizaga (2008) describes slope debris cascades from upper to lower slope locations within the Hindu Kush and Karakoram sectors. Bedrock slopes are covered with scree debris over 1000 m in vertical height range. The greatest sediment yield to valley floors comes from reworking of perched glacial moraines and kame terraces and RSFs farther up the valley. Iturrizaga (2008) notes that such paraglacial slope debris is found below the elevation of glacial trimlines. Several studies describe the timing and extent of different glacial phases across different Himalayan sectors (e.g. Seong et al. 2009; Bisht et al. 2015). Importantly, there is no consistent pattern of ice advance or retreat phases, but that these vary spatially and temporally largely in response to the strength and penetration of the summer monsoon, but also due to local-scale factors including aspect and thus microclimate. Likewise, paraglacial sediment system responses to these ice advance–retreat patterns also vary, related to local geologic structure, tectonic uplift, seismic ground shaking, and the interplay between primary sediment production by frost shattering and rock falls, and secondary reworking from unstable slopes. Petley et al. (2007) showed that from a regional landslide database that there is no consistent spatial pattern of landslides in Nepal in the period 1978–2005, although the number of events has tended to increase over time. Within individual valleys, however, glacial and paraglacial landforms vary in their properties and timing, related to the local-scale factors listed above (Seong et al. 2009). It is also notable that talus cones/fans and fluvial terraces are located in different places within mountain valleys, related to the direction in which sediment is being reworked by slope (downslope) and fluvial (along-valley) processes (Barnard et al. 2006a; Seong et al. 2009; Bisht et al. 2015). Thus, different paraglacial landforms exist within different process domains, which determines where, when and how they are likely to form within the context of a retreating valley glacier. Barnard et al. (2006b), based on [10]Be dating, show that only paraglacial fans, landslides and terraces younger than 5 ka are preserved adjacent to valley floors in the Langtang Himal, Nepal, because of the very high denudation rates experienced in the region (~33 mm year^{-1}). It may be that older paraglacial landform remnants are buried by younger reworked debris, however.

17.4.1   Implications of Paraglacial Processes for Himalayan Glaciers

In detail, modification of rock slopes through RSFs dominate the rock slope paraglacial system, as high mountain systems respond to deglacial unloading or debuttressing following the exposure of glacially steepened rockwalls by glacier downwastage and retreat (Ballantyne 2002). Enhanced paraglacial debris production driven by deglaciation may increase the accumulation of supraglacial debris which, depending on debris cover thickness, can limit ice ablation and increase the longevity of the glacier (e.g., Lambrecht et al. 2011; Pellicciotti et al. 2014). Thick supraglacial debris cover on the order of decimetres to metres (Bosson and Lambiel 2016) reverses the mass balance gradient, with comparatively higher ablation rates upglacier than at the debris-covered terminus. Furthermore, this significantly influences glacier dynamics and behaviour of the glacier snout (Benn et al. 2005), and can also lead to the development of rock glaciers (Shroder et al. 2000; see Fig. 17.5). Large glacier-rock glacier composite landforms represent the most conspicuous geomorphological expression of the transition of debris-covered glaciers into rock glaciers (e.g., Ribolini et al. 2007; Janke et al. 2015; Monnier and Kinnard 2015; Seppi et al. 2015; Monnier and Kinnard 2017). Rock glaciers and debris-covered glaciers exhibit distinctive characteristics that can enable them to be distinguished (Fig. 17.6). Rock glaciers are cryospheric landforms formed by gravity-driven creep of super-saturated accumulations of rock debris (Fig. 17.7), incorporating a perennially frozen mixture of poorly sorted angular rock debris and ground ice (Haeberli et al. 2006). Rock glaciers are also covered by a continuous overlying debris blanket 0.5–5 m thick that seasonally thaws each summer (known as the active layer) (Bonnaventure and Lamoureux 2013) (Fig. 17.1). Commonly, rock glaciers are characterised by "cohesive flow-evocative features" (Monnier and Kinnard 2017), i.e. spatially organized features such as defined furrow-and-ridge topography, steep (~ >30–35°) and sharp-crested frontal and lateral slopes, and individual lobes (Harrison et al. 2008; Fig. 17.6). These features reflect the viscoplastic flow properties of the rock glacier as its ice and debris move downslope under gravity.

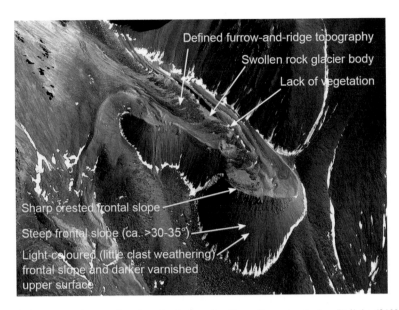

Fig. 17.6 Annotated example of rock glacier: the Caquella rock glacier, Bolivia (21°29′S, 67°55′W). Image data: Google Earth (version 7.1.5.1557, Google Inc., California, USA), DigitalGlobe; imagery date: 20 July 2010

Fig. 17.7 Dughla rock glacier, Nepal [27°54′N, 86°47′E] showing characteristic large blocky clasts covering the surface. Photograph by Darren Jones

Conversely, debris-covered glaciers characteristically have a discontinuous superficial debris layer, typically no more than several decimetres thick, with thickness increasing towards the terminus and towards the lateral margins of the glacier. The flow properties of debris-covered glaciers are devoid of viscous flow morphology; a chaotic distribution of relatively rapidly appearing and disappearing surficial features such as ice-cliffs, supraglacial ponds, hummocks, crevasses, meandering furrows, and thermokarst depressions reflect surface instability.

Paraglacial processes can take place in and can affect the dynamic behaviour of both glaciers and the land surfaces surrounding glaciers. Moreover, it is also clear that important feedback loops can be set up, affecting both heat balance and land surface stability. Both these latter elements are not fully understood, either from field studies or in numerical models, and thus the workings of paraglacial processes are a critical unknown factor in predictions of future environmental change in glaciated mountains generally and in the Himalayas in particular, where debris covered glacier snouts appear to be particularly common. Indeed, 14–18% of total glacierized area in the Himalayas is debris-covered (Kaab et al. 2012).

17.5 Future Environmental Changes in the Himalayas and Their Wider Implications

Ice is stored within the interstices of rock glaciers and beneath surficial debris in debris-covered glaciers. Thermal conduction and cold-air circulation through the debris means that ice melt is reduced in both rock and debris-covered glaciers (Gruber et al. 2016). Therefore, water storage in rock glaciers occurs at long, intermediate and short time-scales, corresponding to water in the solid phase (snow and ice) transitioning towards the liquid phase (water) (Fig. 17.8), and that rock glaciers can be considered as important aquifers in otherwise dry continental interiors.

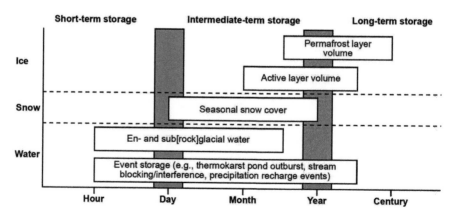

Fig. 17.8 Diagram showing the different forms of rock glacier water storage and their associated time-scales.

Rock glaciers, therefore, form climatically resilient high-altitude frozen water stores of potentially significant hydrological value (e.g., Brenning 2005; Azócar and Brenning 2010; Rangecroft et al. 2015). Jones et al. (2017) showed that >6000 rock glaciers exist within the Nepalese Himalaya, and estimate these cover 1371 km² and contain 20.90 ± 4.18 km³ of water. Across the Nepalese Himalaya, rock glacier to glacier water volume equivalent is 1:9. Therefore, in the context of continued climate warming, glacier-rock glacier relationships may enhance the resilience of the mountain cryosphere (Bosson and Lambiel 2016). Despite this, whereas much has been written on the role of glaciers in maintaining water supplies (e.g. Bradley et al. 2006), that of rock glaciers remains poorly known. Therefore, of critical importance is to better understand glacier-rock glacier relationships, particularly determining which glaciers will 'fully transition' and which will simply downwaste, in order to enable effective water resource management and form adaptation strategies in the context of future climate change.

17.6 Conclusions

In conclusion it is clear that the glaciers in most regions of the Himalaya are in long-term decline and that this will have enormous implications for the people who rely on these ice masses for water supplies. What is less known is the precise ways in which these glaciers and snow packs will respond to warming; the assumption that they will all melt in a predictable way is likely to be simplistic. Better understanding of the evolution of these ice masses is hampered by the uncertainties in climate and glacier modeling and the absence of long-term instrumental data sets with which to reconstruct climate variability and trends. Many Himalayan glaciers will evolve into debris-covered glaciers, and many of these will further evolve to form rock glaciers whose response to warming is poorly studied. Glacier recession will also produce a range of hazards and these can be seen within the context of paraglaciation. Relatively little research has focused on these issues and we argue that this is a significant omission.

Acknowledgements We thank Professor A.P. Dimri for editorial support and to two anonymous reviewers whose contributions significantly improved the paper.

References

Anders AM, Roe GH, Hallet B, Montgomery DR, Finnegan NJ, Putkonen J (2006) Spatial patterns of precipitation and topography in the Himalaya. Geol Soc Am Spec Pap 398:39–53

Azam MF, Wagnon P, Berthier E, Vincent C, Fujita K, Kargel JS (2018) Review of the status and mass changes of Himalayan-Karakoram glaciers. J Glaciol 64(243):61–74

Azócar GF, Brenning A (2010) Hydrological and geomorphological significance of rock glaciers in the dry Andes, Chile (27°–33°S). Permafr Periglac Process 21(1):42–53

Ballantyne CK (2002) Paraglacial geomorphology. Quat Sci Rev 21(18):1935–2017

Barnard PL, Owen LA, Finkel RC (2004) Style and timing of glacial and paraglacial sedimentation in a monsoon-influenced high Himalayan environment, the upper Bhagirathi Valley, Garhwal Himalaya. Sediment Geol 165:199–221

Barnard PL, Owen LA, Finkel RC (2006a) Quaternary fans and terraces in the Khumbu Himal south of Mount Everest: their characteristics, age and formation. J Geol Soc Lond 163:383–399

Barnard PL, Owen LA, Finkel RC, Asahi K (2006b) Landscape response to deglaciation in a high relief, monsoon-influenced alpine environment, Langtang Himal, Nepal. Quat Sci Rev 25:2162–2176

Barry RG (2008) Mountain weather and climate. Cambridge University Press, Cambridge

Benn DI, Kirkbride M, Owen LA, Brazier V (2005) Glaciated valley landsystems. In: Evans DJA (ed) Glacial landsystems. Hodder Education, Oxford, pp 372–406

Benn DI, Bolch T, Hands K, Gulley J, Luckman A, Nicholson LI, Quincey D, Thompson S, Toumi R, Wiseman S (2012) Response of debris-covered glaciers in the Mount Everest region to recent warming, and implications for outburst flood hazards. Earth Sci Rev 114(1–2):156–174

Bhambri R, Bolch T (2009) Glacier mapping: a review with special reference to the Indian Himalayas. Prog Phys Geogr 33(5):672–704

Bhutiyani MR, Kale V, Pawar NJ (2007) Long-term trends in maximum, minimum and mean annual air temperatures across the Northwestern Himalaya during the twentieth century. Clim Chang 85(1–2):159–177

Bisht P, Ali SN, Shukla AD, Negi S, Sundriyal YP, Yadava MG, Juyal N (2015) Chronology of late Quaternary glaciation and landform evolution in the upper Dhauliganga valley, (Trans Himalaya), Uttarakhand, India. Quat Sci Rev 129:147–162

Bliss A, Hock R, Radic V (2014) Global response of glacier runoff to twenty-first century climate change. J Geophys Res Earth 119:717–730

Bolch T, Pieczonka T, Benn DI (2011) Multi-decadal mass loss of glaciers in the Everest area (Nepal Himalaya) derived from stereo imagery. Cryosphere 5(2):349–358

Bolch T, Kulkarni A, Kääb A, Huggel C, Paul F, Cogley JG, Frey H, Kargel JS, Fujita K, Scheel M, Bajracharya S, Stoffel M (2012) The state and fate of Himalayan glaciers. Science 336(6079):310

Bolch T, Pieczonka T, Mukherjee K, Shea J (2017) Brief communication: glaciers in the Hunza catchment (Karakoram) have been nearly in balance since the 1970s. Cryosphere 11(1):531–539

Bonnaventure PP, Lamoureux SF (2013) The active layer: a conceptual review of monitoring, modelling techniques and changes in a warming climate. Prog Phys Geogr 37(3):352–376

Bookhagen B, Burbank DW (2006) Topography, relief, and TRMM-derived rainfall variations along the Himalaya. Geophys Res Lett 33(8):L08405

Bookhagen B, Burbank DW (2010) Toward a complete Himalayan hydrological budget: spatio-temporal distribution of snowmelt and rainfall and their impact on river discharge. J Geophys Res Earth 115(F3):F03019

Bosson J-B, Lambiel C (2016) Internal structure and current evolution of very small debris-covered glacier systems located in alpine permafrost environments. Front Earth Sci 4:39

Bradley RS, Vuille M, Diaz HF, Vergara W (2006) Threats to water supplies in the tropical Andes. Science 312(5781):1755–1756

Brenning A (2005) Geomorphological, hydrological and climatic significance of rock glaciers in the Andes of Central Chile (33–35°S). Permafr Periglac Process 16(3):231–240

Brun F, Berthier E, Wagnon P, Kääb A, Treichler D (2017) A spatially resolved estimate of High Mountain Asia glacier mass balances, 2000–2016. Nat Geosci 10(9):668

Chaturvedi RK, Joshi J, Jayaraman M, Bala G, Ravindranath NH (2012) Multi-model climate change projections for India under representative concentration pathways. Curr Sci 103(791):802

Church M, Ryder JM (1972) Paraglacial sedimentation: a consideration of fluvial processes conditioned by glaciation. Geol Soc Am Bull 83(10):3059–3072

Collins DN, Davenport JL, Stoffel M (2013) Climatic variation and runoff from partially-glacierised Himalayan tributary basins of the Ganges. Sci Total Environ 468–469(Supplement):S48–S59

Dhar ON, Nandargi S (2004) Rainfall distribution over the Arunachal Pradesh Himalayas. Weather 59(6):155–157

Forsythe N, Kilsby CG, Fowler HJ, Archer DR (2012) Assessment of runoff sensitivity in the Upper Indus Basin to interannual climate variability and potential change using MODIS satellite data products. Mt Res Dev 32(1):16–29

Fowler HJ, Archer DR (2006) Conflicting signals of climatic change in the Upper Indus Basin. J Clim 19(17):4276–4293

Fujita K, Nuimura T (2011) Spatially heterogeneous wastage of Himalayan glaciers. Proc Natl Acad Sci U S A 108(34):14011–14014

Gardelle J, Berthier E, Arnaud Y (2012) Slight mass gain of Karakoram glaciers in the early twenty-first century. Nat Geosci 5(5):322–325

Gardner AS, Moholdt G, Cogley JG, Wouters B, Arendt A, Wahr JA, Berthier E, Hock R, Pfeffer WT, Kaser G, Ligtenberg SRM, Bolch T, Sharp MJ, Hagen JO, van den Broeke MR, Paul F (2013) A reconciled estimate of glacier contributions to sea level rise: 2003 to 2009. Science 340:852–857

Gleick P, Palaniappan M (2010) Peak water limits to freshwater withdrawal and use. Proc Natl Acad Sci 107(25):11155–11162

Gruber S et al (2016) Review article: inferring permafrost and permafrost thaw in the mountains of the Hindu Kush Himalaya region. Cryosphere Discuss 2016:1–29

Haeberli W et al (2006) Permafrost creep and rock glacier dynamics. Permafr Periglac Process 17(3):189–214

Harrison S (2009) Climate sensitivity: implications for the response of geomorphological systems to future climate change. Geol Soc Lond, Spec Publ 320(1):257–265

Harrison S, Whalley B, Anderson E (2008) Relict rock glaciers and protalus lobes in the British Isles: implications for Late Pleistocene mountain geomorphology and palaeoclimate. J Quat Sci 23(3):287–304

Hewitt K (2005) The Karakoram anomaly? Glacier expansion and the 'elevation effect,' Karakoram Himalaya. Mt Res Dev 25(4):332–340

Hewitt K (2011) Glacier change, concentration, and elevation effects in the Karakoram Himalaya, Upper Indus Basin. Mt Res Dev 31(3):188–200

Huss M, Zemp M, Joerg PC, Salzmann N (2014) High uncertainty in 21st century runoff projections from glacierized basins. J Hydrol 510:35–48

Immerzeel WW, van Beek LPH, Bierkens MFP (2010) Climate change will affect the Asian water towers. Science 328(5984):1382–1385

Immerzeel WW, Beek LPHV, Konz M, Shrestha AB, Bierkens MFP (2012) Hydrological response to climate change in a glacierized catchment in the Himalayas. Clim Chang 110:721–736

Immerzeel WW, Pellicciotti F, Bierkens MFP (2013) Rising river flows throughout the twenty-first century in two Himalayan glacierized watersheds. Nat Geosci 6(9):742–745

Immerzeel WW, Kraaijenbrink PDA, Shea JM, Shrestha AB, Pellicciotti F, Bierkens MFP, De Jong SM (2014) High-resolution monitoring of Himalayan glacier dynamics using unmanned aerial vehicles. Remote Sens Environ 150:93–103

Iturrizaga L (2008) Paraglacial landform assemblages in the Hindukush and Karakoram Mountains. Geomorphology 95:27–47

Jacob T, Wahr J, Pfeffer WT, Swenson S (2012) Recent contributions of glaciers and ice caps to sea level rise. Nature 482(7386):514–518

Janke JR, Bellisario AC, Ferrando FA (2015) Classification of debris-covered glaciers and rock glaciers in the Andes of central Chile. Geomorphology 241:98–121

Jones DB, Harrison S, Anderson K, Selley HL, Wood JL, Betts RA (2017) The distribution and hydrological significance of rock glaciers in the Nepalese Himalaya. Glob Planet Chang 160(2018):123–142

Kaab A, Berthier E, Nuth C, Gardelle J, Arnaud Y (2012) Contrasting patterns of early twenty-first-century glacier mass change in the Himalayas. Nature 488(7412):495–498

Kansakar SR, Hannah DM, Gerrard J, Rees G (2004) Spatial pattern in the precipitation regime of Nepal. Int J Climatol 24(13):1645–1659

Knight J, Harrison S (2012) The impacts of climate change on terrestrial earth surface systems. Nat Clim Chang 3(1):24

Kothawale DR, Rupa Kumar K (2005) On the recent changes in surface temperature trends over India. Geophys Res Lett 32(18):L18714

Kothawale DR, Revadekar JV, Rupa Kumar K (2010) Recent trends in premonsoon daily temperature extremes over India. J Earth Syst Sci 119(1):51–65

Kraaijenbrink PDA, Bierkens MFP, Lutz AF, Immerzeel WW (2017) Impact of a global temperature rise of 1.5 degrees Celsius on Asia's glaciers. Nature 549(7671):257

Krishna Kumar K, Patwardhan SK, Kulkarni A, Kamala K, Koteswara Rao K, Jones R (2011) Simulated projections for summer monsoon climate over India by a high-resolution regional climate model (PRECIS). Curr Sci 101(3):312–326

Kumar KK, Rajagopalan B, Hoerling M, Bates G, Cane M (2006) Unraveling the mystery of Indian monsoon failure during El Niño. Science 314(5796):115–119

Kumar KK, Kamala K, Rajagopalan B, Hoerling MP, Eischeid JK, Patwardhan SK, Srinivasan G, Goswami BN, Nemani R (2011) The once and future pulse of Indian monsoonal climate. Clim Dyn 36(11):2159–2170

Kumar P, Wiltshire A, Mathison C, Asharaf S, Ahrens B, Lucas-Picher P, Christensen JH, Gobiet A, Saeed F, Hagemann S, Jacob D (2013) Downscaled climate change projections with uncertainty assessment over India using a high resolution multi-model approach. Sci Total Environ 468–469(Supplement(0)):S18–S30

Lambrecht A et al (2011) A comparison of glacier melt on debris-covered glaciers in the northern and southern Caucasus. Cryosphere 5(3):525–538

Lutz AF, Immerzeel WW, Shrestha AB, Bierkens MFP (2014) Consistent increase in High Asia's runoff due to increasing glacier melt and precipitation. Nat Clim Chang 4(7):587–592

Mathison C, Wiltshire A, Dimri AP, Falloon P, Jacob D, Kumar P, Moors E, Ridley J, Siderius C, Stoffel M, Yasunari T (2013) Regional projections of North Indian climate for adaptation studies. Sci Total Environ 468–469(Supplement):S4–S17

Mathison C, Wiltshire AJ, Falloon P, Challinor AJ (2015) South Asia riverflow projections and their implications for water resources. Hydrol Earth Syst Sci 19(12):4783–4810

Mattson LE, Gardner JS, Young GJ (1993) Ablation on debris covered glaciers: an example from the Rakhiot glacier, Panjab, Himalaya. IAHS Publ 218:289–296

Monnier S, Kinnard C (2015) Reconsidering the glacier to rock glacier transformation problem: new insights from the central Andes of Chile. Geomorphology 238:47–55

Monnier S, Kinnard C (2017) Pluri-decadal (1955–2014) evolution of glacier–rock glacier transitional landforms in the central Andes of Chile (30–33 ° S). Earth Surf Dyn 5(3):493–509

Nandargi S, Dhar ON (2011) Extreme rainfall events over the Himalayas between 1871 and 2007. Hydrol Sci J 56(6):930–945

Owen LA, Sharma MC (1998) Rates and magnitudes of paraglacial fan formation in the Garhwal Himalaya: implications for landscape evolution. Geomorphology 26:171–184

Owen LA, Benn DI, Derbyshire E, Evans DJA, Mitchell WA, Thomson D, Richardson S, Lloyd M, Holden C (1995) The geomorphology and landscape evolution of the Lahul Himalaya, Northern India. Z Geomorphol 39:145–174

Pellicciotti F, Carenzo M, Bordoy R, Stoffel M (2014) Changes in glaciers in the Swiss Alps and impact on basin hydrology: current state of the art and future research. Sci Total Environ 493:1152–1170

Pepin N, Bradley RS, Diaz HF, Baraer M, Caceres EB, Forsythe N, Fowler H, Greenwood G, Hashmi MZ, Liu XD, Miller JR, Ning L, Ohmura A, Palazzi E, Rangwala I, Schöner W, Severskiy I, Shahgedanova M, Wang B, Williamson SN, Yang DQ (2015) Elevation-dependent warming in mountain regions of the world. Nat Clim Chang 5(5):424–430

Petley DN, Hearn GJ, Hart A, Rosser NJ, Dunning SA, Oven K, Mitchell WA (2007) Trends in landslide occurrence in Nepal. Nat Hazards 43:23–44

Rangecroft S, Harrison S, Anderson K (2015) Rock glaciers as water stores in the Bolivian Andes: an assessment of their hydrological importance. Arct Antarct Alp Res 47(1):89–98

Ribolini A, Chelli A, Guglielmin M, Pappalardo M (2007) Relationships between glacier and rock glacier in the Maritime Alps, Schiantala Valley, Italy. Quat Res 68(3):353–363

Ridley J, Wiltshire A, Mathison C (2013) More frequent occurrence of westerly disturbances in Karakoram up to 2100. Sci Total Environ 468–469(Supplement(0)):S31–S35

Rupa Kumar K, Sahai AK, Kumar KK, Patwardhan SK, Mishra PK, Revadekar JV, Kamala K, Pant GB (2006) High-resolution climate change scenarios for India for the 21st century. Curr Sci 90(3):334–345

Sakai A, Nakawo M, Fujita K (1998) Melt rate of ice cliffs on the Lirung glacier, Nepal Himalayas, 1996. Bull Glacier Res 16:57–66

Sakai A, Takeuchi N, Nakawo M (2000) Role of supraglacial ponds in the ablation process of a debris-covered glacier in the Nepal Himalayas. IAHS Publ 255:199–132

Scherler D, Bookhagen B, Strecker MR (2011) Spatially variable response of Himalayan glaciers to climate change affected by debris cover. Nat Geosci 4(3):156–159

Schmidt S, Nüsser M (2012) Changes of high altitude glaciers from 1969 to 2010 in the Trans-Himalayan Kang Yatze Massif, Ladakh, Northwest India. Arct Antarct Alp Res 44(1):107–121

Seong YB, Bishop MP, Bush A, Clendon P, Copland L, Finkel RC, Kamp U, Owen LA, Shroder JF (2009) Landforms and landscape evolution in the Skardu, Shigar and Braldu Valleys, Central Karakoram. Geomorphology 103:251–267

Seppi R et al (2015) Current transition from glacial to periglacial processes in the Dolomites (South-Eastern Alps). Geomorphology 228:71–86

Shroder JF, Bishop MP, Copland L, Sloan VF (2000) Debris-covered glaciers and rock glaciers in the Nanga Parbat Himalaya, Pakistan. Geogr Ann Ser B 82(1):17–31

Sorg A, Huss M, Rohrer M, Stoffel M (2014) The days of plenty might soon be over in glacierized Central Asian catchments. Environ Res Lett 9(10):104018

Turner AG, Annamalai H (2012) Climate change and the South Asian summer monsoon. Nat Clim Chang 2(8):587–595

Viste E, Sorteberg A (2015) Snowfall in the Himalayas: an uncertain future from a little-known past. Cryosphere 9(3):1147–1167

Watson CS, Quincey DJ, Smith MW, Carrivick JL, Rowan AV, James MJ (2017) Quantifying ice cliff evolution with multi-temporal point clouds on the debris-covered Khumbu glacier, Nepal. J Glaciol 63:823–837. https://doi.org/10.1017/jog.2017.47

Wiltshire AJ (2014) Climate change implications for the glaciers of the Hindu Kush, Karakoram and Himalayan region. Cryosphere 8(3):941–958

Chapter 18
Sensitivity of Glaciers in Part of the Suru Basin, Western Himalaya to Ongoing Climatic Perturbations

Aparna Shukla, Siddhi Garg, Vinit Kumar, Manish Mehta, and Uma Kant Shukla

Abstract Temporal and spatial climate variability acts as the major driving force which induces changes in glacier response. However, variation in the glacier behavior could also be introduced due to the influence of non-climatic factors. Therefore, in order to assess the influence of these climatic and non-climatic factors on the response of 15 major glaciers of the Suru basin, western Himalayas, Jammu and Kashmir, a multiparametric study has been carried out involving estimation of dimensional (area and length changes) parameters, snowline altitude (SLA)/accumulation area ratio (AAR) and non-climatic factors (debris cover and topographic). Satellite data from the Landsat series sensors (MSS/TM/ETM$^+$/OLI) during the period 1977–2016 along with the Shuttle Radar Topographic Mission (SRTM) Global Digital Elevation Model version-3 (GDEM v-3) constitute the primary datasets used. Results indicate towards an overall negative health of the glaciers with $6.25 \pm 0.0012\%$ loss in glacier area and increase in the average retreat rate from 16 ± 3.4 (1977) to 23 ± 3.4 m/y (2016). This glacier degeneration was accompanied by a debris cover increase of ~80% and mean snow line altitude (SLA) upshift of 116 ± 17 m over the span of 39 years. The observed glacier changes exhibit strong correlation with long-term temperature variability (average $r^2 = 0.481 \pm 0.06$; maximum $r^2 = 0.925$), however, sensitivity to precipitation trends (average $r^2 = 0.143 \pm 0.07$) is not found to be significant. Besides, disparity in glacier response can be partly explained by the spatial variability in meteorological parameters, with glaciers of the Ladakh Range (LR) shrinking (area loss: 9%) and accumulating more debris cover (debris increase: 116%) as compared to those in the Greater Himalayan Range (GHR) (6% and 78%, respectively). However, SLA rise was more pronounced in the GHR glaciers (average of 141 ± 97 m). The differential

A. Shukla (✉) · S. Garg · V. Kumar · M. Mehta
Wadia Institute of Himalayan Geology, Dehradun, Uttarakhand, India
e-mail: aparna@wihg.res.in

U. K. Shukla
Department of Geology, Banaras Hindu University, Varanasi, Uttar Pradesh, India

© Springer Nature Switzerland AG 2020
A. P. Dimri et al. (eds.), *Himalayan Weather and Climate and their Impact on the Environment*, https://doi.org/10.1007/978-3-030-29684-1_18

behavior of glaciers in both the ranges can be attributed partly to the impact of the non-climatic factors such as glacier size, length, maximum elevation and mean slope.

18.1 Introduction

Glaciers act sensitively to the climate change and hence their variable response has implications on the climatic perturbations and vice versa (Oerlemans 1994). Being one of the major freshwater resources, any variation in the glacier parameters as a consequence of the climatic warming would directly affect the runoff distribution, sea level changes and freshwater availability to the downstream communities (Pritchard 2017) that are important for domestic usage, irrigation and hydroelectric power production (Bolch et al. 2012; Immerzeel et al. 2010). Glaciers show variable response to the climatic warming by change in length (retreat), areal extent (degla-ciation), annual mass balance (MB), alteration in ice flow velocity (IFV), surface elevation change (SEC) (thinning), hydrological parameters (discharge/runoff) or change in the extent of glacial lakes (proglacial/moraine-dammed/periglacial/supra-glacial) (Rabatel et al. 2005; Pandey et al. 2011; Kamp et al. 2011; Sakai 2012; Shukla and Qadir 2016). Hence, repeated and continuous monitoring of these gla-cial parameters is vital to comprehend the glacier response. Also, remote location of the Himalayan glaciers, rugged terrain, under-developed road system and harsh cli-mate makes it difficult to continuously monitor them in the field. Considering these issues, remote sensing can be used as a best tool to complement field studies spatially.

Glaciological studies concerning the western Himalayas have shown that sizable latitudinal variations exist in the glacier response (Scherler et al. 2011; Kääb et al. 2012), with the glaciers in the Karakoram region exhibiting either advancement or stability in last few decades (Hewitt 2005; Kääb et al. 2015; Cogley 2016). In con-trast to this, glaciers in the Greater Himalayan Range (GHR) have largely degener-ated, with more than 65% glaciers retreating during 2000–2008 (Scherler et al. 2011). However, there are two views regarding the glaciers in the Trans Himalayan Range/Ladakh Range (THR/LR), with one view suggesting their affinity either towards the Karakoram or the GHR glaciers (Schmidt and Nusser 2017), while the other connoting their response in transition between the two (Chudley et al. 2017).

So far, studies pertinent to the glaciers of the GHR have also shown a contrasting response to the climate change with their focus either on the dimensional parame-ters (Pandey et al. 2011; Kamp et al. 2011; Mir and Majeed 2016), mass fluctuations (Ghosh and Pandey 2013) or consideration of debris cover to show the differential response of the glaciers (Shukla and Qadir 2016). However, multiparametric studies of the glaciers on a basin wide scale showing their relation with the variation in meteorological (temperature and precipitation data), non-climatic factors are still scarce and need to be explored. Therefore, this study is conducted on 15 major gla-ciers of the Suru basin, western Himalayas, Jammu and Kashmir in order to understand the sensitivity of the glaciers to the ongoing climatic changes and the heterogeneous response of the glaciers due to the non-climatic factors.

18.2   Study Area

This study covers 15 major (>5 km^2) glaciers of the Suru basin (Fig. 18.1). The basin extends between latitude and longitude range of 33°50' to 34°40' N and 75°40' to 76°30'E, respectively. Interestingly, this basin cover parts of two main ranges, i.e., GHR and LR, with the glaciers G-1 to G-11 lying in the former and the G-12 to G-15 in the latter (Fig. 18.1). This demarcation could also be visualized from the variation in size and the debris cover extent of the glaciers, with the glaciers present in the GHR to be larger and more debris covered as compared to the LR glaciers.

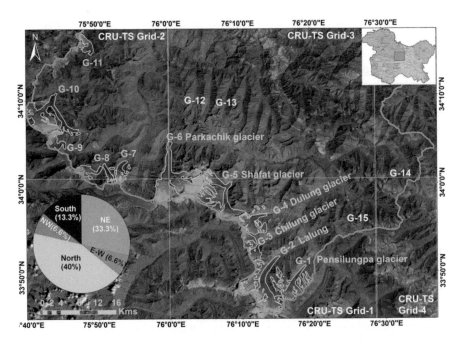

Fig. 18.1 Location map of the study area. 15 major glaciers (G-1 to G-15), situated in the Suru basin were studied for their climatic and non-climatic influence during the period 1977–2016. The glacier boundaries are demarcated using band combination (SWIR1, NIR, Red) on the Landsat ETM$^+$ image with date of acquisition 4 Sep 2000. Suru basin boundary (green outline) encloses the glaciers in GHR (yellow outline) and LR (red outline). White boxes (Grid-1, 2, 3 and 4) are the Climate Research Unit (CRU)-time series (TS) grids and show the microclimatic variability (discussed in Sect. 18.5.1) in the basin. Note the percentage distribution of glaciers based on their aspect as depicted in the pie chart inset. N-E, N-W and E-W represents the north-east, north-west and east-west directions, respectively

Within the basin, Parkachik (G-6) forms the largest (~53 km^2), while G-12 the smallest (4.82 km^2) glacier. Glaciers G-4, G-8 and G-12 have proglacial lakes at their snouts which has implications on the enhanced melting and retreat of these glaciers (Sakai 2012). Based on the percentage of glacier area covered by debris, the studied glaciers were broadly classified as debris covered (>13%), partially debris covered (8–13%) and clean (<8%) glaciers (considering debris cover mapped on the Landsat Operational land imager (OLI) imagery of 2016). In this respect, G-5, G-7, G-10 and G-12 were characterized as debris covered; G-1, G-2, G-8, G-9 and G-11 as partially debris covered and G-3, G-4, G-6, G-13, G-14 and G-15 are clean glaciers.

These glaciers contribute their meltwater to the Suru River, which is a tributary of the Indus River. The river emerges from Pensilungpa glacier and after flowing for nearly 24 kms northward, takes a westward turn from Rangdum. Flowing for about 54 kms and after crossing the townships of Tangole and Panikhar, it flows further north and finally merges with Indus River at Nurla (Negi 2002).

Lying in the western Himalayan region, this basin experiences winter precipitation predominantly as a consequence of Westerlies (Dimri 2013), making the region accessible during July, August and September (Kamp et al. 2011). The weather station at Leh records an annual precipitation of only 93 mm (Klimes 2003; Archer and Fowler 2004) and the mean annual maximum and minimum temperatures being 23.7 °C in August and −15.6 °C in January, respectively (Singh 1998). Snowfall data ranges between 2.05 and 6.84 m at Padum and 7.25–12 m in Durung drung glacier valley. The mean maximum temperature fluctuates between 18.23 and 7.96 °C during summer while mean minimum ranges between 1.32 and −7.8 °C in the Durung drung valley (Raina and Koul 2011). Though the meteorological data for the adjoining regions were available, the region specific trends are not known. Therefore, in order to assess the long term and microclimatic variability within the basin, we have used the CRU-TS 4 dataset for past 115 years (1901–2015), which will be discussed later in the chapter (Sect. 18.5.1) (Fig. 18.2).

18.3 Datasets and Methods

18.3.1 Datasets Used

Primary dataset utilized in the study involves multi-temporal satellite imageries including Landsat series sensors (MSS/TM/ETM+/OLI) for the period 1977–2016 along with the SRTM GDEM-v3 (Table 18.1).

The pre-requisites i.e., peak ablation months (August/September 1st week), minimal snow and cloud cover were taken care off during image acquisition. These conditions have assisted in the identification and accurate demarcation of the glacier boundaries and in the demarcation of SLA. The SRTM GDEM-v3 was incorporated primarily for the extraction of SLA and topographic parameters involving mean

Fig. 18.2 Field photographs of the snouts of (**a**) Pensilungpa glacier, (**b**) Parkachik glacier, (**c**) Shafat glacier and (**d**) Chilung glacier, Suru basin, Jammu and Kashmir during the year 2017. The photographs depict unique characteristics of their snouts and various depositional features present in the glacial environment

Table 18.1 Details of the Landsat series of sensors and other satellite images used along with their acquisition date and co-registration error during the study period

Serial no.	Satellite sensors	Product id	Date of acquisition	RMSE error	Registration accuracy
1.	MSS	LM21590371977213FAK04	1 Aug1977	0.6	48
2.	TM	LT51480361994239ISP00	27 Aug1994	0.3	9
3.	ETM+	LE71480362000248SGS00	4 Sep2000	Base image	
4.	TM	LT51480372011238KHC00	26 Aug2011	0.2	6
5.	OLI	LC81480362016252LGN00	8 Sep2016	0.15	4.5
6.	Sentinel	S2A_OPER_MSI_L1C_TL_MTI	16 Sep2016	0.12	1.2

slope, maximum elevation, altitudinal range and hypsometry. Besides, long term climate data for the duration 1901–2015 was obtained from CRU-TS 4, in which station anomalies from 1961 to 1990 were interpolated into 0.5° latitudinal and longitudinal grid cells (Harris et al. 2013). It is a gridded climate dataset obtained from the monthly meteorological observations done at different meteorological stations of the World and includes six independent climate variables (mean temperature, diurnal temperature range, precipitation, wet-day frequency, vapour pressure and cloud cover). However, in this study only two variables, i.e., (mean annual temperature and mean annual precipitation) were taken into consideration.

18.3.2 Methodology Adopted

In this study, the satellite images were initially pre-processed and the database thus generated was utilized for the extraction of glacier parameters such as area, length, SLA, AAR and other non-climatic factors including debris cover, some topographic factors such as mean slope, maximum elevation, altitudinal range, aspect, compactness ratio and hypsometry and other factors such as glacier boundary characteristics (presence of glacial lake) and glacier size.

In the initial step, the satellite images were co-registered with respect to the base image of Landsat ETM+ image at sub-pixel accuracy with the Root Mean Square Error (RMSE) to be less than 1 (Table 18.1). The tributary glaciers contributing the ice mass to the main trunk were considered as a single glacier entity. The glaciers were then mapped manually utilizing varying band combinations Near Infra-red (NIR), red, green and Short Wave Infra-red (SWIR), NIR, red. Also, high resolution Google Earth™ imagery was referred for the 3-D visualization of the scene under study. In addition to this, presence of proglacial lakes at the snout of some of the glaciers, emergence of stream, ice wall were also examined for the identification of the glacier boundary. Length of individual glacier was measured along the central flow line (CFL) drawn from bergschrund to the snout. Snout fluctuations during different time periods were measured using the parallel line method, in which parallel strips of 50 m spacing were taken on both sides of the CFL and their average values were used to determine the frontal retreat of the glaciers (Shukla and Qadir 2016; Garg et al. 2017a, b). Mean SLA was also estimated at the end of the ablation season which in turn can be used as a reliable proxy for equilibrium line altitude (ELA) and mass balance estimation for the hydrological year (Guo et al. 2014). The maximum difference of spectral signature of snow and ice in the SWIR and NIR bands helped in delineation of the snow line separating the two facies. Thereafter, in order to yield the mean SLA, a buffer of 15 m was created on both sides of the snow line. The debris cover boundary was delineated on the previously marked glacier boundary using Normalized Difference Snow Index (NDSI). It is an image transformation technique, which distinguishes the snow and ice region from the surrounding non-glacier features and hence, could be effectively used to separate the debris cover polygons from the snow and ice within the glacier boundary.

Aspect is a factor exhibiting the orientation of the glacier and controls the amount of solar insolation received by it. The mean aspect is determined based on the direction of the central flow line of the main trunk of the glacier and was thus, classified into 5 major categories, i.e., north, north-east, north-west, east-west and south (Fig. 18.1; Table 18.3) in our study. Compactness ratio, which is a measure of the ratio of area and perimeter of the glacier body to the ratio of area and perimeter of the circle having the same area as the glacier was also estimated (Allen 1998). Other topographic parameters such as hypsometry, mean slope, maximum elevation and altitudinal range were extracted with the help of SRTM GDEM v-3. Hypsometry is the measure of glacier mass distribution over a range of altitudes. It is largely influenced by Snow line altitude (SLA) because if a large part of glacier has an elevation similar to the SLA, a small change in the SLA significantly alters the distribution of glacier mass over ACZ and ABZ (Garg et al. 2017a).

These non-climatic factors were subsequently correlated with the change in glacier dimensional parameters, i.e., area change and retreat using some statistical tests. In the statistical analysis, the variables were initially tested for normality, which involves Shapiro-Wilk's test (Shapiro and Wilk 1965) and visual inspection of the histogram, normal Q-Q plots and box plots. The test revealed normal distribution for nearly all the variables, however, variables which did not exhibit a normal distribution were subsequently log transformed. Thereafter, these parameters were regressed with respect to the topographic parameters through a stepwise process. In this process, it was observed that with the successive addition of a topographic parameter (according to the correlation coefficient (r)), the coefficient of determination (R^2) increased, while root mean square error (RMSE) decreased. Based on this analysis, correlation ranks were assigned to the topographic factors in order of their influence (Table 18.4) on the area change and retreat (Garg et al. 2017b).

18.3.3   Sources of Errors

The study involves extraction of various glacial parameters utilizing satellite data of variable resolutions and specifications. Therefore, estimation of errors and their propagation is essential as these glacier parameters are dependent on the quality of data used for their derivation. Different levels of error may arise due to variance in the remote sensing data such as locational/ positional, pre-processing and processing, data quality and interpretational/conceptual errors (Wu et al. 2014; Shukla and Qadir 2016). As a standard procedure of uncertainty estimation, glacier outlines are compared directly with the ground truth data such as differential global positioning system (DGPS) measurements (Racoviteanu et al. 2008). However, due to the unavailability of the ground control data, high resolution dataset could be used for comparing the medium to coarse resolution with the high resolution imagery (Paul et al. 2013). In this study, high resolution Sentinel imagery (spatial resolution of 10 m), was used for validating the glacier mapping results for the year 2016. The length and area mapping accuracy for the glacier boundaries delineated during 2016 was found to be ±6 m and 0.05 km², respectively. In our study, the locational/positional or conceptual error may have resulted on account of miss-registration of the satellite images or due to the mapping errors, respectively. The locational errors were corrected to an extent by estimation of sub-pixel co-registration RMSE (Sect. 18.3.2). The multi-temporal datasets were assessed for glacier length and area change uncertainty as per methods given by Hall et al. 2003 and Granshaw and

Table 18.2 Terminus and Area change uncertainty associated with satellite dataset as defined by Hall et al. 2003

Serial no.	Satellite sensor	Terminus uncertainty $U_T = \sqrt{a^2 + b^2} + \sigma$	Area change uncertainty $U_A = 2$ $U_T * x$
1.	Landsat MSS	133.44 m	0.02 km²
2.	Landsat TM	51.42 m	0.003 km²
3.	Landsat ETM⁺	48.43 m	0.002 km²
4.	Landsat OLI	46.92 m	0.002 km²

U_T terminus uncertainty, U_A area change uncertainty, x spatial resolution, σ registration accuracy

Fountain 2006. Following formulations (Hall et al. 2003) were used for estimation of the said parameters:

$$\text{Terminus uncertainty}\left(U_T\right) = \sqrt{a^2 + b^2} + \sigma \tag{18.1}$$

where, 'a' and 'b' are the pixel resolution of image 1 and 2, respectively and 'σ' is the registration error. The terminus and areal uncertainty estimated are given in Table 18.2.

$$\text{Area change uncertainty}\left(U_A\right) = 2UT * x \tag{18.2}$$

where, 'x' is the spatial resolution of the sensor.

In the buffer method of estimating the area mapping uncertainty, a buffer size equal the registration error of the satellite image was taken into consideration (Granshaw and Fountain 2006; Bolch et al. 2012; Garg et al. 2017a, b) and the error was estimated to be 2.08, 7, 4.57 and 3.41 km² for the 1977 (MSS), 1994 (TM), 2011 (TM) and 2016 (OLI) imageries. Since the debris extents were delineated within the respective glacier boundaries, the proportionate errors are likely to have propagated in debris cover estimations which were estimated accordingly (Garg et al. 2017b).

Uncertainty in SLA estimation was done in X, Y and Z directions. In this context, error in X and Y direction should equal to the distance taken for creating the buffer on either side of the snow line demarcating the snow and ice facies. Since, the buffer size was taken to be 15 m in this study, therefore, error in X and Y direction was considered as ±15 m. However, uncertainty in Z direction would be similar to the SRTM GDEM-v3, i.e., ±17 m (Garg et al. 2017b).

18.4   Results

This study involved investigation of glacier (area, length changes, SLA and debris cover) parameters in detail in order to assess the sensitivity of the glaciers to the ongoing climatic perturbations.

18.4.1   Area Changes

Results reveal the area change of 15 major glaciers from 288.13 ± 18.78 km^2 (1977) to 270.12 ± 1.49 km^2 (2016), exhibiting a significant loss of 18 ± 0.02 km^2 during the period 1977–2016. To assess the variability of glacier area loss on decadal scale, the total time period was sub-divided into four time frames, i.e., 1977–1994 (16 years), 1994–2000 (6 years), 2000–2011 (11 years), 2011–2016 (5 years). Results show the highest pace of deglaciation during 2000–2011 (8.37 ± 3.5 km^2) followed by 1977–1994 (6.79 ± 18.78 km^2) and a significantly lower deglaciation during 1994–2000 (1.43 ± 3.5 km^2) and 2011–2016 (1.41 ± 2.28 km^2).

The loss in individual glacier area varied from 0.10 ± 1.67 km^2 (G-6) to 4.39 ± 2.43 km^2 (G-4) during the period 1977–2016. In contrast to the area loss, glacier G-6 have also gained mass (0.07 km^2) during 1994–2011 (Fig. 18.3).

18.4.2   Length Changes

Results show that all the studied glaciers have retreated during the entire study period, i.e., 1977–2016, however, with a heterogeneous rate varying from 31.9 ± 3.4 m/y (G-4) to 3.13 ± 3.4 m/y (G-6). Besides, phases of advancement are

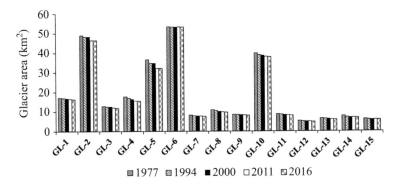

Fig. 18.3 Annual area changes (km^2) of glaciers in the study area during the period 1977–2016

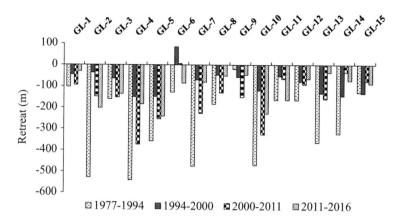

Fig. 18.4 Period wise (1977–1994, 1994–2000, 2000–2011 and 2011–2016) retreat (m) of the glaciers in the study area. Note the advancement shown by the Parkachik glacier (G-6) during 1994–2000 and 2000–2011

also observed in glaciers G-6 during 1994–2000 (14.05 ± 8.6 m/y) and 2000–2011 (0.6 ± 4.4 m/y). Decadal observations show highest rate of retreat during 2011–2016 (22.85 ± 9.4 m/y), with the maximum and minimum rates shown by glaciers 48.28 ± 9.4 m/y (G-4) and 5.9 ± 9.4 m/y (G-1), respectively. Relatively lower rate of retreat is observed during 1977–1994 (16.13 ± 8.3 m/y), followed by 2000–2011 (13.82 ± 4.4) and 1994–2000 (12 ± 8.4 m/y) (Fig. 18.4).

18.4.3   SLA and AAR

Temporal variations in SLA during the study period 1977–2016 show an average increase by 116 ± 17 m. However, this rise in SLA varies from 18 to 333 m for the glaciers G-10 and G-6, respectively. In contrast, a decrease in the SLA by 44 m is observed in G-13. Although a significant increase in SLA is seen in the glaciers during the time interval of 39 years, a considerable fluctuation is observed on the decadal scale. A notably higher average upward shift of SLA is observed during 1994–2000 (263 m), followed by 1977–1994 (80 m). However, SLA has shown a decrease of 197 m and 29 m during 2000–2011 and 2011–2016, respectively (Fig. 18.5a).

Further, the derived SLAs were also used for estimating the AAR of the studied glaciers for respective years. An overall trend in the AAR values during the period 1977–2016 reveals an average increase of 0.09. The average values of AAR of the glaciers initially decreased from 0.59 in 1977 to 0.25 in 2000 and subsequently increased to 0.67 in 2016 (Fig. 18.5b).

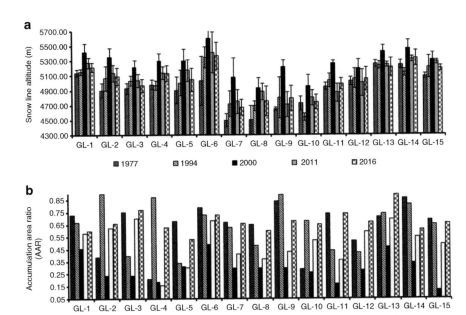

Fig. 18.5 Snow line altitude (SLA) (m) and accumulation area ratio (AAR) variations of studied glaciers during the period 1977–2016. (**a**) Changes in SLA from 1977 to 2016. Note the upward shift of SLA during the period 1977–2000 and a decrease thereafter. The error bars reflect the elevation range of individual glacier snowlines. (**b**) Annual variation in AARs of the glaciers from 1977 to 2016. Note the lowest AAR values in 2000

18.4.4 Debris Cover

Results reveal an overall increase in the debris cover extent by 80% during the period 1977–2016. Decadal variation for four different time frames, i.e., 1977–1994 (16 years), 1994–2000 (6 years), 2000–2011 (11 years), 2011–2016 (5 years) reveal maximum increase of debris cover by 19% during 2011–2016 followed by an increase of 17%, 15% and 12% during 1977–1994, 1994–2000 and 2000–2011, respectively. (Fig. 18.6).

18.5 Discussion

Present study involving the multi-parametric approach of assessing the glacier status suggests collective deterioration in the health of the major glaciers in the Suru basin during the period 1977–2016. However, these glacier parameters show wide variance in their rate of change when considered for a particular time frame. Therefore, differential behavior of the glaciers in the basin during different time

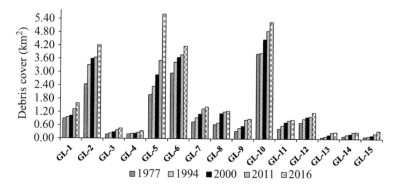

Fig. 18.6 Debris cover (DC) area (km²) of the studied glaciers during the period 1977–2016

periods have been discussed in detail in the following section in order to understand the influence of climatic and non-climatic factors on the same.

18.5.1 Climatic Variability in the Region

Station interpolated data obtained from CRU-TS 4 shows long term fluctuation in the temperature and precipitation records during the period 1901–2015, with the mean annual temperature and precipitation anomaly varying from −1.43 to 1.64 °C and − 1.75 to 2.80 mm/day, respectively. The temperature anomaly pattern for the entire Suru basin shows an initial decreasing trend till 1937 followed by an increase to the maximum in 1941, decreasing thereafter till 1997 and finally tend to increase continuously till 2015. However, precipitation anomaly shows a near negative trend till 1955, with a minimum in 1946 and finally attaining a positive trend reaching a maximum in 2015.

Apart from these generalized climatic variations, grid-wise analysis of the meteorological parameters reveals existence of local climate variability within the basin, which could be manifested in the differential response of the glaciers. Observations show that the glaciers covered in grid 2 have been experiencing a warmer climatic regime with the maximum annual mean temperature of 1.39 °C as compared to the other glaciers in the region (grid 4 = 0.72 °C, grid 1 = 0.63 °C, grid 3 = 0.44 °C). Spatial variability in precipitation data shows that the grid 2 received the maximum mean rainfall amounting to 37.39 mm/day followed by grid 1 (36.83 mm/day), grid 4 (31.95 mm/day) and minimum in grid 3 (31.06 mm/day). These observations suggest that local climatic variability does exist in the basin for the entire duration of 115 years, with the mean temperature and precipitation data showing that the GHR glaciers (2.02 °C/74.22 mm/day) experiencing a warmer and wetter climate as compared to the LR glaciers (1.17 °C/63.02 mm/day).

In this study, some correlations have been discussed in detail (Sect. 18.5.2) between the glacier parameters (area/length/SLA and debris cover) and mean annual temperature and precipitation of a particular year in order to understand the influence of these climatic factors on the response of each glacier. Due to the unavailability of meteorological data for 2016, these climatic relations have been derived for 4 years only (1977, 1994, 2000 and 2011) (Fig. 18.7).

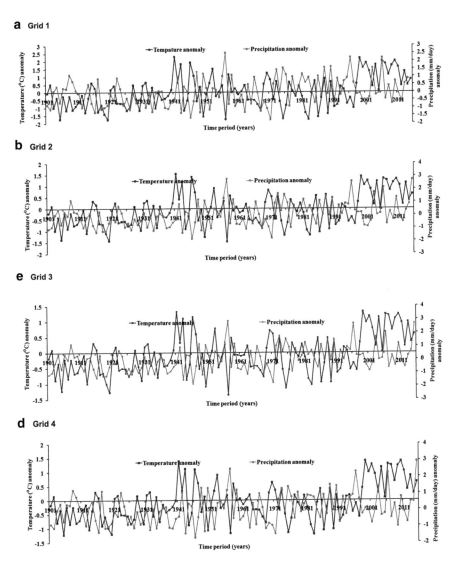

Fig. 18.7 CRU-TS 4 temperature (°C) and precipitation (mm/day) anomalies recorded for the period 1901–2015 showing long term variability in the micro-climatic data. **(a)**, **(b)**, **(c)** and **(d)** are the temperature and precipitation anomalies for grid 1, 2, 3 and 4 respectively

18.5.2   Sensitivity of the Glaciers to Climate Change

Results show an overall deglaciation of ~6.25% (i.e., 18 ± 0.02 km²) during the period 1977–2016. This estimate is significantly less when compared to other studies carried out in the western Himalayas such as Shukla and Qadir 2016, who report a deglaciation of 15% during the period 1977–2013, while Pandey et al. 2011 report an area loss of 18% in glaciers of Zanskar valley between 1962 and 2001. Decadal observations of area loss show that in the entire time duration of 39 years, nearly 47% area loss occurred during 2000–2011, followed by 37% loss during 1977–1994. These results differed from the study conducted on glaciers of Zanskar region which report a maximum area loss during 1992–2000 (Shukla and Qadir 2016). However, our results were synchronous with the meteorological data that show maximum mean annual temperature and minimum mean annual precipitation (1.55 °C and 31.84 mm/day, respectively) during the period of maximum area loss, i.e., 2000–2011. In this study, glaciers in the GHR (G-1 to G-11) constitute 73% of the investigated glaciers and cover an area of ~262 km², with G-6 covering maximum area of 53.39 km² as compared to just 26 km² occupied by LR glaciers, with G-14 covering maximum area of 7.76 km² in 1977. However, the results reveal greater area loss of the LR (9%) glaciers as compared to the ones in GHR (6%). These results were compatible with that of Bhambri et al. 2011; Basnett et al. 2013, who report that the smaller glaciers usually experience greater deglaciation. Also, glacier area of LR shows good negative correlation with the average temperature (r^2 = 0.488 to 0.596) as compared to the GHR glaciers (r^2 = 0.257 to 0.582), with G-6 exhibiting a fair positive correlation (r^2 = 0.222). In contrast, glaciers in LR (r^2 = 0.029 to 0.295) and GHR (r^2 = 0.017 to 0.855) show no significant correlation with precipitation. It means that temperature has a stronger control over the area shrinkage of glaciers in the region (Fig. 18.8). A significantly higher area loss was noted in G-5: Shafat (4.39 ± 2.43 km²) in contrast to other debris covered glaciers. This could also be the reason for the disintegration of Shafat into two, i.e., Shafat-1 & Shafat-2 after 2000 (Fig. 18.9b).

Monitoring the glacier recession for the study period 1977–2016 shows an average rate of 16 ± 3.4 m/y. However, decadal variations show the highest rate of retreat of 23 ± 9.4 m/y during 2011–2016, which is consistent with Shukla and Qadir 2016, who report a retreat rate of 16 m/y during 2009–2013. Glaciers in LR and GHR also show variability in terms of retreat, with the glaciers in the former retreating more (14%) as compared to the latter (8%). These results show that the glaciers in the LR have degenerated more irrespective of their smaller average length (4 km) as compared to the GHR (8.3 km). Moreover, length of the glaciers in LR show good negative correlation with the average temperature (r^2 = 0.458 to 0.838), as compared to the GHR (r^2 = 0.238 to 0.543) glaciers, except G-6, which display a positive correlation (r^2 = 0.260) (Fig. 18.10). In contrast, glaciers in LR (r^2 = 0.003 to 0.157) and GHR (r^2 = 0.002 to 0.838) show no significant correlation with precipitation. This means that temperature has a strong control over the length changes of the glaciers in the LR glaciers. An anomalous advancement was seen in Parkachik glacier (G-6)

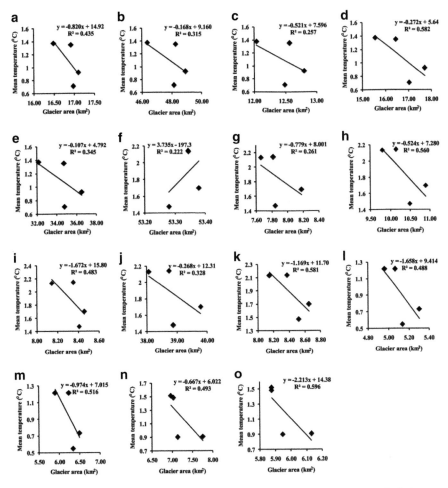

Fig. 18.8 Scatter plots displaying the relation between mean annual temperature and glacier area during the period 1977–2016. (**a–o**) represents the relation between glacier area (G-1 to G-15) during four different years (1977, 1994, 2000 and 2011) and the mean temperature of the particular grid in which that glacier lie. All the relationships were found to be significant at 90% confidence level, i.e., $\alpha = 0.1$

for a period 1994–2011, similar to the results reported by Kamp et al. 2011 during the period 1990–2003 who have attributed it to be due to the high altitudinal range of the glacier. Detailed investigation of the debris covered, partially debris covered and comparatively clean glaciers show the average rate of retreat was significantly high for debris covered glaciers (23 ± 3.4 m/y) followed by clean glaciers (15 ± 3.4 m/y) and partially debris covered glaciers (12 ± 3.4 m/y) during the period 1977–2016. These results are not in line with the previous studies which have obtained a higher retreat of clean glaciers as compared to the debris covered ones. However, the probable reason of such contradictory results could be the presence of patchy debris cover which was not thick enough to have an insulating effect of the glacier snout.

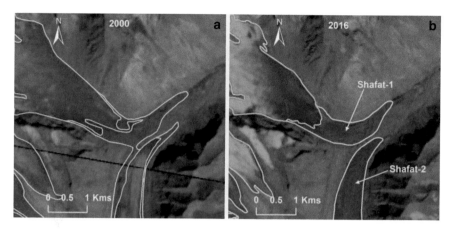

Fig. 18.9 Satellite imageries depicting the fragmentation of glacier G-5. (**a**) The united boundary of glacier G5 in 2000 (ETM⁺ image), while (**b**) The fragmented boundaries of the glacier G-5 into Shafat-1 and Shafat-2 in 2016 (OLI image)

Besides, the higher retreat of G-12 and G-8, which are debris and partially debris covered glaciers, respectively, is probably due to the presence of a proglacial lake at their snout. This inference is well in agreement with the fact that the glacial lakes (supraglacial or proglacial) associated with the debris covered glaciers can significantly alter its response inducing considerable changes in the enhancement of the ablation mechanism (Basnett et al. 2013).

Debris cover on the glaciers increased by 80% during the study period, with the highest increase noted during 2011–2016. The increment in the debris cover during the study period is accompanied with the overall degeneration (area loss:18 ± 0.02 km², retreat rate:16 ± 3.4 m/y) of the glaciers during the same period. The pronounced degeneration might lead to the melting of the glaciers which consequently may increase the debris cover over the glaciers as documented in many parts of the world (Bolch et al. 2008; Shukla et al. 2009; Schmidt and Nusser 2009). However, differential accumulation of debris over was seen in the LR and GHR regions, with 116% increase in the former and 78% in the latter during the period 1977–2016. Also, debris cover extent of the LR glaciers shows a good positive correlation ($r^2 = 0.479$ to 0.713) with the mean temperatureas compared to the GHR ($r^2 = 0.325$ to 0.850) in the respective year (Fig. 18.11). However, precipitation does not show a significant relation with LR ($r^2 = 0.004$ to 0.037) and GHR ($r^2 = 0.008$ to 0.220) glaciers. This shows that temperature fluctuations in the basin had a direct and strong control over the increase in the debris cover on the glaciers.

Temporal and spatial variations in SLAs are an indicator of ELAs which in turn can be used as a reliable proxy for mass balance (Kulkarni et al. 2004; Hanshaw and Bookhagen 2014). SLA is a dynamic parameter which changes frequently and depends on the prevailing temperature and precipitation received during a particular year. In our analysis, it is found that all the glaciers except G-13 experienced a steady increase in SLA values during the period 1977–2016, with an overall upward

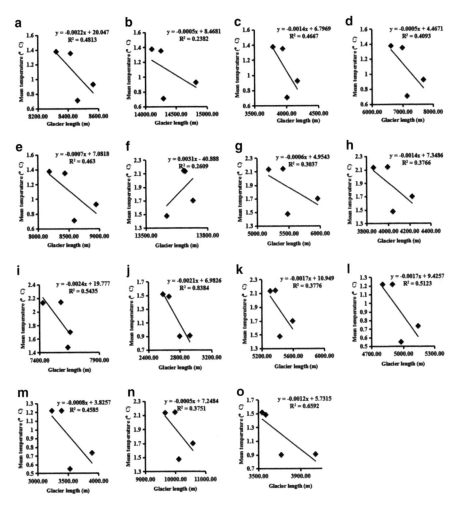

Fig. 18.10 Scatter plots displaying the relation between mean annual temperature and glacier length during the period 1977–2016. (**a–o**) represents the relation between glacier area (G-1 to G-15) during four different years (1977, 1994, 2000 and 2011) and the mean temperature of the particular grid in which that glacier lie. All the relationships were found to be significant at 90% confidence level, i.e., $\alpha = 0.1$

movement of SLA during 1977–2000 and a decrease thereafter. Also, in our results, we find a good positive correlation of mean SLA of the glaciers during 4 different time periods (1977, 1994, 2000 and 2011) with the mean temperature changes in the basin during the respective time periods. Considering these, it appears that a general rise in SLA can be attributed to regional climatic warming while that of individual SLA variation in glaciers may be related to their unique topography (Shukla and Qadir 2016). These results are also in agreement with the fact that the glacier mass fluctuations give an immediate response to the climatic changes. SLA rise in the

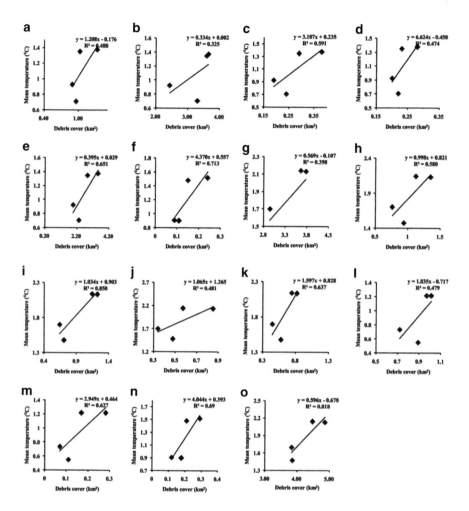

Fig. 18.11 Scatter plots displaying the relation between mean annual temperature and debris cover of the glaciers during the period 1977–2016. (**a–o**) represents the relation between debris cover of glaciers (G-1 to G-15) during four different years (1977, 1994, 2000 and 2011) and the mean temperature of the particular grid in which that glacier lie. All the relationships were found to be significant at 90% confidence level, i.e., $\alpha = 0.1$

glaciers of GHR and LR shows discrepancy spatially and temporally, with a greater upward shift observed in GHR (141.8 ± 97 m) as compared to the LR (44.8 ± 67 m) glaciers. However, SLA of the GHR glaciers shows good positive correlation with the mean temperature ($r^2 = 0.207$ to 0.835) as compared to the LR ($r^2 = 0.095$ to 0.70) in the respective year (Fig. 18.12). On the contrary, precipitation does not show a significant relation with LR ($r^2 = 0.214$ to 0.119) and GHR ($r^2 = 0.566$ to 0.021) glaciers. This shows that temperature fluctuations in the basin had a direct control over the SLA increase of the GHR glaciers.

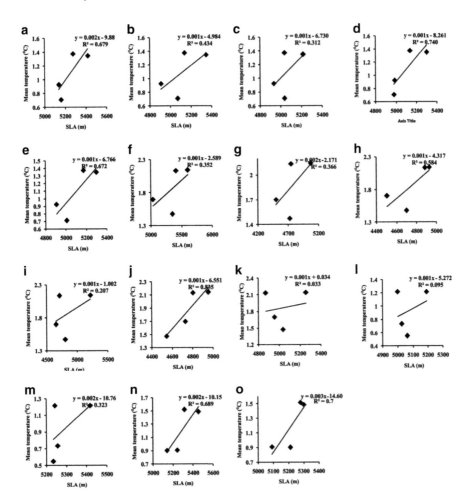

Fig. 18.12 Scatter plots displaying the relation between mean annual temperature and mean SLA of studied glaciers during the period 1977–2016. (**a–o**) represents the relation between SLA of glaciers (G-1 to G-15) during four different years (1977, 1994, 2000 and 2011) and the mean temperature in the particular grid in which that glacier lie. All the relationships were found to be significant at 90% confidence level, i.e., $\alpha = 0.1$

Results of AAR change during the period 1977–2016 shows that the glaciers have suffered maximum loss during 1977–2000 with a gain thereafter (2000–2016). AAR variations also show strong negative correlation with mean temperature in the GHR ($r^2 = 0.003$ to 0.925) and in LR ($r^2 = 0.007$ to 0.84) glaciers. However, its correlation with precipitation was not so significant in GHR ($r^2 = 0.007$ to 0.85) and LR ($r^2 = 0.007$ to 0.027).

Characterized by the contrasting climatic regimes, the glaciers of GHR and LR exhibit a differential response, with the LR glaciers shrinking at a greater pace and accumulating more debris cover as compared to the GHR glaciers. However, a greater rise in SLA was seen in GHR glaciers than their counterparts. In future studies, more glaciers would be considered in LR in order to have a clear understanding of the differential response of the glaciers to the ongoing climate perturbations in the two contrasting regions considered in this study.

18.5.3 Influence of Non-climatic Factors

Glaciers under study have shown wide variance temporally as a consequence of ongoing climatic fluctuations (Sect. 18.5.2). However, the spatial variation and uniqueness in response of the individual glaciers mandate the investigation of non-climatic factors, i.e., debris cover, topographic and other snout characteristics as well, in order to understand their influence on glacier response. In this regard, debris cover, compactness ratio, maximum elevation, altitudinal range, glacier hypsometry, mean slope, aspect, glacier size and other snout characteristics (presence of glacial lakes) were studied in detail to understand their correlation with the glacier parameters, i.e., deglaciation and retreat (Figs. 18.13 and 18.14). These correlations show the presence of few outliers (not removed in this study) as could be seen in Fig. 18.13a. In this correlation graph between glacier size (km^2) and absolute area change G-4, G-5 and G-6 are the outliers. The possible reason for G-4 to be one of the outliers in this case is the presence of an MDL at its snout which has led to an abrupt increase in the area loss of the glacier. On the other hand, G-6 is the largest glacier of the Suru-basin, yet has undergone less area loss probably due to the high altitudinal range (3578–6115 m) and northerly aspect of the glacier. This causes its response to deviate from the general trend of area loss. We also observe that after removing these outliers, the correlation improved from $r^2 = 0.407$ to $r^2 = 0.878$ (Fig. 18.13a).

In our observation, it is found that glacier size, both in terms of length and area plays a dominant role influencing the response of glaciers. Glacier length shows a fair negative correlation with deglaciation ($r = -0.492$) and retreat rate ($r = -0.366$) of the glaciers (Table 18.4), which suggests that smaller glaciers have undergone more deglaciation and retreat as compared to the larger ones. Also, glacier size (km^2) has a fair positive correlation with absolute area change ($r = 0.661$), while a negative correlation with the percentage area loss ($r = -0.512$). Contrary to this, glacier size ($r = 0.268$) and percentage length change ($r = -0.334$) reflected no direct correlation with retreat rate (Table 18.4). This indicates that the large sized (in terms of area) glaciers lose more area, however, undergo a lesser amount of percentage area loss simultaneously. This relationship is similar to the one obtained from central Himalayas (Garg et al. 2017b).

Maximum elevation of the glaciers in this region also influences their response to a larger extent. It shows a fair negative correlation with deglaciation ($r = -0.464$) and no significant relation with the retreat ($r = -0.006$). These observations show that the glaciers which extend to a higher elevation undergo less deglaciation and retreat as compared to the one located at lower elevation. This is so because at

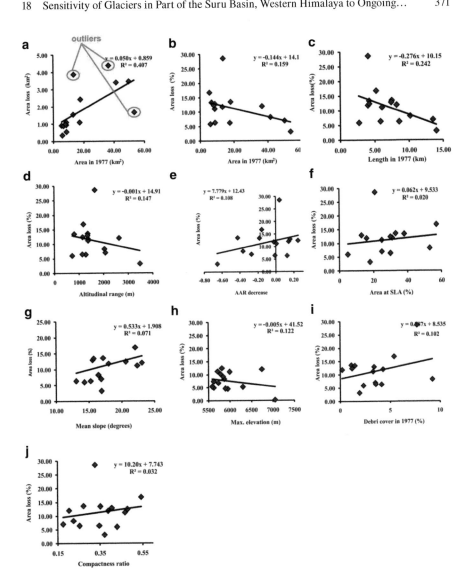

Fig. 18.13 Scatter plots displaying the relation between topographic factors with deglaciation during the period 1977–2016. All the relationships were found to be significant at 90% confidence level, i.e., α = 0.1. Note the presence of outliers depicted in (**a**)

higher elevations, glaciers are more likely to receive more precipitation in form of snow and have lower air temperature simultaneously. This inference is in line with Bhambri et al. 2011 and Pandey and Venkataraman 2013, who found similar negative association of elevation range with the glacier area changes in the central and western Himalaya, respectively. A significant positive correlation is also found between mean slope and the deglaciation (r = 0.372) and with retreat rate (r = 0.330).

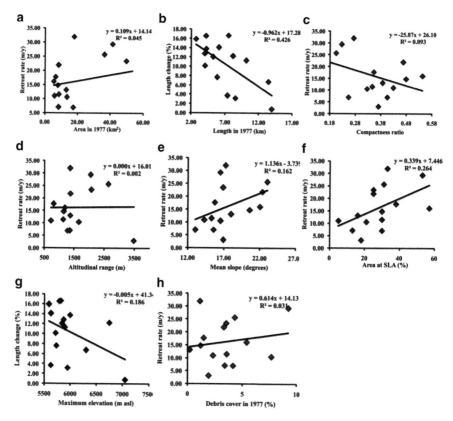

Fig. 18.14 Scatter plots displaying the relation between topographic factors with retreat during the period 1977–2016. All the relationships were found to be significant at 90% confidence level, i.e., $\alpha = 0.1$

This analysis clearly suggests that the glaciers having steeper slope retreat and loose area at a faster pace than those having relatively gentle slopes. Glacier hypsometry is the measure of its mass distribution over a range of altitudes. Depending upon the geometry of the glaciers, even a slight shift in the position of the SLA could result into marked alteration in the accumulation or ablation if a large portion lies at an elevation similar to the SLA. In this study, a good positive correlation is found between glacier area at SLA and retreat ($r = 0.514$) and low correlation with the area loss ($r = 0.141$). It is also found that larger glacier have more area at SLA (31%) as compared to the medium (25%) and small (28%) sized glaciers. The higher average retreat of larger glaciers (20 ± 3.4 m/y) may possibly be due to their larger area similar to SLA. Further, considering the aspect of the glaciers, it is quite peculiar that there are very few south trending glaciers in the basin while the field observations reveal existence of many south trending deglaciated valleys (Fig. 18.1). Poor representation of glaciers with all the aspects precluded a proper investigation of their control on present glacier response. Results show that the maximum length and area

Table 18.3 Glacier area (km²) and retreat rate (m/y) of the glaciers during the period 1977–2016 and topographic parameters such as area at mean SLA (%), altitudinal ranges (m asl), mean slope (°) and aspect of the studied glaciers

Glaciers	Glacier area (km²) 1977–2016	Retreat rate (m/y) 1977–2016	Area at mean SLA (%)	Altitudinal range (m asl)	Mean slope (°)	Aspect
G1	0.77 ± 1.36	7 ± 3.4	68	5949–4670	13.32	NE
G2	2.52 ± 3.31	23.43 ± 3.4	45	6301–4228	17.34	NE
G3	1.06 ± 0.83	13.13 ± 3.4	40	5869–4504	18.26	NE
G4	2.21 ± 1.24	31.97 ± 3.4	38	5818–4456	17.21	E
G5	4.39 ± 2.43	25.68 ± 3.4	47	6747–4116	23.05	NE
G6	0.10 ± 1.67	3.13 ± 3.4	37	6115–3578	16.93	N
G7	0.60 ± 0.49	21.9 ± 3.4	38	5629–4254	22.39	NE
G8	1.22 ± 0.81	10.55 ± 3.4	43	5717–4059	16.56	NW
G9	0.42 ± 0.57	7.06 ± 3.4	40	5624–4273	15.58	S
G10	1.81 ± 2.39	29.38 ± 3.4	51	5896–3859	16.49	N
G11	0.59 ± 0.69	11.51 ± 3.4	39	5730–4593	15.30	S
G12	0.49 ± 0.50	16.12 ± 3.4	56	5839–4653	22.08	N
G13	0.66 ± 0.62	17.76 ± 3.4	38	5784–4972	15.77	N
G14	0.85 ± 0.58	14.79 ± 3.4	41	5996–4843	20.62	N
G15	0.33 ± 0.49	10.98 ± 3.4	28	5590–4864	14.33	N

loss was experienced by north trending glaciers followed by north-east and east facing glaciers. For these reasons, aspect ranks low amongst the non-climatic factor with regard to its impact of glacier response in this region (Table 18.3). The compactness ratio shows a positive correlation with area loss ($r = 0.197$) and a negative correlation with the retreat ($r = -0.313$) (Table 18.4). The positive correlation indicates that more the compactness of the glacier, more will be its deglaciation. However, it does not show a clear relationship with retreat of the glaciers as they mainly involve frontal changes while compactness ratio is a factor which is defined by the entire geometry of the glaciers.

Likewise, the formation and subsequent enlargement of glacial ponds/lakes over or at the snout of the glaciers is known to contribute considerably to rapid wasting and increasing ablation (Sakai and Fujita 2010; Benn et al. 2012). Thus, unique snout characteristics also contribute to the differential behavior of glaciers in the study area. As per our results, relatively higher retreat rates are observed for glaciers G-4 (Dulung) followed by G-10, G-5 (Shafat) and G-2 (Lalung). Dulung glacier, although a debris free glacier, shows the highest retreat rates. Also, a moraine-dammed lake (MDL) is situated at the snout of Dulung and has continuously increased its size from 0.15 km² in 1977 to 0.56 km² in 2016. This significant increase in the size of MDL has possibly influenced the enhanced retreat rate of the glacier. These results are in agreement with the previous research works emphasizing on the fact that the presence of a proglacial lake at the snout of the glacier significantly enhances the rate of retreat by increasing the calving processes (Sakai 2012; Basnett et al. 2013).

Table 18.4 Correlation coefficient (r) and coefficient of determinations (r²) computed between various topographic factors and glacier changes

Influential Factors	Area loss (%)		Order of influence	Retreat rate (m /y)		Order of influence
	r	r²		r	r²	
Glacier size (km²)	0.637 (area loss in km²)	0.407 (area loss in km²)	8	0.268	0.072	6
Glacier size (%)	−0.398	0.159		–	–	
Glacier length (km)	−0.492	0.242	7	−0.366	0.134	8
Compactness ratio (full glacier)	0.179	0.032	3	−0.304	0.093	1
Altitudinal range (m)	−0.383	0.147	4	0.045	0.002	3
Maximum elevation (m asl)	−0.349	0.122	6	−0.109	0.011	5
Mean slope (°)	0.266	0.071	5	0.574	0.109	7
Glacier area at SLA (%)	0.141	0.020	2	0.514	0.264	2
Aspect			1	0.109	0.012	4

These relationship were found to be significant at 90% confidence level i.e. α =0.1

Order of influence (9) has the highest and (1) has least influence of topographic factors on changes in glacier parameters

ACZ accumulation zone, *SLA* snowline altitude, *AAR* area accumulation ratio, *GA* glacier area

In order to understand the degree of influence of each factor on deglaciation and retreat we have tried to arrange these factors in order of their influence (Table 18.4). These ranks have been assigned on the basis of preliminary analysis of the individual correlations of the non-climatic factors with glacier response followed by a thorough step-wise multi-variate analysis to gauge their cumulative influence on glacier changes.

18.6 Conclusions

This study monitored the status of 15 major glaciers of the Suru basin, western Himalayas, Jammu and Kashmir and evaluated the changes in different glacier parameters with respect to the variability in the climatic and non-climatic factors over the time period 1977–2016. The major inferences drawn from the study involves:

1. Trends apparent from the long-term meteorological records show recent increase in temperature (after 1997) and precipitation (after 1946) in the study area. However, the spatial variability of climate parameters reveals a progressively warmer and wetter climatic regime for glaciers hosted in the GHR as compared to the LR.

2. The differential climate regimen characterizing the GHR and LR also influences the glacier behavior in these regions. The glaciers in the LR are smaller in size (~5 to 8 km^2) and cleaner (average debris: ~10.13%) as compared to their GHR (size:~8 to 53 km^2 and maximum debris: ~19%)counterparts. It has been observed that LR glaciers have been shrinking faster (area loss: 9%) and accumulating more debris cover (debris increase: 116%) as compared to the GHR glaciers (6% and 78%, respectively). The GHR glaciers have, however, experienced greater rise in SLA (141 ± 97 m) in comparison to the LR ones (44.8 ± 67 m).

3. Apart from the long-term and local climate variability, the heterogeneous glacier response is also governed by the non-climatic factors. Analysis reveals that the glacier size, length, maximum elevation, slope and altitudinal range are important non-climatic factors controlling glacier area change. The retreat of the studied glaciers is majorly influenced by glacier length, size, slope, maximum elevation and aspect.

Acknowledgement Authors are grateful to the Director, Wadia Institute of Himalayan Geology, Dehradun for providing all the facilities and support for successful conduction of our research work. Authors thank the two anonymous referees for their valuable comments and suggestions for improving the original article and the editorial team for effectively processing the chapter.

References

Allen TR (1998) Topographic context of glaciers and perennial snowfields, glacier National Park, Montana. Geomorphology 21(3–4):207–216. https://doi.org/10.1016/S0169-555X(97)00059-7

Archer DR, Fowler HJ (2004) Spatial and temporal variations in precipitation in the Upper Indus Basin, global Teleconnections and hydrological implications. Hydrol Earth Syst Sci 8:47–61. https://doi.org/10.5194/hess-8-47-2004

Basnett S, Kulkarni AV, Bolch T (2013) The influence of debris cover and glacial lakes on the recession of glaciers in Sikkim Himalaya, India. J Glaciol 59:1035–1046

Benn DI, Gulley J, Thompson S, Bolch T, Hands K, Luckman A, Nicholson LI, Quincey D, Toumi R, Wiseman S (2012) Response of debris-covered glaciers in the Mount Everest region to recent warming, and implications for outburst flood hazards. Earth-Sci Rev 114(1–2):156–174. https://doi.org/10.1016/j.earscirev.2012.03.008

Bhambri R, Bolch T, Chaujar RK (2011) Mapping of debris-covered glaciers in the Garhwal Himalayas using ASTER DEMs and thermal data. Int J Remote Sens 32(23):8095–8119

Bolch T, Buchroithner M, Pieczonka T et al (2008) Planimetric and volumetric glacier changes in the Khumbu Himal, Nepal, since 1962 using Corona, Landsat TM and ASTER data. J Glaciol 54:592–600

Bolch T, Kulkarni A, Kääb A et al (2012) The state and fate of Himalayan glaciers. Science 336:310–314

Chudley TR, Miles ES, Willis IC (2017) Glacier characteristics and retreat between 1991 and 2014 in the Ladakh Range, Jammu and Kashmir. Remote Sens Lett 8(6):518–527. https://doi.org/10.1080/2150704X.2017.1295480

Cogley (2016) Glacier shrinkage across High Mountain Asia. Ann Glaciol 57(71):41–49. https://doi.org/10.3189/2016AoG71A040

Dimri AP (2013) Interseasonal oscillation associated with the Indian winter monsoon. J Geophys Res Atmos 118(3):1189–1198. https://doi.org/10.1002/jgrd.50144

Garg PK, Shukla A, Tiwari RK et al (2017a) Assessing the status of glaciers in parts of the Chandra basin, Himachal Himalaya: a multiparametric approach. Geomorphology 284:99–114. https://doi.org/10.1016/j.geomorph.2016.10.022

Garg PK, Shukla A, Jasrotia AS (2017b) Influence of topography on glacier changes in the central Himalaya, India. Glob Planet Chang 155:196–212. https://doi.org/10.1016/j.gloplacha.2017.07.007

Ghosh S, Pandey AC (2013) Estimating the variation in glacier area over the last 4 decade and recent mass balance fluctuations over the Pensilungpa Glacier, J&K, India. Glob Perspect Geogr 1(4):58–65

Granshaw FD, Fountain AG (2006) Glacier change (1958–1998) in the north cascades national park complex, Washington, USA. J Glaciol 52(177):251–256. https://doi.org/10.3189/172756506781828782

Guo Z, Wanga N, Kehrwald NM et al (2014) Temporal and spatial changes in Western Himalayan firn line altitudes from 1998 to 2009. Glob Planet Chang 118:97–105. https://doi.org/10.1016/j.gloplacha.2014.03.012

Hall DK, Bayr KJ, Schöner W et al (2003) Consideration of the errors inherent in mapping historical glacier positions in Austria from the ground and space (1893–2001). Remote Sens Environ 86:566–577. https://doi.org/10.1016/S0034-4257(03)00134-2

Hanshaw MN, Bookhagen B (2014) Glacial areas, lake areas, and snow lines from 1975 to 2012: status of the Cordillera Vilcanota, including the Quelccaya Ice Cap, northern Central Andes, Peru. Cryosphere 8:359–376. https://doi.org/10.5194/tc-8-359-2014

Harris I, Jones PD, Osborna TJ et al (2013) Updated high-resolution grids of monthly climatic observations – the CRU TS3.10 Dataset. Int J Climatol 52(3–4):591–611doi. https://doi.org/10.1002/joc.3711

Hewitt K (2005) The Karakoram anomaly? Glacier expansion and the elevation effect, Karakoram Himalaya. Mt Res Dev 25(4):332–340

Immerzeel WW, Beek LPH, Bierkens MFP (2010) Climate change will affect the Asian water towers. Science 328:1382–1385

Kääb A, Berthier E, Nuth C et al (2012) Contrasting patterns of early twenty first century glacier mass change in the Himalayas. Nature 488:495–498. https://doi.org/10.1038/nature11324

Kääb A, Treichler D, Nuth C et al (2015) Brief Communication: contending estimates of 2003–2008 glacier mass balance over the Pamir–Karakoram–Himalaya. Cryosphere 9:557–564. https://doi.org/10.5194/tc-9-557-2015

Kamp U, Byrne M, Bolch T (2011) Glacier fluctuations between 1975 and 2008 in the Greater Himalaya Range of Zanskar, southern Ladakh. J Mt Sci 8:374–389

Klimes L (2003) Life-forms and clonality of vascular plants along an altitudinal gradient in E Ladakh (NW Himalayas). Basic Appl Ecol 4(4):317–328

Kulkarni AV, Rathore BP, Suja A (2004) Monitoring of glacial mass balance in the Baspa basin using accumulation area ratio method. Curr Sci 86(1):101–106

Mir RA, Majeed Z (2016) Frontal recession of Parkachik glacier between 1971–2015, Zanskar Himalaya using remote sensing and field data. Geocarto Int. https://doi.org/10.1080/10106049.2016.1232439

Negi S (2002) Cold deserts of India. Indus Publishing, New Delhi

Oerlemans J (1994) Quantifying global warming from the retreat of glaciers. Science 264:243–245

Pandey P, Venkataraman G (2013) Changes in the glaciers of Chandra–Bhaga basin, Himachal Himalaya, India, between 1980 and 2010 measured using remote sensing. Int J Remote Sens 34(15):5584–5597

Pandey A, Ghosh S, Nathawat MS (2011) Evaluating patterns of temporal glacier changes in greater Himalayan range, Jammu & Kashmir, India. Geocarto Int 26:321–338. https://doi.org/10.1080/10106049.2011.554611

Paul F, Barrand NE, Baumann S et al (2013) On the accuracy of glacier outlines derivedfrom remote-sensing data. Ann Glaciol 54(63):171–182

Pritchard HD (2017) Asia's glaciers are a regionally important buffer against drought. Nature 545:169–187. https://doi.org/10.1038/nature22062

Rabatel A, Dedieu JP, Vincent C (2005) Using remote-sensing data to determine equilibrium-line altitude and mass-balance time series: validation on three French glaciers,1994–2002. J Glaciol 51(175):539–546. https://doi.org/10.3189/172756505781829106

Racoviteanu AE, Arnaud Y, Williams MW et al (2008) Decadal changes in glacier parameters in the Cordillera Blanca, Peru, derived from remote sensing. J Glaciol 54(186):499–510

Raina RK, Koul MN (2011) Impact of climatic change on agro-ecological zones of the Suru-Zanskar Valley, Ladakh (Jammu and Kashmir), India. J Ecol Nat Environ 3(13):424–440

Sakai A (2012) Glacial lakes in the Himalayas: a review on formation and expansion process. Glob Environ Res:23–30

Sakai A, Fujita K (2010) Correspondence. Formation conditions of supraglacial lakes on debris-covered glaciers in the Himalaya. J Glaciol 56(195):177–181. https://doi.org/10.3189/002214310791190785

Scherler D, Bookhagen B, Strecker MR (2011) Spatially variable response of Himalayan glaciers to climate change affected by debris cover. Nat Geosci 4(3):156–159. https://doi.org/10.1038/ngeo1068

Schmidt S, Nusser M (2009) Fluctuations of Raikot glacier during the past 70 years: a case study from the Nanga Parbat massif, northern Pakistan. J Glaciol 55(194):949–959. https://doi.org/10.3189/002214309790794878

Schmidt S, Nusser M (2017) Changes of high altitude glaciers in the Trans-Himalaya of Ladakh over the past five decades (1969–2016). Geosciences 7(2):27. https://doi.org/10.3390/geosciences7020027

Shapiro SS, Wilk MB (1965) An analysis of variance test for normality

Shukla A, Qadir J (2016) Differential response of glaciers with varying debris cover ex- tent: evidence from changing glacier parameters. Int J Remote Sens 37(11):2453–2479

Shukla A, Gupta RP, Arora MK (2009) Estimation of debris cover and its temporal variation using optical satellite sensor data: a case study in Chenab basin, Himalaya. J Glaciol 55(191):444–452. https://doi.org/10.3189/002214309788816632

Singh H (1998) Economy, society and culture: dynamics of change in Ladakh. In: Stellrecht I (ed) Karakoram-Hindu Kush-Himalaya, dynamics of change. Culture area. Karakoram scientific studies 4.2. Cologne, Germany, pp 351–366

Wu Y, He J, Guo Z et al (2014) Limitations in identifying the equilibrium-line altitude from the optical remote-sensing derived snowline in the Tien Shan, China. J Glaciol 60(224):1093–1100. https://doi.org/10.3189/2014JoG13J221

Chapter 19
Glacio-Hydrological Degree-Day Model (GDM) Useful for the Himalayan River Basins

Rijan Bhakta Kayastha and Rakesh Kayastha

Abstract This chapter describes a Glacio-hydrological Degree-day Model (GDM) which uses degree-day factors for estimating snow and ice melt that calculates total discharge from a river. It is a physically based gridded glacio-hydrological model which is useful for the Himalayan river basins. The GDM is successfully used in the Marsyangdi River basin (MRB) and Trishuli River basin (TRB). The model is first calibrated and validated by using observed discharge over the period of 2004–2014. A long-term continuous simulation is then carried out for 2020–2100 in both basins. Results show that the model simulations are good. The Nash-Sutcliffe Efficiency (NSE) are 0.79 and 0.83 for the period of 2004–2007 in MRB and from 2007 to 2010 in TRB, respectively during the calibration period and 0.81 and 0.76, for the period of 2008–2010 in MRB and from 2011 to 2014 in TRB, respectively. The snow melt and ice melt contributions to total discharge in MRB are 15% and 13%, respectively whereas 12% and 16% in TRB for the calibration period. The Representative Concentration Pathways (RCPs) 4.5 W m^{-2} scenario for the period of 2020–2100 shows an average increase of simulated discharge by 1.43 m^3 s^{-1} per year and 0.25 m^3 s^{-1} per year for MRB and TRB, respectively. Similarly, in RCP 8.5 the discharge increases by 0.71 m^3/s per year and 0.94 m^3 s^{-1} per year in MRB and TRB, respectively. The model can be used as a promising tool for the study of hydrological system dynamics and potential impacts of climate change on the Himalayan river basins.

R. B. Kayastha (✉) · R. Kayastha
Himalayan Cryosphere, Climate and Disaster Research Center (HiCCDRC),
Department of Environmental Science and Engineering, School of Science,
Kathmandu University, Dhulikhel, Nepal
e-mail: rijan@ku.edu.np

© Springer Nature Switzerland AG 2020
A. P. Dimri et al. (eds.), *Himalayan Weather and Climate and their Impact on the Environment*, https://doi.org/10.1007/978-3-030-29684-1_19

19.1 Introduction

The Himalayas with a length of 2400 km are inhabited by 52.7 million people and are spread across five countries: Bhutan, China, India, Nepal and Pakistan (Apollo 2017). Some of the world's major rivers, the Indus, the Ganges and the Brahmaputra, rise in the Himalayas, and their combined drainage basin is home to roughly 700 million people. The Himalayas together with Tibetan Plateau have a profound effect on the climate of the region, helping to keep the monsoon rains on the Indian plain and limiting rainfall on the Tibetan Plateau. The Himalayan range encompasses about 15,000 glaciers, which store about 12,000 km^3 of fresh water (IPCC 2007). Furthermore, the Himalayan river system is supplying water resources for more than 700 million people for their drinking, irrigation, navigation, industrial and hydropower uses which is ever increasing due to increase in economic activities in the region (Eriksson et al. 2009). The hydrology of the Himalayan region is attracting attention of more scientists and general public mainly because of contribution of snow and ice melt in the discharge of perennial rivers of the Himalayas in the changing climate scenarios and its behavior in future. More than 65% of monsoon influence glaciers are retreating (Scherler et al. 2011), regional variation on mass balance of Himalayan glaciers is reported and is less negative after 2000 than the average global mass loss till 2000 (Azam et al. 2018; Brun et al. 2018). Many authors have suggested that shrinking of glaciers in response to climate change might change the hydrological regime in these regions (Bolch et al. 2012; Kääb et al. 2012; Immerzeel et al. 2012). Decrease in glacier volume and area tends to influence the intensity of the seasons and the inter-annual variation on runoff (Juen et al. 2007). Glacier melt contribution is projected to increase until 2050 and then decrease on sub-basins (Immerzeel et al. 2013) and regionally, it is expected to increase discharge till 2050 and then decrease (Lutz et al. 2016). In western Himalaya, glacier melt contribution on runoff is projected to increase by 16–50% with 1–3 °C increase in temperature (Singh and Kumar 1997; Tahir et al. 2011; Sam et al. 2016).

The variation in discharge and contribution of hydrological components such as snow melt, ice melt, ground water and rain to total discharge in many rivers are still not well understood due to glacier and snow melts complexities in the high mountains. Therefore, many studies have carried out in the recent past in order to know the exact hydrologic regime of the Himalayan river systems. A study by Alford and Armstrong (2010) indicated that in Nepal the glacier contribution to sub-basin stream flow varies from approximately 20% in the Budhi Gandaki basin to approximately 2% in the Likhu Khola basin, averaging approximately 10% across nine basins. This discharge volume represents approximately 4% of the total mean annual estimated volume of 200,000 million cubic meters for the rivers flowing out of Nepal.

19.1.1 Glacio-Hydrological Modeling

Glacio-hydrological models are used because they can be tailored to fit the characteristics of available data. Many studies are carried out in the region using simple to complex glacio-hydrological modelling approaches to fill the large data gaps prevalent in the Himalayan basins. Generally, two melt modelling approaches: energy balance and temperature index model are widely used in most part of the world today in order to calculate discharge of glacierized river basins. The energy balance approach which explicitly models melt as residual in the surface energy balance equation by accounting sums of energy fluxes within the atmosphere and glacier boundary (Reid and Brock 2010) and temperature-index-model which derives melt from empirical relationship between air temperatures and melt rates (Braithwaite 1995; Hock 2003). Although the energy balance approach best describes melt totals (Hock 1999, 2003), this approach is not always feasible for Himalayan glaciers that are located in remote places and input data availability is a major constraint in these regions (Kayastha et al. 2000). A number of studies have used temperature index models in data daunted Himalayan basins (Takeuchi et al. 1996; Kayastha et al. 2000, 2005) to estimate river discharge at different temporal scales due to four main reasons: (1) wide availability of air temperature data; (2) relatively easy interpolation and forecasting possibilities of air temperature; (3) generally good model performance despite their simplicity; and (4) computational simplicity (Hock 2003). Several studies on the other hand have modified the simple temperature index model by incorporating different parameters such as albedo, shortwave radiation and melt factors to improve model performance (Cazorzi and Dalla Fontana 1996; Hock 1999).

Fukushima et al. (1991) used a conceptual runoff model called HYCYMODEL in the Langtang River basin to estimate streamflow change by global warming. Braun and Renner (1992) used the conceptual precipitation-runoff model in the same basin for better understanding of hydrological processes and efficient planning and operation of water resources. Similarly, Rana et al. (1996) used the same HYCYMODEL and empirical relation for melting of snow and ice for modeling runoff from the basin with inclusion of effect of debris on melting of underlying ice. Hock (1999) used a distributed temperature-index ice- and snowmelt model including potential direct solar radiation to estimate hourly melt and discharge of Storglaciaren, a small glacier in Sweden. The classical degree-day method yielded a good simulation of the seasonal pattern of discharge, but the pronounced melt-induced daily discharge cycles were not captured. Rango (1992) used the Snowmelt Runoff Model (SRM) for the Rio Grande and Kings River basins to study the changes in snowmelt runoff under warmer climate scenarios. HBV Light model developed by Uppsala University has been used in many parts of the world for river runoff simulation since its development. The degree-day methods have been use in many variants for more than a century and perform well in Alps (Braun and Renner 1992), Greenland (Braithwaite 1995), Scandinavia (Hock 1999), the Himalayas (Singh and Kumar 1997; Kayastha et al. 2005; Pradhananga et al. 2014), New

Zealand (Woo and Fitzharris 1992). Khadka et al. (2016) used SRM in five sub-river basins of Koshi River basin in Nepal, in which there was a high possibility of increase in mean stream flow for the period of 2041–2060 in comparison to the baseline period (2000–2008) in three sub-basins Tamor, Tamakoshi and Sunkoshi while decline in the flows in Arun and Dudhkoshi sub-basins in future under the impact of climate change.

Many researches have been carried out to investigate the sensitivity of glaciers to climate change (Laumann and Reeh 1993; Braithwaite and Zhang 1999; De Woul and Hock 2005; Fujita et al. 2006; Pradhananga et al. 2014; Ragettli et al. 2015). For the upper Indus Basin, Immerzeel et al. (2009) found that glacier melt contributed substantially to stream flow – 32% in a reference situation, peaking in July (with snow melt providing 40% of the total with a peak in June, and rain comprising 28% with a peak in July). Bookhagen and Burbank (2010) combined the snowmelt model with the Tropical Rainfall Measuring Mission (TRMM) – derived rainfall and MODIS-derived evapotranspiration measurements into annual and seasonal river flow estimated of the entire Himalayan region. They found that snowmelt accounts for ~50% in the annual runoff budget in the western Himalaya (Indus, Sutlej). In contrast, the generally smaller central and eastern Himalaya catchments receive more than 80% of their annual runoff from rainfall and less than 20% from snowmelt (except the large Brahmaputra catchment which receives ~34% of its annual discharge from snowmelt). Despite the eastward decrease in the contribution of snowmelt to annual runoff, snowmelt is significant and important for all catchments in the pre- and early monsoon months from March to June. During this time, more than 40% of the discharge was derived from snowmelt and is vital to agriculture, hydropower, and water quality throughout the Himalayan foreland. Immerzeel (2010) have applied their modeling approach (Normalized melt index) across the greater Himalayan region and concluded that glacier melt water is extremely important in the Indus Basin and reasonably important for the Brahmaputra, but only plays a modest role for the Ganges, Yangtze, and Yellow rivers. Preliminary results indicated that the snow and glacier melt contribution, compared to total runoff generated below 2000 m, was the following: Indus, 15.1%; Brahmaputra, 27%; Ganges, 10%; Yangtze, 8% and Yellow, 8%. This showed the much higher contribution of glacier melt to the Indus than to other rivers.

Immerzeel et al. (2013) had studied hydrological regime of the Baltoro and Langtang watersheds that drain into the Indus and Ganges Rivers, respectively using a high resolution glacio-hydrological model. They showed that the largest uncertainty in future runoff was a result of variations in projected precipitation between climate models. In both watersheds, strong, but highly variable, increases in future runoff were projected and, despite the different characteristics of the watersheds, their responses were surprisingly similar. In both cases, glaciers will recede but net glacier melt runoff was on a rising limb at least until 2050. In combination with a positive change in precipitation, water availability during this century was not likely to decline. Huss and Hock (2018) computed global glacier runoff changes for 56 large-scale glacierized drainage basins to 2100 and analyzed the glacial impacts on stream flow. In roughly, half of the investigated basins, the modelled annual glacier runoff continues to rise until a maximum ('peak water') is reached, beyond which runoff steadily declines. In the remaining basins, this tipping point has already been passed. Peak water timing in Indus, Ganges and Brahmaputra occurs later in basins with larger glaciers and higher ice-cover fractions. Although most of the 56 basins have less than 2% ice coverage, by 2100 one-third of them might have experienced runoff decreases greater than 10% due to glacier mass loss in at least 1 month of the melt season, with the largest reductions in central Asia and the Andes.

As a continuing research on how future climate will impact on hydrologic regime of glacierized river basins of the Himalayas, we have developed a Glacio-hydrological Degree-day Model (GDM) suitable to use in such river basins. The model is used in two river basins in Nepal and future discharge variations are studied using future predicted climate data.

19.2 Study Area

This study is carried out in two river basins of Central Himalaya, Marsyangdi and Trishuli River basins (Fig. 19.1) (hereafter MRB and TRB), sub-basin of Narayani River basin in Nepal Himalaya. MRB elevation ranges from 355 to 7819 m above

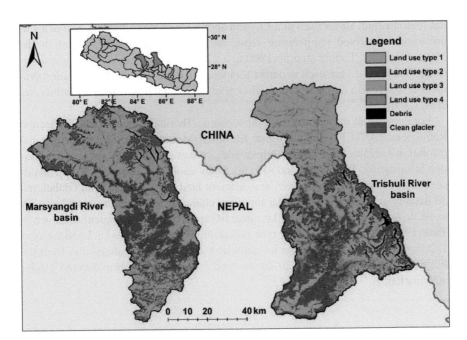

Fig. 19.1 Location map of MRB and TRB (map of Nepal in the inset). The map showing the land use types with clean and debris covered glaciers

sea level (a.s.l), covering an area of 4059 km^2. About 13.3% (543 km^2) of total basin area of MRB is glaciated, within that 5.15% (28 km^2) is covered by debris. TRB covers an area of 4601 km^2, 62% (2852 km^2) lies in Tibetan Plateau (TP), China and 38% (1748.65 km^2) in Nepal extending from 566 to 7659 m a.s.l. Out of total basin area, 13.5% (623 km^2) is glaciated including (544 km^2) debris free and (79 km^2) debris-covered. Glacier area in both of the basins lies in the elevation of 4000 to 6500 m a.s.l, debris covered and clean ice glacier area are comparatively higher in TRB at elevation 4500–5000 m a.s.l and 5000–5500 m a.s.l than in MRB (Fig. 19.2).

The climate of both basins is mostly dominated by Indian Summer Monsoon (ISM) from June to September and occasionally by the westerly disturbance during post- monsoon (October–January). Most of the MRB lies in the southern flanks of Central Himalayas and North Western part of this basin lies in the leeward of Annapurna Massif. Less than 40% of TRB lies in Nepal and is influenced by ISM and orographic effects, and rest of the basin lies in the TP where leeward phenomena prevails.

19.3   Input Data

An input data such as daily air temperature, precipitation and stream flow data are obtained from Department of Hydrology and Meteorology (DHM), Government of Nepal (Table 19.1). Air temperature and precipitation measured at climatological station Khudi Bazar (823 m a.s.l.) and another station at Chame (2680 m a.s.l.) are used to derived temperature lapse rate and precipitation gradient in the MRB. Similarly, Timure station (1900 m a.s.l.) and Kyangjing climatological station (3862 m a.s.l.) are used to derive temperature lapse rate and precipitation gradient in TRB. The derived precipitation gradients in TRB is similar to Immerzeel et al. (2014).

For geo-spatial dataset, Advance Spaceborne Thermal Emission and Reflection Radiometer (ASTER) Global Digital Elevation Model Version 2 of 30 m spatial resolution available from USGS (https://gdex.cr.usgs.gov) is used for the grid elevation information. The GlobeLand30 of 30-meter resolution (http://www.globalland-cover.com) is used for land cover. Ten different land cover classes from GlobeLand 30 dataset are merged with similar topology character and surface runoff behavior to create six land classes for similar range of rainfall runoff coefficient as shown in Table 19.2. In this study, land use type is classified into six classes i.e. Land use type 1, 2, 3, 4, debris-covered glacier ice and clean glacier ice. The shape files from the ICIMOD Glaciers Inventory (2010) are used for clean and debris-covered glacier information.

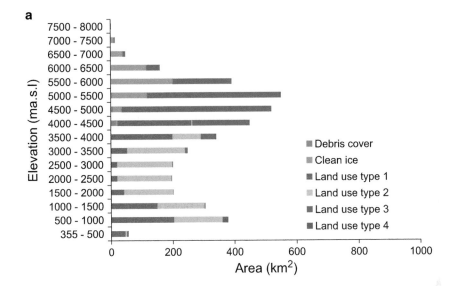

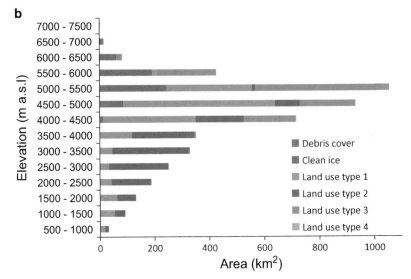

Fig. 19.2 Area – altitude distribution of debris covered and clean ice along with other land use types in MRB (**a**) and TRB (**b**) based on the ASTER GDEM of 30 m resolution, GlobeLand30 and ICIMOD Glacier Inventory (2010)

Table 19.1 List of climatological and hydrological stations in the MRB and TRB

River basin	Stations (m a.s.l.), type	Hydrological stations (m a.s.l.)	Year	Reference station
MRB	Khudi bazar (823), climatological	Bimal Nagar (354)	2004–2010	Khudi Bazar
	Chame (2680), climatological			
TRB	Timure (1900), Climatological	Betrawati (600)	2007–2015	Timure
	Kyangjing (3862), Climatological			

Table 19.2 Re-classification of land cover classes from GlobeLand30 dataset in MRB and TRB

GlobeLand30 Land classes	Land use type	MRB	TRB
Agriculture land	Land use type 1	29.21%	34.88%
Grass land		(1186 km²)	(1605 km²)
Shrub land	Land use type 2	27.39%	27.38%
Forest		(1112 km²)	(1260 km²)
Barren land	Land use type 3	29.12%	24.10%
		(1182 km²)	(1109 km²)
Artificial surface	Land use type 4	0.88%	0.08%
Water bodies		(36 km²)	(4 km²)
Permanent snow and ice	Clean ice	12.69%	11.81%
		(515 km²)	(544 km²)
	Debris covered	0.69%	1.72%
		(28 km²)	(79 km²)

19.4 Model Setup

The Glacio-hydrological Degree-day Model (GDM, Version 1.0) is a gridded and distributed glacio-hydrological model capable to simulate the contribution of hydrological components in the river discharge. GDM simulates four different runoff components in total discharge: glacier ice melt, snow melt, rainfall and baseflow at daily time step (Fig. 19.3). A melt module is based on degree-day approach, a simplification of complex process (Braithwaite and Olesen 1989) to estimate the glacier ice and snow melts with minimal data requirements (Kayastha et al. 2005). The concept of two reservoir based modelling approach of Soil and Water Assessment Tool (SWAT) (Luo et al. 2012) is adopted to simulate the hydrological response of baseflow and rainfall runoff contribution in river discharge. The basin is divided into 3×3 km grids and classified land class information from GlobeLand30 is extracted to each grid. Daily extrapolated temperature and precipitation from the reference station to each grid are used to force the model for the discharge simulation. The threshold temperature (T_T) determines whether the precipitation is in the form of snow or rain in each grid in the respective time step as:

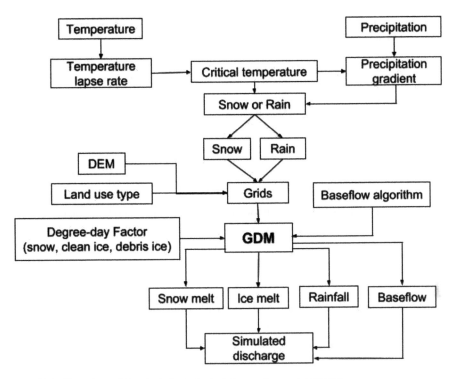

Fig. 19.3 Flow chart of the glacio-hydrological degree-day model (GDM)

$$Precipitation = \begin{cases} rain, if & T \geq T_T \\ snow, if & T < T_T \end{cases} \tag{19.1}$$

where, T is extrapolated daily air temperature for grids and T_T is threshold temperature, both in °C.

In each grid, daily ice melt from debris free and debris-covered ice and snow melt from glacierized and glacier free areas is calculated as:

$$M = \begin{cases} K_d \text{ or } K_s \text{ or } K_b \times T & if, T > 0 \\ 0 & if, T \leq 0 \end{cases} \tag{19.2}$$

where, M is the ice or snow melt in mm day^{-1} in each grid, T is daily air temperature in °C, K_d, Ks, and K_b are the degree-day factors for debris-covered ice, snow and clean glacier ice in mm °C^{-1} day^{-1}. The model takes an account for the multilayer melting of the snow above clean ice and debris-covered ice.

Baseflow is calculated using a baseflow simulation approach as in SWAT. Two aquifer system: shallow and deep aquifer system concept is applied to simulate the baseflow in glacier and snow melt dominated basin (Luo et al. 2012; Zhang et al. 2015). The advantage of two reservoir system over single reservoir system is that it releases the discharge in recession period and assures the level of discharge much more similar to observed discharge during the recession period. Hence, Luo et al. (2012) baseflow scheme is adopted for the baseflow simulation in this study. The major components accounted on baseflow simulation are the amount of percolation of surface water, W_{seep}, the delay time for the overlying geological formation for shallow aquifer percolation, $\delta_{gw, sh}$, the recession constant for the shallow aquifer, $\alpha_{gw, sh}$, the seepage constant for deep water percolation, β_{dp}, delay time for the deep aquifer percolation, $\delta_{gw, dp}$ and recession constant for deep aquifer, $\alpha_{gw, dp}$. The details of baseflow approach and parameter values are given in Neitsch et al. (2011) and Luo et al. (2012).

The surface runoff, Q_G consists of runoff from rainfall, snowmelt and ice melt runoff from each grid. The surface runoff component is calculated grid-wise based on the following equation:

$$Q_G = Q_r * C_r + Q_s * C_s + Q_i \qquad (19.3)$$

where, Q_r is discharge from rain, Q_s is discharge from snowmelt and Q_i is discharge from clean and debris-covered icemelt in $m^3 \ s^{-1}$, C_r and C_s are the rain and snow coefficients, Q_G is surface runoff component from each grid in $m^3 \ s^{-1}$. Then the total surface runoff contribution, Q_R from all grids and the total baseflow contributions, Q_B from all grids are expressed as:

$$Q_R = \sum_{G=1}^{n} Q_G \qquad (19.4)$$

$$Q_B = \sum_{b=1}^{n} Q_b \qquad (19.5)$$

where Q_b is the baseflow contribution from each grid and n is the number of grids. Total surface discharge, Q_R is then routed along with the baseflow contribution, Q_B towards the outlet through the following equation:

$$Q = Q_R * (1 - k) + Q_{R(d-1)} * k + Q_B \qquad (19.6)$$

where, k is recession coefficient, Q is total discharge in $m^3 \ s^{-1}$ and d is the dth day.

19.5 Hydrologic Simulation

19.5.1 Simulation Experiment Design

The performance of GDM is first calibrated over a four-year period (2004–2007) and evaluated over a period of another 4 years (2007–2010) by comparing the simulated and observed discharges at Betrawati hydrological station in TRB and Bimal Nagar hydrologic station in MRB. The melt module parameter such as degree-day factors for snow and ice melt are based on field observations in Nepal Himalayas carried out by Kayastha et al. (2000, 2003). Degree-day factor for ice melt under a debris layer is assumed to be around half of the clean ice melt based on the field observation on Khumbu and Lirung Glaciers in Nepal Himalayas.

19.5.2 Performance Indices

An assessment efficiency of model performance is carried out by comparing the daily time series observed and simulated discharges. The Nash-Sutcliffe Efficiency (NSE) index (Nash and Sutcliffe 1970) is used to assess the model result between the simulated and observed discharge as shown in Eq. 19.7.

$$NSE = 1 - \frac{\sum_{i=1}^{d} \left(Q_{obs} - Q_{sim} \right)^2}{\sum_{i=1}^{d} \left(Q_{obs} - Q_{avg} \right)^2} \tag{19.7}$$

where Q_{obs} is the daily observed discharge and Q_{sim} is the daily simulated discharge and Q_{avg} is the average observed discharge.

Similarly, volume difference is also used to determine the model accuracy and is calculated by using the following equation:

$$VD = \frac{V_R - V'_R}{V_R} \times 100 \tag{19.8}$$

where V_R and V'_R is the measured and the simulated discharge, respectively.

19.6   Results and Discussion

19.6.1   Model Calibration and Validation

The main calibrating parameters of the GDM are positive degree-day factors, snow and rain coefficients and recession coefficient. The model is calibrated with different positive degree-day factors and a set of degree-day factors are adopted for different months which are within the range of estimated degree-day factors on different glaciers of Nepal Himalayas. Similarly, the model is also calibrated with snow and ice coefficients and recession coefficients. All these calibrated parameters and coefficients and used in the study are listed in Table 19.3.

Daily simulated discharge is compared with the observed hydrographs of MRB (Fig. 19.4.1) and TRB (Fig. 19.4.2) for the both calibration and validation period. Both high and low discharges of the simulated discharge are consistent with the observed discharge. The model overestimated slightly in the pre-monsoon or low flow period that might have associate with the precipitation distribution. Precipitation distribution to higher elevations from the reference station at lower elevation may not well representative and precipitation at glacier altitude may be two to ten time higher than the valley bottoms (Bocchiola et al. 2011; Immerzeel et al. 2015). Barry (2012) also said that the precipitation gradient in the mountainous environment is considered to vary vertically and horizontally. The high Himalayan region with

Table 19.3 Parameter and coefficients used for calibrating the model in MRB and TRB

Parameters		MRB	TRB
Critical temperature		2 °C	2 °C
Temperature lapse rate		0.6 °C 100 m^{-1}	0.58 °C 100 m^{-1}
Precipitation gradient		+ 20% (~2680 m) + 30% (> 2680 m)	+ 15% (~ 3862 m) + 39% (> 3862 m)
Recession coefficient		0.90 and 0.008	0.91 and 0.008
Runoff coefficient	Land use type 1	0.5	0.13
	Land use type 2	0.3	0.22
	Land use type 3	0.3	0.26
	Land use type 4	0.95	0.95
Degree day factor	Snow melt	7–8.5 mm °C^{-1} day^{-1}	7.5–9.0 mm °C^{-1} day^{-1}
	Ice melt	6–10.5 mm °C^{-1} day^{-1}	6–8.5 mm °C^{-1} day^{-1}
	Ice under debris	3 mm °C^{-1} day^{-1}	3 mm °C^{-1} day^{-1}
Rain coefficient		0.05–0.1	0.01–0.03
Snow coefficient		0.15–0.3	0.2–0.4
$\delta_{gw,sh}$		30 days	10 days
$\alpha_{gw,sh}$		0.9	0.1
$\delta_{gw,dp}$		180 days	125 days
$\alpha_{gw,dp}$		0.2	0.3
β_{dp}		0.8	0.9
Initial recharge		8 mm	20 mm

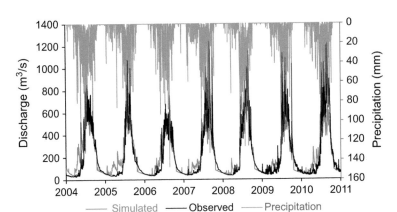

Fig. 19.4.1 Precipitation distribution and observed vs. simulated discharges for the calibration (2004–2007) and validation (2008–2010) periods for the MRB

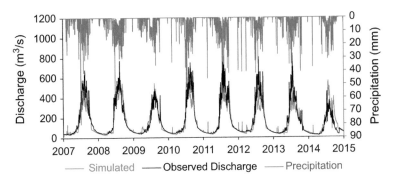

Fig. 19.4.2 Precipitation distribution and observed vs. simulated discharge for the calibration (2007–2010) and validation period (2011–2015) for the TRB

complex topography within the basin also effects on spatial and temporal distribution of precipitation which might be a constraint to represent the precipitation pattern or distribution. In this study, climatological stations within each respective basin are used to derive the precipitation gradient and distributed to higher elevations. Even with such limitations the model simulates the daily discharge with a good NSE and R^2 values and within 10% volume difference in spite of less input data.

The model performance is evaluated with the best performing parameter for the respective basins. Nash-Sutcliffe Efficiency (NSE) values for the calibration and validation period for the both river basin is >0.75 where as volume difference is in between ±10%. The coefficient of determination (R^2) value is also found to be 0.8 for the both of the cases. Model performance is satisfactory for both calibration and validation periods in both basins as listed in Table 19.4.

Table 19.4 NSE and volume difference for the calibration and validation periods

	MRB		TRB	
	Calibration	Validation	Calibration	Validation
NSE	0.79	0.81	0.83	0.76
Vol. diff %	−9.22	0.42	4.93	7.11
R^2	0.80		0.80	

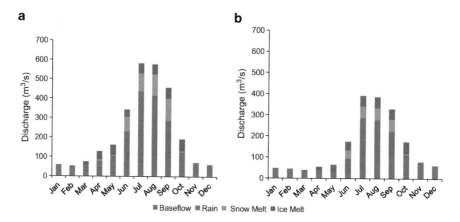

Fig. 19.5 Contribution of baseflow, rain, snow melt and ice melt for the calibration and validation periods in MRB (**a**) and TRB (**b**)

19.6.2 Contributions of Snow Melt, Ice Melt, Rain and Baseflow

The results of the model are then analyzed in detail for different components of discharge. Figure 19.5 shows the mean annual contributions of snow, ice melts (clean and debris), rain and baseflow for calibration and validation periods for the MRB and TRB. In MRB, 15% of total annual discharge is contributed from snow melt, where as 13% from clean ice and ice melt under debris, 44% from rain and 28% from the baseflow during calibration period. Similarly, 12% from snow melt, 16% from the ice melt from clean and debris covered ice, 36% from the rain and remaining from baseflow contributed to the annual discharge in TRB in calibration period. Bookhagen and Burbank (2010) found that the contribution of snow melt to discharge of MRB with an area or 4805 km² was 15.1% which was very close to the result of the present study. Also Racoviteanu et al. (2013) found that the contribution of glacier ice melt to measured discharge at Betrawati was 9.5%. The higher contribution of ice melt from the present model compared to their result may due to different modeling approach and difference in river basin area consideration. They used elevation dependent ice ablation model based on glacier mass balance gradient and considered only the river basin area within Nepal (3250.6 km²). Rain contribution on discharge is dominant on both river basins, during the monsoon period (June–Sept), while the baseflow contribute throughout the year with the maximum in the late monsoon period. The ice melt contribution on discharge begins from the pre-monsoon period with the maximum contribution in June to October in both basins which is attributed to the temperature and precipitation. The ice melt contribution on TRB with debris-covered area of ~79 km^2 contributes only ~3% more in annual discharge than that of MRB having debris cover glacier area of 28 km^2. The debris cover glacier in TRB lies in the higher elevation extent than that of MRB resulting comparatively less ice melt contribution and also attribute by the temperature and precipitation distributions.

19.6.3 Future Hydrological Regimes Under Different Climate Scenarios

In this study, daily temperature and precipitation data for the future climate scenarios are obtained from the World Climate Research Program (WCRP) sponsored Coordinate Regional Downscaled Experiments (CORDEX) for South Asia, driven by CNRM Regional Climate Model (RCMs) from the Commonwealth Scientific and Industrial Research Organization (CSIRO) of 50 km resolution, to run the model for the future climatic scenarios. The two Representative Concentration Pathways (RCPs) for the greenhouse gas concentration trajectories values of +4.5 W m^{-2} and + 8.5 W m^{-2} up to 2100 is used to simulate future discharges. The daily precipitation and temperature series data for each basin are bias corrected using the methods given by (Weiland Sperna et al. 2010) for temperature and precipitation as given below (Eqs. 19.9 and 19.10):

$$T_{corrected\_MOD} = T'_{MOD} + \left(T_{OBS} - T_{MOD}\right) \tag{19.9}$$

where $T_{corrected\_MOD}$ is the value obtained after calibration; T_{OBS} and T_{MOD} are the observed and modeled daily temperature, respectively; and T'_{MOD} is the mean daily value of modeled temperature.

$$P_{corrected\_MOD} = P_{MOD} \frac{\dot{P}_{OBS}}{\dot{P}_{MOD}} \tag{19.10}$$

where P_{MOD} is the daily modeled precipitation, and $\dot{P}_{OBS}$ and $\dot{P}_{MOD}$ are the average observed and modeled precipitation used for bias correction.

Temperature and precipitation projection of RCP 4.5 and RCP 8.5 climatic scenarios from the CORDEX is used for model simulation to estimate the future hydrological regime. The daily time series temperature and precipitation data from the CORDEX for two different scenarios are extracted at the reference station and distributed to grids in respect to elevation using temperature lapse rate and precipitation gradient. The daily time series data from RCM dataset is bias corrected with the observed data and fitted as forcing dataset for model simulation. The projected discharge till 2100 in MRB are increasing at a rate of 1.4 and 0.71 m³ s⁻¹ and similarly, in TRB are also increasing at a rate of 0.25 and 0.94 m³ s⁻¹ in RCP4.5 and 8.5, respectively (Fig. 19.6.1). Variations of different components of discharge snowmelt, icemelt, rain and baseflow are also shown in Fig. 19.6.2. The contribution of snow melt in MRB after 2050s is decreasing which may be due to increase in temperature; converting snow to rain and both snowmelt and icemelt in TRB after 2050s are increasing which may be due to increase in temperature and precipitation.

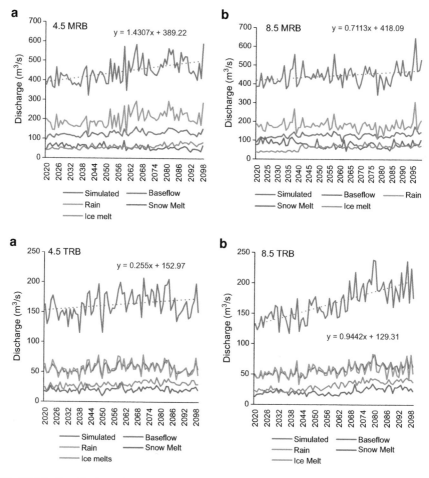

Fig. 19.6.1 Contribution of each component on total discharge for 2020–2099 in projected future climate scenarios for RCP 4.5 (left) and RCP 8.5 (right) in MRB (upper) and TRB (lower)

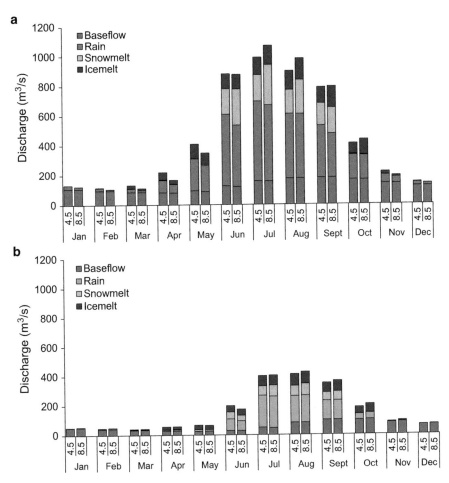

Fig. 19.6.2 Monthly variation of hydrological components contribution on discharge in future projected scenarios for RCP4.5 and RCP8.5 from 2020 to 2099 in MRB (**a**) and TRB (**b**)

19.7 Conclusions

The Glacio-hydrological Degree-day Model (GDM) is successfully setup to estimate discharge from a glacierized Himalayan river basins. As the model provides information on snow melt, ice melt, rain and baseflow, it gives insights into the hydrological system dynamics of the basin. Although the model is simple, it is able to simulate the discharge with a reasonable accuracy in the calibration as well as the validation periods. For the calibration and validation periods, NSE in both river basins are in between 0.76 and 0.83 and R^2 0.8 which are very good. Future water scenarios from this model suggest that the discharge will be higher in both river basins in both scenarios and the maximum increase in discharge is 1.42 m^3 s^{-1} in

MRB in RCP 4.5 scenario and lowest increase in discharge is 0.25 m^3 s^{-1} in TRB in RCP 4.5 scenario. Therefore, the model can be used as a promising tool for the study of hydrological system dynamics and potential impacts of climate change on the Himalayan river basins.

Acknowledgments Authors would like to thank the CHARIS (Contribution to High Asia Runoff from Ice and Snow) Project funded by United State Agency for International Development (USAID) for the financial support. We would also like to thank the Department of Hydrology and Meteorology (DHM), Government of Nepal for providing the hydro-meteorological data and the reviewers Rajesh Kumar and Mohd. Farooq Azam of the manuscript.

References

Alford D, Armstrong R (2010) The role of glaciers in stream flow from the Nepal Himalaya. Cryosphere Discuss 4:469–494

Apollo M (ed) (2017) The population of Himalayan regions – by the numbers: past, present and future. In: Efe R, Öztürk M (eds) Contemporary studies in environment and tourism, vol 9. Cambridge Scholars Publishing, Newcastle upon Tyne, pp 143–159

Azam MF, Wagnon P, Berthier E et al (2018) Review of the status and mass changes of Himalayan-Karakoram glaciers. J Glaciol 64:61–74. https://doi.org/10.1017/jog.2017.86

Barry RG (2012) Recent advances in mountain climate research. Theor Appl Climatol 110:549–553. https://doi.org/10.1007/s00704-012-0695-x

Bocchiola D, Diolaiuti G, Soncini A et al (2011) Prediction of future hydrological regimes in poorly gauged high altitude basins: the case study of the upper Indus, Pakistan. Hydrol Earth Syst Sci 15:2059–2075. https://doi.org/10.5194/hess-15-2059-2011

Bolch T, Kulkarni A, Kaab A et al (2012) The state and fate of Himalayan glaciers. Science 336:310–314. https://doi.org/10.1126/science.1215828

Bookhagen B, Burbank DW (2010) Toward a complete Himalayan hydrological budget: spatio-temporal distribution of snowmelt and rainfall and their impact on river discharge. J Geophys Res Earth Surf 115:1–25. https://doi.org/10.1029/2009JF001426

Braithwaite RJ, Olesen OB (1989) Calculation of glacier ablation from air temperature, West Greenland. In: Oerlemans J (ed) Glacier fluctuations and climatic change. Kluwer Academic Publishers, Dordrecht, pp 219–233

Braithwaite RJ (1995) Positive degree-day factors for ablation on the Greenland ice sheet studied by energy-balance modelling. J Glaciol 41:153–160. https://doi.org/10.1017/S0022143000017846

Braithwaite RJ, Zhang YU (1999) Modelling changes in glacier mass balance that may occur as a result of climate changes. Geogr Ann Ser A Phys Geogr 81:489–496. https://doi.org/10.1111/j.0435-3676.1999.00078.x

Braun JN, Renner CB (1992) Application of a conceptual runoff model in different physiographic regions of Switzerland. Hydrol Sci J 37:217–231. https://doi.org/10.1080/02626669209492583

Brun F, Berthier E, Wagnon P et al (2018) Correction: a spatially resolved estimate of High Mountain Asia glacier mass balances from 2000 to 2016 (Nature Geoscience DOI: 10.1038/ngeo2999). Nat Geosci 11:543. https://doi.org/10.1038/s41561-018-0171-z

Cazorzi F, Dalla Fontana G (1996) Snowmelt modelling by combining air temperature and a distributed radiation index. J Hydrol 181:169–187. https://doi.org/10.1016/0022-1694(95)02913-3

De Woul M, Hock R (2005) Static mass-balance sensitivity of Arctic glaciers and ice caps using a degree-day approach. Ann Glaciol 42:217–224. https://doi.org/10.3189/172756405781813096

Eriksson M, Jianchu X, Shrestha AB et al (2009) The changing Himalayas: impact of climate change on water resources and livelihoods in the greater Himalayas. International Centre for Integrated Mountain Development, Kathmandu

Fujita K, Thompson LG, Ageta Y et al (2006) Thirty-year history of glacier melting in the Nepal Himalayas. J Geophys Res Atmos 111:3–8. https://doi.org/10.1029/2005JD005894

Fukushima Y, Watanabe O, Higuchi K (1991) Estimation of streamflow change by global warming in a glacier-covered high mountain area of the Nepal Himalaya. In: Snow, hydrology forest high alpine areas. IAHS, Wallingford, pp 181–188

Hock R (1999) A distributed temperature-index ice- and snowmelt model including potential direct solar radiation. J Glaciol 45:101–111. https://doi.org/10.1111/j.0435-3676.1999.00089.x

Hock R (2003) Temperature index melt modelling in mountain areas. J Hydrol 282:104–115. https://doi.org/10.1016/S0022-1694(03)00257-9

Huss M, Hock R (2018) Global-scale hydrological response to future glacier mass loss. Nat Clim Chang 8:135–140. https://doi.org/10.1038/s41558-017-0049-x

Immerzeel WW, Droogers P, de Jong SM, Bierkens MFP (2009) Large-scale monitoring of snow cover and runoff simulation in Himalayan river basins using remote sensing. Remote Sens Environ 113:40–49. https://doi.org/10.1016/j.rse.2008.08.010

Immerzeel WW Ludovicus PH, Bierkens MFP (2010) Corrected 30 July 2010; see last page. Sci Mag 328

Immerzeel WW, Bierkens MFP, Konz M et al (2012) Hydrological response to climate change in a glacierized catchment in the Himalayas. Clim Chang 110:721–736. https://doi.org/10.1007/s10584-011-0143-4

Immerzeel WW, Pellicciotti F, Bierkens MFP (2013) Rising river flows throughout the twenty-first century in two Himalayan glacierized watersheds. Nat Geosci 6:742–745. https://doi.org/10.1038/ngeo1896

Immerzeel WW, Kraaijenbrink PDA, Shea JM et al (2014) High-resolution monitoring of Himalayan glacier dynamics using unmanned aerial vehicles. Remote Sens Environ 150:93–103. https://doi.org/10.1016/j.rse.2014.04.025

Immerzeel WW, Wanders N, Lutz AF et al (2015) Reconciling high-altitude precipitation in the upper Indus basin with glacier mass balances and runoff. Hydrol Earth Syst Sci 19:4673–4687. https://doi.org/10.5194/hess-19-4673-2015

IPCC (2007). Summary of policy makers, climate change 2007. The physical science basis. Contribution of working Group I to the fourth assessment report of the intergovernmental panel on climate change. Cambridge University Press

Juen I, Kaser G, Georges C (2007) Modelling observed and future runoff from a glacierized tropical catchment (Cordillera Blanca, Perú). Glob Planet Change 59:37–48. https://doi.org/10.1016/j.gloplacha.2006.11.038

Kääb A, Berthier E, Nuth C et al (2012) Contrasting patterns of early twenty-first-century glacier mass change in the Himalayas. Nature 488:495–498. https://doi.org/10.1038/nature11324

Kayastha RB, Takeuchi Y, Nakawo M, Ageta Y (2000) Practical prediction of ice melting beneath various thickness of debris cover on Khumbu Glacier, Nepal, using a positive degree-day factor. In: Debris-covered glaciers. IAHS Publ, Wallingford, pp 71–81

Kayastha RB, Ageta Y, Nakawo M et al (2003) Positive degree-day factors for ice ablation on four glaciers in the Nepalese Himalayas and Qinghai-Tibetan Plateau. Bull Glaciol Res 20:7–14

Kayastha RB, Ageta Y, Fujita K (2005) Use of positive degree-day methods for calculating snow and ice melting and discharge in Glacierized basins in the Langtang Valley, Central Nepal. Clim Hydrol Mt Areas:5–14. https://doi.org/10.1002/0470858249.ch2

Khadka A, Devkota LP, Kayastha RB (2016) Impact of climate change on the snow hydrology of Koshi River basin. J Hydrol Meteorol 9:28. https://doi.org/10.3126/jhm.v9i1.15580

Laumann T, Reeh N (1993) Sensitivity to climate change of the mass balance of glaciers in southern Norway. J Glaciol 39:656–665. https://doi.org/10.1017/S0022143000016555

Luo Y, Arnold J, Allen P, Chen X (2012) Baseflow simulation using SWAT model in an inland river basin in Tianshan Mountains, Northwest China. Hydrol Earth Syst Sci 16:1259–1267

Lutz AF, Immerzeel WW, Kraaijenbrink PDA et al (2016) Climate change impacts on the upper Indus hydrology: sources, shifts and extremes. PLoS One 11:1–33. https://doi.org/10.1371/journal.pone.0165630

Nash J, Sutcliffe J (1970) River flow forecasting through conceptual models part I—a discussion of principles. J Hydrol 10:282–290

Neitsch S, Arnold J, Kiniry J, Williams J (2011) Soil & water assessment tool theoretical documentation version 2009. Texas Water Resour Inst:1–647. https://doi.org/10.1016/j.scitotenv.2015.11.063

Pradhananga NS, Kayastha RB, Bhattarai BC et al (2014) Estimation of discharge from Langtang River basin, Rasuwa, Nepal, using a glacio-hydrological model. Ann Glaciol 55:223–230. https://doi.org/10.3189/2014AoG66A123

Racoviteanu AE, Armstrong R, Williams MW (2013) Evaluation of an ice ablation model to estimate the contribution of melting glacier ice to annual discharge in the Nepal Himalaya. Water Resour Res 49:5117–5133. https://doi.org/10.1002/wrcr.20370

Ragettli S, Pellicciotti F, Immerzeel WW et al (2015) Unraveling the hydrology of a Himalayan catchment through integration of high resolution in situ data and remote sensing with an advanced simulation model. Adv Water Resour 78:94–111. https://doi.org/10.1016/j.advwatres.2015.01.013

Rana B, Fukushima Y, Ageta Y, Nakawo M (1996) Runoff modeling of a river basin with a debris-covered glacier in Langtang Valley, Nepal Himalaya. Bull Glacier Res 14:1–6

Rango A (1992) Worldwide testing of the snowmelt runoff model with applications for predicting the effects of climate change. Hydrol Nord 23:155–172. https://doi.org/10.2166/nh.1992.011

Reid TD, Brock BW (2010) An energy-balance model for debris-covered glaciers including heat conduction through the debris layer. J Glaciol 56:903–916. https://doi.org/10.3189/002214310794457218

Sam L, Bhardwaj A, Singh S, Kumar R (2016) Remote sensing flow velocity of debris-covered glaciers using Landsat 8 data. Prog Phys Geogr 40:305–321. https://doi.org/10.1177/0309133315593894

Scherler D, Bookhagen B, Strecker MR (2011) Spatially variable response of Himalayan glaciers to climate change affected by debris cover. Nat Geosci 4:156–159. https://doi.org/10.1038/ngeo1068

Singh P, Kumar N (1997) Impact assessment of climate change on the hydrological response of a snow and glacier melt runoff dominated Himalayan river. J Hydrol 193:316–350. https://doi.org/10.1016/S0022-1694(96)03142-3

Tahir AA, Chevallier P, Arnaud Y, Ahmad B (2011) Snow cover dynamics and hydrological regime of the Hunza River basin, Karakoram Range, Northern Pakistan. Hydrol Earth Syst Sci 15:2275–2290. https://doi.org/10.5194/hess-15-2275-2011

Takeuchi Y, Naruse R, Skvarca P (1996) Annual air-temperature measurement and ablation estimate at Moreno Glacier, Patagonia. Bull Glacier Res 14:23–28

Weiland Sperna FC, Van Beek LPH, Kwadijk JCJ, Bierkens MFP (2010) The ability of a GCM-forced hydrological model to reproduce global discharge variability. Hydrol Earth Syst Sci 14:1595–1621. https://doi.org/10.5194/hess-14-1595-2010

Woo M, Fitzharris BB (1992) Reconstruction of mass balance variations for Franz Josef glacier, New Zealand, 1913–1989. Arct Alp Res 24:281–290

Zhang Y, Hirabayashi Y, Liu Q, Liu S (2015) Glacier runoff and its impact in a highly glacierized catchment in the southeastern Tibetan Plateau: past and future trends. J Glaciol 61:713–730. https://doi.org/10.3189/2015JoG14J188

Chapter 20
Hydrology of the Cold-Arid Himalaya

Renoj J. Thayyen

Abstract Hydrological characteristics of the Cold-arid trans-Himalayan region is least known due to lack of studies in the region. Hydrology of a region having around 60 mm of mean annual precipitation can be thought as a paradox. But the area have numerous glaciers constituting more than 75% of the Indian glacier resource and significant snow cover area mainly constrained over the top of the mountain range. These glaciers and snow sustain the livelihood of immediate lowland of arid valley bottom and contribute to the flow of Indus and its major tributaries like Shyok and Zanskar. These big rivers sustain livelihood of millions further downstream. Along with snow and glaciers; large extent of permafrost areas makes the hydrology of the cold-arid regions unique as compared to other regions of Indian Himalaya. Present study focus on the Ladakh range of the cold-arid system. Small glaciers with <1 km^2 area is characteristic of the Ladakh and Zanskar ranges. Discharge in the stream reach between glacier and foothill is restricted around 43% days as the mountain reach of the stream freezes in winter. Summer discharge show significant reduction during the lean snow years when ground ice melt component probably contributing to the stream flow. Steep precipitation and temperature gradients are another key feature of the cold-arid system. The temperature gradient during summer months surpasses 9.8 K/km and play a key role in sustaining the mountain cryospheric system in the region. Water related disasters are also very common in the area and have varied genesis such as Cloudburst, Glacial Lake Outburst Floods, Landslide Dam Outburst Flood and glacial surge dam outburst floods. More studies are imperative to comprehend the hydrology of this area in a better way. Mass and Energy balance of bigger glaciers, Permafrost thaw modeling, Groundwater dynamics and recharge areas and climate modeling at valley- ridge scale are some of the key topic to be pursued in greater details in future.

R. J. Thayyen (✉)
National Institute of Hydrology, Roorkee, Uttarakhand, India

© Springer Nature Switzerland AG 2020
A. P. Dimri et al. (eds.), *Himalayan Weather and Climate and their Impact on the Environment*, https://doi.org/10.1007/978-3-030-29684-1_20

399

20.1   Introduction

Cold-arid region of the Himalaya located beyond the Great Himalayan range is one of the three major hydrologic regimes of the Himalaya. Owing to the harsh climate and limited water resources, these areas are sparsely populated. Studies on the hydrology of these areas are also limited, due to the perceived limited contribution of these areas to the downstream discharge of major rivers. However, the limited water resources of the area is sustaining the livelihood of the local population and gaining better knowledge of the system has become imperative for managing the scarce water resources, especially under the changing climate of the Himalaya. Better understanding of the hydrology of each of the major climate regimes of the Himalaya is also important to develop a holistic understanding of climate change impact on Himalayan hydrology. Perceived asynchronous response of the Himalayan cryospheric systems and glaciers to the changing climate, espoused by the Karakorum anomaly in the west (Hewitt 2005; Gardelle et al. 2012) to the significant melting of glaciers in the eastern Himalaya has added an urgency in this regard. The role of mountains as water towers and its contribution to sustaining the low land population is highly acknowledged (Alford 1985; Viviroli and Weingartner 2004; Messerli et al. 2004). It is suggested that the world's major rivers could be classified into four different groups (Viviroli et al. 2007), based on the relative contribution of mountain runoff to the low land flows. Therefore, significance of a mountain range is determined by assessing the climate and hydrology of the low land regions. The low lands experiencing significant precipitation have limited significance of the mountain runoff while desert low lands almost completely dependent on the mountain water resources (Messerli et al. 2004). The same approach holds true for local mountain- valley system as well. In the cold-arid region, the drier valley bottom is almost entirely fed by the melt water from the cryospheric sources along the high mountain ranges. Therefore, cryospheric resources of the cold-arid region could acquire higher value as compared to the similar resources in the wet monsoon regions of the Himalaya. The importance of the Himalayan mountain chain stems from the fact that a quarter of the world's population live in its surrounding plains and major rivers originating from the Himalaya support this huge population. Glaciers and snow cover of the Himalaya influence the runoff regime of these rivers, but the nature of this influence vary significantly across various glacio-climatic setup in the Himalayas (Kaser et al. 2010). A large swath of the Himalaya experience monsoon rains in summer months simultaneously with peak glacier melt and significantly reduce the impact of glacier melt at the downstream reaches. However, river discharge during April to mid –June in this region is dominated by the snow-melt and have high impact on the mountain eco-system. Contrary to this, cryospheric resources have year round impact on the cold-arid region eco-system of the trans-Himalaya. A large area of the Himalayas north- west of the great Himalayan range comes under the climatic classification of cold-arid region (Fig. 20.1). This Cold - arid regions in the country spread across the trans- Himalayan mountains and constitutes a unique climate regime of the Himalayan orogeny (Thayyen and Gergan

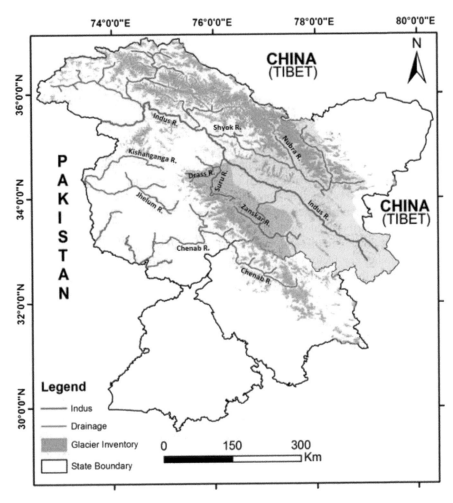

Fig. 20.1 Study area. Major part of the cold-arid region is in the upper Indus basin and have some of the heavily glacierised regions in the HKH region

2010). Most of this region comes under the administrative unit of Ladakh Autonomous Hill Development Council (LAHDC) in the state of Jammu & Kashmir. These high altitude plateau regions lie in the rain shadow zones of western disturbances and SW Indian monsoon seldom reach in this area. However, high elevation region of these mountains experience considerable snowfall and therefore, main water resource of this region are snow and glacier melt. Cold-arid regions of the country is part of the upper Indus basin and have a glacier cover of about 29,119 km² constituting more than 75% of the Indian glacier reserve (Raina and Srivastava 2008). Paradoxically, this region remain as water scarce and water stressed region owing to very little precipitation occurrence over the large part, especially along the valley bottom region. Majority of the villages in the Ladakh settled at the foothill zones of the mountains along the small mountain rivulets where water and alluvial soil is available for agricultural activities. Rest of the area remain barren and non-productive. For example, Ladakh region of Jammu and Kashmir state including Zanskar, Indus, Shyok, Nubra and Suru valleys cover an area of 58,000 km² in which potentially cultivable land constitute around 500 km² and actually cultivated land is only around 160 km² (Osmaston 1990). Water for most of this cultivated land comes from the small mountain streams fed by small glacier/snow cover. Even though there are big glaciers in the region, especially in the Karakorum Mountains and along the northern slopes of the great Himalayan range, population is concentrated along the small streams. Effectively huge amount of water from snow and glacier melt in the bigger rivers such as Shyok, Indus, Suru and Zanskar are not utilised by the hill population. One of the key water resources management challenge in this region is caused by the topographical constraints. The cultivable alluvial land is distributed along the foothills of the mountain ranges away from the major rivers flowing in this area which also lie significantly elevated from the river valley. Another key factor is the lack of research including weather and discharge monitoring in this area, which limit the ability to understand the hydrological processes of the area. There are 227 villages settled along both sides of the river Indus and spread over southern slopes of the Ladakh and northern slopes of the stok (Zanskar) mountain ranges (Census 2011). Majority of these villages sustain on the mountain streams, but none of these streams are gauged. Quantifying the available water resources, its seasonal distribution and dynamics is imperative for building better water management solutions for this area in future. The resource – habitat equilibrium built up over a long period is also under challenge recent times. The region is witnessing a tourism explosion since 2004 with annual tourist footfall crossing 200,000 per annum recently.

Aridity in the Ladakh region is the result of the very low precipitation occurrence in this region. Long term mean annual precipitation at Leh station, representing most of the valley region is only 60 mm (1978–1990) (Chevuturi et al. 2018) and combined with persistent low temperatures driven by comparatively long winter season from October to March/April classifies it as a cold arid region (Bhatt et al. 2015). The arid conditions also reflected in the land use as large area of the region is completely barren. The aridity is determined by the moisture index given by Thornthwaite and Mather (1955) and modified by the Venkataraman and Krishanan (1992) as

$$Mi = \left\{ \frac{(P - PE)}{PE} \right\} \times 100$$

Where Mi is the moisture index, P the average annual rainfall and PE is the average annual potential evaporation.

Based on the value derived for the moisture Index (Mi) the climate classification has been made as per the criteria of Table 20.1.

Table 20.1 Climatic classification based on Moisture Index (Mi)

Value of Mi	Climatic Zone
< −66.7	Arid
−66.6 to −33.3	Semi-arid
−33.3 to 0	Dry sub-humid
0 to +20	Moist sub- humid
+20.1 to +99.9	Humid
100 or more	Per- humid

Raju et al. (2013) undertook an exercise to revisit the climatic zones of India by using the gridded rainfall data of IMD (Rajeevan et al. 2008) at 0.5° × 0.5° for a period of 1971–2005 and suggested that the Ladakh district fall into dry sub-humid category. However, IMD gridded data gives an annual precipitation of 300 mm as compared to 60 mm from Leh station data (Chevuturi et al. 2018). Hence, this assessment need further evaluation but paucity of data is the major constrain.

20.2 Study Area: Ladakh

Ladakh region account for more than half of the landmass of the Jammu & Kashmir state in India and administered under two districts of Leh and Kargil. Leh is the largest district in the country in terms of area (45,100 km^2) and lies between 32 to 36° N latitude and 75 to 80°E longitude (Fig. 20.1). The district is bounded by Pakistan in the West and China in the north and eastern part and Lahul Spiti of Himachal Pradesh in the South East. Leh district comprises of Leh town and 112 inhabited villages and one un-inhabited village. The total population of Leh district is 1.33 lakhs. Kargil district has a population of 1.4 lakhs scattered over an area of 14,036 km^2 (Census 2011). Valley of Indus is situated at a high altitude around 3300 m a.s.l. and mountain ranges passes through 2750 m a.s.l. high at Kargil to 7672 m a.s.l. at Saser Kangri, in the Karakoram Range. High altitude and arid climate resulted into sparse vegetation and reduced level of oxygen, making it a formidable challenge to the visitors. Encircled by the Karakoram in the north and the Great Himalayas in the south, it presents a picturesque sight. Even though Ladakh is a high-altitude desert, it had number of extensive lake system at one point of time and many big lakes are still found on its southeast plateau of Rupshu and Chushul, or lakes of Tso-moriri, Tso-Kar and Pangong-Tso. Ladakh's culture is evolved through centuries and highly influenced by its climate. Ladakh's earliest inhabitants consisted of a mixed Indo-Aryan population of Mons and Dards. Buddhism came to western Ladakh through Kashmir in the second century when much of eastern Ladakh and western Tibet was still practicing the Bon religion. In the past, Ladakh gained importance from its strategic location at the crossroads of important trade routes. Since 1974 the Indian Government has encouraged tourism in Ladakh and fast emerging as a major Himalayan tourist destination.

20.3 Methodology

National Institute of Hydrology, Roorkee has initiated a study focusing on a major cold-arid catchment near Leh to build better understanding of the hydrology of the system in 2010. The study is carried out in the Upper Ganglass catchment covering 15.7 km² area with Phuche glacier in the upper reaches covering an area of 0.62 km². This paper discuss the data collected during 2010–14 period (Fig. 20.2). Meteorological forcing including its altitudinal response is studied by installing two automatic water stations at 4700 m a.s.l. and 5600 m a.s.l. The valley bottom meteorological data is sourced from India meteorological Department (IMD). Proglacial stream gauge is monitored using a Radar water level recorder and velocity by a handheld current meter. Discharge at this station is calculated from the rating curve established by area- velocity method. Study of winter and summer glacier mass balance is carried out by glaciological method during the period to generate data on glaciers role in the hydrologic system response. Stream originating from the high altitude cryospheric systems of the Ladakh range are interrupted streams. As per the definition, interrupted streams have perennial discharge in their upper reaches and intermittent flow in their reaches at lower elevations. However, small streams originating from the Ladakh range have intermittent flow at the higher reaches and perennial flow at the lower reaches. This clearly indicate that the small cirque glaciers have limited thickness and do not undergo pressure melting at the bottom during the winter months as happens in the case of bigger glaciers in the region. From its origin at Phuche glacier, the Phuche stream travels around 19 km before it joins with the River Indus. The intermittent section of the stream is around 11.5 km long

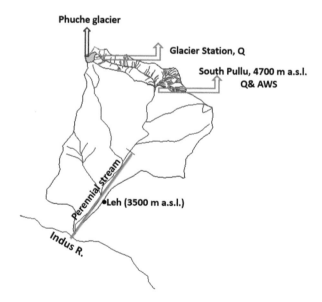

Fig. 20.2 Perennial lower reaches below the intermittent flow regime of the Phuche stream

constituting around 60% of the total stream length from the glacier (Fig. 20.2). The perennial stretch of the river starts 1–2 km offset to the foothill zone of the catchment.

20.4 Glacier Resources of the Cold-Arid Region

Glaciers are the core water resources in this region covering 9810 km² (Table 20.2). This is far more than the glacier reserve in the Indian part of the Ganga-Brahmaputra river systems. Indian part of the Ganga basin cover an area of 2857 km² mainly in the Uttarakhand Himalaya and Himalayan tributaries of the Brahmaputra have glacier cover of 929 km². (Sangewar et al. 2009). However, characteristics of the glaciers and its distribution vary across the cold-arid region. The northern slopes of the great Himalayan range feeding Suru and Zanskar rivers has large number of medium (5–7 km) and big (>10 km) sized glaciers (Raina et al. 2008). However, glaciers of Zanskar and Ladakh mountain ranges are characteristically small with 79% of the glaciers in these region have area less than 0.75 km² and experienced significant area loss during the past five decades (Schmidt and Nüsser 2012, 2017). Further north, the Karakoram mountains are home to some of the biggest and largest glaciers in the entire HKH region. Overall, the cold-arid mountain system is endowed with huge glacier resources (Fig. 20.1). An area of significant interest is the Ladakh and stok range with small glaciers feeding to the small streams sustaining the livelihood of the population.

20.5 Hydrology of the Cold Arid Regime: A Case Study of Ganglass Catchment Ladakh

Climate of Leh (3500 m a.s.l.) is characterized by high annual range of temperature between (+) 34.8 °C and (−)28.3 °C during 1951–1980 period (IMD). Contemporary weather of the region is assessed during 2010–2012 period show annual temperature range between (+) 33.8 °C to (−) 23.4 °C. Highest temperatures recorded in the months of July and August and mean monthly temperature during these months was 17.8 and 17.7 °C during 1951–1980 period while it increased to 20.8–23.0 °C during the present study period. Minimum mean monthly temperature recorded was

Table 20.2 Glacier distribution in the Upper Indus basin

Basin	Catchment area	Glacier Area (RGI 4.0) Km²
Shyok	38,545	7658
Indus (Suru, Zanskar and Main Indus)	46,450	2153

RGI Randolph Glacier Inventory

(−)14.4 °C during 1951–1980 period while during the study period it has risen to (−)8.5 °C in January 2011. As compared to this, higher elevation station at South Pullu (4700 m a.s.l.) recorded the lowest mean monthly temperature of (−)16.3 °C in January 2012 and highest mean monthly temperature of 11.2 °C is recorded in July 2011. Annual range of temperature recorded at South Pullu station at 4700 m a.s.l. varied from (−)27.3 °C to (+)21.9 °C. Further higher up over the mountain ridge at 5600 m a.s.l. annual range temperature observed between (−)27.4 °C and (+)13.6 °C and the mean monthly temperature ranges between (−)21.6 °C and (+)2.8 °C.

In the cold-arid region, dry valley bottom with very little precipitation is fed by the water from wet upper mountain zone sustaining the livelihood of the people in the region. The discharge regime of the high elevation mountain slopes are mainly governed by the snowfall amount and time of occurrence. The stream flow initiates at the gauging site at South Pullu (4700 m a.s.l.) during May 1–15 period depending on the prevailing weather conditions. Out of five years of observations in the Ganglass catchment, 2 years experienced early flow in the first week of May (2011 & 2013) and other two years (2010 & 2012) experienced late initiation of discharge around 15 May. The river ceases to flow at the South Pullu between 1–15 October. Hence, on an average stream flow at 4700 m a.s.l. is limited to around 158 days (5th May to 10th October) constituting 43% of days during cold-arid system water year (1st October to 30th September). In other words, it indicate that the upper mountain region remain frozen around 200 days in a year.

Inter annual variation of discharge is high in the region. Discharge from the catchment reduced to half during the years experiencing low snowfall (Figs. 20.3 and 20.4) characterized by the early disappearance of snow. The small glaciers play a very important role in sustaining the flows during such years with contribution rising up to 34%. Three distinct runoff response were observed during the study period, governed by the snow cover and precipitation. Years with early disappear-

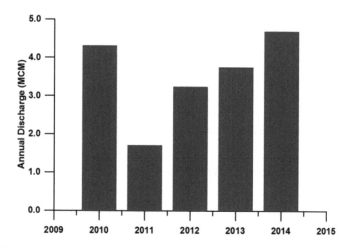

Fig. 20.3 Annual discharge flux variations at 4700 m a.s.l. of the cold-arid system catchment

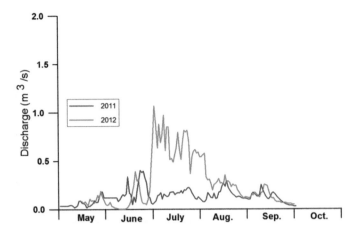

Fig. 20.4 Example of varying hydrograph characteristics of the same catchment (Ganglass) during high snow year followed by a low snow year

ance of the snow cover by mid of May produce annual runoff, distributed more or less equally through the summer months (M_{6-8}) experiencing 31–23% of runoff in each month. Years with normal winter snowfall with persistence of snow in the catchment till first week of June experienced more than 50% of the annual discharge in the month of July alone. The third category are the years experiencing persistent snowfall in the month of May and June which facilitate the extended snow cover in the catchment till mid-July and produce higher discharge in the months of July and August constituting more than 75% of the annual discharge. July discharge showed highest year-to-year variability ranging from 23 to 57% of the annual totals and managing this kind of variability is the biggest challenge in this region. Water demand is highest during these months due to most favorable conditions for tourists as well as for agriculture. Knowledge of seasonal discharge fluctuation based on prevailing snow regime will be extremely helpful in managing the scarce water resources in the region effectively. During the low snow years, as mentioned before, contribution from glacier melt increases to 34% of the annual flow as observed in the year 2011 and demonstrate the key role being played by these small glaciers in the Ladakh range. The monthly distribution of discharge during the low snow period and its response to electrical conductivity (EC) point towards the possibility of ground ice melt contribution in the region (Fig. 20.5). This observation has added a new dimension to the hydrology of the cold-arid regime. (Thayyen 2015).

There are number of geomorphological features present in the catchment signalling seasonal freeze and thaw activites in the higher elevations of the cold -arid regions in the form of patterned ground and exposed subsurface ice formations (Fig. 20.6). Intial modeling exercise by the researchers of the Univeristy of Zurich also indicate significant permafrost areas in the cold-arid region of the Himalaya (Gruber 2012; Gruber et al. 2017). Signature of ground ice melt, immediately after the snow melt in the higher elevations of the catchment is observed as the higher

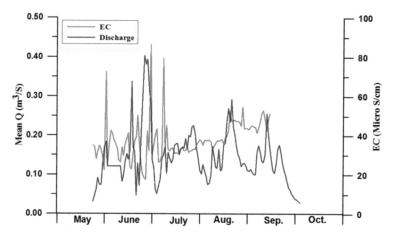

Fig. 20.5 Discharge- EC relationship during a low snow year showing typical inverse relationship during snow melt period. In the absence of snow from mid-July onwards EC shows gradual increase associated with steady low discharge probably influced by ground ice melt

Fig. 20.6 Patterned ground and exposed sub surface ice formation in the Ganglass catchment

ionic pulses in the meltwater. These ionic pulses during the late melt periods are dominated by higher calcium and bicarbonate concentrations (Namrata Priya et al. 2016). Permafrost processes is identified as a key knowledge gap, which could have significant implication for cryosphere-livelihood nexus in the cold-arid region. Cold-arid climate found to be facilitating sublimation in the glacier regimes, probably affecting the hydrology of the glaciers in the region (Rai et al. 2016). Even in the low elevation glaciers of cold-arid region of the Zanskar basin show significant deviation in the stable isotope values of snout ice as compared to the isotopic characteristics of glacier ice in the Kashmir region (Rai et al. 2016). This suggest that number of key knowledge gaps exists and significant effort is required to understand the cold-arid system hydrological process in a better way.

20.5.1   Precipitation Gradients in the Cold-Arid Regime

Snow is the dominant precipitation component in the catchment. 4 years of precipitation monitoring shows that the precipitation recorded at 4700 m a.s.l. at South Pullu is more than 4–5 times higher than the valley bottom at 3500 m a.s.l. (Leh). While South Pullu experienced around 209 mm of winter precipitation in place of 47.5 mm at Leh (Fig. 20.7). Compared to this, glacier elevations experienced further high values up to 620 mm w.e as evident from the winter mass balance of Phuche glacier. This demonstrate the steep precipitation gradient in the catchment during winter months. It clearly suggest that the hydrological assessment of the region based on the valley bottom measurements is insufficient for understanding the cold-arid region hydrology holistically.

Relationship of elevation gradient of precipitation is developed between the altitude index and the cumulative precipitation. The altitude index AI = Ax/Ab, Where Ax is the higher elevation stations and Ab is the base station. The winter precipitation between these altitudes are related through the power relationship.

The relationship have given a correlation on 0.92 and equation governing the precipitation distribution across these altitudes as

$$Ph = Pb\left(Ax / Ab\right)^{6.9}$$

Where Ph is the precipitation at desired higher altitude (h), Pb is the known precipitation at the base station, Ah is the altitude of the desired altitude and Ab is the altitude of the base station.

Similar relationship is found for the summer precipitation between 3500 m a..s.l. and 4700 m a.s.l. as shown in the Fig. 20.8. The relationship has returned the same exponent and slightly varying intercept. This suggest that the relationship between 3500–4700 m a.s.l. for winter as well as summer period remain same. However, it may not be prudent to assume the same relationship extending up to the glacier elevations in summer. The routine field observations during the mass balance data

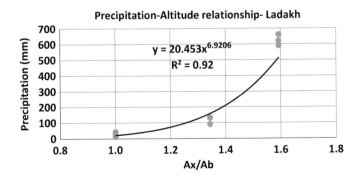

Fig. 20.7 Relationship between winter precipitation between 3500, 4700 and 5575 m a.s.l. in the study catchment in Ladakh

Fig. 20.8 Relationship of
summer precipitation
(June–September) between
3500–4700 m a.s.l

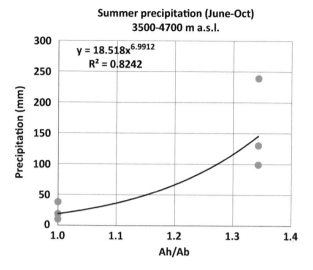

collection suggests that the summer precipitation between 4700 m a.s.l. and that of
glacier did not vary significantly or even may be less. However, this issue need to be
resolved by proper precipitation data collection at the glacier altitudes in future.
Bhutiyani (1999) developed the relationship for Siachen glacier in the Karakorum
Himalaya with an exponent of 4.78. This suggests that the Ladakh range have
steeper precipitation gradient in winter months.

20.5.2 Temperature Gradients in the Cold-Arid Regime

Temperature play a key role in determining the runoff regime of a cryospheric system as well as glacier mass balance. The annual temperature at 4700 m a.s.l ranges
between (−)27.3 °C and 21.9 °C during 2010–2014 period. The lowest temperature
occur during the last week of January and the highest in the month of June or July.
Prolonged sub-zero temperature regime of the catchment play an important role in
regulating the catchment hydrology. However, short period of positive temperatures
during July and August months produces abundant melting of the cryospheric components. Whenever fresh snowfall occurred during May & June months, temperature showed a sharp decline leading to significant reduction in discharge. Years with
precipitation free May & June months showed higher number of positive degree
days forcing higher melting during these two months as well.

Near surface temperature, lapse rate or Slope Environmental Lapse Rate (SELR)
(Thayyen and Dimri 2014) of the cold-arid regime is found to be unique. Temperature
lapse rate data of 2010, 2011 and 2012 between 3500 m and 4700 m a.s.l. is studied.
The SELR response between 4700 and 5600 m a.s.l representing the nival regime
also studied for one year (2012–2013). The foremost observation from these data is

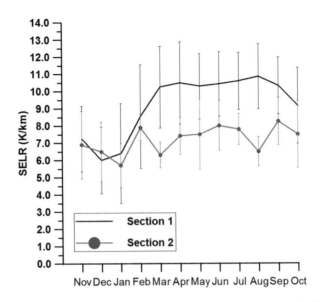

Fig. 20.9 Differing SELR of various sections of the mountain slope of the cold-arid system governed by elevation. Lapse rate of nival section 2 (4700–5600 m a.s.l.) (Pink line) is lower than the lapse rate of lower section 1 (3500–4700 m a.s.l.)

the higher annual range of daily SELR of the cold-arid system ranging from 2.8 to 17.0 K/km with consistently steep SELR during summer months clearly reflecting the characteristics of the arid conditions. In fact, lower section (3500–4700 m a.s.l.) started experiencing higher mean monthly SELR from March onwards (9.5–11.0 K/ km). However, SELR of core winter months (November–January) showed striking similarity with that of the monsoon regime with winter lowering of the SELR (Thayyen and Dimri 2014). Daily SELR during this period ranged between 5.8 and 7.5 K/km (Fig. 20.9), which clearly indicates the role of moisture and low temperature combination controlling the SELR of cold-arid regime mountain slopes during these months. Comparable observations were reported earlier from far end of the cold-arid system at the Chandra basin as well. SELR derived from average daily temperature between 3330 m a.s.l. and 3940 m a.s.l. near Hamtah glacier in Chandra basin during July, August, September months reported as 11.1 K/km, 10.9 K/km, 12.5 K/km, 11.3 K/km and 11.1 K/km in 2002, 2003, 2004, 2005 and 2006 respectively (Siddiqui and Maruthi 2007). However, SELR of higher elevation nival/glacier section is characterized by lower daily values ranging between 7.4 and 6.5 K/ km during summer months as compared to the consistently higher values of the lower section >9.8 K/km. Moreover, specific humidity values at nearby Chhota Shigri glacier meteorological station mimics that of Leh stations ranging between 4–7 g/kg during summer months. Hence, it is suggested that the higher summer SELR of temperature is a characteristic of the cold-arid regime of Ladakh.

20.5.3 Groundwater Resources

During the last decade, resource-habitat equilibrium has come under severe stress in the region due to heightened tourism activities. Tourist arrival at Leh has increased from 20,000 in 2004 to 1.8 Lakh in 2012 (Harjit Singh 2013) and become a major economic activity in the region surpassing the traditional agrarian economy. In the urban centres, this has forced people to explore the alternate source of water which has led to a trend of rampant ground water exploration in this region. However, the recharge area of groundwater in this area is lying the high elevation nival-glacier areas and sustainability of ground water as an assured long-term source is in serious doubt. More research is required to establish the recharge mechanism of groundwater in this region and its flow paths and sustainability.

20.6   Water Related Disasters in the Cold-Arid Region

Water related disasters have become an essential component of Cold- arid hydrology in the past decade. The region experienced couple of highly disastrous flood events in the recent past and left a deep scar on the society. Managing these floods could achieve twin objectives of saving the populations from disastrous consequences of the floods and turning it into viable source of water for the communities in a water scarce region through further research and development. Large areas of cultivable land as well as shelters were destroyed by these floods leading to financial stress for the communities. Loss of large number of life and livelihood create emotional trauma for the communities, which take long time to overcome. Hence, one of the future thrust area in research on cold-arid hydrology could be the flood management. There are four major types of floods experienced in the region such as (1). Cloudburst floods (2). Landslide Dam Outburst Floods (LDOF) (3). Glacial lake Outburst floods (GLOF) and (4). Glacial surge dam outburst floods. A brief description of these floods are given in the following section.

Irrespective of the low precipitation occurrence disasters triggered by extreme rainfall is not alien to the Cold-arid regions of IHR. Leading cause of concern in this regard are the cloudbursts (Dimri et al. 2016). The scale of devastation of the 2010 Leh cloudburst has brought the focus sharply to this phenomenon. Past records of flood occurrence in the region show number of such events since 2005 (Thayyen et al. 2013). The definition of cloudburst by the India Meteorological Department (IMD) suggest a high rate of rainfall at 100 mm/h covering an area of 20–30 km^2 (Ashrit,2010). However, study of cloudbursts in the Ladakh region suggests much shorter duration up to 15 min and much higher intensity ranging from 100 to 300 mm (Thayyen et al. 2013; Dimri et al. 2017). Floods resulting from such a short duration storm falling over the steep barren slopes of the mountain is often devastating with peak flood discharge of 545 ± 35% m^3/sec for a catchment with 2.83 km^2 area and 1070 ± 35% m^3/sec for a bigger catchment with 64.95 km^2 area (Thayyen et al. 2013). As it occur in the mountain slopes, cloudburst often impact the settlements in the first order stream catchment. In the city centers like Leh, the migrant locals and labourers are increasingly occupying these areas making them vulnerable to such events. Tourism is the main stay of economic activities in Ladakh and recurring cloudburst in the region will have negative impact the economic prosperity of the region, if not better managed.

Landslide Dam Outburst Flood (LDOF) is another flood source in the cold-arid Himalaya. LDOF occur due to landslide into the river leading to river blockage. The "Great Indus Flood" of 1841 (Mason 1929) is the biggest such flood occurred in the region. This particular landslide was triggered by an earthquake on the west side of the Lechar spur of Nanga Parbat. By June 1841 the lake has grown into 64 km in length and 300 m deep. The lake was emptied in 24 h, devastating hundreds of villages and killing thousands of people (Falconer 1841). Most recent occurrence of LDOF in the cold-arid region is recorded in the Phutkal River in Zanskar which got blocked by a heavy landslide that occurred on 15 January 2015. This lake had busted on 07 May 2015, affecting around 40 villages downstream. Factors triggering landslides in this non-monsoon region need better understanding. The role of permafrost thaw, persistent frost action and snow melt water seepage need greater focus.

Glacial lake outburst flood (GLOF) is another water related disaster in the region. As glaciers retreat in the cold-arid Himalaya, many glacier lakes are formed in the recent decades. Many of these lakes are dammed by the ice-cored moraine, which is susceptible for thawing with the rise in temperature. Much of the damage created during the GLOF events are associated with large amounts of debris that accompany the floodwaters. Due to extreme hazard potential, it is necessary to take into account of GLOF, while planning, designing and constructing any infrastructure projects such as bridges and culverts as they are located on the path of the flood. Cloudbursts and GLOF occur in the remote mountain areas and often confuse people and authorities about exact cause of the flood. Proper investigation is necessary to reveal the flood source, but often not done due lack of administrative mechanism to monitor and record such events in the most part of the cold-arid region. In September 2007 NIH team has got a rare opportunity to witness a GLOF event near the Pangong Lake (Fig. 20.10). The flood occurred in the Pagal Nalla, 3 km enroute to Pangong Lake. As its name suggests, this stream experiences frequent flash floods. The GLOF happened on 15 September 2007 between 2.30 and 3.00 p.m. It is informed by the local people that small instances of flooding reported earlier on 10 September 2007and local police temporarily suspended the tourist movement towards the Pangong lake area. NIH team visited the place from Tangse, a place 26 km away from the site. A rough on field estimate suggested as discharge of 30m^3/sec at 6.15 p.m. Electrical conductivity of the food water was 204 µS/cm, clearly indicating that the flood water had been stored for long time allowing ionic enrichment. Water sample collected from washed off bridge site on following day (16 September) showed reduced EC of 22 µS/cm, suggesting complete evacuation of stored water from the glacial system. We also carried out isotopic analysis of flood and post flood water and δO^{18} values of the both the samples showed $-14.09^0/_{00}$ and $14.59^0/_{00}$ respectively suggesting the same source of the water. δO^{18} values of the glacier ice samples in the Ladakh region vary between -16.26 to $-13.86^0/_{00}$ (Rai et al. 2016).

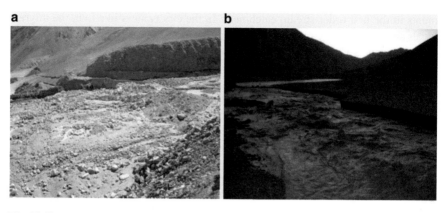

Fig. 20.10 (**a**) Washed off bridge (**b**) Flood water gushing down through the broad valley reducing its impact

Later, we identified the glacial lake responsible for this flood on the satellite imagery of 5 July 2007 (Thayyen et al. 2010).

Glacial surge dam outburst floods are another unique feature of the cold-arid region because occurrence of the surging glaciers in the Indian Himalayan region (IHR) is limited to the Karakorum Mountains of the cold- arid Himalaya. Such floods are rare and probably have high recurrence interval. Surging glaciers flow at exceptional rates during limited durations ranging from a few months to a few years resulting into sudden advances in the front (Meier and Post,1969; Hewitt,1969). Detailed recording of a glacier surge dam outburst floods in the region date back to the eighteenth century at the headwaters of Shyok River. This region have three surging glaciers, Chong Kumdan, Kichik Kumdan and Aktash. Mason (1929) presented a detailed account of the snout position of these glaciers and floods that originated from the river blockade by these glaciers. He listed number of such floods spanning through 1780, 1835,1839,1842,1873, 1903, 1926 and 1929. However, the flood of 1929 was the most devastating surge dam burst flood. Kumdan glacier surged into the Shyok riverbed and blocked the river flow. By 1928 it was grown into a huge lake having 16 km length and 120 m depth and stored around $1.5 \times 10^9 \, m^3$ of water blocked by 2.4 km ice barrier. The dam breached in 1929 (Mason 1930a, b) and entire lake water was evacuated in 48 h with an estimated peak discharge of 22,650 m^3/s (Hewitt 1982). This flood resulted in a huge devastation in the Shyok valley and Indus up to 1194 km downstream (Attock) of its origin. As glacier surge is cyclic phenomena (Murray et al. 2000; Fowler et al. 2001) the region is susceptible for similar floods in future.

20.7 Conclusions

Hydrology of the Cold-Arid region of the Himalaya is unique in many ways. Steep hydrological gradients supported by steep precipitation and temperature gradients are characteristics of the cold-arid region. More than tenfold increase in winter precipitation along the ridges than the valley bottom signifies its impact on the hydrological system. Similarly, very high Slope Environmental Lapse Rate (SELR) of temperature during the March to September also pay a critical role defining the cold-arid system hydrology. Snowmelt in the high elevation region is the major water source but year-to-year variations are also significant. Observed that the contributions from the glacier is up to 34% during the lean flow years. Study suggest that the ground ice melt could also be significant in such years. However, further studies on permafrost thaw is important for the region. Some areas of Ladakh, especially the urban centers experience groundwater mining and this resource may not be sustainable in the long run. However, self-reliance in energy generation by solar and hydro - power development will help the region to use the water from the River Indus, which is currently underutilized. Optimizing the huge power generation potential of the area is key to it sustenance. Big glaciers in the region ensure certain stability in river flow and these resources should be leveraged for the development of the region. Even though Ladakh is rainfall deprived, this region is highly susceptible for floods of various genesis. Cloudbursts floods are emerging as most common threat to the life and livelihood of the people. The threat from GLOF is also very real as there are number of glacier lakes are developed the region. Hence it is important to monitoring the status of this glacier lakes on a regular basis. Present study is based on the data generated from a single catchment in the entire region as there is no other such database linking local mountain process between valley bottom and mountain ridge is available in the region. This show urgent need for scaling up the hydrological research in the region to refine and improve the assessment presented in this paper.

Acknowledgement Sincere thanks to Dr. Sharad K. Jain, Director, NIH and Dr. Sanjay K. Jain, Head, WRS Division for their constant support and encouragement. Financial assistance for the Ladakh Glacier project from SERB, DST (No.SR/DGH/PK-1/2009) and NIH, Roorkee is duly acknowledged.

References

Alford D (1985) Mountain hydrological systems. Mt Res Dev 5(4):349–363

Ashrit R (2010) Investigating the Leh 'cloudburst'. National Centre for Medium Range Weather Forecasting, Ministry of Earth Sciences, India

Bhatt RK, Raghuvanshi MS, Kalia RK (2015) Achieving sustainable livelihood in cold arid regions of India through multienterprise options. Ann Arid Zone 54(3&4):1–12

Bhutiyani MR (1999) Mass-balance studies on Siachen Glacier in the Nubra valley, Karakoram Himalaya, India. J Glaciol 45(149):112–118

Census (2011) Census of India, 201, Registrar General & Census Commissioner, Govt. of India

Chevuturi A, Dimri AP, Thayyen RJ (2018) Climate change over Leh (Ladakh), India. Theor Appl Climatol 131(1–2):531–545

Dimri AP, Thayyen RJ, Kibler K, Stanton A, Jain SK, Tullos D, Singh VP (2016) A review of atmospheric and land surface processes with emphasis on flood generation in the southern Himalayan rivers. Sci Total Environ 556:98–115

Dimri A, Chevuturi A, Niyogi D, Thayyen RJ, Ray K, Tripathi SN, Pandey AK, Mohanty UC (2017) Cloudbursts in Indian Himalayas: a review. Earth Sci Rev

Falconer H (1841) Letter to the secretary of the Asiatic society, on the recent cataclysm of the Indus. J Asia Soc Bengal 10:615–620

Fowler AC, Murray T, Ng FSL (2001) Thermally controlled glacier surging. J Glaciol 47(159):527–538

Gardelle J, Berthier E, Arnaud Y (2012) Slight mass gain of Karakoram glaciers in the early twenty-first century. Nat Geosci 1–4. https://doi.org/10.1038/NGEO1450

Gruber S (2012) Derivation and analysis of a high-resolution estimate of global permafrost zonation. Cryosphere 6(1):221

Gruber S, Fleiner R, Guegan E, Panday P, Schmid MO, Stumm D, Wester P, Zhang Y, Zhao L (2017) Inferring permafrost and permafrost thaw in the mountains of the Hindu Kush Himalaya region. Cryosphere 11(1):81–99

Harjit Singh (2013) Tourism and climate change: their effects and impacts, Workshop Odyssey. National Institute of Hydrology, Roorkee, pp 20–22

Hewitt K (1969) Glacier surges in the Karakoram Himalaya (Central Asia). Can J Earth Sci 6(4):1009–1018

Hewitt K (1982) Natural dams and outburst floods of the Karakoram Himalaya. IAHS 138:259–269

Hewitt K (2005) The Karakoram anomaly? Glacier expansion and the "elevation effect" Karakoram Himalaya. Mt Res Dev 25:332–340

Kaser G, Großhauser M, Marzeion B (2010) Contribution potential of glaciers to water availability in different climate regimes. P Natl Acad Sci USA 107:20223–20227. https://doi.org/10.1073/pnas.1008162107

Mason K (1929) Indus floods and Shyok glaciers. Himalayan J 1:10–29

Mason K (1930a) The glaciers of the Karakoram and neighbourhood. Geological Survey of India, Bangalore

Mason K (1930b) The Shyok flood: a commentary. Himalayan J. 2:40–47

Meier MF, Post A (1969) What are glacier surges? Can J Earth Sci 6(4):807–817

Messerli B, Viviroli D, Weingartner R (2004) Mountains of the world: vulnerable water towers for the 21st century. Ambio:29–34

Murray T, Stuart GW, Miller PJ, Woodward J, Smith AM, Porter PR, Jiskoot H (2000) Glacier surge propagation by thermal evolution at the bed. J Geophys Res 33(B6):13491–13507

Namrata Priya, Thayyen RJ, Ramanathan AL, Singh VB (2016) Hydrochemistry and dissolved solute load of meltwater in a catchment of a cold-arid trans-Himalayan region of Ladakh over an entire melting period. Hydrol Res. https://doi.org/10.2166/nh.2016.156

Osmaston H (1990) Environmental determinism and economic possibilism in Ladakh. Wissenschaftsgeschichte und gegenwärtige Forschungen in Nordwest Indien. Staatliches Museum für Völkerkunde, Dresden, pp 141–150

Rai SP, Thayyen RJ, Kumar B, Purushothaman P (2016) Isotopic characteristics of cryospheric waters in parts of Western Himalayas, India. Environ Earth Sci 75(7):1–9. https://doi.org/10.1007/s12665-016-5417-8

Raina VK, Srivastava D (2008) Glacier Atlas of India. Geological Society of India, Bangalore. 315 p

Raina VK, Srivastava D, Singh RK, Sangewar CV (2008) Glacier regimen, fluctuation, hydrometry and mass transfer studies in the Himalayas. In: Raina VK, Srivastava D (eds) Glacier Atlas of India. Geological Society of India, Bangalore, pp 206–216

Rajeevan M, Bhate J, Jaswal AK (2008) Analysis of variability and trends of extreme rainfall events over India using 104 years of gridded daily rainfall data. Geophys Res Lett 35(18)

Raju BMK, Rao KV, Venkateswarlu B, Rao AVMS, Rao CR, Rao VUM, Ra BB, Kumar NR, Dhakar R, Swapna N, Latha P (2013) Revisiting climatic classification in India: a district-level analysis. Curr Sci 105(4):492–495

Sangewar CV, Shukla SP, Singh RK (2009) Inventory of the Himalayan glaciers. Geological Survey of India Publication, Bangalore

Schmidt S, Nüsser M (2012) Changes of high altitude glaciers from 1969 to 2010 in the trans-Himalayan Kang Yatze massif, Ladakh, Northwest India. Arct Antarct Alp Res 44(1):107–121

Schmidt S, Nüsser M (2017) Changes of high altitude glaciers in the trans-Himalaya of Ladakh over the past five decades (1969–2016). Geosciences 7(2):27

Siddiqui MA, Maruthi KV (2007) Detailed glaciological studies on Hamtah glacier, Lahaul and Spiti district, H. P, Geological Survey of India, 140(8):92–93

Thayyen RJ (2015) Ground ice melt in the catchment runoff in the Himalayan cold-arid system. In: International symposium on glaciology in High-Mountain Asia Kathmandu, Nepal, 2–6 March 2015. International Glaciological Society, Bangalore

Thayyen RJ, Dimri AP (2014) Factors controlling slope environmental lapse rate (SELR) of temperature in the monsoon and cold-arid glacio-hydrological regimes of the Himalaya. Cryosp Dis 8:5645–5686. https://doi.org/10.5194/tcd-8-5645-2014

Thayyen RJ, Gergan JT (2010) Role of glaciers in watershed hydrology: a preliminary study of a "Himalayan catchment". Cryosphere 4:115–128. https://doi.org/10.5194/tc-4-115-2010

Thayyen RJ, Goel MK, Kumar N (2010) Estimation of aerial and volumetric changes for selected glaciers in Jammu & Kashmir, project report. National Institute of Hydrology, WHRC, Jammu, pp 1–92

Thayyen RJ, Dimri AP, Kumar P, Agnihotri G (2013) Study of cloudburst and flash floods around Leh, India during 4–6 August 2010. Nat Hazards 65:2175–2204. https://doi.org/10.1007/s11069-012-0464-2

Thornthwaite CW, Mather JR (1955) Publications in climatology. Water Bal 8:1–104

Venkataraman S, Krishanan A (1992) Crops and weather. Indian Council of Agricultural Research, New Delhi

Viviroli D, Weingartner R (2004) The hydrological significance of mountains: from regional to global scale. Hydrol Earth Syst Sci Discuss 8(6):1017–1030

Viviroli D, Durr HD, Messerli B, Meybeck M, Weingartner R (2007) Mountains of the world, water towers for humanity: typology, mapping, and global significance. Wat Reso Res 43:W07447. https://doi.org/10.1029/2006WR005653

Chapter 21
Hydrology of the Himalayas

Nuzhat Q. Qazi, Sharad K. Jain, Renoj J. Thayyen, Pravin R. Patil, and Mritunjay K. Singh

Abstract The Himalayan Mountain chain is the third-largest deposit of ice and snow in the world, serves as an important source of freshwater for the 1.3 billion population living in the lowlands of river basins of Indus, Ganga and the Brahmaputra (IGB) covering eight countries (Afghanistan, Bangladesh, Bhutan, China, India, Myanmar, Nepal, and Pakistan). Influence of Himalayan cryosphere is very significant in headwater tributaries of these river basins and also plays a significant role in the livelihood of the people through river runoff. Understanding of the timing and relative contribution of individual components of the hydrological cycle and water resources characteristics across the Himalayas is limited and is due to inadequate investigations and lack of synthesis of existing information. This chapter presents outcome of an extensive review of available knowledge and discuses knowledge gaps in the current understanding of hydrology of IGB river basins. Many factors that are considered important in managing Himalayan water resources have been identified and discussed in this chapter.

21.1 Introduction

High mountain areas around the world have served as an important source of freshwater for the population living in the adjacent lowlands, center of biodiversity, source regions for important natural resources and ecosystem services (Bandyopadhyay et al. 1997; Barnett et al. 2005; Schickhoff et al. 2016). The Himalayan mountain is one of the largest mountain chains on Earth, serving as a source of fresh water for the major river systems (Ganges, Brahmaputra, Indus, Irrawady, Mekong, Amu Darya, Salween, Tarim, Yangtze, and Yellow rivers) of the world, extends from 15.95° to 39.31° N and 60.85° to 105.04° E, covering an area of more than 4,192,000 km^2 (Bajracharya and Shrestha, 2011). Together these rivers support the drinking water, irrigation, energy, industry and sanitation needs of 1.3

N. Q. Qazi · S. K. Jain (✉) · R. J. Thayyen · P. R. Patil · M. K. Singh
National Institute of Hydrology, Roorkee, Uttarakhand, India

© Springer Nature Switzerland AG 2020
A. P. Dimri et al. (eds.), *Himalayan Weather and Climate and their Impact on the Environment*, https://doi.org/10.1007/978-3-030-29684-1_21

billion people living in the mountains and downstream (Shrestha et al. 2015). Himalayan mountain range directly influences the economic prospect of eight countries; Afghanistan, Bangladesh, Bhutan, China, India, Myanmar, Nepal, and Pakistan. It contains the third-largest deposit of ice and snow in the world, after the Antarctica and the Arctic (Ibsen 2018) and aptly called as 'Third Pole' or 'Water Tower of Asia' (Bajracharya and Shrestha 2011).

Three major river systems, trans-boundary in nature, originate from the Himalayas. Discussion in this chapter is limited to the Indus, Ganga and Brahmaputra (IGB) river systems, which are flowing into the Indian sub-continent. The Indus river basin (IRB) is distributed across Pakistan, India, China and Afghanistan. Ganga river basin (GRB) occupies the central part of the Himalayas and outspreads in India, Tibet (China), Nepal and Bangladesh. The third, the Brahmaputra river basin (BRB) covers China, India, Bangladesh and Bhutan.

Influence of Himalayan cryosphere is very significant in headwater tributaries of all the three river basins. A major share of the Himalayan glaciers lies in the IRB covering an area of 18,195 km^2 followed by BRB, 10,593 km^2 and GRB,7939 km^2. While glaciers occupy high elevation un-inhabited areas and water runs through deep gorges, seasonal snow cover spread over a much large mountain area also plays a significant role in the livelihood of the people through river runoff. Maximum snow cover in the IRB could extend up to 2,74,686 km^2, 67,708km^2 in the GRB and 1,22,905 km^2 in the BRB. Figure 21.1 and Table 21.1 illustrate the glacier and snow cover distribution in these basins.

Our understanding of the timing and relative contribution of individual components of the hydrological cycle across the Himalayas is limited (Bookhagen 2012) and water resources characteristics in the Himalayan region have many knowledge gaps (Bajracharya and Shrestha 2011). For example, many studies conducted in the IGB only focused on the downstream parts of the basin (e.g. Hirabayashi et al. 2013; Gain et al. 2011) and did not take into account the processes that are relevant in mountainous basins (e.g. ice and snowmelt). In the upstream domains of the IGB, where mountain-hydrological processes are important, the number of studies on extremes is very limited. The study conducted by Lutz et al. (2016); is only about high flows and does not take the effects of climate change on low flows into consideration. Available estimates of the ice volume and thickness showed large difference in the IGB region (Azam et al. 2018). Advances in knowledge of Himalayan springs, geology, sediment transfer etc. is limited due to inadequate investigations and lack of synthesis of existing information in published and gray literature. Therefore, the present chapter presents outcome of an extensive review of available knowledge of hydrology of Indus, Ganges and Brahmaputra river basins and has identified gaps in the current understanding of hydrology of these basins. Many factors that are considered important in managing Himalayan water resources have been identified.

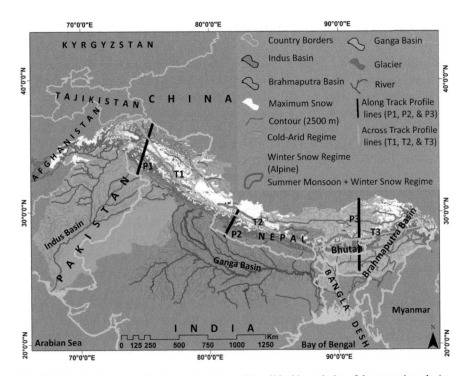

Fig. 21.1 Indus, Ganga and Brahmaputra basins with political boundaries of the countries, glacier distribution and maximum snow cover. Glacier area is based on Randolph Glacier Inventory, RGI6.0

Table 21.1 Country wise distribution of Indus, Ganga and Brahmaputra basin area, its glaciers and corresponding maximum snow cover areas

Basin	Country	Glacier Area km²	Max. Snow cover area km²	Total basin area km²	%
Indus	China	1216	46,480	83,669	08.85
	India	14,880	165,284	321,289	34.00
	Pakistan	1892	31,268	469,355	49.67
	Afghanistan	207	31,655	70,687	07.48
	Total	**18,195**	**274,687**	**945,000**	**100**
Ganga	China	2613	17,189	34,992	03.33
	Nepal	3099	33,566	147,181	14.02
	India	2227	16,953	861,452	82.04
	Bangladesh	0	0	6375	0.61
	Total	**7939.2**	**67,708**	**1,050,000**	**100**
Brahmaputra	China	8770	95,604	289,609	49.93
	Bhutan	1019	6461	46,500	8.02
	India	805	20,840	194,413	33.52
	Bangladesh	0	0	49,478	8.53
	Total	**10,594**	**122,905**	**580,000**	**100**

21.2 Hydrology of the Himalayas

Hydrology of the Himalayas is complex because of the impact of two circulation systems, the Asian Summer Monsoon (ASM) and Western Disturbances (WD) (Bookhagen and Burbank 2010). ASM declines in strength from east to west along the Himalayas, whereas WDs decline as they move from west to east (Gautam et al. 2013). To the west (Indus), there is general pattern of winter accumulation and summer melt, similar to glaciers of North America and Europe. However, part of IRB (Western Himalaya) is a transition region receiving precipitation from both the ASM and WD (Azam et al. 2016). Conversely, in east (Ganges and Brahmaputra), where glaciers experience maximum accumulation in the summer due to high monsoonal precipitation and high elevations, periods of summer time ablation punctuate overall summer-long snow accumulation (Ageta and Higuchi 1984). Further, the Himalaya is a barrier to monsoon winds, causing maximum precipitation on southern slopes with a regional east to west decrease in the monsoon intensity (Shrestha et al. 1999) and hence, large local orographic control on climate. For instance, ASM provided low precipitation (21% of the annual sum) on the leeward side of the orographic barrier at Chhota Shigri Glacier (western Himalaya) and high precipitation (51% of annual total) on the windward side at Bhuntar city (~50 km south from Chhota Shigri) (Azam et al. 2014). Depending on their geographical position and regional orography, the glaciers in the Himalaya (IGB) are subjected to different climates. This variability of precipitation regimes along the Himalaya begets varying types and behaviors of glaciers over short distances (Maussion et al. 2014).

The intensity, timing, and magnitude of the monsoon precipitation vary from east to west, with the longest duration of monsoon and greatest amounts of precipitation in the east. In the eastern part of the region, more than three-quarters of all precipitation falls during the summer monsoon months from June to September (Nepal 2012), whereas the western area receives more than one-third of total precipitation in winter (Shrestha 2008). Snow is an important component of the hydrology of the Himalaya. Snow-cover dynamics in the high Himalayas and on the Tibetan Plateau, influence the water availability and timing in downstream basins, specifically in the spring at the onset of the growing season, but also in the fall after the monsoon season (Barnett et al. 2005; Kundzewicz et al. 2007; Lemke et al. 2007; Viviroli et al. 2007; Immerzeel et al. 2009; Bookhagen and Burbank 2010). Hydrological characteristics of IGB is presented in Table 21.2.

21.2.1 Glaciers and Snow

Himalayan mountains contain important natural reserves of frozen fresh water in the form of glaciers and snow. These glaciers are unique as they are located in tropics, high altitude regions, predominantly valley type and many are covered with debris. Today there are about 30 million km^3 of ice on our planet that covers an almost 10% of the World's land area. In the Himalayas, the glaciers cover

Table 21.2 Hydrological characteristics of Indus, Ganga and Brahmaputra

Parameter	Indus	Ganges	Brahmaputra
Catchment area (km$^{2)}$	**945,000**	**1,050,000**	**580,000**
River length (km)	3200	2500	2900
Average annual rainfall (mm)	415	1125	1350
Average annual discharge (m^3/sec)	7610	11,600	19,300
Annual sediment load at river mouth (million tons/ year)	291	599	580–650
Sediment yield (million tons/km^2/year	0.3	0.61	0.85–1.12
No. of glaciers/area (km^2)	18,495/21000	7963/9000	11,497/14000
Snow coverage (annual Avg. %)	13.5	5	20
Contribution of snow melt (%)	>50	22	<25
Snow and glacier melt index	150	10	27

Source: Modified from Hasson et al. (2013), Sinha and Tandon (2014), and others

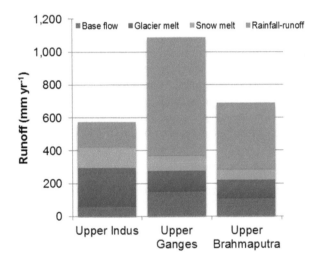

Fig. 21.2 Hydrological regime 1998–2007 in the upper IGB. (Lutz and Immerzeel 2013)

approximately 33,000 km^2 area and this is one of the largest concentrations of gla-cier-stored water outside the Polar Regions. IGB are originating from the glaciated Himalayan region where these rivers are fed by seasonal snow- and glacier- melt water. IGB is having 18,495, 7963 and 11,497 glaciers covering an area of 21,000 km^2, 9000 km^{2}and 14,000 km^2 (Bajracharya and Shrestha 2011) and 13.5%, 5% and 20% of average snow coverage (Gurung et al. 2011). Model suggested that Indus has the highest meltwater index as compared to Ganges and Brahmaputra (Table 21.2), (Immerzeel et al. 2010). The total annual glacier runoff (IGB) for the period of 1961–1990 was 41 km^3, 16 km^3 and 17 km^3, respectively. However, in the recent periods of 2001–2010, total glacier runoff was reduced to 36 km^3, 15 km^3 and 16 km^3, respectively (Savoskul and Smakhtin 2013). As is clear from the Fig. 21.2, glacier melt has highest importance in the upper Indus basin, while the contribution of glacier melt to total flow is very small in the upper Ganges and upper Brahmaputra. GRB and BRB are dominated by rainfall-runoff, followed by baseflow, however, baseflow contribution is just 10.8% of total flow in IRB. Continuous changes in surface temperatures and precipitation may have serious consequences glacier melt and rainfall-runoff relationships, e.g. drought and floods.

Role of glaciers in basin hydrology varies significantly in different mountain hydrological systems. Most of the interpretations are based in the conceptual framework of 'Alpine catchment' where most of the annual precipitation occurs as snow in the winter months. The melting of the winter snow persists through the summer season and together with the glacier melt, produces the seasonal high discharges. In such systems, glaciers are critical for the river flow, especially during the peak summer months (Fig. 21.3a). A 'Himalayan catchment' is significantly different from 'Alpine catchment' with the presence of monsoon coinciding with the temperature, melt and discharge peaks of the annual hydrograph. Hence the 'Himalayan catchment' is characterized by the peak glacier runoff contributing to the crest of the annual stream flow hydrograph from monsoon in July and August months (Thayyen and Gergan 2010), (Fig. 21.3b). Consequently, higher discharges in the headwater streams of the monsoon catchment could occur during the positive mass balance regimes of the glacier and reduced discharges associated with the negative glacier mass balance regimes. Third important glacio-hydrological regime of the Himalaya is the cold-arid systems, geographically situated in the Ladakh region of the trans-Himalaya (Fig. 21.3c). The seasonal precipitation distribution of these three dominant hydrological regimes of the Himalaya is shown in Fig. 21.4.

21.2.2 Discharge Characteristics of IGB Basins

The *discharge* of the *IGB rivers* are governed by a strong precipitation. The mean annual discharge of the Ganges at Bay of Bengal is more than 51,176 $m^3 s^{-1}$ and thus, occupies ninth position among the world's largest rivers. Whereas annual discharge of Indus and Brahmaputra are 6600 $m^3 s^{-1}$ and 21,261 $m^3 s^{-1}$, respectively. Almost 80% of the mean annual flows of Indus and Ganges are confined to the summer months (April–September) with a peak in the month of August, due to snow and glacier melt as well as the monsoonal rainfall. However, the analysis of the observed discharge of Brahmaputra shows that more than 90% of the flows are confined to the high flow period (April-early November) with a mean maximum during mid-July. During the winter (October–March), the rivers of IGB experiences low flow conditions, because contributions mainly come from the snowmelt, winter rainfall and base flow. Water availability of different tributaries of IGB basins is shown in Table 21.3.

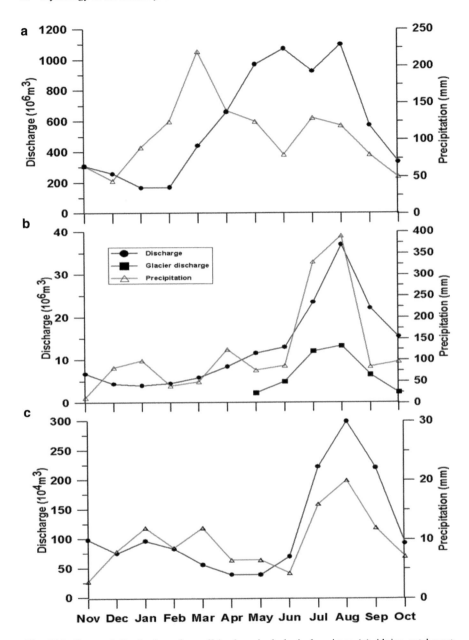

Fig. 21.3 Seasonal distribution of runoff in three hydrological regimes (**a**) Alpine catchment (peak glacier runoff contributes to other wise low flow period of annual stream hydrograph governed by lower precipitation in summer), (**b**) Himalayan catchment (characterized by the peak glacier runoff contributing to the crest of the annual stream flow hydrograph from monsoon in July and August months) and (**c**) Cold-arid catchment (annual discharge peak occurs in the month of July and August mainly from glacier/permafrost/snow melting at higher reaches during the period e.g. Ladakh region

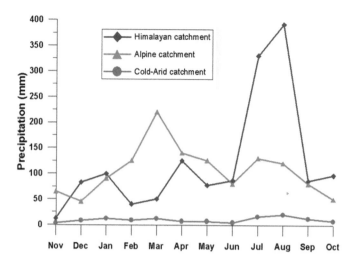

Fig. 21.4 Seasonal distribution of precipitation in the major hydrological regimes of the Himalaya. (Thayyen and Gergan 2010)

21.2.3 Sediment Transfer Characteristics of IGB Basins

Rivers and streams in the Himalayas carry sediment in their flows and transports from one part of the system to other (Lupker et al., 2012). Table 21.3 presents quantitative estimation of erosional budget/sediment yield for individual tributary of IGB and are useful in understanding the dynamics of movement, storage, or removal of water and sediment in systems. These estimates provide a framework for scientific analysis and a basis for policy decision and management. In case of a vast and dynamic alluvial river like the Brahmaputra, which exhibits a high variability in flow and sediment yield, changing boundary conditions of the channel, complex river morphology, and recurrent seismic instability in the basin region, the uncertainties involved in the estimation are significantly high. Yet, the studies compiled (Table 21.3) indicate useful results and prospects for further development.

Sediment in the Indus system is preferentially eroded from the western Tibetan Plateau and Karakoram (Clift et al. 2000, 2001). During summer, when snow melts, water discharge increases by 20–50 times and sediment load by 500–1000 times (Ferguson 1984). Eastern portion of the Himalayan Range is eroding faster than the western portion, which contributes to the Brahmaputra having a higher suspended load than the Ganges (Galy and France-Lanord 2001). The higher erosion in the eastern region is likely caused by higher precipitation in the eastern region (Fluteau et al. 1999; Galy and France-Lanord 2001). The sediment budget mentioned in Table 21.3 is mainly based on sediment yields at selected reaches on the main stem and at confluence points of tributaries. The Table 21.3 indicate that the Indus carries the highest amount of sediment per unit drainage area followed by Brahmaputra.

Table 21.3 Characteristics of annual average discharge and sediment yield of different tributaries of Indus, Ganga and Brahmaputra basin

Basin	Rivers/Station	Basin Area (km²)	Avg. Annual discharge (m³/s)	Sediment Yield (tons/km²/year)
INDUS	Shyok (Yugo)	33,670	347.1[a]	754
	Shigar (Shigar)	6610	NA	2547
	Hunza (Dainyor)	13,157	338.6[a]	3375
	Gilgit (Gilgit)	12,095	281.9[a]	1008
	Gilgit (Alam Br.)	26,159	644[a]	2108
	Astore (Doyian)	4040	136.8[a]	401
	Gorband (Karora)	635	NA	1406
	Indus (Kharmong)	67,858	489.1[a]	496
	Indus (Kachura)	112,665	1069.1[a]	705
	Indus (Parlab Br.)	142,709	1775.8[a]	973
	Indus (Shatial Br.)	150,220	NA	709
	Indus (Besham Q.)	162,400	2412.2[a]	1345
	Kabul (Kabul City)	NA	247	4700
	Jehlum (Chinari)	13,775	330[b]	NA
	Jehlum (Kohala)	25,000	828[b]	NA
	Chenab (Akhnoor)	NA	8001	NA
	Ravi (Muksar)	NA	268	NA
	Beas (Mandi plain)	NA	499	NA
	Satluj (Ropar)	NA	500	NA
GANGES	Ganga (Garhmukteshwar)	29,709	660	766
	Ganga (Fatehgarh)	40,096	576	444
	Ganga (Ankinghat)	82,209	1015	366
	Ganga (Kanpur)	87,650	895	376
	Ramganga (Dabri station)	23,919	274	213
	Garra (Husepur)	6155	82	1262
	Ganga (Farraka barrage)	907,000	16,648	1235
	Gomti (Saidpur)	NA	234	NA
	Ghaghara (Revelganj)	NA	2990	NA
	Gandak (Sonepur)	NA	1760	NA
	Koshi (Kursela)	NA	2166	NA
	Tons (Dehradun)	NA	2827	NA
	Sone (Indrapuri)	NA	33,045	NA

(continued)

Table 21.3 (continued)

Basin	Rivers/Station	Basin Area (km²)	Avg. Annual discharge (m³/s)	Sediment Yield (tons/km²/year)
Brahmaputra	Subansiri	27,400	755,771	959
	Ranganadi	2941	74,309	1598
	Burai	791	20,800	5251
	Bargang	550	16,000	1749
	Jia Bharali	11,300	349,487	4721
	Gabharu	577	8450	520
	Belsiri	51	9300	477
	Dhansiri (north)	10,240	26,577	379
	Noa Nadi	907	4450	166
	Nanoi	860	10,281	228
	Bamadi	739	5756	323
	Puthimuri	1787	26,324	2887
	Pagladiya	383	15,201	1887
	Manas-Aie-Beki	36,300	307,947	1581
	Chamramati	1038	32,548	386
	Gaurang	1379	22,263	506
	Tipkai	1.364	61,786	598
	Gadadhar	610	7000	272
	Burhi Dehang	4923	1,411,539	1129
	Disang	3950	55,101	622
	Dikhow	3610	41,892	252
	Jhanzi	1130	8797	366
	Bhogdoi	920	6072	639
	Dhansiri (south)	10,240	68,746	379
	Kopili	13,556	90,046	230
	Kulsi	400	11,643	135
	Krishnai	1615	22,452	13 I
	Jinari	594	7783	96

Note: [a]daily mean, [b]monthly mean

The conjoined Ganges-Brahmaputra River carries 80% of the sum of the loads and remaining 20% of sediment is diverted from the main river by the tributaries and deposited along the main river channel (Rice 2007). Sediment that reaches the Bay of Bengal is dominated by silt and clay, with 15–20% of the total discharge being fine to very fine sand (Thorne et al. 1993). An additional study shows that more than 76% of the bed sediments are within the fine to very fine sand class, and have a mean grain size between 177 μm and 62.5 μm (Datta and Subramanian 1997). The suspended load of a river carries the majority of the sediment, while bedload transport accounts for approximately 10% of the suspended load (Walling 1987).

21.3 Topography

The Himalayas are not a single continuous chain of mountains, but a series of several more or less parallel, or converging ranges, intersected by enormous valleys and extensive plateaus. Their width is between 250 km and 300 km (Sorkhabi 2010) and these comprise of many minor ranges. The individual ranges generally present a steep slope towards the plains of India and a more gently inclined slope towards Tibet. The Eastern Himalayas of Nepal and Sikkim rise very abruptly from the plains of Bengal and Uttar Pradesh and suddenly attain their great elevation above the snow-line within strikingly short distances from the foot of the mountains. But in the Western Himalayas, the rise in elevation is gradual across many cascading mountain ranges of lesser altitudes.

The topography of the Himalayas in the upper IGB is characterized by sudden rise of Shivalik from the Ganga plains which further rise up to Lesser Himalayas and connect to the Great Himalayan range with a steep slope. Some of the highest peaks of Earth such as the Mount Everest (8848 m), K2 (8611 m), Kanchenjunga (8586 m), Dhavalagiri (8167 m), Nanga Parbat (8126 m), Gasherbum (8035 m), Gosainthan (8013 m), Nanda Devi (7816 m), etc. are situated in the Great Himalayan belt with average elevation extending to about 7000 m. The average elevation in the Lesser Himalayas lies between 3500 m and 4600 m whereas lower foothills of Himalayas seldom exceeds the elevation range 900–1200 m (Fig. 21.5).

The varying elevations and slope profiles of each basins play significant role in determining the regional climate and hydrological responses as it critically influences both air and water circulations. As climate and hydrology of the Himalayas is dictated by the orographic nuances of the Asian monsoon in summer and westerly disturbances in winter months, precipitation dominant and shadow zones lay interspersed in the Himalayas. Sudden rise in elevation facilitates lifting of the moist air along the frontal band of Lesser and Great Himalayas resulting in excessive orographic precipitation. As a result, two discrete bands of monsoon precipitation exist which stretches along the length of Great Himalayan Range. The first orographic barrier is created through southward thrusting of Lesser Himalayas over the northernmost proximal edge of the Himalayan foreland basin resulting in a zone of high rainfall at a mean elevation of 0.9 ± 0.4 km and a relief of 1.2 ± 0.2 km (Bookhagen and Burbank 2006). A steep slope front along the boundary of the lesser and Greater Himalayas controls the location of another inner rainfall band at an average elevation of 2.1 ± 0.3 km and relief of 2.1 km.

There are numerous rain shadow zones in the Himalayas and the complex topography of that region create wet and dry regions side by side. Several studies (e.g., Bookhagen and Burbank 2006; Bookhagen and Strecker 2008; Shrestha et al. 2012) conducted on the distribution of rainfall with elevation in various regions of the Himalayas indicate a strong relationship between rainfall and elevation. Some studies (Dhar and Rakhecha 1981) show that no linear relationship exists between rainfall and altitude. However, other studies (e.g., Singh et al. 1995; Singh and Kumar 1997) have reported that there may be a continuous increase in precipitation with

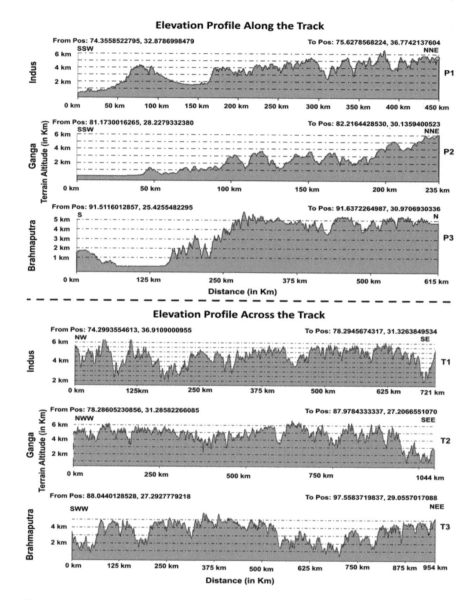

Fig. 21.5 Cross sectional slope progression of the Himalayas from the southern front of Indus, Ganga and Brahmaputra basins. (Slope progressions were extracted using hydrologically conditioned 3 arc-second SRTM DEM freely downloadable from HydroSheds (Hydrological data and maps based on SHuttle Elevation Derivatives at multiple Scales) website (https://hydrosheds. cr.usgs.gov/dataavail.php)

altitude in the Himalayan region but precipitation begins to decrease above a certain altitude. Clearly, this relationship varies considerably with time and place.

21.4 Geology of IGB Basins

Active tectonics in any basin is an important forcing factor affecting the river channel dynamics, its sediment load, and morphology. Himalayas have common geological history but sharply varying geology across the elevations and latitudes. The geology of the Indus drainage is largely shaped by the collision between the Indian Plate with mainland Asia, starting at around 50 million years ago (Inam et al. 2007). Continued tectonic activity and erosion from the valleys has allowed the surrounding ranges to be uplifted to great heights. The drainage of the Indus is dominated by the Western Tibetan Plateau, Karakoram and tectonic units of the Indus Suture Zone. The Indus basin has variety of rock formations, which are continuously undergoing disintegration through glacio-fluvial action. Main central thrust separates lesser Himalayan rocks from higher Himalayas in north and main boundary thrust separates it from Siwalik in the south. Fracture zones, which are developed due to thrust, faults and other lineaments, are good locations for springs and promote groundwater recharge as well as control the drainage pattern. The Ganga basin in the Himalayan foreland forms one of the largest plains of the world over with ~450 million people dwelling in it. The plain evolved by deposition of huge piles of sediments over the basement within the down-warp between Himalayan Orogenic Belt in the north and Precambrian rocks of the Indian craton in the south (Shukla and Raju 2008; Sinha et al. 2009). The basement is marked with linear furrows, ridges and faults, lying obliquely to the Himalayan structural grains (Thakur et al. 2009; Jayangondaperumal et al. 2010; Goswami 2012). These oblique trending basement structures divide the basin into several tectonic blocks (Hazarika et al. 2010; Dasgupta et al. 2013). In many cases, the blocks act as half-grabens with differential uplift and bending. BRB shows significant difference in fluvial dynamics. The vast tract of Brahmaputra valley developed between the Arunachal Himalaya, the Naga-Patkai range of hills and the Shillong plateau is a typical fluvial terrain with a mosaic of numerous minor fluvial landform features. The valley has developed through alleviation over a sequence of Cenozoic sediments which again overlie a basement that has developed a structural high nearly coinciding the flow path of the Brahmaputra. The BRB consists mainly of Holocene floodplain deposits, underlain by poorly consolidated or unconsolidated tertiary and quaternary sediments.

21.5 Major Landuse in the IGB Basins

The forest cover in the IGB Basins are only 4%, 8% and 15% whereas major part of basins are covered with snow and glacier (75%, 53% and 74%), respectively. Moreover, second major part of the IRB and GRB are agriculture, however, forest cover is the second landuse in BRB (Table 21.4). Forest cover is declining in all three basins; overgrazing and deforestation to make room for cultivation and human settlements are common in Indus and Ganges Basin, however, shifting cultivation, is widely being practiced in the hills and foot hill regions of Brahmaputra basin. Nowadays, some part of land in Indus and Ganges are used for other economic uses which consists of mineral exploitation or construction of human settlements, industrial structures, roads, railways, airports and other civil works. On the other hand, grasslands are today broken up by vast tracts of tea plantations and human habitation in Brahmaputra basin.

Quantifying effect of land-use change on water resources is a challenge in hydrological science. The land use/cover changes from forest to other uses have been widespread in the past several decades in the Himalayan region (Batar et al. 2017). Such changes in land use/cover lead to environmental degradation through soil erosion, change in discharge, evapotranspiration etc. Agricultural land area has increased considerably over the past four decades in the Himalaya at the costs of other land uses, particularly forests (Sharma et al. 2007). The forest-dominated watersheds are consequently converted into agrarian watersheds where discharge, sediment and nutrient losses are accelerated. The study conducted by Younis and Ahmad (2017), concluded that as built-up area increase up to 40% from the year 2000 to 2010, which also increases the discharge to 33%, and confirms that LULC change affects discharge values of watersheds of Indus basin. Similarly, Chawla and Mujumdar (2015), analyzed the change in landuse land cover in the Ganges basin and observed that from 1973 to 2011, there was an increase in crop land and urban area by 47% and 122% and decline in dense forest from 14% in 1973 to 11% in 2000), a slight increase in dense forest (11.44–14.8%) between 2000 and 2011, respectively. The results showed that the change in landuse significantly increased

Table 21.4 Landuse landcover distribution of IGB

Landuse	Indus		Ganga		Brahmaputra	
	km^2	%	km^2	%	km^2	%
Built up land	6701	0.6	36,908	2.1	3163	0.4
Agricultural	115,005	10.1	564,866	31.4	50,375	6.9
Forest	48,481	4.2	137,817	7.7	107,854	14.7
Grassland	13,473	1.2	7324	0.4	8531	1.2
Wasteland	91,651	8.0	76,604	4.3	10,117	1.4
Waterbodies	5935	0.5	29,877	1.7	11,266	1.5
Snow/glacier	860,160	75.4	944,978	52.5	543,725	74.0
Total	1,141,406	100.0	1,798,374	100.0	735,031	100.0

Source: CWC and NRSC (2014)

peak discharge and evapotranspiration by 77% and 42% from 2011 to 1973. Likewise, Tsarouchi et al. (2014), also found effect of land cover change in change in hydrology of Ganges basin from 1984 to 2010. On the other hand, the Brahmaputra basin (especially upper Brahmaputra basin) is one of the worst flood-affected areas in India. River dynamics of Brahmaputra basin changes landuse land cover (LULC) significantly especially agriculture due to flooding, leads to changes in the sediment flux and other hydrology of the basin (Hazarika et al. 2015). Therefore, number of studies on forested watersheds provided situations from macro to micro scales of IGB. Sustenance of watershed functioning in the IGB needs substantial support from various approaches and promotion of forests and agroforestry practices in combination with rehabilitation of degraded lands and better land husbandry could provide hydrological benefits for both upstream and downstream users.

21.6 General Climate of the IGB Basins

The climate of IGB ranges from tropical at the base of the mountains to permanent ice and snow at the highest elevations. The average annual precipitation in the IGB basins are 415 mm year^{-1}, 1125 mm year^{-1} and 1350 mm year^{-1}, respectively (CRU 2012; Hasson et al. 2013). Variation in rainfall across the basins are considerable, particularly between the upper and lower basins (Turner and Slingo 2009; Bera 2017). The upper IRB receives nearly 500 mm year^{-1}, whereas the lower basin receives just under 300 mm year^{-1}. Although there is not much difference between the annual amount of precipitation in the lower and upper parts of the GRB. In contrary to IRB, the lower BRB receives approximately 2216 mm year^{-1} of rain annually, which is over three times more than the upper basin. Overall, the upper IRB and GRB basin experience more rainy days (132 and 179) in a year, compared to their lower basins (84 and 152 days). However, upper Brahmaputra basin experiences less number of rainy days (164) in a year as compared to lower basin (214 days). There are two sources of precipitation in the IGB; monsoon (July–September) and westerly (December–March), (Bookhagen and Burbank 2006). The hydrological regimes of the IGB is dominated by the monsoon system and its contribution is about 55% (I), 84% (G), and 70% (B) (Kripalani et al. 2007; Sabade et al. 2011; Hasson et al. 2013; CWC-NRSC 2014) and westerly contributes about 45% (I), 16% (G) and 30% (B), respectively. Deka et al. (2013), Bera (2017), Latif et al. (2018), observed large spatial and temporal variability in the annual and seasonal rainfall trends in IGB. The increase in rainfall intensity, changes in rainfall patterns, and greater frequency of extreme rainfalls (Turner and Slingo 2009) likely to aggravate flooding problems within IGB whose early signs are visible now.

Temperature plays important role in IGB in several ways. Over the past decades and across the IRB, winters are getting warmer, but summers are getting cooler. However, extreme hot days are getting hotter and extreme cold days are getting milder. On the other hand, in GRB, winters are getting warmer, but summer average temperatures have remained constant. Summer extremes are becoming more intense, while winter extremes are showing mixed trends across the basin. Nevertheless, in BRB, temperatures are changing over time and showing mixed trends across the seasons and in different areas of the basin. The average maximum temperatures of IGB are about 30 °C, 30.3 °C and 19.6 ° C in summer and 13 °C, 21.1 °C and 9.2 °C in winter. Average minimum temperatures range from 18 °C, 21.5 °C and 18.3 °C in summer to −0.3 °C, 6.4 °C and − 0.3 °C in winter. Over last decades, average maximum temperatures of Indus basin have slightly decreased (0.5 °C), while minimum temperatures have increased (1.2 °C) in the winter. On the other hand, there has been no significant trend in terms of changes in maximum temperatures, but there has been rise of 0.7 °C and 0.5 °C in average minimum winter temperature across the GRB and BRB.

21.7   Precipitation (P), Evaporation (E) and Potential Evapotranspiration (PET) over the Basins

The mean annual cycle of the P, E, and PET over the IRB, GRB, and BRB is presented in Fig. 21.6. It can be observed that monthly P values from ERAI tend to be slightly greater than those computed from CRU, but the annual cycle is the same. These differences are best appreciated in the annual cycle of P over the BRB. In the IRB, the P annual cycle is characterized by two maximum peaks in February–March and July–August (Fig. 21.6a). The E approximately follows this cycle but with lower values. In IRB basin, the PET remains higher than the P and E across the year; in fact, Cheema (2012), argue that the major part of this basin is dry and located in arid to semiarid climatic zones. Laghari et al. (2012), also found for the climatology from 1950 to 2000 that PET exceeds P at the IRB across the year. PET is enhanced after maximum precipitation; maximum values occur in May–June. Over the GRB maximum P occurs between May and October and is greater than over the IRB (Fig. 21.6b, c). The PET and E annual cycles over this basin differ, and as expected, PET > E. The PET annual cycle is mainly like for the IRB. Indeed, both variables reflect close but different information. The E annual cycle agrees with that obtained by Hasson et al. (2014), for the three river basins. Over the BRB, the monthly average precipitation both from CRU (Climatic Research Unit) and ERA-I increases abruptly from March until a maximum (> 11.0 mm day^{-1}) in July and later falls until a minimum is reached in December (Fig. 21.6c). The PET and E are very close and do not surpass 4 mm day^{-1} in the annual climatology. In particular, the PET annual cycle is notable for being lower than what was obtained for the IRB and GRB. The annual cycles of P (from CRU and ERA-I) and E for the IRB, GRB, and BRB follow the same annual cycle as those obtained by Hasson et al. (2014). These authors analysed the seasonality of the hydrological cycle over the same basins for the twentieth century climate (1961–2000 period), utilizing PCMDI/CMIP3 general circulation models (GCMs) and observed precipitation data.

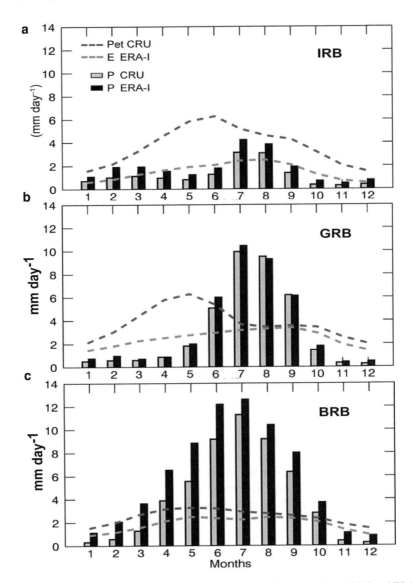

Fig. 21.6 The 1981–2015 annual cycle of precipitation (gray, black bars from CRU and ERA-I, mm day−1) and potential evapotranspiration (blue line from CRU, mm day−1) and evaporation (green line from ERA-I, mm day−1) over the Indus (**a**), Ganges (**b**), and Brahmaputra (**c**) river basins from CRU 3.24.01. (Sori et al. 2017)

21.8 Climate Change Impact on IGB Basins

Impacts of Climate change (CC) on hydrological regimes vary from basin to basin. CC have potential impact on hydrological processes including precipitation, evapo-transpiration, overland flow, streamflow (volume, timing, frequency and magnitude), infiltration, groundwater flow, soil erosion and transport, water temperature, snow and glacier change. The results of such hydrological changes would affect almost every aspect of life i.e., agricultural productivity, water supply for urban and industrial use, power generation, wildlife and biotic ecosystems, sedimentation, plant growth and nutrient flow into water bodies (Zhang et al. 2007). Climate change is also expected to have significant consequences on snowmelt and glacier runoff across the Himalayan region. Our knowledge of high-altitude snow/ice and its response to climate is still incomplete (Azam et al. 2018; Brun et al. 2017). Understanding the present hydrological, climatological and glaciological processes of high elevation catchments is thus vital. This requires better insights into the present composition of runoff and interactions between climate, glaciers, snow and soil. Climate change impacts of IGB basins is divided into following two sections:

21.8.1 Observed Impacts

The cryosphere of the Karakoram and the Greater Himalayas, source of the headwaters of all major rivers of IGB, is highly susceptible to climate change. Degradation of perennial snow covers by thinning, retreat, and negative mass balance of glaciers or ice losses have been widely observed in the Himalayas (Azam et al. 2016, 2018; Brun et al. 2017). Compared to such prevalent losses of the cryosphere in the Himalayas, several reports suggest glacial stability or even positive glacier mass balance in the Karakoram (Hewitt 2005; Gardelle et al. 2012, 2013; Zhou et al. 2017; Bolch et al. 2017). However, glacial growth and stability is not ubiquitous throughout Upper Indus sub-basin (Kaab et al. 2012). Glacierized watersheds of upper IRB show that in the central and eastern Karakoram glacier, mass balance is negative whereas in the western Karakoram it is positive (Mukhopadhyay and Khan 2014a, b, 2015). Such findings have made the prediction and assessment of impacts of climate change on the future of the Upper IRB and associated flows downstream very uncertain.

Climate change is also altering ecosystems of the GRB largely through changes in water quantity, water quality as well as changes in biodiversity (Hosterman et al. 2009). Reduced flows to the Sundarbans wetland ecosystem have resulted in salinity intrusion in the south-western part of Bangladesh, loss of biodiversity, and loss of ecosystem functionality (Islam 2016). Due to increase in temperatures of GRB, resulting in retreat of glaciers, increase variability in precipitation, increased magnitude and frequency of droughts and floods; and leads to sea level rise (Hosterman et al. 2009). Droughts, floods and other extreme events results in scarcity of water and food which further leads to displacement of populations, loss of livelihoods, communicable disease and malnutrition (Menne and Bertollini 2000). Increased runoff from glacier retreat and ice/snow melt could increase annual discharge into the GRB in the short term, followed by the reduction of runoff in long term (Barnett et al. 2005). Changes in precipitation pattern, glacial retreat and increased sedimentation may adversely affecting dam storage capacity and hydropower generation in GRB and its tributaries (Hosterman et al. 2009). Climate change also has significant impact on fresh water storage (glacial snow covered regions) at high elevations and fresh water runoff to low elevations (Jianchu et al. 2007).

Numerous studies also have assessed climate change impacts on hydrological processes in the BRB, e.g. temperature (Immerzeel 2008; Shi et al. 2011), precipitation (Kripalani et al. 2007), snow (Shi et al. 2011), streamflow (Gain et al. 2011; Jian et al. 2009), groundwater (Tiwari et al. 2009), runoff (Ghosh and Dutta 2012; Mirza 2002), extreme events (Rajeevan et al. 2008; Webster and Jian 2011), and even water quality (Huang et al. 2011). However, a few studies have assessed how projected changes in climate and land use and land cover could impact long-term patterns in the basin's hydrological components. Some studies e.g. Flugel et al. (2008) and Immerzeel (2008), suggests a consistent rise in average and seasonal temperatures over the last 50 years, and the projections indicate that the temperature will continue to rise, although the magnitudes of the projected changes differ depending on the driving models.

21.8.2 Future Implications/Changes

The potential impact of climate change will be more evident in future in the Himalayan region, where the runoff is dominated, largely, by glacier melt and snow-melt (Viviroli et al. 2007; Immerzeel et al. 2013; Lutz et al. 2014b). Climate change and rising temperature will increase annual runoff by 7–12% by 2050 due to accelerated melt in the upper IRB together with an increase in precipitation (Lutz et al. 2014a). However, the projected future hydrology depends on the precipitation projections which have a large uncertainty and large variation between annually averaged and seasonal projections among the General Circulation Models (GCMs). IRB may lose up to 8.4% of its total water resources by 2050 and Pritchard (2017), showed that glacier melt is a critical buffer at the time of drought in the whole Asia including all Himalayan basins. Nepal et al. (2014), estimated that the contribution of snow melt to river flow in the Dudh Koshi catchment would decrease by 31% with a 2 °C rise in temperature, and by 60% with a 4 °C rise, changing the river from 'snow-dominated' to 'rain-dominated'. Wiltshire (2014), suggested that under a warming climate, the volume of glaciers in the eastern Himalayas (Nepal and Bhutan) will decline over the twenty-first century, despite increasing precipitation, as a result of less precipitation falling as snow as well as increased ablation. Application of the water balance model in the Tamor catchment suggested an annual decrease in runoff up to 8% for a 5 °C temperature increase (Sharma et al. 2000), however, the study did not take glacier melt into account. All these studies suggest that a rise in temperature will affect the snow/glacier melt pattern and annual runoff. The predicted decrease in the water flows of the IRB will have serious consequences for India and Pakistan which receive 63% and 36% of its water. Per capita availability of water in the IRB has suffered a 70% decline during the past few decades.

Demographic trends in both India and Pakistan are fraught with ominous consequences for the water sector with serious impacts on their food and energy security.
In India, per capita water availability stood at 1539 m^3 in 2011. Demographic
changes during the future will further aggravate water scarcity.

A study by Immerzeel et al. (2010), projected a decrease of 17.6% in mean
upstream water supply in the GRB, with the reduction in melt runoff partly compensated for by increased upstream rainfall (+8%). Immerzeel et al. (2013), suggested
that under projected climate change, the glacier area in the Langtang catchment will
be reduced by 54% by the end of the century and the ice volume by 60%. Initially,
net glacier melt runoff will increase, with a peak in 2045 and 2048 for RCP 4.5 and
RCP 8.5, respectively, after which it will decrease. However, water availability is
not likely to decline during this century as the reduction in runoff will be offset by
an increase in precipitation. Lutz et al. (2014b), found that the runoff (Koshi River
basin, eastern Nepal) is likely to increase up to 2050, primarily due to an increase in
precipitation in upstream areas, with the maximum increase during the pre-monsoon
period, but the hydrograph remains unchanged. These studies indicate that the
future reduction due to melt runoff of GRB will be offset by increased
precipitation.

Many tributaries of BRB are flashy in nature, which increase the chance of significant damages due to extreme events. During the last 10 years, the river has seen
some of the most destructive floods in its history. During 2009 and 2012 flooding,
thousands of people died and 2.2 million people were forced to evacuate their homes
as monsoon rains inundated large areas. Ghosh and Dutta (2012), revealed that
although the number of flood events would decrease in future (2010–2100), the
peak discharge and duration of the floods would increase. Gain et al. (2011), predicted a very strong increase in annual peak flow for the river, which may have
severe impact on flooding. Immerzeel et al. (2010), estimated that the discharge
generated by snow and glacier melt is 27% of the total discharge naturally generated
in the downstream areas of the BRB. The study by Immerzeel et al. (2010), for
2046–2065 projected a decrease of 19.6% in mean upstream water supply, with the
reduction in melt runoff partly compensated by increased upstream rainfall (+25%).
Prasch, et al. (2011), suggested that glacier ice melt will accelerate from 2011 to
2040 due to the increase in air temperature and longer melting periods and that as
the amount of glacier ice is reduced, ice melt will decrease. Lutz et al. (2014b), in
their investigation projected an increase in total runoff in the upper BRB up to 2050,
primarily due to an increase in precipitation and accelerated melt runoff, with the
increase occurring throughout the year. Mirza (2002), projected a substantial
increase in mean peak discharge in the BRB (although less than in the Ganges),
based on climate change scenarios from four GCMs, which could lead to more frequent flooding of different magnitudes. Gain et al. (2011), indicated that there will
be a strong increase in peak flows, both in size and frequency, although dry-season
conditions are likely to increase. Thus, number of authors have looked at intra-
annual variation and the impact of climate change on flooding and other negative
impacts on basin. The results from various studies and their quantitative analysis

(hydrological impact on climate change) are helpful for better understanding of potential hydrological risks for future water management planning.

21.9 Challenges in Water Resources Management in IGB Basins

21.9.1 Methodological Challenges

There are number of methods to analyze the runoff components from rainfall, melting of snow and glaciers including water balance analysis (Thayyen et al. 2005; Kumar et al. 2007), glacier degradation from observation or modelling as a contribution to runoff (Kotliakov 1996; Kaser et al. 2010), isotopic investigations (Dahlke et al. 2013) and hydrological modelling (Hagg et al. 2007; Naz et al. 2013). However, the water balance method can only estimate the effects of glacier and snow at monthly or larger time scales; additionally this method cannot separate snowmelt and ice melting on glaciers. Isotopic investigation cannot be widely used as it demands large financial and laboratory support. However, the application of hydrological models to understand the glacier effects in hydrology is relatively new (Hagg et al. 2007; Huss et al. 2008; Koboltschnig et al. 2008; Prasch 2010; Nepal et al. 2013). A practical challenge is the lack of long-term good quality data to represent the hydrological dynamics of Himalayan rivers. Moreover, the rainfall-runoff models rarely describe the snow and ice melting (Singh and Singh 2001) and glacier dynamics (Naz et al. 2013) at basin scale. Physically based models, are likely to produce more realistic results because they depend less on parameter calibration and their parameters have a physical basis. All models, including physically based ones, have parameters that need to be estimated or identified through calibration (Foglia et al. 2009). Appropriate calibration is a key issue in modern hydrological science, and much attention has been recently devoted to it.

In the Himalayas, fieldwork is difficult due to the tough terrain, remoteness of glaciers as well as logistical, financial and political obstacles. For this reason, in recent years the focus has been on remote sensing approaches used to reconstruct snow cover, frontal and areal changes of glaciers and ice volumetric changes (Gardelle et al. 2013; Kaab et al. 2012; Shangguan et al. 2014). However, in the light of possible changes in the snow-glacier-energy balance due to climatic changes, there is a strong call for more in-situ measurements across the Himalayas and models that integrate those data in space and time (Azam et al. 2018; Cogley 2012; Reid and Brock 2010). Local processes and effects that are difficult to study using remotely sensed data could explain regional differences and temporal changes in glacier mass balance across the region, such as the glacier expansion in the central Karakorum known as the 'Karakorum anomaly' (Hewitt 2005). Glacio-hydrological models are indispensable tools to study these effects and to understand the characteristics of a catchment and its response to climate change. Their applicability in high elevation regions is restricted due to: (i) the lack of representative data to force the models (Huss et al. 2014; Pellicciotti et al. 2014), (ii) simplifications in model structure due to insufficient process understanding and the scarcity of detailed information about glacio-hydrological processes (Huss et al. 2014) and (iii) parametric uncertainty due to insufficient quality or paucity of data for model calibration and validation (Ragettli et al. 2013).

21.9.2   Disaster Challenges

Due to its physical setting, the IGB region is prone to various water-induced hazards (landslides, floods, glacial lake outburst floods, and droughts). Every year, during the monsoon season, floods wreak havoc on the mountains and the plains downstream. These floods are often trans-boundary. Globally, 10% of all floods are transboundary, and they cause over 30% of all flood casualties and account for close to 60% of all those displaced by floods. The social and economic setting of the region makes its people more vulnerable to natural hazards. Lack of supportive policy and governance mechanisms at the local, national and regional levels, and the lack of carefully planned structural and non-structural measures of mitigation lead to increased vulnerability.

21.9.3   Energy Challenges

Energy is one of the most important pillars of sustainable development and hydropower is one of the most promising environmentally friendly sources of energy of the IGB. The varying estimated potential Hydro-electric Power (HEP) figures for India is 45,635 MW, Nepal is 83,000 MW and for Bhutan is 21,000 MW. Moreover, innovative solutions such as electric transportation and a clean source of domestic and industrial energy supply would significantly improve the deteriorating environmental condition of the region. However, many countries in the region have been able to tap only a small fraction of their available potential. Still, people in these countries face many hours of scheduled power cuts. Major causes are lack of cooperation among nations, politicization of the water resource development and management aspects, prolonged negotiations, disagreements on the location of dams, reservoir safety, resettlement and rehabilitation issues, environmental concerns, cost and benefits sharing, etc.

21.9.4 Water Quality Challenges

Water quality has witnessed progressive deterioration due to growing urbanization and industrialization. The increased use of agrochemicals, discharge of untreated domestic sewage and poor sanitation facilities have aggravated the problem of water pollution. The optimum utilization of the water resource, effective management to meet the multi-sectoral uses, enhancing the efficiency of water utilization, technological modernization, checking pollution and inter countries cooperation are the major challenges for maintaining water quality in the IGB basins.

21.9.5 Environmental Challenges

Water plays a vital role in maintaining different ecosystem services in riparian areas. Freshwater ecosystems in particular largely depend on the specific flow regime of rivers passing through them. However, due to intervention of infrastructure development, the flow regime changes in the downstream areas, where, in many cases, communities depend on water resources for livelihoods such as fishing. A major concern is how to make sure that a certain minimum flow is maintained so as to sustain freshwater supply and support dependent ecosystems. There is very weak monitoring of the minimum flow requirement in the region.

21.9.6 Food Challenges

Water and food share a strong nexus, both being essential ingredients for human survival and development. Agriculture is a major contributor to many countries of IGB. The Indus river system is a source of irrigation for about 144,900 hectares of land, whereas the Ganges basin provides irrigation for 156,300 hectares of agricultural land. Access to water resources for food production and their sustainable management is a concern from the local to national level. Amid rapid environmental and socio-economic changes, the growing population will require more water and food, and equitable access to vital resources has become a major question. Sustainable solutions to these problems require efficient use of water resources for agricultural use in which technological innovation plays a vital role.

21.9.7   Other Challenges

Construction of new projects is becoming progressively more tedious due to concerns about submergence of forests, siltation of reservoirs, fragmentation of rivers, loss of biodiversity, displacement of population etc.

21.10   Floods and Droughts

Floods and drought have been observed to cause a gradual increase in human suffering and damage to property as well as increased loss of life and economic costs (Doocy et al. 2013). The frequent occurrence of droughts and floods in the IGB regions also has had huge impacts on regional food and energy security (Webster et al. 2011). During the summer monsoon months (June–September), the IGB experiences severe floods; and frequency and magnitude of floods generated in one country affect another country, and erosion in one country can deposit sediment in another. Therefore, floods are not only linked to single country/stateproblem but is also linked to trans-boundary internationally. Future hydrological extremes, such as floods and droughts, may pose serious threats for the livelihoods in the upstream domains of the IGB (Mirza 2011; Lutz et al. 2016). A recent study, assessing the impacts of climate change on hydrological regimes and extremes in the Upper IRB, showed that, in general, summer peak flow will likely shift to other seasons, and projected an increase in the frequency and intensity of extreme discharge conditions (Lutz et al. 2016). Another study projected increases in heavy precipitation indices during monsoon period, accompanied with extended periods of no precipitation during the winter months, in the GRB (Mittal 2014). Hence, the cited study (Mittal 2014) indicated an increase in the incidence of extreme weather events over the first half of the twenty-first century. Studies performed on global flood risk show similar patterns (Hirabayashi et al. 2008, 2013; Pechlivanidis et al. 2016). Significant increasing trends in high flows (i.e. 10 percentile exceedance discharge) were found in the GRB with relative increases up to about 100% (Pechlivanidis et al. 2016). Thereby, the changes in high flows were projected to be more significant than the changes in low flows (i.e. 90 percentile exceedance discharge). Assessments on future flood and drought frequencies shows that a future 100-year flood will occur once in 26.1 years and 3.8 years, respectively, at the end of the twenty-first century in the IGB basins (Hirabayashi et al. 2008). Furthermore, the average number of drought days were found to increase by a factor 1.17 and 4.05 in the IGB basins, respectively (Hirabayashi et al. 2008). It is difficult to compare the magnitude of absolute and relative changes in discharge levels because different studies have used different climate forcing and approaches to investigate impacts of climate change, precipitation, temperature, landuse cover etc. on hydrological extremes (i.e. floods and drought).

21.11 Research Gaps

- One of the greatest areas of uncertainty in Himalayan science remains how changes in glacier melt will affect river discharge over coming decades. Naturally, there will be variability within and between river systems in each basin; but these data are in agreement with previously published findings that allay water shortage fears, at least in the short term. Perhaps what they do less well, simply because of the broad-scale nature of the assessment, is to assess how these projections change with increasing distance from the source. This remains a major data gap for the discipline to address in coming years.
- The last decade has seen a proliferation of research focusing on Himalayan climate, glaciers, water resources, and related policy, but rarely these disciplines are considered together in a single volume.
- The need for more research in the Himalayas can hardly be overstated. It is known the mountain range is being badly affected by deforestation, habitat loss and global warming, and its ability to act as Asia's water tower over the long term is increasingly in doubt. But there are serious data gaps whenever scientists and policymakers look at the ecology of the world's highest mountain range.
- Research methodology is a way to systematically solve the research problem, but ongoing research gaps are also due to methodological challenges e.g. (a) lack of standardized image analysis, (b) limited field validation data (c) lack of accurate elevation data for remote glacierized areas (d) algorithms for automatically discerning debris-covered ice from non-ice areas with debris, etc.

21.12 Summary

The Himalayan Mountain chain which is the third-largest deposit of ice and snow in the world, serves as an important source of freshwater for the 1.3 billion population living in the lowlands of river basins of Indus, Ganga and the Brahmaputra (IGB) covering eight countries (Afghanistan, Bangladesh, Bhutan, China, India, Myanmar, Nepal, and Pakistan). Influence of Himalayan cryosphere is very significant in headwater tributaries of these river basins. While glaciers occupy high elevation uninhabited areas and water runs through deep gorges, seasonal snow cover spread over a much large mountain area also plays a significant role in the livelihood of the people through river runoff. Understanding of the timing and relative contribution of individual components of the hydrological cycle and water resources characteristics across the Himalayas is limited. In the upstream domains of the IGB, where mountain-hydrological processes are important, the number of studies on extremes is very limited. Although there are some studies conducted about high flows but they does not take the effects of climate change on low flows into consideration. Estimates of the ice volume and ice thickness is mostly lacking for the glaciers in this region. Advances in knowledge of Himalayan springs, geology, sediment transfer etc. is limited due to inadequate investigations and lack of synthesis of existing information. Therefore, this chapter presents outcome of an extensive review of available knowledge about the hydrology of IGB river basins. The chapter also identifies and discuss the knowledge gaps in the current understanding of hydrology for this region. Many factors that are considered important in managing Himalayan water resources have been identified and discussed. Observed and future implications of climate change impacts in IGB has also been discussed in detail. **Major findings discussed in the chapter are as:**

- Temperatures across the Himalayan region will increase by about 1–2 °C (in some places by up to 4–5 °C) which will result in increase of annual runoff by 7–12% and IRB may lose up to 8.4% of its total water resources by 2050.
- Glaciers will continue to suffer substantial ice loss, with the main loss in the Indus basin.
- Precipitation will change with the monsoon expected to become longer and more erratic.
- Extreme rainfall events are becoming less frequent, but more violent and are likely to increase in intensity.
- Communities living immediately downstream from glaciers are the most vulnerable to glacial changes.
- Despite overall greater river flow projected, higher variability in river flows and more water in pre-monsoon months are expected, which will lead to a higher incidence of unexpected floods and droughts, greatly impacting on the livelihood security and agriculture of river-dependent people.
- Changes in temperature and precipitation will have serious and far-reaching consequences for climate-dependent sectors, such as agriculture, water resources and health.

References

Ageta Y, Higuchi K (1984) Estimation of mass balance components of a summer-accumulation type glacier in the Nepal Himalaya. Geogr Ann Ser Phys Geogr 66:249–255

Azam MF, Wagnon P, Vincent C, Ramanathan AL, Favier V, Mandal A, Pottakkal JG (2014) Processes governing the mass balance of Chhota Shigri glacier (western Himalaya, India) assessed by point-scale surface energy balance measurements. Cryosphere 8(6):2195–2217

Azam MF, Ramanathan AL, Wagnon P, Vincent C, Linda A, Berthier E, Sharma P, Mandal A, Angchuk T, Singh VB, Pottakkal JG (2016) Meteorological conditions, seasonal and annual mass balances of Chhota Shigri glacier, western Himalaya, India. Ann Glaciol 57(71):328–338

Azam MF, Wagnon P, Berthier E, Vincent C, Fujita K, Kargel JF (2018) Review of the status and mass changes of the Himalayan-Karakoram glaciers. J Glaciol 64(243):61–74. https://doi.org/10.1017/jog.2017.86

Bajracharya SR, Shrestha BR (2011) The status of glaciers in the Hindu Kush-Himalayan region. International Centre for Integrated Mountain Development (ICIMOD)

Bandyopadhyay J, Kraemer D, Kattelmann R, Kundzewicz ZW (1997) Highland waters: a resource of global significance. In: Messerii B, Ives J (eds) Mountains of the world: a global priority. Parthenon, New York, pp 131–155

Barnett TP, Adam JC, Lettenmaier DP (2005) Potential impacts of a warming climate on water availability in snow-dominated regions. Nature 438(17):303–309. https://doi.org/10.1038/nature04141

Batar AK, Watanabe T, Kumar A (2017) Assessment of land use/land cover change and forest fragmentation in the Garhwal Himalayan region of India. Environments 4:34

Bera S (2017) Trend analysis of rainfall in Ganga Basin, India during 1901-2000. Am J Clim Chang 6:116–131. https://doi.org/10.4236/ajcc.2017.61007

Bolch T, Pieczonka T, Mukherjee K, Shea J (2017) Brief communication: glaciers in the Hunza catchment (Karakoram) have been nearly in balance since the 1970s. Cryosphere 11(1):531–539. https://doi.org/10.5194/tc-11-531-2017

Bookhagen B (2012) Hydrology: Himalayan groundwater. Nat Geosci 5:97–98

Bookhagen B, Burbank DW (2006) Topography, relief, and TRMM-derived rainfall variations along the Himalaya. Geophys Res Lett 33(L08405). https://doi.org/10.1029/2006GL026037

Bookhagen B, Burbank DW (2010) Toward a complete Himalayan hydrological budget: spatio-temporal distribution of snowmelt and rainfall and their impact on river discharge. J Geophys Res Earth 115(F03019). https://doi.org/10.1029/2009JF001426

Bookhagen B, Strecker MR (2008) Orographic barriers, highresolution TRMM rainfall, and relief variations along the eastern Andes. Geophys Res Lett 35:L06403. https://doi.org/10.1029/2007GL032011

Brun F, Berthier E, Wagnon P, Kaab A, Treichler D (2017) A spatially resolved estimate of High Mountain Asia glacier mass balances from 2000 to 2016. Nat Geosci 10:668–673. https://doi.org/10.1038/ngeo2999

Chawla I, Mujumdar PP (2015) Isolating the impacts of land use and climate change on stream-flow. Hydrol Earth Syst Sci 19:3633–3651

Cheema MJM (2012) Understanding water resources conditions in data scarce river basins using intelligent pixel information, case: Transboundary Indus Basin. PhD thesis, TU Delft, University, Delft, the Netherlands, 209 pp.

Clift PD, Shimizu N, Layne GD, Gaedicke C, Schluter HU, Clark MK, Amjad S (2000) Fifty five million years of Tibetan evolution recorded in the Indus fan. EOS Trans Am Geophys Union 81(25):277–281

Clift PD, Shimizu N, Layne G, Gaedicke C, Schluter HU, Clark MK, Amjad S (2001) Development of the Indus fan and its significance for the erosional history of the western Himalaya and Karakoram. Geol Soc Am Bull 113:1039–1051

Cogley JG (2012) Himalayan glaciers in the balance. Nature 488(7412):468. https://doi.org/10.1038/488468a

CRU (2012) University of East Anglia Climatic Research Unit: Phil Jones, Ian Harris: CRU Time Series (TS) high resolution gridded data version 3.20, [Internet], NCAS. British Atmospheric Data Centre

CWC and NRSC (2014) Ganga Basin report. Ministry of Water Resources, New Delhi

Dahlke HE, Lyon SW, Jansson P, Karlin T, Rosqvist G (2013) Isotopic investigation of runoff generation in a glacierized catchment in northern Sweden. Hydrol Process. https://doi.org/10.1002/hyp.9668

Dasgupta S, Mukhopadhyay B, Mukhopadhyay M, Nandy DR (2013) Role of transverse tectonics in the Himalayan collision: further evidences from two contemporary earthquakes. J Geol Soc India 81(2):241–247

Datta DK, Subramanian V (1997) Texture and mineralogy of sediments from the Ganges-Brahmaputra-Meghna River system in the Bengal Basin, Bangladesh and their environmental implications. Environ Geo 30(3/4):181–188

Deka RL, Mahanta C, Pathak H, Nath KK, Das S (2013) Trends and fluctuations of rainfall regime in the Brahmaputra and Barak basins of Assam, India. Theor Appl Climatol 114:61–71

Dhar ON, Rakhecha PR (1981) The effect of elevation on monsoon rainfall distribution in the Central Himalayas. In: Lighthill J, Pearce RP (eds) Monsoon dynamics. Cambridge University Press, Cambridge, pp 253–260

Doocy S, Daniels A, Murray S, Kirsch TD (2013) The human impact of floods: a historical review of events 1980-2009 and systematic literature review. PLOS Currents Disasters. 2013 Apr 16. Edition 1. https://doi.org/10.1371/currents.dis.f4deb457904936b07c09daa98ee8171a

Ferguson RI (1984) Sediment load of the Hunza River. The international Karakoram project 2: 581–598. Cambridge University Press

Flügel WA, Pechstedt J, Bongartz K, Bartosch A, Eriksson M, Clark M (2008) Analysis of climate change trend and possible impacts in the upper Brahmaputra River basin – the BRAHMATWINN project. In: 13th IWRA world water congress 2008, Montpelier, France

Fluteau F, Ramstein G, Besse J (1999) Simulating the evolution of the Asian and African monsoons during the past 30 Myr using an atmospheric general circulation model. J Geophys Res Atmos 104(D10):11995–12018

Foglia L, Hill MC, Mehl SW, Burlando P (2009) Sensitivity analysis, calibration, and testing of a distributed hydrological model using error-based weighting and one objective function. Water Resour Res 45(6):1–18. https://doi.org/10.1029/2008WR007255

Gain AK, Immerzeel WW, Sperna Weiland FC, Bierkens MFP (2011) Impact of climate change on the stream flow of the lower Brahmaputra: trends in high and low flows based on discharge-weighted ensemble modelling. Hydrol Earth Syst Sci 15(5):1537–1545

Galy A, France-Lanord C (2001) Higher erosion rates in the Himalaya: geochemical constraints on riverine fluxes. Geology 29(1):23–26

Gardelle J, Berthier E, Arnaud Y (2012) Slight mass gain of Karakoram glaciers in the early twenty-first century. Nat Geosci 5(5):322–325

Gardelle J, Berthier E, Arnaud Y, Kaab A (2013) Region-wide glacier mass balances over the Pamir-Karakoram-Himalaya during 1999-2011. Cryosphere 7(6):1885–1886. https://doi.org/10.5194/tc-7-1263-2013

Gautam R, Hsu NC, Lau WKM, Yasunari TJ (2013) Satellite observations of desert dust-induced Himalayan snow darkening. Geophys Res Lett 40:988–993. https://doi.org/10.1002/grl.50226

Ghosh S, Dutta S (2012) Impact of climate change on flood characteristics in Brahmaputra basin using a macro-scale distributed hydrological model. J Earth Syst Sci 121(3):637–657

Goswami PK (2012) Geomorphic evidences of active faulting in the northwestern, ganga plain, India: implications for the impact of basement structures. Geosci J 16(3):289–299

Gurung DR, Kulkarni AV, Giriraj A, Aung KS, Shrestha B, Srinivasan J (2011) Changes in seasonal snow cover in Hindu Kush-Himalayan region. Cryosphere 5:755–777. https://doi.org/10.5194/tcd-5-755-2011

Hagg W, Braun LN, Kuhn M, Nesgaard TI (2007) Modelling of hydrological response to climate change in glacierized central Asian catchments. J Hydrol 332(1–2):40–53

Hasson S, Lucarini V, Pascale S (2013) Hydrological cycle over south and southeast Asian river basins as simulated by PCMDI/CMIP3 experiments. Earth Syst Dynam 4:199–217

Hasson S, Lucarini V, Pascale S, Böhner J (2014) Seasonality of the hydrological cycle in major south and southeast Asian river basins as simulated by PCMDI/CMIP3 experiments. Earth Syst Dynam 5:67–87. https://doi.org/10.5194/esd-5-67-2014

Hazarika P, Rav IKM, Srijayanthi G, Raju PS, Rao NP, Srinagesh D (2010) Transverse tectonics in the Sikkim Himalaya: evidence from seismicity and focal mechanism data. Bull Seismol Soc Am 100(4):1816–1822. https://doi.org/10.1785/0120090339

Hazarika N, Das AK, Borah SB (2015) Assessing land-use changes driven by river dynamics in chronically flood affected upper Brahmaputra plains, India, using RS-GIS techniques. Egypt J Remote Sens Space Sci 18(2015):107–118

Hewitt K (2005) The Karakoram anomaly? Glacier expansion and the "elevation effect" Karakoram Himalaya. Mt Res Dev 25:332–340

Hirabayashi Y, Kanae S, Emori S, Oki T, Kimoto M (2008) Global projections of changing risks of floods and droughts in a changing climate. Hydrol Sci J 53:754–772. https://doi.org/10.1623/hysj.53.4.754

Hirabayashi Y, Mahendran R, Koirala S, Konoshima L, Yamazaki D, Watanabe S, Kim H, Kanae S (2013) Global flood risk under climate change. Nat Clim Chang 3:816–821. https://doi.org/10.1038/nclimate1911

Hosterman HR, McCornick PG, Kistin EJ, Pant A, Sharma B, Bharati L (2009) Water, climate change, and adaptation: focus on the Ganges River basin. Working paper no: NI WP 09-03. Nicholas Institute for Environmental Policy Solutions, Duke University, Durham, North Carolina

Huang X, Sillanpaa M, Gjessing ET, Peräniemi S, Vogt RD (2011) Water quality in the southern Tibetan plateau: chemical evaluation of the Yarlung Tsangpo (Brahmaputra). River Res Appl 27(1):113–121

Huss M, Farinotti D, Bauder A, Funk M (2008) Modelling runoff from highly glacierized alpine drainage basins in a changing climate. Hydrol Process 22(19):3888–3902

Huss M, Zemp M, Joerg PC, Salzmann N (2014) High uncertainty in 21st century runoff projections from glacierized basins. J Hydrol 510:35–48. https://doi.org/10.1016/j.jhydrol.2013.12.017

Ibsen T (2018) The Arctic cooperation, a model for the Himalayas – third pole? In: Goel P, Ravindra R, Chattopadhyay S (eds) Science and geopolitics of the white world. Springer, Cham

Immerzeel WW (2008) Historical trends and future predictions of climate variability in the Brahmaputra basin. Int J Climatol 28:243–254. https://doi.org/10.1002/joc.1528

Immerzeel WW, Droogers P, De Jong SM, Bierkens MFP (2009) Large-scale monitoring of snow cover and runoff simulation in Himalayan river basins using remote sensing. Remote Sens Environ 113(1):40–49. https://doi.org/10.1016/j.rse.2008.08.010

Immerzeel WW, Van Beek LP, Bierkens MFP (2010) Climate change will affect the Asian water towers. Science 328(5984):1382–1385

Immerzeel WW, Pellicciotti F, Bierkens MFP (2013) Rising river flows throughout the twenty-first century in two Himalayan glacierized watersheds. Nat Geosci 6(8):1–4. https://doi.org/10.1038/ngeo1896

Inam A, Clift PD, Giosan L, Tabrez AR, Tahir M, Rabbani MM, Danish M (2007) The Geographic, geological and oceanographic setting of the Indus River. In: Gupta A (ed) Large rivers: geomorphology and management. Wiley, pp 333–346

Islam SN (2016) Deltaic floodplains development and wetland ecosystems management in the Ganges–Brahmaputra–Meghna Rivers Delta in Bangladesh. Sustain Water Resour Manag 2(3):237–256. https://doi.org/10.1007/s40899-016-0047-6

Jayangondaperumal R, Dubey AK, Sen K (2010) Structural and magnetic fabric studies of recess structures in the western Himalaya: implications for 1905 Kangra earthquake. In the Structural geology – classical to modern concept, edited by Mamtani MA. J Geol Soc India 75: 212–225

Jian J, Webster PJ, Hoyos CD (2009) Large scale controls on Gnages and Brahmaputra river discharge on intraseasonal and seasonal time scale. QJR Meterol Soc 135(639):353–370

Jianchu Xu, Shrestha A, Vaidya R, Erickson M, Hewitt K, 2007. The melting Himalayas, Technical paper. International Centre for Integrated Mountain Development (ICIMOD)

Kaab A, Berthier E, Nuth C, Gardelle J, Arnaud Y (2012) Contrasting patterns of early twenty-first century glacier mass change in the Himalayas. Nature 488(7412):495–498. https://doi.org/10.1038/nature11324

Kaser G, Großhauser M, Marzeion B (2010) Contribution potential of glaciers to water availability in different climate regimes. In: Barry RG (ed) National Academy of Sciences of the United States of America, pp 20223–20227

Koboltschnig GR, Schoner W, Zappa M, Kroisleitner C, Holzmann H (2008) Runoff modelling of the glacierized alpine upper Salzach basin (Austria): multi-criteria result validation. Hydrol Process 22(19):3950–3964

Kotliakov VM (1996) Variations of snow and ice in the past and at present on a global and regional scale, UNESCO

Kripalani RH, Oh JH, Kulkarni A, Sabade SS, Chaudhari HS (2007) South Asian summer monsoon precipitation variability: coupled climate model simulations and projections under IPCC AR4. Theor Appl Climatol 90(3–4):133–159. https://doi.org/10.1007/s00704-006-0282-0

Kumar V, Singh P, Singh V (2007) Snow and glacier melt contribution in the Beas River at Pandoh dam, Himachal Pradesh, India. Hydrol Sci J 52(2):376–388. https://doi.org/10.1623/hysj.52.2.376

Kundzewicz ZW, Mata LJ, Arnell NW, Doll P, Kabat P, Jimenez B, Miller K, Oki T, Zekai S, Shiklomanov I 2007. Freshwater resources and their management. In: Parry ML et al (ed) Climate Change 2007: impacts, adaptation and vulnerability. Contribution of Working Group II to the Fourth Assessment Report of the Intergovernmental Panel on Climate Change, Cambridge University Press, Cambridge, UK

Laghari AN, Vanham D, Rauch W (2012) The Indus basin in the framework of current and future water resources management. Hydrol Earth Syst Sci 16:1063–1083. https://doi.org/10.5194/hess-16-1063-2012

Latif Y, Yaoming MA, Yasin M (2018) Spatial analysis of precipitation time series over the upper Indus Basin. Theor Appl Climatol 131(1–2):761–775

Lemke P, Ren J, Alley RB, Allison I, Carrasco J, Flato G, Fujii Y, Kaser G, Mote P, Thomas RH, Zhang T (2007) Observations: changes in snow, ice and frozen ground. Cambridge University Press, Cambridge

Lupker M, Blard PH, Lave J, France-Lanord C, Leanni L, Puchol N, Charreau J, Bourles D (2012) [10] Be-derived Himalayan denudation rates and sediment budgets in the ganga basin. Earth Planet Sci Lett 333–334:146–156

Lutz AF, Immerzeel WW (2013) Water availability analysis for the UpperIndus, Ganges, Brahmaputra, Salween and Mekong river basins. Future Water, Costerweg 1V, 6702 AA Wageningen, The Netherlands

Lutz AF, Immerzeel WW and Kraaijenbrink PDA (2014a) Gridded meteorological datasets and hydrological modelling in the upper Indus Basin. Report Future Water 130, Wageningen, The Netherlands

Lutz AF, Immerzeel WW, Shrestha AB, Bierkens MFP (2014b) Consistent increase in high Asia's runoff due to increasing glacier melt and precipitation. Nat Clim Chang 4:587–592. https://doi.org/10.1038/nclimate2237

Lutz AF, Immerzeel WW, Kraaijenbrink PDA, Shrestha AB, Bierkens MFP (2016) Climate change impacts on the upper indus hydrology: sources, shifts and extremes. PLoS One 2016:11. https://doi.org/10.1371/journal.pone.0165630

Maussion F, Scherer D, Molg T, Collier E, Curio J, Finkelnburg R (2014) Precipitation seasonality and variability over the Tibetan plateau as resolved by the high Asia reanalysis. J Clim 27(5):1910–1927

Menne B, Bertollini R (2000) The health impacts of desertification and drought. Down to earth: the newsletter to the convention to combat desertification 14: 4–6 (December)

Mirza MQ (2002) Global warming and changes in the probability of occurrence of floods in Bangladesh and implications. Glob Environ Chang 12(2):127–138

Mirza MMQ (2011) Climate change, flooding in South Asia and implications. Reg Environ Chang 11:95–107. https://doi.org/10.1007/s10113-010-0184-7

Mittal N, Mishra A, Singh R, Kumar P (2014) Assessing future changes in seasonal climatic extremes in the Ganges river basin using an ensemble of regional climate models. Clim Change 123:273–286. https://doi.org/10.1007/s10584-014-1056-9

Mukhopadhyay B, Khan A (2014a) A quantitative assessment of the genetic sources of the hydrologic flow regimes in upper Indus Basin and its significance in a changing climate. J Hydrol 509:549–572

Mukhopadhyay B, Khan A (2014b) Rising river flows and glacial mass balance in Central Karakoram. J Hydrol 513:192–203

Mukhopadhyay B, Khan A (2015) A reevaluation of the snowmelt and glacial melt in river flows within upper Indus Basin and its significance in a changing climate. J Hydrol 527:119–132

Naz B, Frans C, Clarke G, Burns P, Lettenmaier D (2013) Modeling the effect of glacier recession on stream flow response using a coupled glacio-hydrological model. Hydrol Earth Syst Sci Discuss 10(4):5013–5056

Nepal S (2012) Evaluating upstream-downstream linkages of hydrological dynamics in the Himalayan region. PhD thesis, Friedrich Schiller University, Jena

Nepal S, Krause P, Flügel W-A, Fink M, Fischer C (2013) Understanding the hydrological system dynamics of a glaciated alpine catchment in the Himalayan region using the J2000 hydrological model. Hydrol Process 28:1329–1344. https://doi.org/10.1002/hyp.9627

Nepal S, Flügel WA, Shrestha AB (2014) Upstream-downstream linkages of hydrological processes in the Himalayan region. Ecol Process 3(1):19. https://doi.org/10.1186/s13717-014-0019-4

Pechlivanidis IG, Arheimer B, Donnelly C, Hundecha Y, Huang S, Aich V, Samaniego L, Eisner S, Shi P (2016) Analysis of hydrological extremes at different hydro-climatic regimes under present and future conditions. Clim Chang 141(3):467–481. https://doi.org/10.1007/s10584-016-1723-0

Pellicciotti F, Ragettli S, Carenzo M, McPhee J (2014) Changes of glaciers in the Andes of Chile and priorities for future work. Sci Total Environ 493:1197–1210. https://doi.org/10.1016/j. scitotenv.2013.10.055

Prasch M (2010) Distributed process oriented modelling of the future impact of glacier melt water on runoff in the Lhasa river basin in Tibet. PhD thesis, Ludwig-Maximilians-University of Munich, Germany

Prasch M, Weber M, Mauser W (2011) Distributed modelling of snow- and ice-melt in the Lhasa River basin from 1971 to 2080. In: Cold regions hydrology in a changing climate, proceedings of an international symposium, H02, held during IUGG 2011 in Melbourne, Australia, July 2011, vol 346, IAHS Publishing, pp 57–64

Pritchard HD (2017) Asia's glaciers are a regionally important buffer against drought. Nature 545(7653):169–174. https://doi.org/10.1038/nature22062

Ragettli S, Pellicciotti F, Bordoy R, Immerzeel W (2013) Sources of uncertainty in modeling the glaciohydrological response of a Karakoram watershed to climate change. Water Resour Res 49(9):6048–6066. https://doi.org/10.1002/wrcr.20450b

Rajeevan M, Bhate J, Jaswal AK (2008) Analysis of variability and trends of extreme rainfall events over India using 104 years of gridded daily rainfall data. Geophys Res Lett 35(18):L18707

Reid TD, Brock BW (2010) An energy-balance model for debris-covered glaciers including heat conduction through the debris layer. J Glaciol 56(199):903–916. https://doi.org/10.3189/002214310794457218

Rice SK (2007) Suspended sediment transport. In the Ganges-Brahmaputra River system, Bangladesh. Master thesis. Texas A&M University, Texas

Sabade SS, Kulkarni A, Kripalani RH (2011) Projected changes in south Asian summer monsoon by multi-model global warming experiments. Ther Appl Climatol 103(3–4):543–565. https://doi.org/10.1007/s00704-010-0296-5

Savoskul OS, Smakhtin V (2013) Glacier systems and seasonal snow cover in six major Asian river basins: hydrological role under changing climate. IWMI Research Report 150. International Water Management Institute (IWMI), Colombo, Sri Lanka, 53p. https://doi.org/10.5337/2013.204

Schickhoff U, Bobrowski M, Böhner J, Bürzle B, Chaudhary RP, Gerlitz L, Lange J, Muller M, Scholten T, Schwab N., 2016. Climate change and treeline dynamics in the Himalaya. In: Climate change, glacier response, and vegetation dynamics in the Himalaya. Springer, Cham, pp 271–306. https://doi.org/10.1007/978-3-319-28977-9_15

Shangguan D, Liu S, Ding Y, Wu L, Deng W, Guo W, Wang Y, Xu J, Yao X, Guo Z, Zhu W (2014) Glacier changes in the Koshi River basin, central Himalaya, from 1976 to 2009, derived from remote-sensing imagery. Ann Glaciol 55(66):61–68. https://doi.org/10.3189/2014AoG66A057

Sharma KP, Vorosmarty CJ, Moore B III (2000) Sensitivity of the Himalayan hydrology to land-use and climatic changes. Clim Chang 47(1-2):117–139. https://doi.org/10.1023/A:1005668724203

Sharma E, Bhuchar S, Xing MA, Kothyari BP (2007) Land use change and its impact on hydro-ecological linkages in Himalayan watersheds. Trop Ecol 48(2):151–161

Shi Y, Gao X, Zhang D, Giorgi F (2011) Climate change over the Yarlung Zangbo–Brahmaputra River basin in the 21st century as simulated by a high resolution regional climate model. Quat Int 244(2):159–168

Shrestha AB (2008) Resource manual on flash flood risk management. ICIMOD Training Manual, ICIMOD, Kathmandu. isbn: 9789291152667

Shrestha AB, Wake CP, Mayewski P A, Dibb JE (1999) Maximum temperature trends in the Himalaya and its vicinity: an analysis based on temperature records from Nepal for the period 1971–94. J Clim 12(9):2775–2786

Shrestha D, Singh P, Nakamura K (2012) Spatiotemporal variation of rainfall over the central Himalayan region revealed by TRMM precipitation radar. J Geophys Res 117:D22106. https://doi.org/10.1029/2012JD018140

Shrestha AB, Agrawal NK, Alfthan B, Bajracharya SR, Maréchal J, Van Oort B (2015) The Himalayan climate and water atlas: impact of climate change on water resources in five of Asia's major river basins. ICIMOD, GRID-Arendal and CICERO

Shukla UK, Raju JN (2008) Migration of the Ganga River and its implication on hydro-geological potential of Varanasi area, UP, India. J Earth Syst Sci 117:489–498

Singh P, Kumar N (1997) Effect of orography on precipitation in the western Himalayan region. J Hydrol 199:183–206. https://doi.org/10.1016/S0022-1694(96)03222-2

Singh P, Singh VP (2001) Snow and glacier hydrology, vol. 37. Kluwer Academic Publishers, Dordrecht/Boston/London. isbn: 0792367677

Singh P, Ramasastri KS, Kumar N (1995) Topographical influence on precipitation distribution in different ranges of western Himalayas. Nord Hydrol 26:259–284

Sinha R, Tandon SK (2014) Indus-ganga-Brahmaputra plains: the alluvial landscape. In: Landscapes and landforms of India, Springer, pp 53–63

Sinha R, Kettanah Y, Gibling MR, Tandon SK, Jain M, Bhattacharjee PS, Dasgupta AS, Ghazanfari P (2009) Craton-derived alluvium as a major sediment source in the Himalayan Foreland Basin of India. GSA Bull 121(11/12):1596–1610. https://doi.org/10.1130/B26431

Sori R, Nietol R, Drumond A, Vicente-Serrano SM, Gimeno L (2017) The atmospheric branch of the hydrological cycle over the Indus, Ganges, and Brahmaputra river basins. Hydrol Earth Syst Sci 21:6379–6399

Sorkhabi R (2010) Geologic formation of the Himalaya. Himal J 66

Thakur VC, Jayangondaperumal R, Suresh N (2009) Late quaternary–Holocene fold and landform generated by morohogenic earthquakes in Chandigarh anticlinal ridge in Panjab SubHimalaya. Himal Geol 30(2):103–113

Thayyen RJ, Gergan JT (2010) Role of glaciers in watershed hydrology: a preliminary study of a "Himalayan catchment". Cryosphere 4(1):115–128

Thayyen RJ, Gergan JT, Dobhal DP (2005) Monsoonal control on glacier discharge and hydrograph characteristics, a case study of Dokriani glacier, Garhwal Himalaya, India. J Hydrol 306(1–4):37–49

Thorne CR, Russell APG, Alam MK (1993) Planform pattern and channel evolution of Brahmaputra river, Bangladesh. Geol Soc Lond Spec Publ 75(1):257–276

Tiwari VM, Wahr J, Swenson S (2009) Dwindling groundwater resources in northern India, from satellite gravity observations. Geophys Res Lett 36(18):1–5. https://doi.org/10.1029/2009GL039401

Tsarouchi GM, Mijic A, Moulds S, Buytaert W (2014) Historical and future land cover changes in the upper Ganges basin of India. Int J Remote Sens 35(9)

Turner AG, Slingo JM (2009) Uncertainties in future projections of extreme precipitation in the Asian monsoon regions. Atmos Sci Lett 10(3):152–158. https://doi.org/10.1002/asl.223

Viviroli D, Durr HD, Messerli B, Meybeck M, Weingartner R (2007) Mountains of the world, water towers for humanity: typology, mapping, and global significance. Wat Resour Res 43(W07447). https://doi.org/10.1029/2006WR005653

Walling DE (1987) Rainfall, runoff and erosion of the land: a global view. In: Energetics of Physical environment: energetic approaches to physical geography. Wiley, Chichester, p 0471913588

Webster PJ, Jian J (2011) Environmental prediction, risk assessment and extreme events: adaptation strategies for the developing world. Phil Trans R Soc A 369(1956):4768–4797

Webster P, Toma V, Kim H (2011) Were the 2010 Pakistan floods predictable? Geophys Res Lett 38(4). https://doi.org/10.1029/2010GL046346

Wiltshire AJ (2014) Climate change implications for the glaciers of the Hindu Kush, Karakoram and Himalayan region. Cryosphere 8:941–958. https://doi.org/10.5194/tc-8-941-2014

Younis SMZ, Ahmad A (2017) Quantification of impact of changes in land use-land cover on hydrology in the upper Indus Basin, Pakistan. Egypt J Remote Sens Space Sci. https://doi.org/10.1016/j.ejrs.2017.11.001

Zhang W, Rickettsb TH, Kremenc C, Carneyd K, Swintona SM (2007) Ecosystem services and dis-services to agriculture. Ecol Econ 64:253–260

Zhou Y, Li Z, Li JIA (2017) Slight glacier mass loss in the Karakoram region during the 1970s to 2000 revealed by KH-9 images and SRTM DEM. J Glaciol 63(238):331–342. https://doi.org/10.1017/jog.2016.142

Part IV
Ecology/Forestry

Chapter 22
The Impact of Climate Change in Hindu Kush Himalayas: Key Sustainability Issues

Surendra P. Singh, Rajesh Thadani, G. C. S. Negi, Ripu Daman Singh, and Surabhi Gumber

Abstract After the observation of IPCC 4th assessment that the Himalayas are "data deficient" with regard to climate change, some progress has been made particularly in the areas of glacier shrinkage, snow cover change, glacial lake outburst flooding, river discharge, treeline advance, phenological shift, climate change mitigation and adaptations, and people's perceptions. This article focuses on complex interactions among climate change impacts on various bio-physical and socio-economic components of the Hindu Kush Himalayan region ecosystems. The magnitude of climate change impacts can be traced from impacts, such as pre-monsoon drought to crop failure and outmigration of people. The interconnectedness of the ecosystem components makes the Climate Change impacts complex, and often has a cascading effect across various environmental systems. We shed light on how climate change effects get intensified by interacting with other anthropogenic factors. There is a need to devote more concerted efforts to generate primary data, and document evidences to understand the complexity and interconnectedness of CC impacts to address sustainable development issues in the Himalayan mountains.

22.1 Introduction

The present climate change (CC) is a highly complicated, deeply politicized and is a matter of huge concern of our time. In a worst scenario, it may result in a diminished human existence (Henson 2006). Though the impact of CC is widespread,

S. P. Singh (✉)
Central Himalayan Environment Association (CHEA), Nainital, India

R. Thadani
Center for Ecology Development and Research (CEDAR), Dehradun, India

G. C. S. Negi
Govind Ballabh Pant National Institute of Himalayan Environment and Sustainable Development (GBPNIHESD), Almora, India

R. D. Singh · S. Gumber
Central Himalayan Environment Association (CHEA), Kumaun University, Nainital, India

© Springer Nature Switzerland AG 2020
A. P. Dimri et al. (eds.), *Himalayan Weather and Climate and their Impact on the Environment*, https://doi.org/10.1007/978-3-030-29684-1_22

some regions such as mountains and oceans are getting more affected. It has been pointed out that development in mountains is going to suffer more (FAO 2015; Schild and Sharma 2011) because mountain people are comparatively poor and depend more on natural ecosystem services, which are particularly sensitive to CC. There is ongoing concern about current and potential CC impacts in the fragile and highly susceptible landscapes of Himalayas, which may include abnormal floods, droughts and landslides (Barnett et al. 2005), loss of biodiversity and shortage of food (Xu et al. 2009). Himalayas are warming at higher rates than global average rate, and a large number of glaciers are shrinking in area and mass (Bajracharya et al. 2015; Singh et al. 2011; Yao et al. 2012). The region is called third pole because it has more snow and ice than other regions outside the two poles. Nearly half of the biodiversity hotspots occur in mountains. Because of high endemism of species, particularly towards mountain summits, Himalayas are great contributor to global biodiversity (Dhar 2002).

Accounting for about 24% of the land surface of the earth, mountains represent important wilderness areas providing several life supporting services. For example, many downstream dry regions heavily depend on water flowing from mountains. They provide 30–60% fresh water downstream in humid regions and 70–95% in arid and semi-arid regions (Kapos et al. 2000). They are highly sensitive to CC, that is why, it is important to assess the impact of CC on mountain ecosystems (Nougues-Bravo et al. 2007). Himalayas along with Hindu Kush (often called Hindu-Kush Himalayas, HKH) are of great geo-ecological importance because of their huge dimensions, heterogeneities, and contributions to life supporting services to about 1.3 billion people living in river basins originating from this mountainous region.

Issues of sustainable development in Himalayan mountains call for a better understanding on the impacts of CC and adaptation measures required to address the same. In IPCC report AR4, Himalayan region was referred to as a "white spot", emphasizing the scarcity of data on CC. Since then, climate change-related information has grown, so much so that it is possible to start analyzing them in the context of sustainable development. Here we have tried to put together the impact of CC in HKH region on various components of mountain ecosystems, ranging from glacial retreat, loss of forest regeneration to social issues like out migration and depopulation of mountain villages.

The objectives of this article are (1) to summarise the facts about CC in HKH, which are fairly well established (such as, rate of warming, pattern of precipitation changes and weather extremes), (2) to analyse the impact of CC on diverse systems both natural (e.g., glaciers) and man-made (e.g., agriculture), and (3) to analyze how CC and anthropogenic factors, such as forest degradation interact or likely to interact to amplify the overall impact, and affect sustainability issues. We have considered local as well as global anthropogenic factors which affect development in HKH.

22.2   Material and Methods

22.2.1   The Study Region

Himalayas along with Hindu-Kush region, occupy 4.2 million km^2 area (Bajracharya and Shrestha 2011), and have hundreds of mountain peaks higher than 6000 m elevation and about 60,000 km^{2}area under glaciers (Miller et al. 2012). It is a relatively more populated mountain region (210 million people live) than others, and serves some 1.3 billion people living in 10 river basins which originate from the HKH region (Fig. 22.1). Called as the water tower, the HKH is the source of water, soil and nutrients to downstream plains (Singh 2017). This has resulted in the establishment of highly productive plains, such as, Gangetic basin where a population density close to 1000/km^2 is quiet common. This region is among the 34 biodiversity hotspots of the globe, harbouring almost entire range of terrestrial ecosystems, from tropical to alpine types. According to an estimate, HKH has 330 globally recognized bird species spots, and nearly 20,000 plant species close to 30% of which are endemic (Chhettri et al. 2008).

In much of the Himalayas slopes in the south of the main ranges are exposed to monsoon thrust, receiving 70–80% of annual precipitation during monsoon months, typically June to September. The annual precipitation which generally ranges from 1000–4000 mm across the Himalayan Arc, declines gradually from east to west. In north-western part, such as Kashmir westerlies are a major contributor to precipitation, accounting for ~40% of the annual precipitation. Some of the inner valleys of HKH are among the driest parts of the world with annual precipitation below 500 mm.

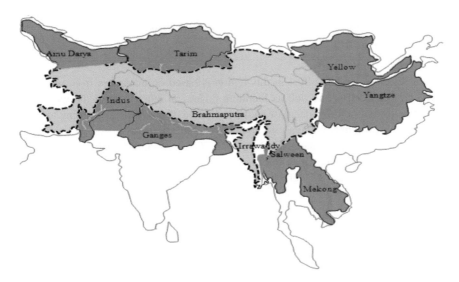

Fig. 22.1 A sketch of 10 river basins originating from Hindu Kush Himalayan (HKH) region. (Map from ICIMOD)

In this region of extraordinary ranges in elevation and precipitation (both amount and seasonality), a great diversity of natural forests occur. Himalayas are a major centre of evergreen oaks (Quercus spp.) and other Fagaceae members (e.g. Castanosis spp.). In much of the Himalayas, the foothills are dominated by Shorea robusta, a dipterocarp. Chir pine (Pinus roxburghii) forms extensive forests between 1000 and 1800 m in much of the Central and Western Himalayas. Between 1500 and 2500 m in eastern part, the species of Quercus and Castanopsis prevail, and in the central and western parts Oak show dominance. Conifers, normally silver fir (Abies spp.), cypress (Cupressus), deodar (Cedrus deodara), spruce (Picea spp.), blue pine (Pinus wallichiana) and Tsuga dumosa make much of the high elevation forests (above 2500 m). Oak, sal and several other Himalayan tree species have desiccation sensitive seeds which germinate with the arrival of monsoon (Singh et al. 2017).

HKH is also being seen as the region of change and stress, nearly 15% of global migrants are from this region. In this region the livelihood of people operates at a subsistence level, and is basically of three types viz., livestock farming, mixed-livestock crop farming, and mixed-crop livestock farming, representing nomadism, semi-nomadism and settled agriculture (Singh et al. 1984). Socially, the HKH is characterized by poverty, inadequate production system, gender inequity, lack of round the year secure water supply, lack of clean energy access, loss of biodiversity, man and wild animal conflicts, diminishing flow of ecosystem services and low adaptation capacity with regard to CC leading to abandonment of agriculture and outmigration (Tewari and Joshi 2012). This has further worsened the problem of food security in the region (Rasul 2011).

22.2.2 Methods: Use of Research Based and Anecdotal Evidences

Our analysis of the CC effect in this region is based on evidences from published research on the HKH, as well as literature review on other parts of the world. For example, to find out how decrease in diurnal range of temperature is going to influence plant growth in Himalayas, we applied generalizations developed for the vegetation of the Northern Hemisphere (Peng et al. 2013). To know whether outmigration from high mountains has CC influence or not, we used general people's perceptions as well as studies carried out on the impact of CC-induced disasters on people's migration. The focus in our approach was shedding light on CC linkages among diverse environmental components of the HKH region and concerned processes. Additionally, it can also be said that it is difficult to separate the effect of CC from those of other anthropogenic factors. For example, abandonment of agricultural land and migration is widespread in Himalayas, but its CC connections are seldom explicit. Climate change induced weather extremes are likely to influence the fate of abandoned lands, but that would also depend on how the farmers left in villages use the abandoned land. Keeping in view the limitations of data and information, we could only give an outline of trends and complexity of interconnections between CC impacts and ecosystem components. There are several limitations in this too. For example, data for a given parameter were not available for all parts of Himalayas, therefore in most cases generalizations were not possible for the whole region. Also data sharing among the HKH countries is still limited. Further, adaptation and people's perceptions are poorly documented and analyzed, and methods used are hardly comparable.

22.3   Results and Discussion

22.3.1   Climate Change- Rate of Warming and Alteration in Precipitation Regime

In the recent decades a number of research papers have unequivocally established that Himalayas are warming at the rates 2–3 times more than global mean rate in the last few decades (Liu and Chen 2000; Schickhoff et al. 2015; Shrestha et al. 1999; Xu et al. 2009), however, in NW Himalayas, Afghanistan and Karakoram regions there has been little or no warming (Yao et al. 2012). There is a growing evidence that rate of temperature rise in mountains is elevation-dependent, that is, it is amplified with elevation. This effect is particularly evident in Tibetan Plateau (Ageta and Kadota 1992). The elevation-dependent amplification of temperature warming could result in a lower temperature lapse rate (TLR) along the elevation gradient. The phenomenon of elevation-dependent warming warrants urgent attention as more rapid changes in high mountain climate have consequences for regional hydrology (Immerzeel et al. 2012; EDW Working Group 2015), and habitats of rare and endangered species. Recently, Ren et al. (2017) have analyzed the temperature and precipitation changes for about last one hundred years (1901–2014) based on climate data set of LSAT-V1.1 and CGP1.0 indicating that temperature has increased in all parts of HKH, though it is more prominent in Tibetan Plateau (TP) and south of Pakistan. The period of 1998–2014 is the warmest of the recorded history (Ren et al. 2017). A summary of published research points out that at some places the annual mean minimum temperature has increased more than twice the mean maximum, resulting in a marked decrease in diurnal temperature range (Table 22.1). This decrease in the diurnal range is far more than that of global mean range. Analysis of extreme temperature events (from 1961 to 2015) indicates that number of cold nights and cold days have decreased (more decrease in cold night), warm nights and warm days have increased (more increase in warm nights), and number of frost days have decreased. Also, rapid urbanization in recent decades has been reported a cause of rise in temperature rapidly (Singh et al. 2010).

Precipitation has decreased, but not significantly when entire period of more than hundred is considered for HKH region as a whole. However, precipitation has significantly increased during recent decades (1961–2013), the rate of increase being about 5.3% per decade in some area (Table 22.1). Change in the number of precipitation days (days with daily rainfall ≥1 mm), however, differs from one

Table 22.1 Observed past changes in certain climatic parameters over HKH

Parameter	Period	Rate
Annual mean surface air T	1901–2014	0.104 °C/decade
Annual mean max. T	1901–2014	0.077 °C/decade
Annual mean min. T	1901-2014	0.176 °C/decade
Diurnal temperature range	1901-2014	-0.101 °C/decade
Total precipitation	1901–2014	−0.360% decade
	1961–2013	3.529% decade
Cold nights T N_{10p}	1961–2015	−0.977 days/decade
Cold days T X_{10p}	1961–2015	−0.511 days/decade

T, temperature

From Ren et al. (2017), Ren and Shrestha (2017), Sun et al. (2017), and Zhan et al. (2017)

region to other. It has increased in Northern India, Northern Tibetan Plateau (TP), but has decreased in Myanmar and Southeast China (Ren et al. 2017). Indian summer monsoon has weakened (Ding and Ren 2008), resulting in decrease in annual precipitation in India and SW China. During this period the events of intense precipitation have increased. Projected future changes include increase in precipitation in most of HKH, except in NW Himalayas (Wu et al. 2017). The summer monsoon precipitation in south eastern Himalayas and TP is projected to intensify by about 22% (Sanjay et al. 2017). Uncertainty in future is projected to be high in precipitation for Central Himalayas (much of Nepal and Kumaun), Karakoram and NW Himalayas. Some of the other changes are decrease in surface wind speed (You et al. 2017) and decline in sunshine in TP, particularly during spring and pre-monsoon season (March–May). It may be pointed out that sparseness of observational data greatly limit the estimates of trends. In brief, consensus about the CC projections for the HKH is rather weak in comparison of the other Asian regions (Ren and Shrestha 2017).

22.3.2 Climate Change Effects and Their Interactions

A schematic representation of CC effects and their complex interaction with various bio-physical and socio-economic components of the HKH region (Fig. 22.1) are divisible into three categories with regard to confidence level: (1) effects which are adequately proven on the basis of published studies in relation to HKH region; (2) those which are very likely to occur on the basis of existing knowledge, but direct HKH based research are yet unavailable; and (3) those which are speculative and anecdotal. It is apparent from Fig. 22.2 that almost all components of mountain ecosystems (physical, biological and social) are affected in one way or the other. Because data are scarce, the relationship between CC and their impacts on various natural and man-made systems and processes are generally not well established. Furthermore, in several cases, CC impact is combined by one or more non-climatic

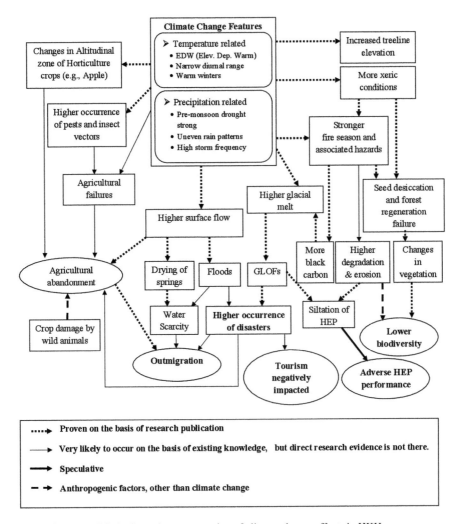

Fig. 22.2 A simplified schematic representation of climate change effects in HKH

impacts. For example, climatic uncertainties and extremes may contribute to the abandonment of agriculture in mountains, but damage of crops by wild animals and economic trivialization of agriculture with increasing GDP (Singh and Singh 2016) may contribute more to it. A study on farmer's perception about CC in Nepal shows that it varies considerably across altitudinal belts (Smadja et al. 2015). Villagers in high mountains notice more clearly the decrease in snowfall and, snow cover, and that severe freezing of ground damages crops. In lower mountain ranges, heightened pre-monsoon drought shows up prominently in people's perception (Smadja et al. 2015). So the schematic representation of CC effects in HKH (in Fig. 22.2) gives only a broad picture of various linkages related to CC.

22.3.3 Glacier Melt, Glacier Hydrology and River Discharge

The most conspicuous and research based impact of CC in Himalayas is glacier melt and glacier shrinkage, which has implications for river discharge. Glacier melt can result in the formation of glacier lakes and, local water availability which in certain circumstances can lead to glacier lake outburst floods (GLOFs), a cause of major disaster. Since, glaciers in Himalayas are remote and located at higher elevations (4800–6200 m) than other regions, they are difficult to be measured manually. A third pole study (based on 30 years observations) surrounding the TP has shown that Himalayan glaciers are depleting more rapidly than those of interior regions, such as Pamir. While in much of western Himalayas and TP where glacier shrinkages are conspicuous, monsoon the principal water source has weakened (Yao et al. 2012). Scherler et al. (2011) has reported that more than 80% glaciers of western Himalayas, northern-Central Himalayas and West Kunlun Shan were retreating, some with the rate upto ~60 m year^{-1}. To conclude, glaciers in HKH, by and large are shrinking/retreating, however, the impact of CC on glaciers can vary depending upon location, precipitation and its source, and debris cover. The increase in precipitation can result in increase in glacier mass, but the effect would be trivial.

The impact of CC on glacier hydrology and river discharge is among the least understood areas of HKH, largely because of the lack of good-quality and long-term observations. In a broad term, the predicted warming will drive rapid glacier shrinkage and resultant initial increases in the amount of melt water, which will be followed by reduced glacier size and drop in melt water discharge from glaciers, continuously diminishing in size (Miller et al. 2012). The contribution of melt water discharge would be higher near the glaciers than away from them in a river basin. The principal time of river discharge varies considerably from one region to other; it is spring for the snow-dominated Kabul, summer for glacier dominated Indus, and monsoon for Ganga, and Brahmaputra. In the last group of rivers, snow-melt dominates during the first monsoon month (July) and rain during later monsoon months (August–September). These rivers, thus will be affected differently because of CC. For example, in the case of Ganga and Brahmaputra rain may combine with snow melt water to cause floods during monsoon months. For Ganga and Brahmaputra the melt water contribution to river discharge is generally estimated far lower (~2–20%) than for Indus (~60%). Hydrology of both Ganga and Brahmaputra basins is monsoon dominated even up to considerable elevations; only in dry periods melt water is a substantial contributor to sustaining flow (Thayyen et al. 2007). Rees et al. (2004) have suggested that in head water catchments mean flow would increase during the first decade and then it would be followed by decline (see also Jain 2008). Because of the scarcity of high altitude precipitation data, spatial variation in flow of melt water, and the lack of good glacio-hydrological models and high complexities associated with HKH region, our understanding of the impact of CC on river discharge is still poor.

22.3.4   Impact of Climate Change on Erosion and Disasters

Being geologically young and geotectonically active, Himalayan mountains are unusually vulnerable to landslides and slope instability. Since the number of days with high intensity rainfall is predicted to increase, CC is likely to increase frequency and intensity of landslides, overflow of rivers and floods. Many areas in the south of the main Himalayan ranges receive between 1500 and 3000 mm rainfall during monsoon months, from June to September. Landslides in Himalayas generally occur when at least 850 mm of rain have accumulated from previous rain storm (Burbank et al. 2012). The threshold for daily rainfall that can trigger a landslide is ~4 cm/day in early season and only 1.1 cm after 2500 mm rainfall. As rainy season progresses and soils get saturated, run off for a given amount of rainfall increases. These factors are also affected by the nature of forest cover. The interception loss of rainfall is distinctly greater in a conifer forest than a broadleaf forest. So the impact of climate change can be affected by several non-climatic factors, such as the time of year and nature of vegetal cover. Mahapatra et al. (2018) have estimated that in Uttarakhand state of India, 48.3% area is losing soil at higher rates than the tolerance limit of 11.2 tha^{-1} year^{-1}. About 33% of the area soil loss is very severe, 40–80 tha^{-1} year^{-1}. The climate change associated new rainfall pattern is likely to increase soil loss from Uttarakhand mountains, which in turn may decrease the life span of water reservoirs, crop soil fertility and hence crop productivity. Uttarakhand alone has been affected by 6–7 major precipitation-induced disasters during last decade or so. Not all disasters might have CC connection, but higher frequency of disasters in recent years is consistent with CC predictions. One of the greatest threats is likely to come from glacier lakes, which when burst, can wreak havoc in downstream areas. For example, in Uttarakhand alone there are 1268 glacier lakes (>2500 m^2 area) between 2900–5850 m elevation. Chorabari glacier lake outburst in combination with excessive rainfall is reported to account for Kedarnath disaster of 2013, the scars of which in terms of infrastructural damages can be seen even in 2018 (Bhambri et al. 2015). Glacier Lake Outburst Floods (GLOFs), cloud burst or excessive rainfall events, all can be highly destructive to infrastructures like roads, bridges and buildings. Such destructions speed up outmigration as well as adversely affect tourism and other economic activities even long years after the disaster. The June 2013 Kedarnath disaster is reported to induce an episodic migration from affected areas, and shift in settlements within mountains.

22.3.5   Widespread Drying of Springs, Intensification of Pre-monsoon Drought

There are now several evidences to suggest that the intensification of pre-monsoon (March to May) drought is becoming a major CC impact in the region. Warmer temperatures without additional water affect plant growth adversely even in a cold

Table 22.2 A summary of information on springs in Himalayas

Dependence on spring water in Himalayan cities
Percentage of Indian Himalayan cities in which people partly or entirely depend on spring water-100% of 45 cities surveyed.
Number of cities which mostly depend on spring water- 17 including famous tourist towns like Mussoorie, Gangtok, Shimla (until recently) and, Shrinagar (Singh and Sharma 2014). 80–90% of rural people in Meghalaya, Sikkim and Uttarakhand depend on spring water (Pradhan 2015).
Drying up and diminishing discharge of springs
In a typical micro-watershed representative of Garhwal Himalaya, water availability in some spring dependent areas now is less than half of the WHO norms of 60 l/capita/day.
Decline in discharge of five springs after rainy season in Garhwal ranges from 36% to 100%.
Decrease of 25–75% in the spring water discharge in the Gaula River Basin, Kumaun Himalayas (Valdiya and Bartarya 1989, 1991).
Complete drying up of 159 springs and conversion of 50 springs from perennial to seasonal in Uttarakhand (Tiwari 2008).
Complete drying up of 39% of the 107 springs in the Kosi headwater, Uttarakhand during last two decades (Grover et al. 2015), resulting in 60% drop in river discharge in about a decade period.
During 1991–2010 the number of rainy days in Kumaun has declined from 60 to 50 and annual rainfall from 1350 mm to 1120 mm (Rawat 2009).

region. An analysis of meteorological data of Mukteshwar (Kumaun region of Uttarakhand) shows that while annual rainfall has not decreased during last three decades or so, pre-monsoon period has become drier (Negi et al. Unpbl.). A warmer and drier pre-monsoon period would cause spring drying and water scarcity. Cascading effects of this water shortage can be traced further to social conflicts over sharing of water resources among villages. Drying up of springs and streams in the region has been mainly connected to land use and inadequate pre-monsoon and erratic rainfall (Negi and Joshi 2004; Valdiya and Bartarya 1989), and the degree of snow/ice melt (Jeelani 2008). Springs and seepages are the major source of potable water in mountain towns (e.g., Gangtok, Shimla, Mussorie, etc.). Singh and Sharma (2014) found that across 45 cities/towns of the Himalayan region, 17 were entirely spring-dependent, and rest of them partially dependent on spring water. Though reliable assessment of spring and ground water status in relation to climate change at regional level is not available, widespread depletion of ground water and spring drying during last 3–4 decades of rapid warming in Himalayas is indicated from several studies (Table 22.2) covering an extensive area of Kumaun. Tiwari (2008) reports that during last decades 159 natural springs have completely dried up and 50 perennial springs have become seasonal. Though land use change is a major cause of spring extinction, the drying effect of warming, particularly during pre-monsoon could be a significant factor. Attempts to revive spring discharge involving bio-engineering measures increased lean period water output by 20% or so (Negi and Joshi 2002; Tambe et al. 2012), and measures such as rainwater harvesting, river water lifting, hand pumps and dug wells, recycling of waste water have come up as alternatives to cope up the water shortages in the region. Unusually higher decline in water level in water bodies, such as that happened in the summer of 2016 in Nainital lake, is a serious threat to the sustainability of tourism in the town. Wild animal mortality is partly related to the water shortage in forests during a fire incident. Thus, pre-monsoon drought can potentially drive a series of aberrations in the rhythm of natural ecosystems and negatively affect the ecosystem services.

22.3.6   Agricultural Abandonment, Loss of Crops and Outmigration

Agriculture is highly dependent on weather conditions, and changes in weather cycle have a major impact on crop yield and food supply. In Kullu valley (Himachal Pradesh), rainfall had decreased by about 7 cm, snowfall by about 12 cm, and mean minimum and maximum temperatures increased by 0.25 and 1 °C, respectively, by 1990s compared to 1880s, severely affecting the gul irrigation systems and other farming activities (Vishvakarma et al. 2003). A revealing example is apple cultivation in Himachal Pradesh, which is often cited as one of the major success stories of its economic growth. Now in Shimla and Kullu valleys, farmers are finding difficulties in sustaining apple cultivation because of warming temperatures and uncertain as well as extreme weather events, resulting in fall in the contribution of apple to their earnings (Table 22.3). Apple requires a chilling period of about 10 weeks below 5 °C for bud-break in springtime (Abbott 1984), which is now available at 2500 m and above because of global warming (Basannagari and Kala 2013). There are evidences to indicate the shift of apple cultivation to higher valleys like Lahul-Spiti, in H.P. (Vedwan and Rhoades 2001). Additionally, alterations in the floral diversity due to land use and land cover change, extinction of local cultivars and use of pesticides have adversely affected pollinators, such as honey bee, causing further decline in apple productivity (Partap and Partap 2003).

Some of the impacts based on anecdotal evidences on mountain agriculture that are linked with CC in the Himalayan region are: (1) Reduced availability of water for irrigation; (2) Extreme drought events and shifts in the rainfall regime resulting into failure of crop germination and seed set; (3) Invasion of weeds in the croplands (e.g., Lantana camara, Parthenium odoratum, Eupatorium hysterophorus etc.); (4) Increased frequency of insect-pest attacks; (5) Decline in crop yield (Negi et al. 2012). These factors have led to loss in agridiversity and change in crops and cropping patterns and many crops are at the brink of extinction, such as Hibiscus sabdar-

Table 22.3   Apple cultivation in three major valleys of Himachal Pradesh

Valley	Mean annual temperature (°C)	Orchard area/household/year		Income from fruit (% of total income	
		1995	2005	1995	2005
Kullu	17	0.55	0.45	69.9	39.6
Shimla	15.4	0.62	0.60	59.3	32.8
Lahul-Spiti	<14	0.48	1.09	17.2	29.1

Source: Rana et al. (2008)

iffa, Panicum miliaceum, Perilla fruitescens, Setaria italica, Vigna spp. etc. (Maikhuri et al. 2001). Studies at the Indian Agriculture Research Institute, New Delhi indicate the possible loss of 4–5 million tons in wheat production in future with every rise of 1 °C temperature throughout the growth periods (Uprety and Reddy 2008). More extreme weather events such as droughts and floods are likely to cause pest and disease outbreaks (Rosenzweig et al. 2001) that could threaten food security.

Nearly 30 million migrants (15% of worldwide 200 million migrants) are from HKH countries (Hoermann and Kollmair 2008). In the foregoing description, water scarcity and disasters due to CC are cited among the recent challenges that farmers encounter. De-intensification and abandonment of agricultural land (up to 30% of total cultivated land) has been reported from several areas of the Nepal Himalayas (e.g., Adhikari 1996, Gautam 2004; Jackson et al. 1998; Khanal 2002; Thapa 2001;). In Garhwal Himalayas, it has been reported upto 60% in some areas (Negi and Joshi 1996). Following abandonment of agriculture, terrace risers are damaged, and land is affected by rills, gullies, sheet-wash, which eventually result in severe soil erosion, landslides and landslips (Khanal and Watanabe 2006; Smadja 1992). One of the consequences of de-intensification of agriculture has been decline in livestock population and reduced supply of farmyard manure. How much is the role of CC in abandonment of agriculture is not known, but the process of land degradation following abandonment can be accelerated because of weather extremes and frequent droughts and extent of excessive rain, particularly when abandoned land is subjected to grazing, burning and infestation with invasive alien species (Khanal and Watanabe 2006). In addition, invasion of abandoned croplands by Lantana and other weeds in the warm valleys of Uttarakhand has provided shelter for wild boar and intensified man-wildlife conflict (Singh 1991).

22.3.7   Forest Degradation May Amplify Climate Change Impact

Of the 22.6 million ha forest area in Indian Himalayas, about 38% is under open forests (with less than 40%canopy density), largely because of degradation (FSI 2011). Baland et al. (2009) have recorded that such forests account for over 50% of forest area in Himachal Pradesh state of India. Agriculture in much of Himalayas heavily depends on forest biomass, as forest floor litter is collected to prepare manure to fertilize the agricultural land (Singh and Singh 1992). However, in some instances, sporadic forest recovery can be seen (Singh and Thadani 2015). Compared to an intact forest in Himalayas, the degraded forest ecosystems have about 50% less biomass and about 80% less carbon sequestration capacity (Singh and Singh 1992; Singh et al. 2019) (Fig. 22.3). In a depauperate forest floor environment even litter decomposition is drastically slowed down, and seeds of trees such as oaks are desiccated much before the arrival of monsoon rain. Increase in the events of heavy rains is likely to accelerate soil erosion from abandoned cropfields,

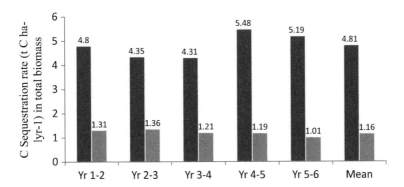

Fig. 22.3 The rate of carbon sequestration in two adjacent forest areas in Uttarakhand (India) over the period from 2003–2010.The dark bars show a well managed community forests; the light bars are unmanaged forest. The well managed forest consists of Banj-oak (Quercus leucotrichophora) and has a high sequestration rate; the degraded forest consists of mixed banj oak and pine (Pinus roxburghii) forest and has a poor sequestration rate. Community management saves an average of 13.4 tonnes CO_2/Ha/year in emissions. (Singh et al. 2011)

landslides, collapse of terraces and overall soil erosion (Singh 2014). The impact of CC on a forest is likely to be far more severe when it is degraded than when it is in a healthy condition. Climate change can increase evapotranspiration loss and drought intensity, accelerate soil erosion, increase pest and diseases infestation and reduce biodiversity. These processes can exacerbate and accelerate land degradation (Webb et al. 2017). In the Himalayan mountains, continued extraction of biomass for firewood, fodder, and NTFPs, and frequent fires lead to forest degradation, which is achronic disturbance (Singh 1998).

Climate change in Himalayas may have several implications for forests, such as forest fires and change in timberline elevation. Increasing incidents of forest fire is an important issue of forest degradation in the western Himalayan region. Though the scale and intensity of fire are driven by several factors, a decline in pre-monsoon precipitation and warmer temperature have increased both frequency and area affected by fire. Fire incidents are fewer in the year's with more pre-monsoon rains (Singh et al. 2016). Fire and smoke pollute air, and deteriorate living condition for local people and tourists. The Third Assessment Report of IPCC (2001) concluded that even with global warming of 1–2 °C, the forest ecosystems could be seriously impacted by future CC through changes in species composition, productivity and biodiversity (Leemans and Eickhout 2004).

Phenological earliness as an effect of global warming has been also well established (e.g., Menzel et al. 2006). In the western Himalayan mountains early flowering of Rhododendron has been attributed to global warming (Gaira et al. 2014). There is a likely impact on natural regeneration of forests trees like sal (Shorea robusta), tilonj oak (Quercus floribunda) and kharsu oak (Q. semecarpifolia) in which seed maturation and seed germination coincide with monsoon rainfall. In wet conditions, these species show vivipary. A rise in temperature and water stress may advance seed maturation, which might result in the breakdown of synchrony between monsoon rains and seed germination (Singh et al. 2010). Impact of pre-monsoon drying due to climate change is also seen on banj oak regeneration. Banj oak seeds mature and fall during winter months, remain on forest floor for several months and germinate only when monsoon arrives or there are pre-monsoon rain showers. But because of global warming they get desiccated well before rains arrive, resulting in a failure of regeneration of banj oak (Quercus leucotrichophora) (Singh 2017). Inevitably, any change in the forest (distribution, density and species composition) would immensely influence households economies and livelihood linked to forestry, agriculture, livestock husbandry, NTFPs and medicinal plants. Better stewardship of land, particularly by taking steps which restore degraded forests can go a long way not only in enhancing several forest based ecosystem services, but also in CC mitigation. Griscom et al. (2017) argue that these natural climate solutions provide a robust basis for global actions to improve ecosystem stewardship to address CC.

22.3.8 Treeline, Climate Change and Pre-monsoon Drought

Since tree growth stops in high mountains because of heat deficiency, climate change is expected to raise treeline elevation. This CC induced upward movement of treeline has been a subject of wider interest globally (Cannone et al. 2007; Kelly and Goulden 2008; Pauli et al. 2003). Upward shift of treeline on a slope in a warming world, partly depends on pre-monsoon conditions. Warming alone may, in fact suppress treelines by causing soil water deficiency (Dawadi et al. 2013; Yadav 2009). Studies on the relationship between tree ring growth and past climate have highlighted the importance of pre-monsoon water (Yang et al. 2014). Changes in treeline elevation could have adverse consequence for herb species diversity, soil carbon and several vital ecosystem processes. Species growing in treeline and near summits can be in problem because of climate change. Singh and Singh (1992) indicate that in long-run a warming of even 1 °C temperature can halve the area under Quercus semecarpifolia forests, a major evergreen oak of high elevation forests distributed in island like fashion near summits.

22.3.9 Climate Change Justice for Himalayas

Very low ratio of per capita CO_2 emission from fossil fuels to the burden of climate change impact is a common feature across the Himalayan region. Per capita fossil fuel CO_2 emission of India is among the lowest in the world, less than 2 t CO_2, compared to global average of ~4.7 t CO_2 or so (Rao and Phadke 2017). It is ridiculously low in Himalayan countries and states (0.4 t CO_2 for Nepal) (Anonymous 2014). Forests over a large area in Himalayas sequester carbon (Singh 2007), and the rivers of the region are tamed to generate hydropower, which is also used by the people living outside Himalayas, which in turn favourably contribute to their carbon budgets. It is quite possible that some Himalayan regions are net carbon sequesters. This means that their CO_2 emissions are less than CO_2 saved by electricity generation and carbon sequestration. However, people in Himalayas suffer most as a consequence of CC which include cloud burst, GLOFs, avalanches, landslides, forest fires, road disruption and road blocks, water scarcity, and crop failures. The cost of infrastructure development in Himalayas is much higher than in plains, and often because of the shortage of finances their quality does not correspond to the requirements in a geologically fragile region. The extra finances required should be made available to mountain states to develop better roads, stable slopes, and measures for CC proofing.

22.4   Conclusion

Climate change affects almost all components of Himalayan environment and people in conjunction with other anthropogenic factors. Despite so much heterogeneity in HKH region and paucity of data, some conclusions are possible to make about CC and its impact on sustainable development. The region is warming at a higher rate than global average rate, and glaciers are, by and large shrinking. Though the impact of rapid retreat of glaciers on river flow is uncertain, most rivers originating from the region are most likely to continue to flow. In the western parts where monsoon is weak, the impact of loss of glaciers and snow, however, can have severe implication on rivers. Pre-monsoon drought, which seems to have intensified under the influence of CC has consequences for several components including drying up of springs, spring water availability, forest fire intensity and frequency, growth and regeneration of important forest tree species. Interactions between CC and forest degradation are likely to reduce ecosystem resilience. Consequences such as invasion of weeds, severity of soil erosion, loss of soil fertility and crop yield, and abandoned of cropfields also impact food security, man-wildlife conflicts and outmigration. However, the form and degree of CC vary both along elevation gradient and the east-to-west Himalayan arc. Consequently, the impact of CC is not the same across a highly heterogenous HKH. It remains to be ascertained that how much of these impacts are due to CC alone or have global change compounding effect, requires further research and analysis of anecdotal evidence to provide coping up strategies for sustainable development. There is a need to strengthen climate data collection in the western Himalayan region. Local climate data are scarce, assessment methods are usually not uniform and the instrumentation is not sufficiently standardized (Negi et al. 2012). The vulnerable mountain ecosystems are likely to face greater risk of CC impacts than other ecosystems. Coordinated efforts are therefore required to develop effective strategies for adaptation and mitigation.

References

Abbott DL (1984) The apple tree: physiology and management. Growers Association, Kullu Valley

Adhikari J (1996) The beginnings of agrarian change: a case study in Central Nepal, TM Publication, Kathmandu. Google Scholar

Ageta Y, Kadota T (1992) Predictions of changes of glacier mass balance in the Nepal Hirnalaya and Tibetan Plateau: a case study of air temperature increase for three glaciers. Ann Glaciol 16:89–94

Anonymous (2014) CO_2 emissions (metric tons per capita)-World Bank Open Data. https://data.worldbank.org/indicator/EN.ATM.CO2E.PC?view=map

Bajracharya SR, Shrestha B (eds) (2011) The status of glaciers in the Hindu Kush-Himalayan region. ICIMOD, Kathmandu

Bajracharya SR, Maharajan R, Shrestha F, Guo W, Liu S, Immerzeel W, Shrestha B (2015) The glaciers of Hindu Kush Himalayas: current status and observed changes from the 1980s to 2010. Int J Water Resour Dev 31:161–173

Baland JM et al (2009) The environmental impact of poverty: evidence from firewood collection in rural Nepal, Economic Development and Cultural Change (forthcoming)

Barnett TP, Adam JC, Lettenmaier DP (2005) Potential impacts of a warming climate on water availability in snow-dominated regions. Nature 438:303–309

Basannagari B, Kala CP (2013) Climate change and apple farming in the Indian Himalayas: a study of local perceptions and responses. PLoS One 8:e77976. https://doi.org/10.1371/journal.pone.0077976

Bhambri R, Mehta M, Dobhal DP, Gupta AK (2015) Glacier lake inventory of Uttarakhand, Wadia Institute of Himalayan Geology. Allied Printers, Dehradun

Burbank DW, Bookhagen B, Gabet EJ, Putkonen J (2012) Modern climate and erosion in Himalaya. C R Geosci 344:610–626

Cannone N, Sgorbati S, Guglielmin M (2007) Unexpected impacts of climate change on alpine vegetation. Front Ecol Environ 5(7):360–364

Chhettri N, Shakya B, Thapa R, Sharma E (2008) Status of a protected area system in the Hindu Kush Himalayas: an analysis of PA coverage. Int J Biodivers Sci Manage 4(3):164–178

Dawadi B, Liang E, Tian L, Devkota LP, Yao T (2013) Pre-monsoon precipitation signal in tree rings of timberline Betula utilis in the Central Himalayas. Quat Int 283:72–77

Dhar U (2002) Conservation of plant endemism in high-altitude Himalaya. Curr Sci 82(2):141–148

Ding YH, Ren GY (2008) An introduction to China climate change science. China Meteorological Press, Beijing

EDW Working Group (2015) Elevation-dependent warming in mountains of the world. Nat Clim Chang 5:424–430

FAO (2015) Mapping the vulnerability of mountain peoples to food insecurity. Food and Agriculture Organization of the United Nations, Rome

Gaira KS, Rawal RS, Rawat B, Bhatt ID (2014) Impact of climate change on the flowering of Rhododendron arboreum in central Himalaya, India. Curr Sci 106:1735–1738

Gautam J (2004) Abandonment of cultivable land: farmer's dependency on imported cereals (in Nepali). Kantipur Daily 29 June 2004, p 4. Google Scholar

Griscom BW et al (2017) Natural climate solutions. PNAS 114(44):11645–11650. https://doi.org/10.1073/pnas.1710465114

FSI (2011) Indian State of Forest Report 2011. Forest Survey of India, Dehradun

Henson R (2006) The rough guide to climate change. Rough Guides Ltd, London, p 325

Grover V, Borsdorf A, Breuste JH, Tiwari PC, Frangetto F (2015) Impacts of global change on mountains: adaptation and responses. CRC Press, Taylor and Francis Group, New York

Hoermann B, Kollmair M (2008) Labour migration and remittances in the Hindu-Kush Himalayan region, ICIMOD background paper. www.books.icimod.org

Immerzeel WW, Van Beek LPH, Konz M, Shrestha AB, Bierkens MFP (2012) Hydrological response to climate change in a glacierized catchment in the Himalayas. Clim Chang 110:721–736

IPCC (2001) 3rd assessment report. http://en.wikipedia.org/wiki/IPCC_Third_Assessment Report

Jackson WJ, Tamrakar RM, Hunt S, Shepherd RK (1998) Land-use changes in two Middle Hills districts of Nepal. Mt Res Dev 18(3):193–212

Jain SK (2008) Impact of retreat of Gangotri glacier on the flow of Ganga river. Curr Sci 95(8):1012–1014. Google Scholar

Jeelani G (2008) Aquifer response to regional climate variability in a part of Kashmir Himalaya in India. Hydrol J 16:1625–1633

Kapos V, Rhind J, Edwards M, Price MF, Ravilious C (2000) Developing a map of the world's mountain 269 forests. In: Price MF, Butt N (eds) Forests in sustainable mountain development: a state-of-knowledge report for 270 2000. CAB International, Wallingford, pp 4–9

Kelly AE, Goulden ML (2008) Rapid shifts in plant distribution with recent climate change. Proc Natl Acad Sci 105(33):11823–11826

Khanal NR (2002) Land use and land cover dynamic in the Himalaya: a case study of the Madi watershed, Western Development Region, Nepal (PhD dissertation). Tribhuvan University, Kirtipur. Google Scholar

Khanal RN, Watanabe T (2006) Abandonment of agricultural land and its consequences: a case study in the Sikles area, Gandaki basin, Nepal Himalaya. Mt Res Dev 26(1):32–40

Leemans R, Eickhout B (2004) Another reason for concern: regional and global impacts on ecosystems for different levels of climate change. Glob Environ Chang 14(3):219–228

Liu X, Chen B (2000) Climate warming in the Tibetan plateau during recent decades. Int J Climatol 20:1729–1742

Mahapatra SK, Reddy GPO, Nagdev R, Yadav RP, Singh SK, Sharda VN (2018) Assessment of soil erosion in the fragile Himalayan ecosystem of Uttarakhand, India using USLE and GIS for sustainable productivity. Curr Sci 115:108–112

Maikhuri RK, Rao KS, Semwal RL (2001) Changing scenario of Himalayan agroecosystems: loss of agrobiodiversity, an indicator of environmental change in Central Himalaya, India. Environmentalist 21:23–39

Menzel A, Sparks TH, Estrella N et al (2006) European phenological response to climate change matches the warming pattern. Glob Chang Biol 12:1969–1976

Miller JD, Immerzeel WW, Rees G (2012) Climatic change impacts on glacier hydrology and river discharge in the Hindu Kush–Himalayas. Mt Res Dev 32(4):461–467. BioOne, Google Scholar

Negi GCS, Joshi V (1996) Land use in a Himalayan catchment under stress: system responses. Ambio 25(2):126–128

Negi GCS, Joshi V (2002) Drinking water issues and hydrology of springs in a mountain watershed in Indian Himalaya. Mt Res Dev 22(1):28–31

Negi GCS, Joshi V (2004) Geohydrology of springs in a mountain watershed: the need for problem solving research. Curr Sci 71(10):772–776

Negi GCS, Samal PK, Kuniyal JC, Kothyari BP, Sharma RK, Dhyani PP (2012) Impact of climate change on the western Himalayan mountain ecosystems: an overview. Trop Ecol 53(3):345–356

Nogués-Bravo D, Araújo MB, Errea MP, Martinez-Rica JP (2007) Exposure of global mountain systems to climate warming during the 21st century. Glob Environ Chang 17(3):420–428

Partap U, Partap T (2003) Warning signals from the apple valleys of the Hindu Kush – Himalayas: productivity concerns and pollination problems. ICIMOD, Kathmandu

Pauli H et al (2003) Assessing the long-term dynamics of endemic plants at summit habitats. In: Nagy L et al (eds) Alpine biodiversity in Europe, Ecological studies 167. Springer, Berlin, pp 195–207

Peng S, Piao S, Ciais P et al (2013) Asymmetric effects of daytime and night-time warming on Northern Hemisphere vegetation. Nature 501(7465):88–92

Pradhan N (2015) An integrated Springshed management approach linking science, policy and practice collaborative applied research in the Kailash sacred landscape (India and Nepal).

International Centre for Integrated Mountain Development (ICIMOD) GPO Box 3226, Kathmandu, Nepal

Rana RS, Bhagat RM, Kalia V, Lal H (2008) Impact of climate change on shift of apple belt in Himachal Pradesh. In: ISPRS Archies XXXVIII-8/W3 workshop proceedings; Impact of climate change on agriculture, pp 131–137

Rao AB, Phadke PC (2017) IOP conference series: earth and environmental science. 76: 012011

Rasul G (2011) The role of Himalayan mountain systems in food security and agricultural sustainability in South Asia. Int J Rural Mgmt 6(1):95–116

Rawat JS (2009) Saving the Himalayan Rivers: developing spring sanctuaries in headwater regions. In: Shah BL (ed) Natural resource conservation in Uttarakhand. Ankit Prakashan, Haldwani, pp 41–68

Rees HG, Holmes MGR, Young AR, Kansaker SR (2004) Recession-based hydrological models for estimating low flows in ungauged catchments in the Himalayas. Hydrol Earth Syst Sci 8(5):891–902

Ren GY, Shrestha AB (2017) Climate change in the Hindu Kush Himalaya. Adv Clim Chang Res 1:4

Ren Y-Y et al (2017) Observed changes in surface air temperature and precipitation in the Hindu-Kush Himalayan region over the last 100-plus years. Adv Clim Chang Res 8(3):148–156

Rosenzweig C, Iglesias A, Yang XB, Epstein PR, Chivian E (2001) Climate change and extreme weather events: implications for food production, plant diseases, and pests. Glob Chang Hum Health 2:90–104

Sanjay J, Krishman R, Shrestha AB et al (2017) Downscaled climate change projection for the Hindu Kush Himalayan region using CORDEX South Asia regional climate models. Adv Clim Chang Res 8(3):185–198

Scherler D, Bookhagen B, Strecker M (2011) Spatially variable responses of Himalayan glaciers to climate change affected debris cover. Nat Geosci 4:156–159

Schickhoff U et al (2015) Do Himalayan treelines respond to recent climate change? An evaluation of sensitivity indicators. Earth Syst Dynam 6:245–265

Schild A, Sharma E (2011) Sustainable mountain development revisited. Mt Res Dev 31(3):237–241

Shrestha AB, Wake CP, Mayewski PA et al (1999) Maximum temperature trends in the Himalaya and its vicinity: an analysis based on temperature records from Nepal for the period 1971–1994. J Clim 12(9):2775–2786

Singh DS (2014) Surface processes during flash floods in the glaciated terrain of Kedarnath, Garhwal Himalaya and their role in the modification of landforms. Curr Sci 106:594–597

Singh SP (1991) Structure and function of the low and high altitude grazing land ecosystems and their impact on live stock component in the central Himalaya. Final report (Department of Environment, Govt. of India, Ref. No. 14/87-ER). Kumaun University Nainital

Singh SP (1998) Chronic disturbance, a principal cause of environmental degradation in developing countries. Environ Conserv 25:1–2

Singh SP (2007) Selling ecosystem services. Samaj Vigyan Shodh Patrika. Special issue (Uttarakhand), p 53

Singh SP (2017) Climate change in Himalayas: research findings, complexities and institutional roles. 23rd Pt. Govind Ballabh Pant memorial lecture, September 10, 2017 at Kosi, Katarmal, Almora

Singh SP, Sharma S (2014) Urbanisation challenges in the Himalayan region in the context of climate change adaptation and disaster risk mitigation stakeholders in the Nainital Lake and its watershed and the benefits/values derived. Background Paper. Centre for Urban Green Spaces, New Delhi

Singh JS, Singh SP (1992) Forests of Himalaya. Gyanodaya Prakashan, Nainital

Singh SP, Singh V (2016) Addressing rural decline by valuing agricultural ecosystem services and treating food production as a social contribution. Trop Ecol 57(3):381–392

Singh SP, Thadani R (2015) Complexities and controversies in Himalayan research: a call for collaboration and rigor for better data. Mt Res Dev 35(4):401–409

Singh JS, Pandey U, Tewari AK (1984) Man and forests: a Central Himalayan case study. Ambio 13(2):80–87

Singh SP, Singh V, Skutsch M (2010) Rapid warming in Himalayas, ecosystem responses and development options. Clim Dev 3:221–232

Singh SP, Khadka IB, Karky BS, Sharma E (2011) Climate change in the Hindu Kush-Himalayas. The state of current knowledge. ICIMOD, Kathmandu

Singh RD, Gumber S, Tewari P, Singh SP (2016) Nature of forest fires in Uttarakhand: frequency, size and seasonal patterns in relation to pre-monsoonal environment. Curr Sci 111(2):398–403

Singh SP, Phartyal SS, Rosbakh S (2017) Tree seed traits' response to monsoon climate and altitude in Indian subcontinent with particular reference to the Himalayas. Ecol Evol 7(18):7408–7419

Singh SP, Pandey A, Singh V (2019) Nature and extent of Forest degradation in Central Himalayas. Springer. (in Press)

Smadja J (1992) Studies of climate and human impacts and their relationship on a mountain slope above Salme in the Himalayan Middle Mountains. Mt Res Dev 121:1–28. Google Scholar

Smadja J et al (2015) Climate change and water resources in the Himalayas. J Alp Res 103:1–33

Sun XB, Ren GY, Shrestha AB et al (2017) Changes in extreme temperature events over the Hindu Kush Himalaya during 1961–2015. Adv Clim Chang Res 8(3):157–165

Tambe S, Kharel G, Arrawatia ML, Kulkarni H, Mahamuni K, Ganeriwala AK (2012) Reviving dying springs: climate change adaptation experiments from the Sikkim Himalaya. Mt Res Dev 32(1):62–72. BioOne, Google Scholar

Tewari PC, Joshi B (2012) Natural and socio-economic factors affecting food security in the Himalayas. Food Sec 4(2):195–207

Tiwari PC (2008) Land use changes in Himalaya and their impacts on environment, society and economy: a study of the Lake region in Kumaon Himalaya, India. Adv Atmos Sci 25:1029–1042

Thapa PB (2001) Land-use/land cover change with focus on land abandonment in Middle Hills of Nepal: a case study of Thumki VDC, Kaski District. M.A. dissertation, Tribhuvan University, Kirtipur. Google Scholar

Thayyen RJ, Gergan JT, Dobhal DP (2007) Role of glacier and snow cover on head water river hydrology in monsoon regime- micro-scale study of din gad catchment, Garhwal Himalaya, India. Curr Sci 92(3):376–382. Google Scholar

Uprety DC, Reddy VR (eds) (2008) Rising atmospheric carbon dioxide and crops. Indian Council of Agricultural Research, New Delhi

Valdiya KS, Bartarya SK (1989) Diminishing discharges of mountain springs in a part of Kumaun Himalaya. Curr Sci 58(8):417–426

Valdiya KS, Bartarya SK (1991) Hydrogeological studies of springs in the catchment of the Gaula River, Kumaun lesser Himalaya, India. Mt Res Dev 11(3):239

Vedwan N, Rhoades RE (2001) Climate change in the western Himalayas of India: a study of local perception and response. Clim Res 19:109–117

Vishvakarma SCR, Kuniyal JC, Rao KS (2003) Climate change and its impact on apple cropping in Kullu Valley, North-West Himalaya, India. In: 7th international symposium on temperate zone fruits in the tropics and subtropics, 14–18 October, Nauni-Solan (H.P)

Webb NP et al (2017) Land degradation and climate change: building climate resilience in agriculture. Front Ecol Environ 15(8):450–459. https://doi.org/10.1002/fee.1530

Wu J, Xu Y, Gao XJ (2017) Projected changes in mean and extreme climate over Hindu Kush Himalayan region by 21 CMIP5 model. Adv Clim Chang Res 8(3):176–184

Xu JC, Grumbine RE, Shrestha A, Eriksson M, Yang XF, Wang Y, Wilkes A (2009) The melting Himalayas: cascading effects of climate change on water, biodiversity, and livelihoods. Conserv Biol 23:520–530

Yadav RR (2009) Tree ring imprints of long-term changes in climate in western Himalaya. India J Biosci 34(5):699–707

Yang B, Qin C, Wang J, He M, Melvin TM, Osborn TJ, Briffa KR (2014) A 3500-year tree-ring record of annual precipitation on the northeastern Tibetan Plateau. Proc Natl Acad Sci U S A 111(8):2903–2908

Yao T, Thompson L, Yang W, Yu W, Gao Y, Guo X, Yang X, Duan K, Zhao H, Xu B, Pu J, Lu A, Xiang Y, Kattel DB, Joswiak D (2012) Different glacier status with atmospheric circulations in Tibetan Plateau and surroundings. Nat Clim ChangNat Clim Change 2:663–667. https://doi.org/10.1038/NCLIMATE1580

You QL, Ren GY, Zhang YQ et al (2017) An overview of studies of observed climate change in the Hindu Kush Himalayan (HKH) region. Adv Clim Chang Res 8(3):141–147

Zhan YJ, Ren GY, Shrestha AB et al (2017) Change in extreme precipitation events over the Hindu Kush Himalayan region during 1961e2012. Adv Clim Chang Res 8(3). https://doi.org/10.1016/j.accre.2017.08.002

Chapter 23
Challenges of Urban Growth in Himalaya with Reference to Climate Change and Disaster Risk Mitigation: A Case of Nainital Town in Kumaon Middle Himalaya, India

Prakash C. Tiwari and Bhagwati Joshi

Abstract Himalaya is the most rapidly urbanizing mountains of the world where the process of urban growth has been fast, but mostly unplanned and unregulated. During the recent decades, Himalaya has experienced rapid urbanization mainly in response to population growth, improved road connectivity, development of tourism and economic globalization. As a result, urbanization has emerged as one of the important drivers of land use and environmental changes depleting natural resources and biodiversity, and disrupting natural drainage and ecosystem services; and increasing vulnerability of anthropogenically modified slopes to a variety of natural risks, particularly under climate change. The paper analyzes the emerging threats of unplanned urban growth in the densely populated Middle Himalayan Ranges with a case illustration of Nainital Town located in Uttarakhand. It was observed that the rapid urbanization is increasing the susceptibility of intensively modified and densely populated fragile slopes to the active processes of mass movement and landslides. Moreover, the rapidly changing climatic conditions, particularly the climate change induced hydrological extremes are posing severe threats to the sustainability of fast-growing urban ecosystem by increasing the frequency, intensity and severity of geo-hydrological hazards in the town and its surrounding region. The city development plan and also the state disaster risk reduction framework and climate change adaptation plan did not make any provision for addressing the emerging risks of climate change in Nainital and other towns of Uttarakhand. A comprehensive climate change vulnerability assessment and mapping of the town should be carried taking into account all the critical parameters of exposure, sensitivity and adaptive capacity of urban ecosystem involving a range of institutions and stakeholders.

P. C. Tiwari (✉)
Kumaun University, Nainital, Uttarakhand, India

B. Joshi
Department of Geography, Government Post Graduate College, Rudrapur, Uttarakhand, India

© Springer Nature Switzerland AG 2020 473
A. P. Dimri et al. (eds.), *Himalayan Weather and Climate and their Impact on the Environment*, https://doi.org/10.1007/978-3-030-29684-1_23

23.1 Introduction

Mountains are highly critical from the view point of marginality, environmental sensitivity, constraints of terrain, inaccessibility and climate change (Meybeck et al. 2001). But, mountains have long been marginalized from the viewpoint of sustainable development of their resources and inhabitants. However, we are experiencing an emergence of responsiveness of the ecological significance of mountain systems and their environmental significance for the sustainability of the global community, particularly after the United Nations Conference on Sustainable Development – the Rios Earth Summit in 1992 (UN). As a result, our understanding about the dilemmas of mountain ecosystems and approach to their development has undergone drastic changes, during the last two decades. Currently, mountain ecosystems, as well as mountain communities, are particularly threatened by the ongoing processes of global environmental change (Borsdorf et al. 2010).

During the recent years, a variety of changes have emerged in the traditional resource use structure in high mountain areas, particularly in developing and underdeveloped regions mainly in response to economic globalization, increased demographic pressure and rapid urban growth (Tiwari 2014). As a result, mountain regions of the world are passing through a process of rapid environmental, socioeconomic and cultural transformation and exploitation and depletion of their natural resources leading to ecological un-sustainability both in upland and lowland areas (Haigh 2002). Moreover, the changing climatic conditions have already stressed mountain ecosystems through higher mean annual temperatures, altered precipitation patterns and frequent extreme weather events (ICIMOD 2010).

Urbanization has emerged as one of the important drivers of global change transforming mountain regions, particularly in developing countries where the process of urban growth has been fast, but mostly unplanned and unregulated (Joshi and Pant 1990; Anbalagan 1993). Urbanization has contributed significantly not only to economic growth through the improvement of infrastructure, development of tourism and the generation of employment opportunities in the mountains; but has also increased community sustainability by strengthening social services, particularly, education, health and communication in their vast rural hinterlands in high mountains (Joshi and Pant 1990). At the same time, rapid and unplanned urbanization is intensifying land use within the cities as well as in their peri-urban zone and unlocking even remote areas of mountains for exploitation of their natural resources by the growing global markets (Anbalagan 1993). These changes are increasing the vulnerability of anthropogenically modified slopes to a variety of natural risks, particularly under rapidly changing climatic conditions (Singh and Gopal 2002). Himalaya representing tectonically alive, and densely populated mountain ecosystems has experienced rapid urbanization during recent decades mainly in response to population growth, improved road connectivity, development of tourism and economic globalization (Ives and Messerli 1990).

The fast expansion of road linkages has facilitated the rapid urbanization through emergence and growth of rural service centers and improved access to markets.

Consequently, more recently, comparatively less accessible areas have also come under the process of speedy urbanization due to the growth of domestic tourism, marketing and increasing popularity of new tourists destinations; and resultant gradual shift from primary resource development practices to secondary and tertiary sectors in the region (Joshi and Pant 1990). This has resulted into tremendous increase in size, area, number, and complexity of urban settlements in Himalaya resulting into the expansion as well as the intensity of urban land use within the towns and their rural hinterlands (Tiwari 2007, 2008).

The sprawling urban growth in fragile mountains and resultant land use intensifications have disrupted the hydrological system of urban areas, and consequently increased the susceptibility of the Himalayan towns to recurrent slope failures, landslides and flash floods (Anbalagan 1993). Moreover, climate change has stressed urban ecosystems by increasing the frequency, severity and intensity of extreme weather events. The natural risks of this unplanned urban sprawl are now clearly discernible in most of the urban centres and their peri-urban zone, all across the densely populated Lesser Himalayan ranges in India (Tiwari 2007, 2008). The urban development in the region is also having long-term impacts on the fragile ecosystem and environment of the urban fringe consisting of natural forests, wildlife habitats, critical headwaters, and prime agricultural land (Tiwari 2014). The natural components of the urban fringe zone are being degraded and depleted steadily and significantly through the expansion of urban land use, deforestation, habitat destruction, mining construction material, waste and sewage disposal, encroachment of productive agricultural land, and changes in the traditional land use and resource management practices under the multiplier effect of urban growth (Tiwari 2007, 2008). The paper analyzes the emerging threats of unplanned urban growth in the densely populated Lesser Himalayan ranges with a case illustration of Nainital Town located in Uttarakhand Himalaya.

Nainital is one of the most important town and popular tourist destination situated in the Lesser Himalayan ranges in Kumaon division of Uttarakhand State (Fig. 23.1). The city encompasses a geographical area of 4.32 km^2 between 1938 m and 2611 m from the mean sea level in the watershed of picturesque Naini Lake (Joshi et al. 1983). Nainital is located in the proximity of the Main Boundary Thrust (MBT) – the tectonic juncture between the Siwaliks mountains (the Outer Himalayan Ranges) situated in the south from the Lesser Himalayan (Middle Himalaya) ranges located in the north. Besides MBT, the town is crisscrossed by several other minor faults which make the geology of the watershed highly complex (Valdiya and Bartarya 1991; Hukku et al. 1977) (Fig. 23.2). A major fault line, called the Nainital fault, passing through the Naini Lake separates the entire watershed into two parts. The morphometric features of Nainital Lake have been presented in Table 23.1 (Rawat 2009).

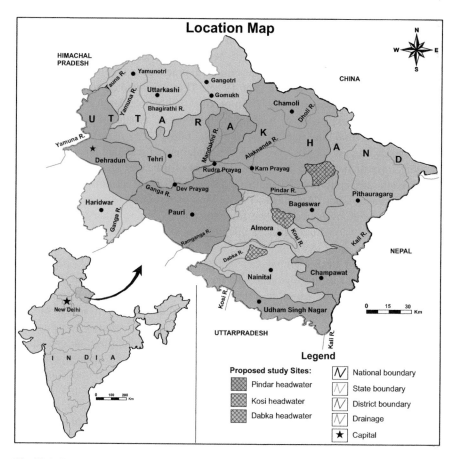

Fig. 23.1 Location map. (Source: Author)

23.2 Objective and Methodology

The main objective of the paper is to analyse the emerging natural and socio-economic risks of unplanned and unregulated urban growth under climate change in densely populated Lesser Himalayan Ranges with a case illustration of Nainital town located in the State of Uttarakhand. The study is mainly based on the analysis and interpretation of secondary data. The data and information used in the present work has been derived from various sources including, Census of India, statistical records of the Government of Uttarakhand, Municipal Council of Nainital, various district level offices, handbooks and reports of both the Central and State Governments department and organizations, and from published and unpublished literature. Besides, necessary qualitative information has also been generated from various primary sources including empirical research and field surveys.

Fig. 23.2 Location of Nainital in Uttarakhand

Table 23.1 Principal morphometric attributes of Naini Lake	Maximum length (m)	1423
	Breadth (m)	253–423
	Maximum depth (m)	27.3 in northern half and 25.5 in southern half
	Mean depth (m)	18
	Surface area (ha)	48
	Lake shoreline (m)	3458
	Volume at maximum level (m³)	8.58

Source: Rawat (2009)

23.3 Results and Discussion

23.3.1 Demographic Growth and Urban Development: A Historical Perspective

Nainital is the most recent town of Kumaon as it was discovered by Mr. P. Barron a European merchant and an enthusiastic hunter in 1841. The habitation in Nainital started towards the end of the first half of the nineteenth century. The building of Nainital Municipal Board was constructed in 1845, and Nainital Municipal Board was formally constituted in the same year. It was the second Municipal Board in the entire North Western Province (Auden 1942; Atkinson 1882). Nainital had become a popular hill resort by 1847 with some 40 houses and buildings. With the growth of town education, administration, real estate, and tourism became the most significant functions of Nainital. The popularity and importance of town was further boosted with the installation of rail link to Kathgodam located at the foot of the mountains in 1884 and the formation of district Nainital in 1891. In 1862, Nainital became the summer seat of the North Western- Provinces (NWP). In 1915, Kathgodam – Nainital road (36 km) was completed, and also electric supply came to the town (Clay 1928). In the latter half of the nineteenth century Nainital witnessed a phenomenal growth in urban functions, and a range of facilities and services emerged to cater the growing needs of the town. Besides, marketing areas, rest houses, recreation centres, dancing halls, rink theatres, cinema halls, holiday centres and military camps; a number of European schools for boys and girls came up in Nainital largely for the children of the British Colonial officials and soldiers; and Nainital became an important centre of education for the British (Joshi et al. 1983). The town enjoyed the status of the summer capital of Uttar Pradesh (U.P.) even after the independence till 1963.

Naini Lake is situated in a densely populated valley in the Kumaon Himalaya and is one of the most popular tourist resorts in Northern India. The picturesque surroundings of valley together with the panoramic beauty of the natural lake, its proximity to plains in the south and salubrious climatic conditions were the main reasons that promoted the development of Nainital a famous health and recreation resort during the British time and afterward (Joshi et al. 1983). Nainital is still one of the most popular tourist destination, and also an important town. The Indo-Pak conflict of 1965 and 1971, and also the continued internal security threats in western Himalaya boosted the tourist economy of the town. However, the tourist influx in the town drastically declined in the year 2013 after the massive disasters that hit Uttarakhand in pre-monsoon months.

The surroundings of Naini Lake which were first inhabited in the 1850s, grew into a town of 6903 people by 1901 and to 41,461 by 2011 (Table 23.2). The population of the town increased by 48.77% during 1901–1911. In the following two decades almost all the mountainous towns of Kumaon were affected by the famine, drought, and epidemic, and consequently, the growth of population was only 9.38% during 1911 and 1921, and following decade the population of Nainital declined by

Table 23.2 Population growth in Nainital (1901–2011)

Census years	Total population	Net change	% Change
1901	6903	–	
1911	10,270	3367	48.77
1921	11,235	965	9.38
1931	10,673	−562	−5.02
1941	11,718	1045	9.72
1951	13,093	1375	11.73
1961	16,080	2987	22.81
1971	25,167	9087	56.51
1981	26,093	926	3.67
1991	29,831	3738	14.32
2001	38,630	8799	29.49
2011	41,461	2831	7.33

Source: Census of India (2011)

5.02% (Table 23.2). The town registered an average death rate of more than 28% per 1000 during 1911–21, however it was reduced to 24.73% per 1000 during 1921–31. At the end of 1931, the growth rate of population was 9.38% which increased slightly (9.77%) in 1941 (Table 23.2). The following decade (1941–51) recorded a low growth rate of the population due to the outbreak of cholera epidemic between 1943 and 1945, and 1947 and in 1948, and plague epidemic from 1946 to 1947 in the region. In spite of this, the decade recorded an increase of 11.73% due to a considerable addition by the migrant population (Table 23.2). The urban population growth, which is positively the bulk of the increase, has taken place after 1951.

Table 23.2 reveals that the region recorded a very high growth rate of population (56.51%) during 1961–71, which was mainly due to the considerable natural increase in population owing to the control over various diseases. The following decades recorded the maximum growth rate of 22.81% (1951–61), 56.51% (1961–1971), 14.32% (1981–91), 29.49% (1991–2001) and 7.33% (2001–2011). During 1971–81 the total growth rate was very low (3.67%). The growth and variation of urban population during 1901 to 1981 had been moderate for Nainital town. The town registered a moderate growth of 7.33% population during 2001–2011 against 29.49% of the previous decade (Table 23.2). The decline of population growth during 2001–2011 is attributed to the establishment of the High Court of Uttarakhand in Nainital and consequent shifting of most of the government offices from Nainital to the growing township of Bhimtal. However, being a major tourist destination and seat of the High Court of Uttarakhand State Nainital has a large floating population, particularly during summer months.

The remarkable feature of the urban demographic evolution of Nainital is that before 1981 the population increased in only those areas which are characterized by gentle slope and lower altitude. However after 1981, higher elevation areas, steep slope and fragile zones and other such environmentally unsafe areas of the town registered phenomenal growth of population. These areas have now also very high concentration and density of population. In 1981, Nainital had an average population density was 211.72 persons/km^2, which increased by over 55 times to 12,148.30 in 2001. The main reasons for the increase in population and its density in Nainital have been the rapid development of tourism, growth of market, service and commercial sectors; and emergence of large number of educational institutions in the town. The increased tourist arrival has not only an impact on the economy of the town, but it equally influenced its evolution and the functional morphology. Conforming to the needs of growing tourism, hotels, restaurants, parks, picnic spots, gardens, shopping centres, parking areas, facilities of recreation, tourist guidance, and transport now constitute important components of the morphology of the town (Joshi et al. 1983).

23.3.2 Seasonal Demographic Flux and Its Implications for Climate Change

As per the Census of India 2011 record, Nainital had a total permanent population of 41,461 persons. In addition to this, the town also hosts a large floating population of approximately 10,000 persons during the peak tourist season from April to June who mostly works as petty vendors, coolies, boatmen, horsemen, waiters in hotels and restaurants. Nainital has been a tourist destination ever since it was discovered in the mid-nineteenth century. The war with Pakistan in 1965 and 1971 further boosted Nainital's tourism industry as the Kashmir valley remained unofficially closed for tourist arrivals for the years till recently. The incidences of high intensity rainfall and the resultant flash floods, landslides and mud and debris flow in 2013 that took the toll of several thousand people across the State of Uttarakhand brought the massive decline in the number of visitors. However, the tourist arrival started picking up in the following years.

In 2003, the floating population of Nainital that mainly include tourists was 0.42 million, which increased to 0.52 million by 2005, recording an increase of 22% with average annual growth rate was 7% over a period of 3 years (2003–2005) (Singh and Gopal 2002). In Nainital, most of the tourists (floating population) arrival is in three summer months (April–June) and 2 months in autumn (October–November) which are known as peak tourist seasons. Taking the tourist population of 2005 (0.52 million) as the base, the average days of stay per tourist as 15, the average tourist load per day works out to 34,533 or say 35,000 (Singh and Gopal 2002). Further, it was that the educational and training institutions and the University together account for at least 20,000 population. Besides, being district headquarters town and location of the district court and High Court, and the office of the Divisional Commissioner, large number of people, visits the town on official business. The estimated number of such visitors is around 7000 (Singh and Gopal 2002). Thus, the total number of floating population in Nainital town has been estimated to be about 129,000 in 2015. Another indicator reflecting the increased tourism activity in recent past is the number of vehicles entering the town. The data shows that the number of light vehicles that entered the town during the peak tourism months has increased by about 46% in the past 3 years (Singh and Gopal 2002). One of the implications of increased tourists vehicle is the rise in vehicular pollution and traffic congestion on the narrow roads with limited carrying capacity. This pollution and traffic congestion is already being felt, and could become a major problem to human health and pedestrians' movement especially during summer months, and particularly in the event of disasters.

23.3.3 Exposure to Climate Change Induced Geo-Hydrological Disasters

Nainital situated in close proximity of the Main Boundary Thrust (MBT) – the tectonic boundary between the Lesser Himalayan Ranges in the north and the Siwaliks Hills in the south – is particularly vulnerable to geo-hydrological hazards (Middlemiss 1880, 1898). Impact of tectonic movements has resulted in intense shearing, faulting, thrusting and fracturing of the rocks observed in the area (Valdiya 1988; Valdiya and Bartarya 1991; Sharma 2006). Consequently, the entire area is tectonically alive, ecologically fragile and highly exposed to a variety of geo-hydrological hazards. The whole township of Nainital is highly vulnerable to landslides and other processes of mass movement, particularly creeping and subsidence (Valdiya 1988; Sharma 2006; Rautelaa et al. 2014). Moreover, the terrain is characterized by the predominance of high relative relief that renders the area highly sensitive to slope failure and processes of the mass movement. Geographically, it is a critical zone, in as much as, it lies within the belt of maximum precipitation (2814.06 mm/year), and also shows relief differences of the highest order (Joshi et al. 1983). As mentioned earlier that due to the presence of Main Boundary Thrust (MBT) and a number of other major and minor faults the town constitutes a tectonically alive domain which is responsible for highly deformed rock conditions in the entire area (Sharma 2006). Besides, during the recent past the rapid urbanization, settlement growth and infrastructural development have been very massive and phenomenal intensifying the anthropogenic process interacting with the fragile environment of the urbanized zone (Tiwari 2014; Tiwari and Joshi 2014, 2015). Moreover, population growth and rapid urbanisation have led to the expansion of construction activities in fragile terrains and has catapulted frequency of landslides to dramatic proportions right since the evolution of the town. The recent observed variability in rainfall pattern has further increased the vulnerability of the settlement to a variety of hydro-geological hazards, particularly the landslides (Sharma 2006).

Nainital has experienced devastating landslides of variable magnitude ever since the evolution and development of town (Oldham 1880; Auden 1942; Nautiyal 1949; Hukku et al. 1977; Pant and Kandpal 1990; Sharma 2006). The five disastrous landslide events occurred in the year 1867, 1880, 1898, 1924 and 1998 caused massive devastation of urban infrastructure and loss of lives in the town and its surrounding areas. These geo-hydrological disasters not only transformed the urban and natural landscape of the town, but also underlined the need to understand the local geo-tectonic and geomorphological conditions before allowing to expand the urbaniza-tion in fragile mountain landscape. It was observed that more than 50% built up area in the township of Nainital is situated on landslide debris deposited over the years (Rautelaa et al. 2014; Disaster Management and Mitigation Centre, Government of Uttarakhand 2011). The geo-hydrological hazards are thus causing colossal damage to the urban infrastructure in the complex geological environment of the urban habi-tation zone of Nainital as well as its surrounding peri-urban areas. Rapid urbaniza-tion in the town may result in increased vulnerability of mountain slopes to mass wasting processes. The urban sprawl of the town during last few decades has been phenomenal consequently exerting pressure on already vulnerable geological envi-ronment (Tiwari 2014; Tiwari and Joshi 2015).

Studies indicated that lack of proper surface drainage and unplanned anthropo-genic intervention emerged as the major reasons for slope instability in Nainital. In view of this detailed network of surface drains were developed in the watershed, and human intervention and construction on unstable slopes were prohibited. In the recent past, the issue of environmental instability around Nainital has also been raised by various civil society groups and individuals in the High Court of Uttarakhand and in the apex court of the country (Supreme Court of India). Both the honourable courts advised against undertaking construction activities on the vulner-able slopes around the lake. In spite of this, the built up area has significantly increased in the town. Although 1.64 km² of total town is demarcated as prohibited area and unsafe for any construction. However, this area is under pressure for fur-ther urban growth without the desired level of planning and development control. During 2005–2010 the built up area in these areas has increased by almost 50% which is much more than average built up area increase of 34% in the town during the period resulting in intensive land use changes (Tables 23.3 and 23.4 and Fig. 23.3). The growth of built up area has been particularly very high in areas pro-hibited for construction where it has increased by more than 56%. The unplanned

Table 23.3 Urban land use changes around Nainital (2005–2010)

Land use classes	Area (in m²)			
	2001	2015	Change in area	% Change
Built up area	630,498.18	844,108.12	213,609.94	33.88
Open area	345,878.41	208,344.41	−137,534.00	−39.76
Trees outside Forest	1,242,649.75	699,390.98	543,258.77	−43.72
Open Forest	2,143,476.85	2,661,936.97	518,460.12	24.19
Dense Forest	7,789,118.89	7,746,524.83	−42,594.06	−0.55
Agriculture	34,849.72	24,333.99	−10,515.74	−30.17
Water-bodies	441,331.48	443,164.00	1832.52	0.42
Total	12,627,803.29	12,627,803.29	–	–

Source: Rautelaa et al. (2014)

Table 23.4 Growth of built up area prohibited zones in Nainital (2005–2010)

Name of the prohibited zone	Area (in m²)			
	2001	2015	Change in area	% Change
Sher ka Danda	62,203.4	97,359.3	35,155.9	56.5
Ayarpatha	31,797.5	48,853.3	17,055.8	53.6
Beyond Lake watershed	57,155.2	82,266.3	25,111.1	43.9
Total	151,156.1	228,478.9	77,322.7	51.2

Source: Rautelaa et al. (2014)

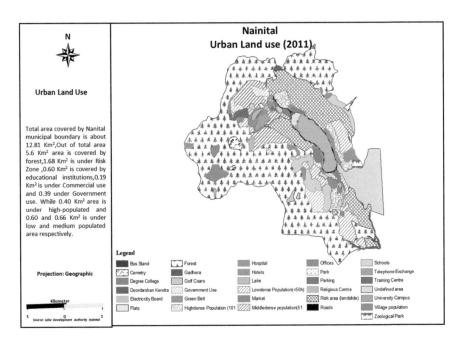

Fig. 23.3 Nainital Town: Urban Land Use [2011]

and unregulated construction on fragile slopes resulted in degradation of forest and biodiversity, and depletion water resources which feed the lake. Studies indicated that steep slopes in this area are presently being levelled for construction of houses and excavated debris is being disposed off along the hill slopes as also along the surface drains. As a result, drainage network is being encroached, obstructed and obliterated increasing the vulnerability of the slopes and large population to climate change induced risks (Disaster Management and Mitigation Centre, Government of Uttarakhand 2011).

Despite rapid urban growth and unplanned constructions of houses and hotels the pressure of heavy influx of tourists and other seasonal population has far exceeded the carrying capacity of urban amenities in Nainital (Table 23.5). Furthermore, the expansion of urbanization and population increase facilitated the emergence and growth of slums in the close proximity of lake, along the drainage channels, fragile

Table 23.5 Urban and tourist
amenties in Nainital

Population and amenities	
Permanent population	39,840
Number of tourists	310,000/year
Number of hotels	120
Number of shops	900
Number of residential houses	8000
Floating population	7000

Source: Singh and Gopal (2002)

Table 23.6 Slum settlements and population in Nainital

S. no.	Slum settlements	Locality	Population
1	Narayan Nagar	Mallital	1947
2	Breysite	Mallital	1360
3	Sardar Line	Mallital	960
4	Committee Line	Mallital	750
5	Bakery Compnd	Mallital	650
6	Mangawali	Tallital	750
7	Harinagar	Tallital	1200
8	Dibhighat	Tallital	780
9	Kathbaas	Tallital	90
10	Rajpura	Mallital	800
11	Sukhatal	Mallital	180
12	Jubleehall	Mallital	200
Total			9667

Source: Urban Development Department, Government of Uttarakhand

slope and other such environmentally unsafe areas in Nainital. As many as 12 slum
pockets have been identified in a small town of 41,000 inhabitants (Table 23.6). The
total population of these slums was 9667 persons in 2011 accounting for about 21%
of the total population of the town. The growth and expansion of slums have further
increased the vulnerability of large population particularly poor and marginalized to
climate change induced risks.

The Naini lake is not only the prime attraction for large number of visitors, but it
constitutes the source of drinking water for most of the population of the town.
Nearly 40% of the total water supply comes directly from the lake (Singh and Gopal
2002). However, due to rapid urban growth and phenomenal magnitude of construction activities in the catchment area the rate of sedimentation has been increasing
progressively posing a serious threat to the quality of water and the life of the lake.
As a result, the mean depth of the lake had reduced by 2.88 m decreasing water
volume by 7682.5 m^3 between a period of 84 years between 1895 and 1979 (Rawat
2009). The bathymetric analysis carried out in Lake Nainital has given a sediment
accumulation rate of 67 m^3/year between 1895 and 1967 and 78 m^3/year during the
period between 1967 and 1979 (Rawat 2009). Sharma (1981) has estimated the life

of the lake to be 314 years based on sedimentation deposit of 0.239 million m^3 during 1960–1977 at the rate of 0.22 m m^3/year. Moreover, carbonate rock lithology which is more susceptible to weathering, high precipitation and frequent landslides may account for a higher sedimentation rate in the lake-bed. The climate change is likely to intensify the magnitude of anthropogenic stress in the lake and affect the availability and quality of water.

23.3.4 Climate Change Adaptation Measures Priorities, Efforts and Gaps

Urbanisation in Uttarakhand has largely been an unplanned process resulting in the lack of civic amenities in proportion to population density. Unplanned urban growth together with rapid urban expansion and increasing inflow of tourists have made severe environmental impacts on the urban ecosystem of Nainital and other urban areas, particularly in view of climate change. Nainital despite being a new town, has grown in a completely unplanned manner causing immense pressure on the limited urban infrastructure and services resulting into the degradation of the urban environmental conditions and increasing vulnerability of the large population to emerging threats of climate change. Major environmental concerns associated with such unplanned urban development are emerging risks of climate change induced geo-hydrological hazards, destruction of forest area, loss of bio-diversity, and depletion of water resources. Despite realizing the increasing vulnerability of urban areas to climate change induced risks no specific climate change adaptation plan has been evolved for any cities of Uttarakhand including Nainital by the State Government (Government of Uttarakhand 2012).

However, Nainital is covered under the Jawaharlal Nehru National Urban Renewal Mission (JNNURM) – an urban development programme sponsored by Government of India – and under which a range of urban development interventions, including the development of city sanitation plan are underway in the town (Urban Development Department, Government of Uttarakhand 2007). However, currently, no detailed climate vulnerability and risk assessments is available for the urban centres of Uttarakhand (Government of Uttarakhand 2012). However, realizing that climate change is likely to have adverse impact on infrastructure and worsen access to basic urban services and quality of life in cities, the State Action Plan on Climate Change (SAPCC) recommended that Urban Development Department (UDD) of Uttarakhand would take necessary steps towards collating available data and information of impacts of climate change on cities, their systems, infrastructure, and people towards improving scientific knowledge and evidence base and understanding of climate change and its impacts (Government of Uttarakhand 2012). The SAPCC says that it will begin the process of developing the necessary systems, databases and protocols for collecting and collating the necessary evidence -based observations and evolve appropriate response strategies.

Towards improving governance mechanisms, institutional decision-making, and convergence, the UDD will initiate the formation of a Climate Cell within the Department. It will take steps to improve understanding of climate change and its effects; education and awareness; and developing and strengthening partnership and cooperation with different stakeholders (Mitchell and Maxwell 2010). It will also initiate processes for developing the necessary coordination mechanisms, sectoral policy initiatives and institutional arrangements to ensure that urban agglomerations and urban populations in the State build their capacity to be resilient to the risks and impacts of climate change through implementing adaptation measures and contributing to mitigation of greenhouse gas emissions (Government of Uttarakhand 2012). The plan further emphasises that 'the UDD will develop and deploy a range of awareness and capacity building programmes for municipal officials for promoting appropriate measures towards climate resilience in their respective Urban Local Bodies (ULBs), as also similar programmes for building awareness on climate change and its impacts for the urban populations. The UDD will also seek to converge such efforts with other sectoral initiatives such as health, education, housing and water, and foster inter and intra departmental coordination' (Government of Uttarakhand 2012).

23.3.5 Climate Change Adaptation and Disaster Risk Mitigation Programs and Gaps

Increasing uncertainty in the precipitation patterns in the state is amply highlighted by the fact that both in the years 2007–08 and 2008–09 the state faced severe drought conditions. In the year 2007–08 nine districts of the state (out of total 13 districts) were officially notified as being drought affected while in 2008–09 ten districts were notified as being drought affected. Whereas, in the years 2010 and 2013 the entire state of Uttarakhand experienced a number of incidences of high intensity rainfall generating flash-floods that triggered devastating landslides and debris flow and caused massive loss of human lives, livelihood assets and infrastructure across the State (Rao et al. 2014; Patra and Kantariya 2014). Under the influence of climate change, the geo-environment of Uttarakhand is increasingly getting more susceptible to some problems that include soil erosion, landslide, prolonged dry spells, glacier recession, erratic precipitation, extreme climate events and rapid loss of habitat and biodiversity. These have a direct implication upon the issues related to the livelihood for people in the state and adjoining regions. During the recent decades, climate change driven fluctuations in the precipitation pattern have shown increasing trends which pose serious threats to ecologically fragile, tectonically active and densely populated urban ecosystems, such as Nainital. During the recent years, the Disaster Mitigation and Management Centre, Dehradun has carried out a number of initiatives in the field of disaster management and mitigation for Nainital town

(Disaster Management and Mitigation Centre, Government of Uttarakhand 2011). These include:

- Land use/land cover change studies using high-resolution satellite data
- Landslide and environmental risk assessment
- Socioeconomic vulnerability assessment
- Preparation of disaster management plans
- Structural vulnerability and risk assessment Vulnerability assessment of the building stock using rapid
- Awareness generation
- Training and capacity building of institutions

However, hazard zone mapping of Nainital Catchment was also carried out by Valdiya (1988), Sharma (2006) and Gupta and Uniyal (2013). However, no detailed climate vulnerability risk assessments so far has been carried out from the view point of disasters, and particularly climate change induced geo-hydrological disasters for any township of Uttarakhand, including Nainital. Nevertheless, in general, the Disaster Management Department has been carrying out a range of related activities for the entire State of Uttarakhand including the urban centres (Disaster Management and Mitigation Centre, Government of Uttarakhand 2011). As per the State Disaster Management Action Plan (SDMAP) evolved by Disaster Management and Mitigation Centre (DMMC), Government of Uttarakhand, the following main initiatives are being taken for disaster risk reduction in Uttarakhand:

- Assessment of the impact of natural disasters upon masses, particularly women;
- Assessment of the people's perception of climate change and documentation of their adaptation strategy through primary data collection in various regions of the state;
- Assessment of the changes being introduced in the geo-environment due to climate change through primary data collection in various regions of the state;
- Study of the impact on natural resources and livelihoods of people due to changing weather patterns and extreme weather events;
- Documentation of best practices in traditional coping methods, possible interventions to meet present demand and promotion of the same
- Documentation of the indigenous technical knowledge of the masses

The City Development Plan (CDP) formulated under Jawaharlal Nehru National Urban Renewal Mission (JNNURM) by Uttarakhand Urban Development Department has made provisions for protection of natural environment, conservation of lake and water resources, and improved sanitation and sewage system in the town. The comparison of the urban land use of 1995 to the proposed land use under City Development Plan (CDP) for 2011 is shown in presented in Table 23.7 (Urban Development Department, Government of Uttarakhand 2007). The land use plan proposed increase in areas under residential and transportation, and marginal increase in the area demarcated as 'prohibited area'. The City Development Plan also proposed increase in the parking area from the existing 0.12 ha in 1995 to 2.24 ha in 2011. However, the areas for parks and open spaces remained unchanged

Table 23.7 Current and proposed land use under City Development Plan (2011) for Nainital

S. no.	Land use categories	Land use 1995		Proposed land use 2011	
		Area (ha)	% to Total NNPP area	Area (ha)	% to Total NNPP area
1	Residential	90.54	7.72	186.00	15.86
2	Rural	15.50	1.31	17.50	1.32
3	Commercial	17.75	1.51	15.75	1.51
4	Institutional	34.00	2.90	34.00	2.90
5	Parks and open spaces	10.64	0.91	10.64	0.91
6	Public utilities	99.02	8.44	89.77	7.65
7	Transportation	14.07	1.20	16.14	1.38
8	Forest areas	508.76	43.37	508.76	43.37
9	Water-bodies	68.90	5.87	68.90	5.87
10	Prohibited areas	135.08	11.52	164.00	13.98
11	Undeveloped open area	177.7	15.15	–	–
12	Others	0.74	1.41	61.16	5.21
	Total	1173.00	100.00	1173.00	100

Source: Urban Development Department, Government of Uttarakhand, Dehradun

(4.02 ha). A project for conservation of Nainital and other Lakes, jointly funded by Government of India and Government of Uttarakhand is currently under implementation in the town. However, these urban development initiatives did not incorporate the climate change impacts on urban ecosystem and a mechanism for adaptation.

23.4 Conclusions and Recommendations

Middle Himalayan mountains are not only the most densely populated range, but it the most rapidly urbanizing mountain tract of Himalaya. Nainital situated in the Himalayan State of Uttarakhand is the most representative sprawling urban area in the Western Himalaya. During the recent years, with the increasing incidences of hydrological disasters, disruption of natural drainage and depletion of the lake and other water sources the Government of India, the State Government, the local bodies and the judiciaries have taken some measures to improve the environmental governance of the town. However, the environmental conditions of the town continued to deteriorate with increasing impacts of climate change. The rapid urbanization is increasing the susceptibility of intensively modified and densely populated fragile slopes to the active processes of mass movement and landslides. Moreover, the rapidly changing climatic conditions, particularly the climate change induced hydrological extremes are posing severe threats to the sustainability of fast- growing urban ecosystem by increasing the frequency, intensity and severity of geohydrological hazards in the town and its surrounding region. The climate change is likely to trigger the slope instability and disrupt the hydrological regime of the lake catchment which is already under stress of increasing urbanization. The city development plan and also the state disaster risk reduction framework and climate change adaptation plan did not make any provision for addressing the emerging risks of climate change, particularly the geo-hydrological disaster in Nainital and other towns of Uttarakhand. In view of this the following recommendations are made:

- A comprehensive climate change vulnerability assessment and mapping of the town should be carried taking into account all the critical parameters of exposure, sensitivity and adaptive capacity of urban ecosystem.
- A detailed and large-scale risk zone mapping of the town should be carried out analyzing the parameters of geology, structure, litho-logy, geomorphology, demography, economy and livelihood, infrastructure and services.
- A comprehensive urban land use policy should be evolved and implemented taking into conservation, developmental, climate change adaptation, disaster risk reduction needs and priorities of the town.
- A participatory framework for the conservation of water resources particularly through reducing anthropogenic intervention in the recharge zone of the Naini Lake and Sukha Tal should be evolved.
- An integrated climate change adaptation governance plan need to be formulated incorporating the above-mentioned points involving a range of institutions and stakeholders (e.g., government line departments, private enterprises, civil society and non-governmental organizations, community based organizations and academic and research institutions).

References

Anbalagan R (1993) Environmental hazards of unplanned urbanization of mountainous terrains: a case study of a Himalayan town. Q J Eng Geol 26:179–184

Atkinson ET (1882) The Himalayan districts of the north-western provinces of India, vol. II. Reprinted as the Himalayan Gazetteer, Cosma, 1973, Delhi

Auden JB (1942) Geological report on the hill side of Nainital. Geological survey of India, unpublished report

Borsdorf A, Tappeiner U, Tasser E (2010) Mapping the Alps. In: Borsdorf A, Grabherr G, Heinrich K, Scott B, Stötter J (eds) Challenges for mountain regions. Tackling complexity. Böhlau, Vienna, pp 186–191

Census of India (2011), Uttarakhand, Government of India, New Delhi

Clay JM (1928) Nainital: a historical and descriptive account, 122. Government Press, United Provinces, Allahabad

Disaster Management and Mitigation Centre, Government of Uttarakhand (2011) Slope instability and geo-environmental issues of the area around Nainital. Disaster Management and Mitigation Centre, Government of Uttarakhand, Dehradun

Government of Uttarakhand (2012) Uttarakhand state action plan for climate change. Government of Uttarakhand, Dehradun

Gupta P, Uniyal S (2013) Landslide hazard zone mapping of the area around Nainital using bivairiate statistical analysis. Int J Adv Technol Eng Res 3(1):60–66

Haigh M (2002) Headwater control: integrating land and livelihoods. paper presented at the International conference on sustainable development of headwater resources, United Nation's International University, Nairobi, Kenya, September

Hukku BM, Srivastava AK, Jaitly GN (1977) Measurement of slope movements in Nainital area. Eng Geol 4:557–467

ICIMOD (2010) Mountains of the world – ecosystem services in a time of global and climate change: seizing opportunities – meeting challenges. Framework paper prepared for the Mountain Initiative of the Government of Nepal by ICIMOD and the Government of Nepal, Ministry of Environment

Ives JD, Messerli B (1990) The Himalayan dilemma. The United Nations University, Routledge

Joshi SC, Pant P (1990) Environmental implications of the recent growth of tourism in Nainital, Kumaon Himalaya, UP, India. Mt Res Dev 10(4):347

Joshi SC, Joshi DR, Dani DD (1983) Kumaun Himalaya: a geographical perspective on resource development. Gyanodaya Prakashan, Nainital

Meybeck M, Green P, Vörösmarty C (2001) A new typology for mountains and other relief classes: an application to global continental water resources and population distribution. Mt Res Dev 21:34–45

Middlemiss CS (1880) Geological sketch of Nainital with some remarks on natural conditions governing the mountain slopes. Rec Geol Surv India 21:213–234

Middlemiss CS (1898) Report on Kailakhan landslip near Nainital of 17th August, 1898. Government Press, Calcutta

Mitchell T, Maxwell S (2010) Defining climate compatible development, CDKN ODI Policy Brief, November 2010/A

Nautiyal SP (1949) A note on the stability of certain hill sides in and around Nainital, U.P. Unpublished report. Geological Survey of India, Calcutta

Oldham RD (1880) Note on the Nainital landslide 18th September 1880. Rec Geol Surv India 13:277–281

Pant GA, Kandpal GC (1990) A report on the evaluation of instability along Balia Nala and adjoining areas, Nainital. Unpublished report, Geological Survey of India

Patra P, Kantariya K (2014) Science-policy Interface for disaster risk management in India: toward an enabling environment. Curr Sci 107(1):39–45

Rao DKHV, Rao VV, Dadhwal VK, Diwakar PG (2014) Kedarnath flash floods: a hydrological and hydraulic simulation study. Curr Sci 106(4):598–603

Rautelaa P, Khanduri S, Bhaisora B, Pande KN, Ghildiyal S, Chanderkala, Badoni S, Rawat A (2014) Implications of rapid land use/land cover changes upon the environment of the area around Nainital in Uttarakhand, India. Asian J Environ Disaster Manag 6(1):83–93

Rawat JS (2009) Saving Himalayan rivers: developing spring sanctuaries in headwater regions. In: Shah BL (ed) Natural resource conservation in Uttarakhand. Haldwani, Ankit Prakshan, pp 41–69

Sharma AK (1981) Structural study of area east of Nainital with special reference to hillside instability. Unpublished PhD thesis, Kumaun University, Nainital

Sharma VK (2006) Zonation of landslide hazard for urban planning case study of Nainital Town, Kumaon Himalaya, India. IAEG2006 paper number 191, The Geological Society of London

Singh SP, Gopal B (2002) Integrated management of water resources of Lake Nainital and its watershed: an environmental economics approach. Final report, EERC, Indira Gandhi Institute for Developmental Research, Mumbai

Tiwari PC (2007) Urbanization and environmental changes in Himalaya: a study of the lake region of district Nainital in Kumaon Himalaya, India. International working paper series ISSN 1935-9160, Urbanization & Global Environmental Change (UGEC), International Human Dimension Programme (IHDP), Working Paper 07-05, pp 1–19

Tiwari PC (2008) Land use changes in Himalaya and their impacts on environment, society and economy: a study of the Lake Region in Kumaon Himalaya, India. Adv Atmos Sci 25(6):1029–1042

Tiwari PC (2014) Urban growth and assessment of its natural and socio-economic risks in high mountain ecosystems: a geospatial framework for institutionalizing urban risk management in Himalaya. In: Hazboun E, Hostettler S (eds) Proceedings of the international conference on technologies for development, 2014, Lausanne, Switzerland. UNESCO Chair in Technologies for Development, Lausanne, p 54

Tiwari PC, Joshi B (2015) Global change and mountains: consequences, responses and opportunities. In: Grover VI, Borsdorf A, Breuste J, Tiwari PC, Frangetto FW (eds) Impact of global changes on mountains: responses and adaptation, science publishers. CRS Press, Taylor and Francis, USA, pp 79–136, International Standard Book Number-13: 978-1-4822-0891-7 (eBook - PDF)

Urban Development Department, Government of Uttarakhand (2007) City development plan: Nainital (revised) under Jawaharlal Nehru National Urban Renewal Mission (JNNURM), Uttarakhand Urban Development Project

Valdiya KS (1988) Geology and natural environment of Nainital hills, Kumaon Himalaya. Gyanodaya Prakashan, Nainital

Valdiya KS, Bartarya SK (1991) Hydrological studies of springs in the catchment of Gaula River, Kumaon Lesser Himalaya, India. Mt Res Dev 11:17–25

Chapter 24
A Relational Vulnerability Analytic: Exploring Hybrid Methodologies for Human Dimensions of Climate Change Research in the Himalayas

Ritodhi Chakraborty, Anne-Sophie Daloz, Tristan L'Ecuyer, Andrea Hicks, Stephen Young, Yanghui Kang, and Mayank Shah

Abstract Vulnerability assessments are critical tools when exploring the Human Dimensions of Climate Change in the Global South. Additionally, Social Ecological Systems research utilizes such assessments to describe and predict potential spaces/tools of policy intervention. However, much of the assessment methodology fails to address the coupled structural processes underlying vulnerability and the experience of climate change. First, most scholarship does not operationalize mixed-methods research using plural epistemologies. Second, it fails to incorporate the communally produced knowledge of marginalized regional populations. Ultimately, power inequalities and their impact on vulnerability within complex adaptive systems, are overwhelmingly ignored. This project attempts to address these issues

R. Chakraborty (✉)
Department of Geography, University of Wisconsin-Madison, Madison, WI, USA

Centre for Excellence (DFPL), Lincoln University, Lincoln, New Zealand
e-mail: rchakrabort5@wisc.edu

A.-S. Daloz
Space and Science Engineering Center & Center for Climatic Research, University of Wisconsin-Madison, Madison, WI, USA

CICERO, Senter for klimaforskning, Oslo, Norway

T. L'Ecuyer
Department of Atmospheric and Oceanic Sciences & Center for Climatic Research, University of Wisconsin-Madison, Madison, WI, USA

A. Hicks
Department of Civil and Environmental Engineering, University of Wisconsin-Madison, Madison, WI, USA

S. Young · Y. Kang
Department of Geography, University of Wisconsin-Madison, Madison, WI, USA

M. Shah
Department of Economics, Kumaun University, Nainital, Uttarakhand, India

© Springer Nature Switzerland AG 2020

A. P. Dimri et al. (eds.), *Himalayan Weather and Climate and their Impact on the Environment*, https://doi.org/10.1007/978-3-030-29684-1_24

through a 'Relational Vulnerability Analytic' (RVA). We utilize a plural epistemological approach to construct an analytic that envisions the various relationships, processes and tools that need to be cultivated and managed in order to empower the community as co-producers of knowledge, while challenging the disciplinary bias in explorations of climate change risk and adaptation. Our method brings top-down spatial analysis tools, mathematical models, grounded ethnographic fieldwork and participatory feminist epistemologies into productive tension to reveal the sources of vulnerability and the agency of subjects, in rural Himalayan households. Additionally, we addresses the appeal for long term, collaborative, multi-dimensional research mobilization in the Himalayas. While the analytic is parameterized for the Himalayan region, it can be implemented in other regions with certain salient customizations. The project concludes that future efforts should be to operationalize this analytic for different regions and populations.

24.1   Introduction

The Social-Ecological Systems (SES) framework has been touted as an analytical tool exploring the complex adaptive systems of our planet, including the varied current and probable impacts of climate change (Vogel et al. 2007). However, the framework has been criticized for its 'ecological bias', and for failing to be epistemically democratic when producing the goals and methods of research (Fabinyi et al. 2014). Majority of SES driven assessments are grounded in static disciplinary boundaries, while there is a need for critical trans-disciplinary work, which holds in view the epistemological and ontological tensions between different ways of conceptualizing a research problem (Nightingale 2003).

In the Himalayan region climate change has been defined as a catastrophic event (Schild 2008; Hartmann et al. 2013). The validity of such claims has been questioned by scholars that have pointed out the lack of ground truthing of model data, the scalar bias of non-regional simulations and the diversity of bio-cultural terrain such reports seek to represent (Singh and Thadani 2015). Additionally, impact assessments of climatic change have failed to incorporate the utility of such assessments to the regional populations. Himalayan small farmers interacting with multiple processes of rural transformation, have a unique perspective on the materialization of climate change impacts (Mathur 2015b; Satyal et al. 2017). The 50 million inhabitants of the region have been overwhelmingly unrepresented as stakeholders in developing a science of impacts that is egalitarian and democratic.

Our work attempts to address some of the methodological concerns raised by scholars about SES research exploring the Human Dimensions of Climate Change (HDCC), in general and in the Himalayan region. We do this through the articulation of a methodological construct we label, 'Relational Vulnerability Analytic'. While we do have an empirical question at the heart of our scholarship, it serves more as an evaluative heuristic for our discussions on methodology. This paper, addresses the operationalization of our framework, through a focus on the often-ignored epistemological and methodological dialogue, critical to a more inter-disciplinary and participative research processes (Murphy 2011). Exploring the conclusions that emerge can provide vital insights into the complex realities of climate change, its interactions with marginalized rural communities of the global south while also creating effective policy interventions (Nightingale 2017).

24.2   Situating the Research

24.2.1   The SES Framework: Limitations and Possibilities

Social Ecological Systems (SES) research emerged as a response to the seemingly catastrophic global environmental challenges which included losses to biodiversity, overexploitation of ecosystems and climatic change (Ostrom 2009). The framework challenged the notion of stable and static equilibrium and proposed that complex adaptive systems are fluid, heterogeneous, non-linear, multiscalar and 'disequilibrium' (Holling 1973; Folke 2006). But, the 'resilience concept' that emerged as the organizing principle of SES research, while applauded for its efforts to construct innovative approaches to deal with the complex relationships between human and natural worlds has been critiqued for the 'narrow topical and theoretical lens' with which it defines and incorporates social dimensions and relationships (Turner et al. 2003; Leach et al. 2010; Stone-Jovicich 2015). Furthermore, as Cote and Nightingale (2012) have critically stated, "the reliance on ecological principles to analyze social dynamics has led to a kind of social analysis that hides the possibility to ask important questions about the role of power and culture in adaptive capacity" (p.479). However, while challenging the SES goal of constructing a grand 'theory of everything' certain scholars have appealed for collaborations between 'resilience ecologists' and social scientists. Importantly, with the caveat that the product of this relationship should not be held responsible for encompassing their complete individual research goals (Turner 2014), allowing for a realistic chance at 'transdisciplinarity'.

Our work is driven by this ethos of collaboration and anchored to the processes underlying community interactions with a changing environment, the organization of communities in dynamic social networks and the movement and distribution of power within communities (see Fabinyi et al. 2014). In doing so we are also guided by Nightingale (2016) prescriptive idea of using a 'socionatural approach' that subverts the nature-society binary and holds the different disciplines in productive tension. We support her claim that even more than the points of 'convergence and complementarity' it is the places of divergence within the process of analysis, which can lead to a truly equitable and realistic research program.

24.2.2 Human Dimensions of Climate Change Research in the Himalayan Region

Global climate change is predicted to have catastrophic effects on the Himalayan region (Rawat et al. 2012; Ren and Shrestha 2017; Zhan et al. 2017). These include changes in precipitation variability, increase in extreme events, shifting seasonal boundaries, change in crop maturation times, increase in winter temperatures and glacial melt rates. These climatic events are projected to have significant effects on land use, livelihood security and over-all stability of SES (Schild 2008; Chaudhary and Bawa 2011; Hartmann et al. 2013). Akin to other mountainous regions of the world, HDCC research in Himalayan spaces has expanded in recent years utilizing a plethora of terminologies and constructs (Kelkar et al. 2008; Sharma et al. 2009; Gentle and Maraseni 2012; Pandey and Jha 2012; Ford et al. 2013; Hoy et al. 2016; Pandey et al. 2017; You et al. 2017).

However, most of this research suffers from four critical flaws. First, most analysis has been overwhelmingly motivated by enduring mountain stereotypes of the Himalayan region as a remote, harsh "riskscape", relatively more disaster prone than other places (Hewitt and Mehta 2012). Second, the two research traditions critical in analyzing HDCC – climate science & land-change science (LCS) and political ecology (PE) have rarely been brought together in conversation, to learn from their points of convergence and divergence (Turner and Robbins 2008). Third, there is a poverty of 'multi-sited, collaborative, long term research' (Singh and Thadani 2015), prompting some to question the validity of top-down, generalized predictions using global models with minimal ground truthing and empirical triangulation. Ultimately, most research has assumed vulnerability to climate change as a function of variables tied to the 'adaptive capacity' of the unit of study and their exposure to a certain hazard. This approach has failed to capture the 'inherently relational' nature of vulnerability and how it manifests differentially across scales and is deeply rooted in the relational matrix of people and places (Turner 2016). These criticisms are not regionally unique and reflect the global conversation around climate change vulnerability and adaptation. Despite such perspectives undergoing a slow progression towards more nuanced analysis, evidenced by the changes between the fourth and fifth assessments reports of the IPCC (2007, 2014), most scientific writing and policy responses have failed to directly engage with the politics of adaptation and more critical studies of vulnerability (Eriksen et al. 2015; Turner 2016; Goldman et al. 2018).

24.3 Research Goals, Tools and Questions

24.3.1 Research Goals

24.3.1.1 Separate But Equal: Utilizing Plural Epistemologies

SES research about the HDCC is contingent on constructing models that utilize static input variables, which while derivative of different perceptions, remains subservient to the idea of uniformity (Cote and Nightingale 2012). The contentious meetings between different epistemological traditions are mined specifically just for quantitative inputs that can be inserted into a schema. But, what about the processes that create the data? The context within which knowledge is produced is critical, as it highlights the intersectionality of the agents or situations involved in the production (Fabinyi et al. 2014). Our goal is to recognize the different ideological backgrounds, without producing a 'hierarchy of methodologies'. Therefore, while we do indulge in data hybridization – evaluating and fitting observations rooted in one epistemology into another, we don't validate such data by the presence of complementarity. Our goal is to push the operational limits of each method to discover moments of rupture, and then use those to identify data gaps and methodological concerns. In doing so we hope to address the concerns about the inherent normativity of climate science – the inclusion of a certain 'valid' research, the overt reliance on and advocacy of knowledge produced by certain global north institutions and the absence of the voices whose futures are being discussed (Eriksen et al. 2015; Lövbrand et al. 2015).

24.3.1.2 Disaggregating Communities: Acknowledging Power

The underlying vulnerability of communities is a product of multiple interacting processes, and is not a simple function of climate change (Cardona et al. 2012; Ribot 2014). However, in most SES research the units of social wellbeing are applied with equity across human populations being studied. The imbalances of power within any scale of enquiry – be it household, village or country is aggregated in functional terms to either an average value, or an essentialist interpretation connected to reductionist variables. Our goal is to advocate for power as a social relation throughout our research process, deconstructing formal institutional scales – village, state, nation, to witness the myriad of exchanges and formations, at these 'intra and inter scalar' spaces. Additionally, we steer clear of the agency-structure dualism when conceptualizing power in our more quantitative analysis and instead explore an equivalency of agents, structures and events (Boonstra 2016). Simplistic correlations of adaptive capacity or resilience to indicators of ownership and access, fail to account for the relative and relational nature of precarity itself (Turner 2016).

24.3.1.3   Challenging the Theory of Himalayan Degradation: Contextualizing Exceptional Precarity in the Anthropocene

The Himalayan region, constituted by the states of Nepal, Bhutan, India and China, is a product of the colonial project of controlling bodies and spaces, through articulations of borders and identities (Gellner 2013; Shneiderman 2010, 2016; Pfaff-Czarnecka and Toffin 2011; Smith 2012; Mathur 2015a, b; Smith and Gergan 2015). Tropes of remoteness and formidable frontier lands established on such colonial imaginations have also colluded with hegemonic transnational politics of resource use and sustainable development, to produce a discourse propagating ecological deterioration and exceptional precarity (Guthman 1997; Ives 2004; Mathur 2015b). The Theory of Himalayan Environmental Degradation (THED), a crisis narrative, is the byproduct of such a cultural and political project. It is a 'scientific' discourse, which attempts to simplify and quantify the causes of apparent SES deterioration (and collapse) and find spaces of optimal intervention. This crisis narrative in recent years has been fortified with the specter of climate change.

These discourses, while critiqued for their scientific validity, rationalist framing, political and historical neutrality and omission of the processes that produce and reproduce inequalities of power (Ribot 2010; Beymer-farris et al. 2012; Singh and Thadani 2015), have generated a powerful material and affective arsenal to address the many points of probable current and future crises in the region (Gupta 2011; Devi et al. 2014; Bhattarai et al. 2015). Our goal is to confront this positivist, rationalist; hazards based approach that fails to provide a holistic understanding into the roots of regional precarity. Furthermore, while the THED is a regionally specific narrative, similar environmentally deterministic explanations are quite popular when describing other people and places (see Taylor 2014 and Davis 2000).

24.3.2   Research Tools

24.3.2.1   Mathematical and Computational Models

Mathematically grounded computer simulations are ubiquitous with incursions into the current and future nature of complex dynamic systems (Clifford 2008; Conte et al. 2012). These include models that incorporate data from remote sensing, point source instruments, census data-sets and historical events. While challenged for representing fluidity through approximations and reducing irregular dynamic entities as fixed objects (Bithell et al. 2008), models allow for vast amounts of spatio-temporal data points to be situated in a controllable framework. Furthermore, they can be harnessed at various steps of the overall analytical process, depending on the questions being explored and the ideological leanings of the overall project.

For our methodological framework, we incorporate deterministic process-based biophysical models, which include – climate models, agro-ecosystem model

(AEM), hydro-geological model (HGM) and agent-based social and cellular ecological models (ABM). While the first three interact with each other in various couplings, with very little structural fluidity, the ABM is constructed ground-up with empirical ethnographic data from the field, census data, outputs from the other model interactions and literature review. The plethora of feedbacks that reflect the relational web tying together disparate elements within the SES world is visualized in the ABM. Contrary to the static and overwhelmingly biophysical positivism driving the ecological and climatic models, ABMs represent a flexibility and design philosophy that supports transdisciplinarity and have become popular in the land use and land cover change (LULCC) scholarship (Schreinemachers and Berger 2011; Wise and Crooks 2012; Berger and Troost 2014). Furthermore, with the development of empirical ABMs there is an attempt to create a space where the collective effects of many interacting individuals can be followed at different scales. But, significant concerns remain about the way ABMs deal with issues of structure and agency, with simplification of human behavior and ultimately their testing (Robinson et al. 2007; Clifford 2008)[1]. We are guided by the philosophy that ABMs are a useful heuristic in visualizing certain narratives about the world, but their outputs of the emerging 'causal' patterns are not explicitly conclusive. Therefore, the outputs are actually inputs in the broader process of analysis. The ABM we propose is deeply rooted in the object-process methodology (OPM), that allows us to confront the agency-structure duality and construct a model universe, where processes and people together, produce the 'units' that describe the function, structure and behavior of systems (Sturm et al. 2010).

24.3.2.2 Political Ecology

Political ecology (PE) emerged as a combination of the concerns of ecology and the analytical framework of political economy, to probe into the structures driving environment and development encounters. However, over time the field has embraced and incorporated a plethora of theoretical approaches that include – feminist theory, post-structuralism and post-colonialism (Turner and Robbins 2008; Peet et al. 2011). PE diverges from the coupled systems theories that concern themselves with locating and evaluating tipping points, using interchangeably the ecological and social vulnerability that trigger such events. Instead, political ecologists are driven by a much more 'value-laden' process that argues "what can be known is prefigured in part by social, political and historical conditions" (Turner and Robbins 2008, p.301). This impetus not just on what the knowledge is, but on how it is produced, makes PE a valuable tool in tempering the goals of conventional 'resilience' driven modeling approaches (Cote and Nightingale 2012; Fabinyi et al. 2014). Additionally it creates space for plural epistemologies that are harbingers of alternative approaches to SES modeling. These include observing transitions in systems, instead of changes

---

[1] While a detailed review of the applicability of ABMs in more critical pedagogy is beyond the scope of this paper (see Bharwani 2004; Ghorbani et al. 2015; Wellman 2016).

in system components (Verburg et al. 2016), telecoupling across scales (Millington et al. 2017) research co-design with a variety of stakeholders (Samson et al. 2017) and the creation of intelligent spatially aware and emotional agents (Molen 2016). Despite recent forays into more critical modeling attempts, very few empirical studies have attempted to combine them with PE (Blythe et al. 2017; Hoque et al. 2017). Our tryst with PE goes beyond the model design and implementation phase and mediates every step of the overall analytical process.

24.3.3   Specific Regional Questions for Parameterization

Vulnerability analysis in the Himalayas has failed to incorporate the role of migrants from rural Himalayan households. The slow demise of agricultural livelihoods, coupled with socio-ecological uncertainties and the increase in industrial infrastructure, have played a role in persuading many young people to leave their villages and move to towns and cities (Bruslé 2008; Jain 2010; Sharma 2013). Urban areas are generally seen as offering better prospects in terms of education and jobs. Youth journeys from the village to the city and back have significant impacts on the urban spaces they chose to inhabit and also for the rural communities they leave behind. We are interested in asking what role, if any, this circular migration of youth plays in helping rural communities adapt to climate change in the Himalayas? The existing research on this topic is very sparse and focuses primarily on the impact of monetary remittances on community wellbeing (Hoermann and Kollmair 2009). However, the myriad relationships these youth have with their rural communities helps to produce a variety of different linkages between the village and the city. These connections have been largely unexplored.

The response of youth to increasing threats to their socio-economic security and livelihoods has recently become a topic of much interest to South Asia scholars (Jeffrey 2010; Smith 2013; Dyson 2015). However, this literature generally overlooks the issue of climate change at different spatial and cultural scales. Consequently, little attention has been paid to connecting climate change with emerging youth-led political movements that transcend the rural-urban divide. In India, there has been a lot of media attention focusing on the migration of educated young people out of rural Himalayas due to the growing threat of environmental catastrophes (Tiwari and Joshi 2015, 2016). In some instances, there are alarmist claims about the growing phenomenon of "ghost villages" that have lost almost their entire population (Pant 2016). However, young people who move to cities are not necessarily severing their ties with their native communities (Smith and Gergan 2015; Korzenevica and Agergaard 2017; Chakraborty 2018). Many of these supposed migrants are better described as "translocal" because they regularly move between urban and rural spaces (Gidwani and Sivaramakrishnan 2003). Moreover, a number of studies have highlighted the importance of the everyday political practices of young people. Some scholars have even argued that young people are naturally inclined to develop innovative ways of approaching social-environmental problems because they experience a "fresh contact" (Cole 2004) with the world around them. It is therefore imperative to understand how educated and mobile young people might play an important role in addressing the vulnerabilities faced by rural households. This could include enabling the household to access climate insurance or compensation programs and to diversify into non-farm economies, as well as remittances.

While not the focus of this paper the specific questions we are exploring using our methodology are:

1. a) What can regional scale modeling tell us about the impacts of climate change on precipitation regimes in the Himalayan region?
   b) How are these changes impacting the salient features of agrarian livelihoods in rural communities?
   c) How are these changes impacting the different processes of vulnerability in these communities?

2. a) What are the important relationships that migrant youth have with their rural communities? How are these relationships produced?
   b) What role, if any, do these relationships play in the different processes of vulnerability in these communities?

3. a) Do the different relationships between migrant youth and their communities react to the changes in climate and agrarian livelihoods?
   b) How do these relationships address the changing nature of vulnerability within rural Himalayan communities? And how do they inform the existing regional knowledge about the HDCC?

Our initial research area was with rural communities in the Indian state of Uttarakhand (western/central Himalayas), however, we are currently exploring possibilities and mobilizing the framework in the Indian state of Sikkim (eastern Himalayas). The current data comes from four villages. While all the four places, Ghargaon, Mana, Inari and Kamu, are situated within the eastern sociocultural region of the state (Kumaon), in the lesser Himalaya (1200–3000 m), they are different in their connectivity to industrial markets, infrastructure access, livelihood spectrum, administrative realities, agrarian practices and ecological characteristics.

While we explore these questions through the framework described in the following section, the focus on this paper is the framework itself since our empirical questions are still a work in progress.

24.4 The Relational Vulnerability Analytic

The Relational Vulnerability Analytic is an assessment tool that builds upon the socionatural approach and attempts to construct a transdisciplinary and democratic methodology to explore the roots of vulnerability in transitioning, rural communities of the Global South. While inspired by our work in the Himalayas, this methodology could be arguably transferred to a different space, however, not without a significant amount of customization based on the realities of the space and the people involved.

The analytic is composed of simultaneously performed tasks in multiple locations by multiple actors. It is not a rapid assessment technique and therefore is not geared towards producing conclusions that can be materialized into short-term technocratic interventions. Instead, it focuses on unearthing the rooted structural processes that produce vulnerability and this requires a significant time investment, informing policies that address such issues. While building an assessment method is the stated end goal of this undertaking, constructing and cultivating relationships both between scholars of different disciplines and between scientists and the communities involved in the process is the real driving force of this framework.

In Fig. 24.1 we have identified different conceptual and operational processes. These are explained below. Each methodological step is deconstructed into the various collaborative spaces of knowledge production. While these spaces interact, intersect and inspire each other, for the purposes of articulation we 'artificially' separate and describe them. These spaces are:

1. The Lab – located usually within a university(ies)/research institutions and peopled by scholars/researchers, administrators and students from various disciplines.
2. The Field – the places, agents and relationships that constitute the space being explored. This is a node encompassing the data production, as well as encounters

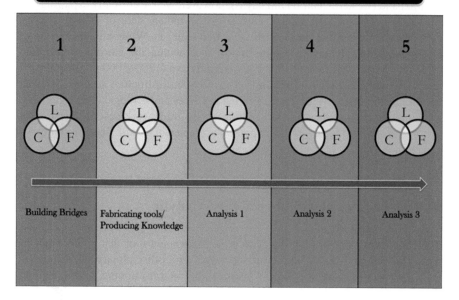

Fig. 24.1 The relational vulnerability analytic (*L* Lab, *C* Community, *F* Field)

with communities and ecologies that work in collaboration with the designs created in the Lab, as well as with community participation.

3. The Community- while in the previous two nodes, the limiting variable is a specific relational agenda focused on research, this node deals more directly with the aspirations of the community.

Explaining the different fundamental steps:

24.4.1 Building Bridges

This is the origin of the relationships both among researchers and with the regional places and people. It begins with assumption-mediated encounters both in the Lab and in the Field.

The Lab Successful trans-disciplinary initiatives are catalyzed through an initial period of trust and relationship building during which epistemologies are openly discussed and the places of complementarity and friction identified. This phase is critical and needs to be accomplished through identifying 'bridging concepts' (Blythe et al. 2017), that allow disparate ideologies to attempt collaboration. However, it is important to keep in mind that, total complementarity is an impossible aim and instead the goal should be epistemic transparency and disclosure. This exercise should help the team create a list of places of convergence and divergence. Furthermore, during this time the focus of the research should be fairly nebulous and abstract, allowing for many degrees of freedom while simultaneously highlighting the researchers' ignorance about processes of vital importance to the community.

Trans-disciplinarity within SES has been attempted through various superficial as well as rooted methods (see Nightingale 2015), however, our opinion is that the focus at this stage should not just be on finding common ground, but should resemble an immersion into different paradigms, as a means of situating individual disciplines, knowing the end goal is collaboration. As an example, for our team this consisted of members taking classes in and attending professional events in collaborating departments. It consisted of trips to the region, without an explicit research agenda and without former exposure, to enter the space not as a researcher but as a traveler or a friend. Finally, it consisted of presenting each other's work in public forums while brainstorming the overall project idea. Along with building collaboration, the most important housekeeping task during the early phase, is identifying and maybe even applying for funding. Given, the significant number of people, fieldwork components, community development tasks, computational power and time investment, critically designed SES projects requires a significant financial investment. While, there are now a growing mobilization of funds for such work, given its 'hybrid' and 'unprecedented' nature, it is still quite limited. As such team members may have to deal with the 'gatekeeping of the institution' when seeking funding (Broadhead and Rist 1976). And should be prepared to string together a plethora of smaller grants that address 'sub-sections' of the project or to reach out to non-traditional funding sources.

The Field Doing SES work in the global south is fraught with a plethora of ethical issues. Among them are: the role of the researcher within the community and the value of the research product to the community as a whole. Furthermore, there is the fear of projects that can have adverse impacts on the community and some that are embedded within dangerous or violent settings (Rodrigues 2014). Simultaneously, there is a persistent call for a 'turn to praxis' in many critical social science fields today that strongly encourages a model of activist research with an agenda of affecting political and social change (Mcguirk and O'Neill 2012). Our strategy during this early phase of the process is similar to our interactions in the Lab, it consists of building trust and transparency with regional communities. Often this resembles an immersion within the relationships that we wish to explore, armed with a reflexivity that remains vigilant for abuses of expert power. In this we are guided by the concepts enshrined in 'feminist epistemology' (Alcoff and Potter 1993), that problematizes the issue of researchers being 'just innocent bystanders' and asks for a critical examination of inter-subjective space between the 'researcher' and the 'researched' (Davids 2014; Van Stapele 2014). To achieve this a significant personal investment is required both from the community and the researchers. This investment can resemble a variety of encounters. Therefore, it can materialize as volunteering labor within community households or working with local governmental and non-governmental institutions.

Initial informal 'pre-surveys' can be conducted during this time to test out the assumptions from the Lab and literature review. Additionally, these 'pre-surveys' can be instrumental in 'unearthing' the probability of getting answers to certain questions, thus setting the structural limits for the research design. As an example, certain members of our team worked as voluntary labor with over a dozen rural communities in western and eastern Himalayas. Furthermore, members also worked as apprentices for certain people within the community, paying them for their educational services, ultimately, we also played supportive roles in setting up youth managed entrepreneurial ventures. Ultimately, members sought out local administrators and scholars, supporting their narratives about the region, their historical and current engagements with the community and their recommendations for building relationships and working in the region.

The Community During this initial phase the community plays an important role in articulating their concerns, fears and aspirations. This process should be supported by a transparency of praxis. In the global south the encounters between the researchers and rural communities are overwhelmingly power misbalanced in the favor of the researcher (Turner 2010; Janes 2016). This notion should accompany the deconstruction of researcher-community relationships and help navigate the 'generative friction' (Cresswell 2015) produced by the socio-cultural differences. It is critical to be upfront about the role of the community, the budget of the research and the probable material impacts if the research process was to proceed. During this phase, it is important to connect democratically with people different people. This prevents the relationship with the community from being hijacked by the intra-community power structure, further marginalizing the most vulnerable. During our initial forays into the communities we were asked for financial and political support. On explaining our inability to provide such support some among the community asked in frustration, "Then why are you here?" This question has been central to our discussions ever since and is brought back for reiteration and acts as a tool of functional evaluation of our complicated utility to the community.

24.4.2 Fabricating Tools/Producing Data

In this stage the research focus moves towards the Field. This phase is when most of the primary qualitative and quantitative data is 'produced' or 'acquired'. Along with the production of data, this is also the stage when the more abstract, ideological dialogue of the previous step, materializes into the fabrication of specific research tools. These include the salient features of research design. For our specific project we created an initial schematic based on step 1 (Fig. 24.2):

The Lab Guided by the pre-survey evaluations, community dialogue and the analytical and capital limitations of the research team, this stage disperses the different team members after bringing them together in the previous stage. This dispersion is necessitated by the division of labor needed to accomplish the various advanced skill based tasks, depending on individual expertise. During this moment it is critical to maintain open channels of communication. Because after the 'disciplinary discomfort' of the previous phase (Head and Gibson 2012) there can be a movement towards the comfort of the familiar, negating much of the hard-fought loosening of disciplinary borders. For our team, this was challenging given the spatial distance between the various members. The people in the Field were also immersed in an environment with unreliable telecommunications access, leading to intermittent conversations with the team. Thus, while Lab mates were dealing with issues of research finance and evaluating model bugs, they were often in the dark about new revelations from the Field and how that impacted their choice of research hypothesis or necessitated their reiteration of analytical tools (Popke 2016). The data generated during this stage was (Table 24.1):

The Field This is arguably the most critical and intensive phase of empirical work for the research team in the Field. To begin with the identity of the researcher becomes a lot more prominent, in the process. Along with this transition in identity comes a dire need for 'checks and balances', since data demands are significant and time and resources are limited. This can lead to unethical behavior and as noted in the Lab location, corrode the hard earned trust and transparency with the community (Kingsley et al. 2010). This should be a time of hyper-vigilance from the

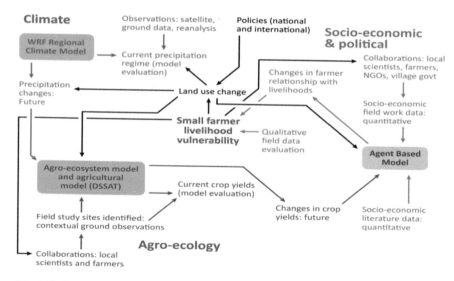

Fig. 24.2 Systems based schematic of research design for model building/coupling

Table 24.1 Model tool development and initial calibration/tests with available data sets

No.	Type	Quantity/ time period
1	Indian Meteorological Division (IMD) long-term rainfall/temperature data for Uttarakhand, IMD rainfall data gridded (0.25 degrees daily product)	1901–2016
2	CMIP5 models were compared against observations from ground stations (APHRODITE), satellites (GPCP, Adler et al. 2012), and atmospheric reanalysis (ERA-Interim, Dee et al. 2014 and MERRA (Rienecker et al. 2011) to document the biases and evaluate objectively the future changes in precipitation	1998–2005
3	Decision Support System for Agrotechnology Transfer (DSSAT) crop model calibrated to create baseline scenarios for staple regional crops (wheat, rice, corn, potatoes)	2005–2010

researcher, and hopefully using a 'buddy system', the production of knowledge should be seen as subservient to the various local relationships. Additionally, given the rather heterogeneous nature of the data produced, the task requires a significant amount of mobility. It is our suggestion that before beginning the process of formal data production on-site teams should be created, favorably with local agents and institutions, using the 'bridges built' in the last stage. These teams should be mobilized simultaneously, in another moment of dispersal, with the administrative head circling through the various sub-teams. The knowledge produced during this stage is often monumental, since it represents daily records of social and ecological systems. While this data may seem overwhelming it is important to begin an initial coding and compartmentalization of the data while in the Field. In our case, the empirical data ark assembled was in the form of video, audio, photographs, social

Table 24.2 Primary socio-economic data collection

No.	Type	Quantity/volume
1	Detailed household surveys (60–90 min to administer)	300 households
2	Semi-structured interviews	1005 individuals
3	Focus group/group meetings	24
4	Travel ethnographies	16
5	Oral history sessions	50
6	Social media data	250 individuals
7	Video/photo diaries	30 individuals

media presence, written testimonies and field journals. Ultimately, addressing the issue of the 'silenced (research) assistant' (Molony and Hammett 2007) we paid our direct collaborators in the Field a wage that was set both by the rate of pay for graduate scholars at local universities and the personal inputs of the assistants. The complete data set generated includes (Table 24.2):

The Community This is the stage during which roots are really deepened in the community. Given the often-invasive requirements of the research process, it is important that the researchers exchange this access into the lives and spaces of the community with their own stories. While being cognizant of potential fallouts of divulging certain personal information, this strategy of 'friendship as research praxis' (Taylor 2011) allows for the researcher to relate to the community in ways that are necessitated beyond just the ones professionally mandated. This allows for the community to become more empowered collaborators of the research process. During this stage community needs also should be addressed that don't directly relate to the research, but which allow for reciprocity. In our case, we were approached by multiple rural youth that wanted help accessing institutional support for entrepreneurship, by community managed Non-profits that needed help fundraising and female wage workers that wanted information about how to improve their career prospects. We were/are engaged in supporting these individuals/institutions within the limitations of our own responsibilities to achieve such goals and have also applied to for more development focused grants to finance the project.

24.5 Analysis 1: Coupling and Comparing

In this the focus moves towards the Lab. It begins with a re-evaluation of research goals after evaluating the data from the Field, the secondary data sources, the initial model outputs and the limitations imposed by financial and temporal realities. After this initial task, the dispersed people and data are brought together, to attempt a coupling of the deterministic models. After this attempt at coupling, the output is

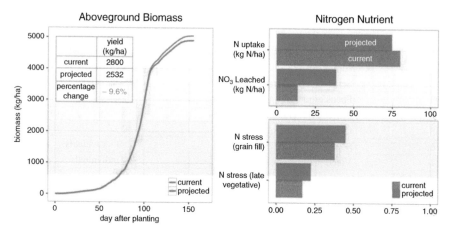

Fig. 24.3 DSSAT simulation results of biomass and nitrogen for a wheat cultivar. Left: comparison of growing season aboveground biomass accumulation with an inset table showing the decrease of final yield. Right: total nitrogen uptake and N stress for baseline and projected scenarios (current: 2016, projected: 2100)

situated next to the Field data, which is thoroughly coded and analyzed through a PE lens, to identify possible convergences and divergences.

The Lab This is the first analytical stage and as such should be done keeping in mind the role of the 'outputs' of such analysis. As suggested by other scholars the 'outputs' of the initial model calibration and projections serve not as conclusions, but as 'inputs' to be utilized for further exploration (Cote and Nightingale 2012). During this stage it is critical to contextualize the points of model coupling, by situating them within the various other points that are not connected. Therefore, in our work, while the agro-ecosystem model receives the input of precipitation regimes from the array of climate models, that represents less than 1% of the total data needed to run the model. The rest of the data is generated by primary field work and literature review. While coupling does occur, there are vast sets of integral variables that do not talk to each other within the two modeling scenarios. Additionally, the limitations of the model architecture need to be corrected through other analytical devices. In our case, the agro-ecosystem model had no spatio-temporally dynamic agricultural pest variable (Jones et al. 2003). This added a significant amount of artificiality to the simulation, since fieldwork and literature had highlighted the massive impact of pests to yields. Figure 24.3 shows the yield predictions without the pest variable.

This model coupling also has to deal with 'mismatch between scales of knowledge' production. Despite SES research advocating for the 'panarchy'[2] model of thinking, in most SES research certain 'scales' of knowledge are privileged more than others

---

[2] For a detailed explanation of this concept see (Folke 2006; Davidson 2010; Miller et al. 2010).

(Ahlborg and Nightingale 2012). Even without straying into the critical epistemologies of PE, our work had to confront this issue, as the regional climate model was producing data on a spatio-temporal scale that effectively flattened all the heterogeneity of the land management and agro-ecological data (Fig. 24.4).

Cycles of systems completed in years and decades had to inform processes that changed daily and hourly. These 'limitations and biases' around scale are further accentuated when the findings from qualitative analysis are brought to bear. While multi-scale assessment is not a panacea to the problem of strategic politics of knowledge production, they reveal the different privileges of the stakeholders in the process (Yeh et al. 2014).

While initial descriptive statistics of regional experiences of climate change produced results such as the ones graphed in Fig. 24.5, they failed to articulate the different relationships that local people had with climatic and ecological transformation. Therefore, using a PE analysis of the overall household vulnerability and the articulations of climatic vulnerability within the community, changed the focus from the climatic hazard and put it on the 'generative structures of vulnerability' (Ribot 2014). In doing so it allowed us to situate the transitions in precipitation regimes and agricultural yield within complex social and political economic history that was being identified by members of the community.

The Field Most of the primary 'data production' for calibrating models and conducting qualitative analysis is done at this point. While, we will have to return to the Field after the first two analytical segments, currently work in the Field consists of 'maintaining social networks' with the community and remotely supporting the procurement of empirical demands that may arise during the early analytical phase. However, that being said, it is important to keep generating data on a scale preferred

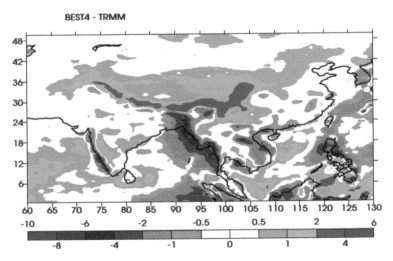

Fig. 24.4 Difference in mean daily precipitation (mm/day) between an ensemble of four CMIP5 models and GPCP satellite observations (1998–2005)

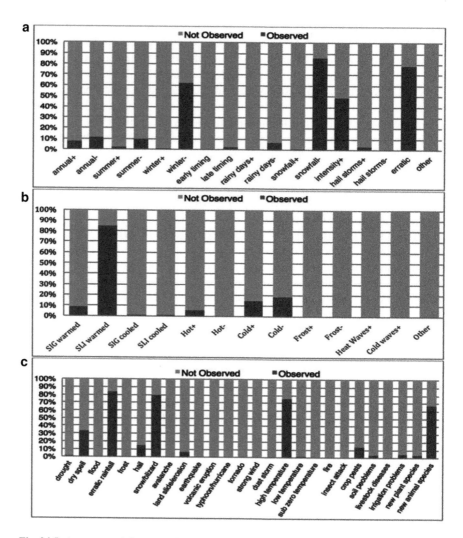

Fig. 24.5 Awareness of disparate biophysical climatic variables (**a**). observations of precipitation events (**b**). observation of temperature events (**c**). observation of unprecedented climatic/ecological events. The percentages represent the number of respondents from the total completed survey set of 188. Therefore, when reduction in snowfall is at 85% in graph **a**, it means 160 families observed the occurrence. (In the third diagram the high responses to snow/blizzard, represents the reduction observed and not the actual occurrence of the event)

by local entities. Therefore, this stage is critical in supporting knowledge production that isn't guided by the time frames and the geographical scales of the formal research design (Ahlborg and Nightingale 2012). Furthermore, with the absence of the researcher within the community, it allows us to expand on the epistemic geographies of climate change (Mahony and Hulme 2016). In our case, this meant a careful chronicling of social media content and telecommunication content, attempted by members of the community. This process was important in formulating an understanding of how the community viewed this project and also of the embodied daily struggles and aspirations, which hinted at the deeper structural processes.

The Community This is a precarious time for the relationships between the researchers and the community. Akin to the problems faced by the relationships between different researchers as they dispersed between the Lab and the Field, the onset of analysis often heralds the physical absence of the researchers from the community, usually after a long period of engagement. This strains the trust and raises certain critical conjectures within the community, about the honest agenda of the research. This is a product of the unequal power relations we mentioned earlier, but also inspired by past negative experiences with various institutional and private agents (Herbert-Cheshire and Higgins 2004; Choudhury and Haque 2016). However, this is the time period during which previous commitments that were made with community members need to be honored and realized. In our case, this meant the slow, but steady mobilization of support for the projects previously identified. While the projects are nowhere near completion, tangible, material steps, such as conversations with loan officers, free marketing on social media and supplies for certain commercial agricultural projects have already been exchanged.

24.6 Analysis 2: Constructing and Investing

This stage consists of building an empirically driven ABM using the directives of 'emic modeling' (Yang and Gilbert 2008) and using inputs from the previously mentioned deterministic models and qualitative analysis. Additionally, it consists of presenting our project and engaging in dialogue with different scholarly and non-scholarly communities, in different locations, to harness the different 'epistemic cultures' (Mahony and Hulme 2016) beyond the Lab and the Field. Our exploratory Himalayan project is currently at this stage.

The Lab This is another critical stage in the lab and arguably the most unpredictable one. Given the massive heterogeneity of applications for ABMs, their limits while structurally resonant within a positivist, quantitative schema, can allow for the visualization of many different processes. While ABMs have come a long way from their nascent iterations within an SES framework (Janssen and Ostrom 2006), they still struggle with constructing agents that are capable of affective behavior (Balke et al. 2010). The goal during this stage should be to incorporate the recent attempts at creating more iterative models that place "objects and processes on the same ontological level" (Molen 2016, p.19), into the model making process.

We suggest building a model using the Object-Process Methodology extended for Multi-Agent Systems (OPM/MAS) (Sturm et al. 2010). This allows for the focus to shift onto fluid processes and events in transition, while representing multiple subjective realities. Furthermore, we advocate for a more grounded theory approach where ecological elements are not represented as entirely discreet entities and are actually deconstructed and embedded as 'actors' within the SES architecture (Molen 2016). Also, eschewing the more 'simplistic' rational agent formulations from the past, we suggest exploring the Brain-Desire-Intention agents (BDI), that supports the integration of cognitive agent modeling platforms with agent-based simulation platforms, and in the process models complex 'humans' moving through non-linear moments of choice (Georgeff et al. 1998; Singh et al. 2016). Ultimately, we support the usage of trans-disciplinary, mixed-methods approaches such as the Knowledge Elicitation Tools (KnETs) process, that attempt to triangulate the rules of conduct for agents through systematic linkages to ethnographic case studies (Bharwani et al. 2015).

The ABM constructed should attempt to capture overall vulnerability for different population units – individuals, households, communities etc. Furthermore, instead of additions of risks as external interventions, they should be constructed as processes themselves, similar to the agents, on temporal pathways that intersect with certain agents only at certain times. Thus the vulnerability experienced by an agent can transition by jumping scale, and may not represent the vulnerability experienced by the different communities that agent belongs to. This brings into sharp focus the salience of relationships (processes) as mediators and barometers of vulnerability (Turner 2016). While the ABM is being constructed and systematized with different inputs, it is important to visit one of our foundational research assumptions – it is impossible to see and visualize the entire problem at once. Modeling, even empirically grounded, just provides us with a partial and one of many articulations of the situation (Nightingale 2016). Therefore, during the ABM construction process, we suggest simultaneous focus on constructing narratives of vulnerability using the PE toolkit. These should be rooted in the critical literature on the Anthropocene, politics of climate change and critical perspectives on 'technocratic, managerial research' that proposes notions like assemblage, post-humanism and hybridity (Castree 2015).

Celebrating and developing this methodological divergence is critical and much needed when operationalizing the socio-natural approach to understanding SES. In our project while the ABM calibrated with more 'reductionist qualitative data' attempted to imagine the futures of Himalayan youth aspiration and conduct, we used a 'mobilities' framework to produce a parallel narrative (see Chakraborty 2018).

The Field A continuation of the process described in the previous step of the analytic.

The Community A continuation of the process described in the previous step of the analytic.

24.7 Analysis 3: Triangulation and Prescription

The final step of analysis consists of triangulation, discussing the generalizability of the analytic, identifying points of future research as well as possible translations into policy interventions. It requires a return to the Field to revisit people and places in an attempt to insert a longitudinal component into the study, while interrogating the results of our analysis with comparisons with actual SES transitions.

The Lab The action during this phase shifts back to the Field with researchers in the Lab setting more engrossed in writing up the results of the outputs, collaborative efforts and methodological and empirical discoveries. This narrative production is vital for both reaching a wider audience as well as identifying and sustaining research funding. Our suggestion is to explore a variety of forms of expression, beyond the scientific article/book chapter (see Nakamura 2013; Smith et al. 2015). This addresses the broader issues around 'ivory tower scholarship' (Singer 1995) while allowing our conclusions to be rendered through a form that reaches a plethora of stakeholders, including the community itself. Furthermore, while most of the analytical goals of the project may have been met, the transformation of theory to praxis is still being negotiated. We believe it is critical to shoulder this responsibility in part. Even though the research team may not control the mechanisms of power through which such a process can occur, it is ethically crucial to stand by our commitments both to the project and the community and pursue feasibility studies symbiotically.

The Field Positivist SES modelers advocate for model transferability and validation as a means to test how closely the model can capture 'key underlying processes' and be used in 'valid comparable contexts' (Sun et al. 2016). But, while triangulation is important for models, divergences between model outputs and results generated from epistemologically different methods are actually recognition of the ideological differences of mixed methods research, and should not be used as a point of value judgement. Furthermore, these encounters need to be moderated for epistemic abuse – where data produced within one paradigm is 'fitted' into analytical mechanisms of a different paradigm (Nightingale 2016). During this stage a return to the field is necessary for two reasons. First, to cultivate the relationships with the community so carefully built over the duration of the work. This is essential both as an ethical prerogative of doing ethical and collaborative research and to grant the community access to the analytical conclusions that were drawn from the knowledge that was co-produced by them. Second, spatio-temporal lives of knowledge are dynamic. That is data collected is just a snapshot of the fluid reality of a space and needs to be embedded within the larger reality. Returning to the communities to observe the stages of their different relational worlds, allows the researcher to add to the narratives of the places and people that were produced through the different analytical tools. Furthermore, it serves as the material weight to a possible theoretical shortsightedness, and as a counterweight to the scalar limitations of the Lab (Ahlborg and Nightingale 2012). Ultimately, the return to the Field allows an intervention in the probable emphasis on 'actionable knowledge' and 'decision relevant science' (Castree 2015). This challenges authoritative attempts to bound and control a space which being part of a much wider system, resists attempts at segregation and insulation (Mahony and Hulme 2016). For our project this meant creating an intensive geo-hydrological watershed health-monitoring project in collaboration with a regional Non-profit Organization. This project while responding to Singh and Thadani's (2015) call for situated multi-sited research with a focus on ground truthing, also engages with issues around irrigation, ground water availability and long term food security which were critical concerns to the communities.

The Community This stage culminates in the ethical litmus test for the research team. While the process of analysis and exploration may have been guided by the dictates of funding resources or institutional demands, the multiple identities of the researcher within the community, can lead to potentially disastrous aftermath with the 'sudden termination' of the research program. Juggling the relationships with the community, both professional and personal, is crucial during this stage, especially since, while advised by some, it is often impossible to keep both of them from bleeding into each other. We imagine the different relationships to have different life expectancies, which are fairly complicated to predict at the outset. Therefore, it is critical that the supportive activities that the team was involved in are seen to their conclusion. This may signify an open-ended engagement, which both the community and the researchers can envision as a calculated risk, with the power still starkly stacked on the researchers' side. Our suggestion is to identify the practical possibilities for extending community collaboration, through exploration of the divergent gaps in the conclusions and then weave them into the non-research activities.

24.7.1 Answering the Empirical Questions

The specific research questions posed in Sect. 24.3 are still in the process of analysis and currently in the stage of Analysis 2 on the RVA. However, there has been a significant amount of analysis already completed and can be read about in Chakraborty (2018) and Chakraborty et al. (in press). Exploring them in detail in this paper is constrained by both the goal of this work and by structural constraints.

24.7.2 Addressing the Gaps in Research

The inadequacies identified in HDCC research in the Himalayan region using a SES framework need to be revisited at this juncture to highlight the interventions proposed by our analytic:

1. <u>The Himalayan Theory of Environmental Degradation</u>: While critical scholars have challenged and repudiated this construct rooted in environmental determinism, using various situated case studies and grounded theory (see Metz 2010) there has been a significant absence of such affective thinking when modeling human-environment relations in the region and doing vulnerability analysis. The RVA complicates the deterministic predictions of future risks, by questioning both how vulnerability is produced and how those modes of production are transformed by global climate change across different spatial and temporal scales.

2. <u>Hybrid Methodologies</u>: A majority of the mixed-methods research on HDCC in the region betrays a disciplinary bias, with usually the more quantitative and positivist epistemologies dominating the design and outcomes (Nightingale 2016). The RVA while resorting to mixing 'biophysical with socioeconomic' data, also categorically connects the knowledge produced to its root paradigm for further analysis and distillation. Furthermore, instead of seeking a holistic complementarity, it probes for moments of divergence.

3. <u>Collaborative regional capacity building</u>: The politics of knowledge production, validation and reproduction in the Himalayan region has been documented by various scholars (Guneratne 2010; Ahlborg and Nightingale 2012; Hewitt and Mehta 2012). The issues of scalar mismatch, over-reliance on top-down data and a marginalization of knowledge produced by both regional communities and scholars is foundational to many regional vulnerability assessment attempts. The RVA attempts to address this issue by creating an analytical equity between different stakeholders, resisting compartmentalizing the information into binaries of local/scientific and simultaneously investing in collaboratively controlled place based assessment systems. In doing so our goal is to create both an open-source, public data set of ground truthing that can help situate the top-down visualization and to support regional scholars and communities to construct reflexive narratives about systems that they are directly embedded within.

4. <u>Relational Vulnerability</u>: In most SES research, proxy variables underlying rationalistic functions are empowered to adequately capture the complex assemblage of processes that produce vulnerability (Rosegrant et al. 2014; Entwisle et al. 2016). These variables are often categorically claimed to be part of artificially segregated 'ecological', or 'social' dimensions (Kelly and Adger 2000). By following socionatural approach of imagining integrated 'social-ecological' processes and nodes, our analytic supports recent attempts at redefining the roots of vulnerability within HDCC research (Yeh et al. 2014; Barnett and Eakin 2015; Carr and Owusu-Daaku 2016). By focusing deeply into the social relations that constitute families and societies, both through the multi-subjective iterations of the ABM as well as PE lens on the relational dynamics within formal social units (e.g. household, village, ward), we attempt to supplement the partiality of reductionist metric driven analyses (Turner 2016).

24.7.3 Challenges and Potential Limitations of the Methodology

24.7.3.1 Choosing an Adequate Level of Complexity

Model parameterization has often relied on bulk datasets (An 2012), which although provide a good starting point, fail to capture the observed heterogeneity of cross-scalar relational networks (Turner 2016). While it may seem logical that the complexity of a model as a heuristic device should echo the processes it is attempting to represent, this comes with a set of problems. Most of them are related to the calibration and transferability of the framework (Sun et al. 2016). Furthermore, these problems are more heightened within a methodology where the model is deemed the overarching analytical method, but this is not the case with the RVA. Therefore, choosing the complexity of the model universe can be a function of its role within a broader analytical frame, freeing the coupled models from the burden of performing the 'god trick'. However, despite this caveat, the volume of data generated through the RVA is significant and as such 'mining' it or 'harvesting' it to observe dynamic, complicated processes that are usually veiled, is a daunting task.

24.7.3.2 Spatio-Temporal Limitations

Unlike other vulnerability assessments frameworks, the RVA requires a significant temporal investment. Furthermore, the analytic encompasses many different spaces, all of which play critical roles in the overall process. Managing and sustaining these heterogeneous set of actors, institutions and activities can be an overwhelming task, especially given the cultural and professional differences and aspirations. While the RVA doesn't make a specific temporal schema, its various stages unfold over years. Therefore, this analytic is not suitable for rapid assessment scenarios and instead is geared towards revealing underlying structural processes which don't require intermediate and long term policy interventions.

24.7.3.3 Committing to the Uncomfortable Search for Divergences and Beyond

The moments of generative friction that will inevitably result from implementing the RVA will situate the researchers in an arguably unprecedented scenario. While venturing past disciplinary borders will challenge deeply ingrained ideological ideas, sharing power with the community and other stakeholders will test the limits of 'scientific' control. These can be potentially fatal to the overall project if not intentionally and democratically mediated. The divergences that emerge from probing knowledge with different epistemologies can be instrumental in highlighting the relationships between different types of data and catalyzing a more robust research design (Nightingale 2016). But it is the divergences in the different relationships that the many agents have to each other and to the overall research process that will need to be carefully mediated. Since, notwithstanding the results of the questions being explored, by going through the process of the RVA, the researchers themselves will embody a materialization of the goals of the analytic.

24.8 Conclusions

Most of the HDCC research in the Himalayan region suffers from a lack of disciplinary and stakeholder collaboration. Additionally, the SES framework has failed to equitably incorporate disparate knowledges within its analysis, inherently depending on quantitative systems literature to articulate and evaluate power relations and cultural values. We address both these situations by proposing a methodology – The Relational Vulnerability Analytic, inspired by the socionatural approach.

The RVA focuses not just on the content of the knowledge but also on the context of its production (Cote and Nightingale 2012). It attempts to produce vulnerability assessment tool with a goal of bringing hybrid epistemologies in dialogue, acknowledging the role of power and challenging 'The Theory of Himalayan Degradation'. In doing so, it utilizes the tools of PE, ABM and deterministic climate, hydro-geological and agro-ecological modeling.   Through different steps undertaken in different spaces by the various team members, the analytic is operationalized as both a method founded in rigor and one providing inclusion to the community. Building trust, both within the interdisciplinary research team and with members of the community is critical for the formalizing of research objectives and the utility of such objectives beyond scholarship. This provides the catalysis for the production of tools and specific methods needed to achieve the research goals. Post empirical data collection there are different analytical stages each with a unique goal. Finally, the conclusions are seen as the beginning points for further exploration of such complex problems and leads to the creation of select community identified projects, which have a tangible impact on their lives.

The RVA is not presented as 'the' definitive solution to the deeper foundational problems with SES and HDCC research, but as a possible alternative that needs to be adequately implemented in the field to evaluate its utility. This analytic is an appeal for foundational collaboration that goes beyond mere superficial encounters among disciplines. Furthermore, beyond scholarly purview, the conclusions of using such a framework can highlight usually overlooked spaces, times and agents that require support from policy instruments. As an example, in our project, an intervention for abatement of climatic risk should include supporting the changes within household management and ownership of land, often catalyzed by young migrants, which advocates for a more representative role of women. Decoupling cultural and political realities from climatic ones and getting trapped inside certain scales is a problem rampant within adaptation planning. The RVA addresses this by creating a dialogue not just between multiple fields of knowledge, but also between the relationships that different groups of people share with each other, and with their transforming worlds.

References

Adler RF, Gu G, Huffman GJ (2012) Estimating climatological bias errors for the Global Precipitation Climatology Project (GPCP). J Appl Meteorol Climatol 51(1):84–99. https://doi.org/10.1175/jamc-d-11-052.1

Ahlborg H, Nightingale AJ (2012) Mismatch between scales of knowledge in Nepalese forestry: epistemology, power, and policy implications. Ecol Soc 17(4). https://doi.org/10.5751/ES-05171-170416

Alcoff L, Potter E (1993) In: Alcoff L, Potter E (eds) Feminist epistemologies. Routledge, London

An L (2012) Modeling human decisions in coupled human and natural systems: review of agent-based models. Ecol Model 229:25–36. https://doi.org/10.1016/j.ecolmodel.2011.07.010

Balke T et al (2010) How do agents make decisions? A survey. Jasss 17(4):1. https://doi.org/10.18564/jasss.2687

Barnett AJ, Eakin HC (2015) "We and us, not I and me": justice, social capital, and household vulnerability in a Nova Scotia fishery. Appl Geogr Elsevier 59:107–116. https://doi.org/10.1016/j.apgeog.2014.11.005

Berger T, Troost C (2014) Agent-based modelling of climate adaptation and mitigation options in agriculture. J Agric Econ 65(2):323–348. https://doi.org/10.1111/1477-9552.12045

Beymer-farris BA, Bassett TJ, Bryceson I (2012) Promises and pitfalls of adaptive management in resilience thinking: the lens of political ecology. Resil Cult Lands:283–299. https://doi.org/10.1017/CBO9781139107778

Bharwani S (2004, December) Adaptive knowledge dynamics and emergent artificial societies: ethnographically based multi-agent simulations of behavioural adaptation in agro-climatic systems, p 369

Bharwani S et al (2015) Identifying salient drivers of livelihood decision-making in the forest communities of Cameroon: adding value to social simulation models. J Artif Soc Soc Simul 18((1)3):1–26

Bhattarai B, Beilin R, Ford R (2015) Gender, agrobiodiversity, and climate change: a study of adaptation practices in the Nepal Himalayas. World Dev Elsevier Ltd 70:122–132. https://doi.org/10.1016/j.worlddev.2015.01.003

Bithell M, Brasington J, Richards K (2008) Discrete-element, individual-based and agent-based models: tools for interdisciplinary enquiry in geography? Geoforum 39(2):625–642. https://doi.org/10.1016/j.geoforum.2006.10.014

Blythe J et al (2017) Feedbacks as a bridging concept for advancing transdisciplinary sustainability research. Curr Opin Environ Sustain Elsevier BV 26–27:114–119. https://doi.org/10.1016/j.cosust.2017.05.004

Boonstra WJ (2016) Conceptualizing power to study social-ecological interactions. Ecol Soc 21(1). https://doi.org/10.5751/ES-07966-210121

Broadhead RS, Rist RC (1976) Gatekeepers and the social control of social research. Soc Probl 23(325):8–23. https://doi.org/10.3868/s050-004-015-0003-8

Bruslé T (2008) Choosing a destination and work. Mt Res Dev 28(3/4):240–247. https://doi.org/10.1659/mrd.0934

Cardona OD, van Aalst MK, Birkmann M, Fordham G, McGregor R, Perez R, Pulwarty RS, Schipper ELF, Singh BT (2012) Determinants of risk: exposure and vulnerability. In: Managing the Risks of Extreme Events and Disasters to Advance Climate Change Adaptation – a special report of working groups I and II of the Intergovernmental Panel on Climate Change (IPCC), Cambridge University Press, Cambridge, pp 65–108. https://doi.org/10.1017/CBO9781139177245.005

Carr ER, Owusu-Daaku KN (2016) The shifting epistemologies of vulnerability in climate services for development: the case of Mali's agrometeorological advisory programme. Area 48:7–17. https://doi.org/10.1111/area.12179

Castree N (2015) Coproducing global change research and geography {The} means and ends of engagement. Dial Hum Geogr 5(3):343–348. https://doi.org/10.1177/2043820615613265

Chakraborty R (2018) The invisible (mountain) man: migrant youth and relational vulnerability in the Indian Himalayas. Available from ProQuest dissertations and theses database. (UMI no. 10829862)

Chakraborty R, Daloz AS, Kumar M, Dimri AP (in press) Does awareness of climate change impacts lead to worries about it? Epistemological pluralism, parallel analysis and community perceptions of climate change in rural Himalayas

Chaudhary P, Bawa KS (2011) Local perceptions of climate change validated by scientific evidence in the Himalayas. Biol Lett 7(5):767–770. https://doi.org/10.1098/rsbl.2011.0269

Choudhury MUI, Haque CE (2016) "We are more scared of the power elites than the floods": adaptive capacity and resilience of wetland community to flash flood disasters in Bangladesh. Int J Dis Risk Reduc Elsevier 19:145–158. https://doi.org/10.1016/j.ijdrr.2016.08.004

Clifford NJ (2008) Models in geography revisited. Geoforum 39(2):675–686. https://doi.org/10.1016/j.geoforum.2007.01.016

Cole J (2004) Fresh contact in Tamatave, Madagascar. Am Ethnol 31(4):573–588. https://doi.org/10.1525/ae.2004.31.4.573

Conte R et al (2012) Manifesto of computational social science. Eur Phys J Spec Top 214(1):325–346. https://doi.org/10.1140/epjst/e2012-01697-8

Cote M, Nightingale AJ (2012) Resilience thinking meets social theory. Prog Hum Geogr 36(4):475–489. https://doi.org/10.1177/0309132511425708

Cresswell T (2015) Afterword – Asian mobilities/Asian frictions? Environ Plan A 48(6):1082–1086. https://doi.org/10.1177/0308518X16647143

Davids T (2014) Trying to be a vulnerable observer: matters of agency, solidarity and hospitality in feminist ethnography. Women Stud Int Forum Elsevier Ltd 43:50–58. https://doi.org/10.1016/j.wsif.2014.02.006

Gellner DN (ed) (2013) Borderland lives in northern South Asia. Duke University Press, Durham & London

Davis M (2000) Late Victorian holocausts: El Niño famines and the making of the third world. Verso, London/New York

Davidson DJ (2010) The applicability of the concept of resilience to social systems: some sources of optimism and nagging doubts. Soc Nat Resour 23(12):1135–1149. https://doi.org/10.1080/08941921003652940

Devi RM, Dimri AP, Dutta J (2014) Uttarakhand disaster: natural or man-made? – a meteorological investigation. J Appl For Ecol 2(September):32–38

Dee DP, Balmaseda M, Balsamo G, Engelen R, Simmons AJ, Thépaut J (2014) Toward a consistent reanalysis of the climate system. Bull Amer Meteor Soc 95:1235–1248. https://doi.org/10.1175/BAMS-D-13-00043.1

Dyson J (2015) Life on the hoof: gender, youth, and the environment in the Indian Himalayas. J R Anthropol Inst 21(1):49–65. https://doi.org/10.1111/1467-9655.12147

Entwisle B et al (2016) Climate shocks and migration: an agent-based modeling approach. Popul Environ Springer Netherlands 38(1):47–71. https://doi.org/10.1007/s11111-016-0254-y

Eriksen SH, Nightingale AJ, Eakin H (2015) Reframing adaptation: the political nature of climate change adaptation. Glob Environ Change Elsevier Ltd 35:523–533. https://doi.org/10.1016/j.gloenvcha.2015.09.014

Fabinyi M, Evans L, Foale SJ (2014) Social-ecological systems, social diversity, and power: insights from anthropology and political ecology. Ecol Soc 19(4). https://doi.org/10.5751/ES-07029-190428

Folke C (2006) Resilience: the emergence of a perspective for social-ecological systems analyses. Glob Environ Chang 16(3):253–267. https://doi.org/10.1016/j.gloenvcha.2006.04.002

Ford JD et al (2013) The dynamic multiscale nature of climate change vulnerability: an Inuit harvesting example. Ann Assoc Am Geogr 103(5):1193–1211. https://doi.org/10.1080/00045 608.2013.776880

Gentle P, Maraseni TN (2012) Climate change, poverty and livelihoods: adaptation practices by rural mountain communities in Nepal. Environ Sci Pol Elsevier Ltd 21:24–34. https://doi.org/10.1016/j.envsci.2012.03.007

Georgeff M et al (1998) The belief-desire-intention model of agency. In: Intelligent Agents V: agents theories, architectures, and languages. 5th international workshop, ATAL'98, pp 1–10. https://doi.org/10.1007/3-540-49057-4_1

Ghorbani A, Dijkema G, Schrauwen N (2015) Structuring qualitative data for agent-based modelling case study: horticulture innovation. J Artif Soc Soc Simul 18(1):1–6

Gidwani V, Sivaramakrishnan K (2003) Circular migration and the spaces of cultural assertion. Ann Assoc Am Geogr 93(1):186–213. https://doi.org/10.1111/1467-8306.93112

Goldman MJ, Turner MD, Daly M (2018) A critical political ecology of human dimensions of climate change: epistemology, ontology, and ethics. Wiley Interdiscip Rev Clim Chang 9:1–15. https://doi.org/10.1002/wcc.526

Guneratne A (2010) Culture and environment in the Himalayas. Routledge, London

Gupta V (2011) A critical assessment of climate change impacts, vulnerability and policy in India. Pres Environ Sustain Dev 5(1):11–22

Guthman J (1997) Representing crisis: the theory of Himalayan environmental degradation and the project of development in post-Rana Nepal. Dev Chang 28(1):45–69. https://doi.org/10.1111/1467-7660.00034

Hartmann DL, Tank AMGK, Rusticucci M (2013) IPCC fifth assessment report, climate change 2013: the physical science basis. IPCC, AR5 (January 2014), pp 31–39. https://doi.org/10.1017/CBO9781107415324.004

Head L, Gibson C (2012) Becoming differently modern: geographic contributions to a generative climate politics. Prog Hum Geogr 36(6):699–714. https://doi.org/10.1177/0309132512438162

Herbert-Cheshire L, Higgins V (2004) From risky to responsible: expert knowledge and the governing of community-led rural development. J Rural Stud 20(3):289–302. https://doi.org/10.1016/j.jrurstud.2003.10.006

Hewitt K, Mehta M (2012) Rethinking risk and disasters in mountain areas. Revue de géographie alpine 100–1:0–13. https://doi.org/10.4000/rga.1653

Hoermann B, Kollmair M (2009) Labour migration and remittances in the Hindu Kush-Himalayan region. International Centre for Integrated Mountain, Kathmandu

Holling CS (1973) Resilience and stability of ecological systems. Annu Rev Ecol Syst 4:1–23. https://doi.org/10.1146/annurev.es.04.110173.000245

Hoque SF, Quinn CH, Sallu SM (2017) Resilience, political ecology, and well-being: an interdisciplinary approach to understanding social-ecological change in coastal Bangladesh. Ecol Soc 22(2). https://doi.org/10.5751/ES-09422-220245

Hoy A et al (2016) Climatic changes and their impact on socio-economic sectors in the Bhutan Himalayas: an implementation strategy. Reg Environ Change Springer Berlin/Heidelberg 16(5):1401–1415. https://doi.org/10.1007/s10113-015-0868-0

Ives JD (2004) The theory of Himalayan environmental degradation: its validity and application challenged by recent research author(s): Jack D. Ives Conference: The Himalaya-Ganges Problem (August, 1987), pp. 189–199. Published by: Int Mt Soc 7(3):189–199

IPCC (2007) Climate change 2007: synthesis report. Contribution of Working Groups I, II and III to the Fourth Assessment Report of the Intergovernmental Panel on Climate Change [Core Writing Team, Pachauri, R.K and Reisinger, A. (eds.)]. IPCC, Geneva, Switzerland, 104 pp

IPCC (2014) Climate change 2014: synthesis report. Contribution of Working Groups I, II and III to the Fifth Assessment Report of the Intergovernmental Panel on Climate Change [Core Writing Team, R.K. Pachauri and L.A. Meyer (eds.)]. IPCC, Geneva, Switzerland, 151 pp

Jain A (2010) Labour migration and remittances in Uttarakhand. International Centre for Integrated Mountain Development, Kathmandu

Janes JE (2016) Democratic encounters? Epistemic privilege, power, and community-based participatory action research. Action Res 14(1):72–87. https://doi.org/10.1177/1476750315579129

Janssen MA, Ostrom E (2006) Empirically based, agent-based models. Ecol Soc 11(2):art37. https://doi.org/10.5751/ES-01861-110237

Jeffrey C (2010) Timepass: youth, class, and time among unemployed young men in India. Am Ethnol 37(3):465–481. https://doi.org/10.1111/j.1548-1425.2010.01266.x

Jones JW et al (2003) The DSSAT cropping system model. Eur J Agron. https://doi.org/10.1016/S1161-0301(02)00107-7

Kelkar U et al (2008) Vulnerability and adaptation to climate variability and water stress in Uttarakhand State, India. Glob Environ Chang 18(4):564–574. https://doi.org/10.1016/j.gloenvcha.2008.09.003

Kelly PM, Adger WN (2000) Theory and practice in assessing vulnerability to climate change and facilitating adaptation. Clim Chang 47(4):325–352. https://doi.org/10.1023/A:1005627828199

Kingsley J, 'Yotti' et al (2010) Using a qualitative approach to research to build trust between a non-aboriginal researcher and aboriginal participants (Australia). Qual Res J 10(1):2–12. https://doi.org/10.3316/QRJ1001002

Korzenevica M, Agergaard J (2017) "The house cannot stay empty": a case of young rural Nepalis negotiating multilocal householding. Asian Popul Stud Taylor & Francis 13(2):124–139. https://doi.org/10.1080/17441730.2017.1303110

Leach M, Scoones I, Stirling A (2010) Governing epidemics in an age of complexity: narratives, politics and pathways to sustainability. Glob Environ Change Elsevier Ltd 20(3):369–377. https://doi.org/10.1016/j.gloenvcha.2009.11.008

Lövbrand E et al (2015) Who speaks for the future of Earth? How critical social science can extend the conversation on the Anthropocene. Glob Environ Chang 32:211–218. https://doi.org/10.1016/j.gloenvcha.2015.03.012

Mahony M, Hulme M (2016) Epistemic geographies of climate change. Prog Hum Geogr:1–30. https://doi.org/10.1177/0309132516681485

Mathur N (2015a) A "remote" town in the Indian Himalaya. Mod Asian Stud 49(2):365–392. https://doi.org/10.1017/S0026749X1300053X

Mathur N (2015b) "It's a conspiracy theory and climate change " of beastly encounters and cervine dissapearances in Himalayan India. HAU J Ethnogr Theor 5(1):87–111

Mcguirk P, O'Neill P (2012) Critical geographies with the state: the problem of social vulnerability and the politics of engaged research. Antipode 44(4):1374–1394. https://doi.org/10.1111/j.1467-8330.2011.00976.x

Metz JJ (2010) Downward spiral? Interrogating narratives of environmental change in the Himalaya. In: Guneratne A (ed) Culture and environment in the Himalayas. Routledge, London

Miller F et al (2010) Resilience and vulnerability: complementary or conflicting concepts? Ecol Soc 15(3):11

Millington J et al (2017) Integrating modelling approaches for understanding telecoupling: global food trade and local land use. Land 6(3):56. https://doi.org/10.3390/land6030056

Molen N (2016) A method for employing qualitative data in the development of spatial agent-based models. Michigan State University, Michigan

Molony T, Hammett D (2007) The friendly financier: talking money with the silenced assistant. Hum Organ 66(3):292–300

Murphy BL (2011) From interdisciplinary to inter-epistemological approaches: confronting the challenges of integrated climate change research. Can Geogr 55(4):490–509. https://doi.org/10.1111/j.1541-0064.2011.00388.x

Nakamura K (2013) Making sense of sensory ethnography: the sensual and the multisensory. Am Anthropol 115(1):132–132. https://doi.org/10.1111/j.1548-1433.2012.01543.x

Nightingale A (2003) A feminist in the forest: situated knowledges and mixing methods in natural resource management. Acme 2(1):77–90. https://doi.org/10.1016/S0016-7185(99)00025-1

Nightingale AJ (2016) Adaptive scholarship and situated knowledges? Hybrid methodologies and plural epistemologies in climate change adaptation research. Area:41–47. https://doi.org/10.1111/area.12195

Nightingale AJ (2017) Power and politics in climate change adaptation efforts: struggles over authority and recognition in the context of political instability. Geoforum Elsevier 84(May):11–20. https://doi.org/10.1016/j.geoforum.2017.05.011

Ostrom E (2009) Social-ecological systems. Science 325(5939):419–422

Pandey R, Jha SK (2012) Climate vulnerability index – measure of climate change vulnerability to communities: a case of rural lower Himalaya, India. Mitig Adapt Strateg Glob Chang 17(5):487–506. https://doi.org/10.1007/s11027-011-9338-2

Pandey R et al (2017) Agroecology as a climate change adaptation strategy for small holders of Tehri-Garhwal in the Indian Himalayan region. Small Scale For Springer Netherlands 16(1):53–63. https://doi.org/10.1007/s11842-016-9342-1

Pant A (2016) Rural tourism a solution for ghost villages of Uttarakhand. Int J New Technol Res 6:52–60

Peet R, Robbins P, Watts M (2011) Global political ecology (Peet R, Robbins P, Watts M (eds)). Routledge, London

Pfaff-Czarnecka J, Toffin G (2011) Introduction: belonging and multiple attachments in contemporary Himalayan societies. The politics of belonging in the Himalayas: local attachments and boundary dynamics, pp xi–xxxviii. https://doi.org/10.4135/9788132107729

Popke J (2016) Researching the hybrid geographies of climate change: reflections from the field. Area 48(1):2–6. https://doi.org/10.1111/area.12220

Rawat PK, Tiwari PC, Pant CC (2012) Climate change accelerating land use dynamic and its environmental and socio-economic risks in the Himalayas. Int J Climate Change Strateg Manage 4(4):452–471. https://doi.org/10.1108/17568691211277764

Ren GY, Shrestha AB (2017) Climate change in the Hindu Kush Himalaya. Adv Clim Change Res (National Climate Center (China Meteorological Administration)) 8(3):137–140. https://doi.org/10.1016/j.accre.2017.09.001

Ribot J (2010) Vulnerability does not fall from the sky: towards multi-scale, pro-poor climate policy. Social dimensions of climate change: equity and vulnerability in a warming world, p 319. https://doi.org/10.1088/1755-1307/6/34/342040

Ribot J (2014) Cause and response: vulnerability and climate in the Anthropocene. J Peasant Stud 41(5):667–705. https://doi.org/10.1080/03066150.2014.894911

Rienecker MM, Suarez MJ, Gelaro R, Todling R, BacmeisterE J, Liu MG, Bosilovich SD, Schubert L, Takacs G, Kim S, Bloom J, Chen D, Collins A, da Conaty A, Silva WG, Joiner J, Koster RD, Lucchesi R, Molod A, Owens T, Pawson S, Pegion P, Redder CR, Reichle R, Robertson FR, Ruddick AG, Sienkiewicz M, Woollen J (2011) MERRA: NASA's modern-era retrospective analysis for research and applications. J Clim 24:3624–3648. https://doi.org/10.1175/JCLI-D-11-00015.1Metz (2010)

Robinson DT et al (2007) Comparison of empirical methods for building agent-based models in land use science. J Land Use Sci 2(1):31–55. https://doi.org/10.1080/17474230701201349

Rodrigues CD (2014) Doing research in violent settings: ethical considerations and ethics committees. DSD Working Papers on Research Security, No.5, pp 1–20

Rosegrant MW et al (2014) Food security in a world of natural resource scarcity: role of agricultural technologies. IFPRI Book. https://doi.org/10.2499/9780896298477

Samson E et al (2017) Early engagement of stakeholders with individual-based modeling can inform research for improving invasive species management: the round goby as a case study. Front Ecol Evol 5(November):1–15. https://doi.org/10.3389/fevo.2017.00149

Satyal P et al (2017) A new Himalayan crisis? Exploring transformative resilience pathways. Environ Model Softw Elsevier Ltd 23(November 2016):47–56. https://doi.org/10.1016/j.envdev.2017.02.010

Schild A (2008) ICIMOD's position on climate change and mountain systems. Mt Res Dev 28(3/4):328–331. https://doi.org/10.1659/mrd.mp009

Schreinemachers P, Berger T (2011) An agent-based simulation model of human-environment interactions in agricultural systems. Environ Model Softw Elsevier Ltd 26(7):845–859. https://doi.org/10.1016/j.envsoft.2011.02.004

Sharma JR (2013) Marginal but modern: young Nepali labour migrants in India. Young 21(4):347–362. https://doi.org/10.1177/1103308813506307

Sharma E et al (2009) Climate change impacts and vulnerability in the eastern Himalayas. Icimod, p 32. https://doi.org/10.1007/978-3-540-88246-6

Shneiderman S (2010) Are the central Himalayas in Zomia? Some scholarly and political considerations across time and space. J Glob Hist 5(2):289–312. https://doi.org/10.1017/S1740022810000094

Shneiderman S (2016) Association for Nepal and afterword|Charting Himalayan histories afterword. Chart Himal Hist 35(2):136–138

Singer M (1995) Beyond the ivory tower: critical praxis in medical anthropology. Published by Wiley on behalf of the American Anthropological Association Stable. URL: http://www.jstor.org/stable/648559. Beyond Ivory Tower Critic Prax Med Anthropol 9(1):80–106

Singh SP, Thadani R (2015) Complexities and controversies in Himalayan research: a call for collaboration and rigor for better data. Mt Res Dev 35(4):401–409. https://doi.org/10.1659/MRD-JOURNAL-D-15-00045

Singh D, Padgham L, Logan B (2016) Integrating BDI agents with agent-based simulation platforms. Auton Agent Multi-Agent Syst Springer US 30(6):1050–1071. https://doi.org/10.1007/s10458-016-9332-x

Smith S (2012) Intimate geopolitics: religion, marriage, and reproductive bodies in Leh, Ladakh. Ann Assoc Am Geogr 102(6):1511–1528. https://doi.org/10.1080/00045608.2012.660391

Smith SH (2013) "In the heart, there"s nothing': unruly youth, generational vertigo and territory. Trans Inst Br Geogr 38(4):572–585. https://doi.org/10.1111/j.1475-5661.2012.00547.x

Smith SH, Gergan M (2015) The diaspora within: Himalayan youth, education-driven migration, and future aspirations in India. Envir Plan D Soc Space 33(1):119–135. https://doi.org/10.1068/d13152p

Smith A, Hall M, Sousanis N (2015) Envisioning possibilities: visualising as enquiry in literacy studies. Literacy 49(1):3–11. https://doi.org/10.1111/lit.12050

Stone-Jovicich S (2015) Probing the interfaces between the social sciences and social-ecological resilience: insights from integrative and hybrid perspectives in the social sciences. Ecol Soc 20(2). https://doi.org/10.5751/ES-07347-200225

Sturm A, Dori D, Shehory O (2010) An object-process-based modeling language for multiagent systems. IEEE Trans Syst Man Cybernet Part C Appl Rev 40(2):227–241. https://doi.org/10.1109/TSMCC.2009.2037133

Sun Z et al (2016) Simple or complicated agent-based models? A complicated issue. Environ Model Softw Elsevier Ltd 86(3):56–67. https://doi.org/10.1016/j.envsoft.2016.09.006

Taylor J (2011) The intimate insider: negotiating the ethics of friendship when doing insider research. Qual Res 11(1):3–22. https://doi.org/10.1177/1468794110384447

Tiwari PC, Joshi B (2015) Climate change and rural out-migration in Himalaya. Change Adapt Socio Ecol Syst 2(1):8–25. https://doi.org/10.1515/cass-2015-0002

Tiwari PC, Joshi B (2016) Gender processes in rural out-migration and socio-economic development in the Himalaya. Migr Dev Routledge 5(2):330–350. https://doi.org/10.1080/21632324.2015.1022970

Turner S (2010) Challenges and dilemmas: fieldwork with upland minorities in socialist Vietnam, Laos and southwest China. Asia Pac Viewp 51(2):121–134. https://doi.org/10.1111/j.1467-8373.2010.01419.x

Turner MD (2014) Political ecology I. Prog Hum Geogr 38(4):616–623. https://doi.org/10.1177/0309132513502770

Turner MD (2016) Climate vulnerability as a relational concept. Geoforum Elsevier Ltd 68:29–38. https://doi.org/10.1016/j.geoforum.2015.11.006

Turner BL, Robbins P (2008) Land-change science and political ecology: similarities, differences, and implications for sustainability science. Annu Rev Environ Resour 33(1):295–316. https://doi.org/10.1146/annurev.environ.33.022207.104943

Turner BL et al (2003) A framework for vulnerability analysis in sustainability science. Proc Natl Acad Sci U S A 100(14):8074–8079. https://doi.org/10.1073/pnas.1231335100

Van Stapele N (2014) Intersubjectivity, self-reflexivity and agency: narrating about "self" and "other" in feminist research. Women Stud Int Forum Elsevier Ltd 43:13–21. https://doi.org/10.1016/j.wsif.2013.06.010

Verburg PH et al (2016) Methods and approaches to modelling the Anthropocene. Glob Environ Change Elsevier Ltd 39:328–340. https://doi.org/10.1016/j.gloenvcha.2015.08.007

Vogel, C. et al. (2007) 'Linking vulnerability, adaptation, and resilience science to practice: pathways, players, and partnerships', Glob Environ Chang, 17(3–4), pp. 349–364. doi: https://doi.org/10.1016/j.gloenvcha.2007.05.002

Wellman MP (2016) Putting the agent in agent-based modeling. Auton Agent Multi-Agent Syst Springer US 30(6):1175–1189. https://doi.org/10.1007/s10458-016-9336-6

Wise S, Crooks AT (2012) Agent-based modeling for community resource management: Acequia-based agriculture. Comp Environ Urban Syst Elsevier Ltd 36(6):562–572. https://doi.org/10.1016/j.compenvurbsys.2012.08.004

Yang L, Gilbert N (2008) Getting away from numbers: using qualitative observation for agent-based modeling. Adv Complex Syst 11(2):1–11. https://doi.org/10.1142/S0219525908001556

Yeh ET et al (2014) Tibetan pastoralists' vulnerability to climate change: a political ecology analysis of snowstorm coping capacity. Hum Ecol 42(1):61–74. https://doi.org/10.1007/s10745-013-9625-5

You QL et al (2017) An overview of studies of observed climate change in the Hindu Kush Himalayan (HKH) region. Adv Clim Change Res Elsevier Ltd 8(3):141–147. https://doi.org/10.1016/j.accre.2017.04.001

Zhan Y-J et al (2017) Changes in extreme precipitation events over the Hindu Kush Himalayan region during 1961–2012. Adv Clim Change Res Elsevier Ltd 8(3):166–175. https://doi.org/10.1016/j.accre.2017.08.002

Chapter 25
Climate Change Trends and Ecosystem Resilience in the Hindu Kush Himalayas

Nakul Chettri, Arun Bhakta Shrestha, and Eklabya Sharma

Abstract During the past few decades, our understanding of the potential risks from climate change to mountain ecosystem has increased. The Hindu Kush Himalayas (HKH) is characterised by diverse climate due to diversity in geology, monsoon influence and ecosystems. Though paucity in studies, it was observed that the HKH ecosystems witnessed changes in climate over the period with evidence of change in phenology and species range shift altering ecosystem functions. During 1901–2014, annual mean surface air temperature significantly increased in the HKH at a rate of about 0.11 °C per decade showing significant upward trend. The intense precipitation also showed increasing trend in annual intense precipitation amount, days and intensity with 5.28 mm per decade, 0.14 day per decade and 0.39 mm/day per decade respectively. The elevation dependent warming has also been prominent in the HKH with higher warming with the increasing elevation. Higher warming is projected during winter and the projected warming differs by more than 1 °C between the eastern and western HKH, with relatively higher values during winter. The highest warming is projected to be over the central Himalaya for the far-future period with the RCP8.5 scenario. The projections made by the study for the near-future and far-future periods for HKH are relatively higher than the seasonal global means. These changes have indicated that rapidly changing climatic conditions could significantly thwart efforts for ecosystem resilience at a national and regional scales. There have been a wide range of interpretation from observed and people's perceptions impacting on a wide range of ecosystems and biodiversity at different scales.. However, there is still a major gap in understanding the cross-linkages among areas of research, for example, linking social-ecological knowledge on resilience contributing to evolutionary adaptation. Although numerous important contributions have emerged in recent years, synthesis of such practices and its consequences has not yet been achieved. This chapter is an attempt to relate the climate change science with ecosystem resilience in the HKH, identify gaps, and understand the social-ecological interaction and contribute towards social-ecological resilience.

N. Chettri · A. B. Shrestha · E. Sharma (✉)
International Centre for Integrated Mountain Development, Kathmandu, Nepal
e-mail: Eklabya.Sharma@icimod.org

© Springer Nature Switzerland AG 2020
A. P. Dimri et al. (eds.), *Himalayan Weather and Climate and their Impact on the Environment*, https://doi.org/10.1007/978-3-030-29684-1_25

25.1 Introduction

The mountain biomes are considered sensitive to climate change (Nogués-Bravo et al. 2008). The climate change affects directly or indirectly different key features (ecosystems, agriculture, biodiversity, snow cover, glaciers, run-off processes, and water availability) of the mountains (Bharali and Khan 2011; Tiwari and Joshi 2012; Joshi et al. 2012; Bhagawati et al. 2017; Lamsal et al. 2017a; Tewari et al. 2017; Bajracharya et al. 2018). There are growing evidences that the rate of warming is amplified with elevation (Shrestha et al. 1999; Sun et al. 2017). As a result, the high-mountain environment is witnessing more rapid changes in temperature than at lower elevations (Pepin et al. 2015). As a result of higher warming and visible changes in cryosphere and ecosystems, the mountains are considered as an early indicators of climate change (Singh et al. 2010) and such elevation-dependent warming can accelerate the rate of change in mountain ecosystems, cryosphere, hydrological regimes and biodiversity (Beniston 2003; Pauchard et al. 2009; Brandt et al. 2013; Pepin et al. 2015). Mountains have been considered as the last bastion for biodiversity with fragile ecosystems and recognized as major sources of ecosystem services contributing for human wellbeing (Messerli and Ives 1997; Viviroli et al. 2011). Climate change is likely impacting food production and security, sustained water supply, biodiversity and other natural ecosystems, human health limiting sustainable development (Xu et al. 2009; Ravindranath et al. 2011, Iizumi et al. 2013; Palomo 2017).

The maintenance of mountain ecosystem resilience is vital for human well-being as 85% of the people living in the mountains depends directly on ecosystem services, yet despite increase in conservation activities, the vulnerability of mountain ecosystems has been a major challenge (Nogués-Bravo et al. 2008; Rodríguez-Rodríguez and Bomhard 2012; Palomo 2017). Although habitat degradation, fragmentation, and destruction, overexploitation, and invasive species have driven recent biodiversity loss, climate change is projected to be a major driver of extinction throughout the twenty-first century, impacting ecosystems directly or indirectly or via synergies with other stressors (Parmesan 2006; Aukema et al. 2017). Climate change impacts over ecosystem degradation have put the Hindu Kush Himalayas (HKH) at the centre of regional and global attention (Brooks et al. 2006; Shrestha et al. 2012; Xu and Grumbine 2014). There is paucity of systematic research and long term analysis in the HKH due to differences on research priorities, accessibility to the remote areas, availability of financial and human resources and political will. However, the anecdotal information and scatted case studies have realized the vulnerability of diverse ecosystems in the context of changing climate in the HKH (Pandey and Jha 2012; Negi et al. 2017). Evidences showed that climate change have major implications on the poor and marginalized communities who exclusively depend on the ecosystem services for their livelihoods (Chettri et al. 2010). Broadly, the interaction between climate change and healthy ecosystems is twofold, on one hand healthy ecosystems is threatened by climate change, and on the other hand, proper management of ecosystems provides significant opportunities to mitigate the impacts of climate change (Lo 2016).

The HKH play a key role on supporting economy of the countries within the region and downstream of ten major river basins, which depend largely on hydropower, water supply, agriculture, and tourism. For example, Bhutan's export revenue from hydropower contributed up to 16.3% to the nominal gross domestic product (GDP) or 39% in terms of total exports in 2009/2010 (RMA 2011). Addressing vulnerability is a key issue in the HKH and analyses of existing knowledge, and gaps on how mountain ecosystems could be impacted under climate change is essential. The fragile landscapes of the HKH are highly susceptible to natural hazards, leading to ongoing concern about current and future climate change impacts in the region (Xu and Grumbine 2014). Climate change concerns in the Himalayas are multifaceted encompassing floods, droughts, landslides (Barnett et al. 2005), human health, biodiversity, endangered species, agriculture livelihood, and food security (Xu et al. 2009). While there are some reviews of existing literature on climate change observations and physical impacts on some of these aspects, a comprehensive review covering the HKH from all dimensions of impacts is still missing. Thus, this study has two specific objectives: (i) to synthesize the current state of knowledge on climate change impacts on the biophysical system (e.g., temperature, precipitation, snow coverage, streamflow, glacier melt, and ecosystem changes) in the Himalayan region and (ii) to review existing literature on resilience building practices to address impacts of climate change in the region. This study will help identify critical research gaps on the impacts of climate change in the Himalayas and strengthen understanding on social-ecological interlinkages necessary to understand the resilience contributing to evolutionary adaptation.

25.2   Hindu Kush Himalaya – The Vulnerable Mountain Ecosystem

The HKH mountain ecosystems is one of the most fragile ecosystems in the world (Ren and Shrestha 2017). Stretched over more than 4.3 million km^2 area includes areas of Afghanistan, Bangladesh, Bhutan, China, India, Myanmar, Nepal, and Pakistan, the HKH is characterized by some of the most complex terrain, and has a substantial influence on the East Asian monsoon, and even on global atmospheric circulation (Chettri and Sharma 2016). The region provide key livelihood resources such as food, timber, fibre, medicine and a wide range of services such as drinking water, water for irrigation, climate regulation, carbon storage, and the maintenance of aesthetic, cultural, and spiritual values (Sharma et al. 2015; Sandhu and Sandhu 2014; Chaudhary et al. 2017). The natural and semi-natural vegetation cover on mountains helps to stabilize headwaters, prevent flooding, landslide and maintain steady year-round flows of water by facilitating the seepage of rainwater into aquifers, vital for maintaining human life in the densely populated areas downstream. As a result, mountains have often been referred to as 'water towers' (Molden et al. 2014; Mukherji et al. 2015). Recognizing the importance of mountains for biodiversity and sustaining ecosystem services, the Convention on Biological Diversity (CBD) in Chapter 13 of Agenda 21 (1992) has recognized the significance of the mountains that supports all forms of living organisms, animals (including humans), and plants (UN 1992).

Driven by plate tectonics, the mountains of the HKH have unique ecosystems with altitudinal variation giving rise to numerous micro climates and diverse ecological gradients. It is the youngest and one of the most diverse ecosystems among the global mountain biomes, with extreme variations in vegetation, climate, and ecosystems resulting from altitudinal, latitudinal, and soil gradients (Xu et al. 2009; Sharma et al. 2010). This diverse biophysical habitat sets the stage for a rich biodiversity and species evolution (Miehe et al. 2014; Hudson et al. 2016). The region is the source of 10 major river systems with productive landscapes and strong upstream downstream linkages (Xu et al. 2009), and includes all or part of four global biodiversity hotspots — Himalaya, Indo-Burma, mountains of Southwest China, and mountains of Central Asia (Mittermeier et al. 2004; Chettri et al. 2010) — which contain a rich variety of gene pools and species with high endemism and novel ecosystem types with one of the highest rate of deforestation and habitat degradation. In addition, the region supports more than 60 different ecoregions, many of them are part of the Global 200 ecoregions (Olson and Dinerstein 2002). The ecosystem services from the HKH sustain 240 million people in the region and benefit some 1.6 billion people in the downstream river basin areas and have been well recognized by many scholars (Quyang 2009; Xu et al. 2009; Molden et al. 2014; Sharma et al. 2015).

However, the HKH region, has been witnessing various direct and indirect pressure through wide range of drivers such as habitat degradation, pollution, invasive species and climate change to name a few (Chettri and Sharma 2016). Climate change trigger higher rate of erosion manifested by change in precipitation pattern (Burbank et al. 2003; Pandit et al. 2007) increasing the frequency of disasters leading to vulnerability of these fragile ecosystems. Forest fragmentation has been identified as one of major drivers leading to vulnerability in both western (Uddin et al. 2015) and eastern Himalayas (Ravindranath et al. 2011). Sensitive ecosystems such as wetlands and high altitude rangelands are more vulnerable to combined effects of various drivers such as climate change, land use change, overexploitations etc. (Chettri et al. 2010; Chaudhary et al. 2017). As a result, the soil retention capacity has been reduced with increased erosions and frequency of floods (Burbank et al. 2003; Joshi and Kumar 2006; Cánovas et al. 2017).

25.3   Climate Change Trend in the Hindu Kush Himalayas

25.3.1   *Weather and Climate of Hindu Kush Himalayas*

The Himalayas are sensitive to climate change and variability (Eriksson et al. 2009; Shrestha and Aryal 2011). The climate of this region is found to be related to several large scale global climatic phenomena. The climate in this region has experienced large scale change in the historical and paleoclimatic time scales owing to natural changes and variability such as solar variability, orbital changes, tectonics and volcanism (Chettri and Sharma 2016). Moreover, topography, seasonality, and variability of weather patterns strongly determine the spatial and temporal pattern of temperatures and precipitation across different geographic regions of the HKH. The average summer and winter temperatures are about 30 °C and 18 °C respectively in the southern foothills. In the middle Himalayan valleys mean summer temperatures range between 15 °C and 25 °C and very cold winters (Shrestha and Aryal 2011). Areas having elevations above 4800 m experience winter temperatures below freezing point and receive precipitation largely in the form of snow. Topography also plays important role in the form of precipitation with a major part of precipitation from the southwest Indian summer monsoon, fall as frozen precipitation at higher elevations and liquid precipitation at lower elevations and adjacent plains of the Himalaya. As revealed by the sparse rain gauge network, the southern slopes of the Himalaya typically experience large annual precipitation totals as high as 400 cm per year during the period of 1998–2007 (Bookhagen and Burbank 2010). Since the influence of summer and winter monsoon circulation is not evenly distributed over the Himalayas, the summer (winter) rainfall is typically the greatest over the southeast (northwest) part of the HKH. Monsoon precipitation is found to be the highest over the Siwalik and Pir Panjal ranges of the lower Himalaya, while it reduces northwards into the Great Himalaya, Zanskar, Ladakh, and Karakoram ranges. However, most of the warming observed during the last few decades of the twentieth century is attributed to the increase in anthropogenic greenhouse gas concentration (IPCC 2007; You et al. 2017).

25.3.2   *Past Climate Changes*

25.3.2.1   **Temperature**

Several studies have been conducted on surface air temperature change for a different areas of the HKH, including the Tibetan Plateau (Liu and Chen 2000; Wang et al. 2008; Liu et al. 2009; Yao et al. 2012; Duan and Xiao 2015; Wang et al. 2016; Fan et al. 2015; Ren et al. 2017; You et al. 2017). These studies show significant warming in recent decades and the last century, in spite of the fact that the warming varies in different part of the HKH. Among them Ren et al. (2017) covered the

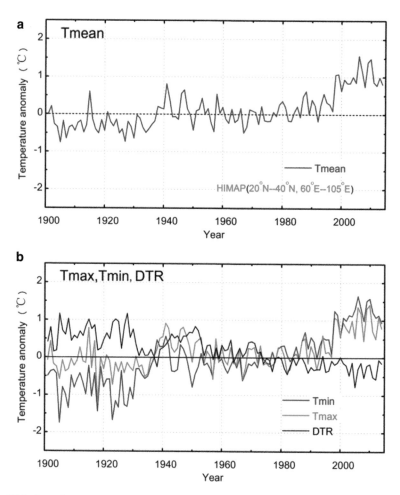

Fig. 25.1 Annual mean temperature anomaly series (°C) relative to 1961–1990 mean values for Tmean (**a**), Tmax, Tmin, and DTR (**b**) for the HKH between 1901 and 2014 (data source: CMA GLSAT; Ren et al. 2017)

whole HKH domain. According to this study, during the period 1901–2014, annual mean surface air temperature significantly increased in the HKH at a rate of about 0.11 °C per decade. For the full period of record (1901–2014), annual temperature trends show significant upward trends ($p < 0.05$), and the increase rates of Tmean, Tmax and Tmin are 0.10 °C per decade, 0.08 °C per decade, and 0.18 °C per decade, respectively (Fig. 25.1; Ren et al. 2017). Diurnal Temperature Range (DTR), the difference between the daily maximum and minimum temperature, shows a significant negative trend of −0.10 °C per decade, due to the much larger rise of minimum temperature than of maximum temperature in the region. Locally, deviations from the general pattern described above have been found in the Karakoram region, where in decreasing (most notably) summer temperature have been measured

(Forsythe et al. 2017). These trends are comparable with or greater than the global land- surface temperature trends.

25.3.2.2   Precipitation

Analysis of past 114 years' (1901–2013) trends of annual precipitation in the entire HKH using Global Land Monthly Precipitation (GLMP) and Global Land Daily (GLDP) data sets recently developed by the China Meteorological Administration (CMA) did not show significant trends (Ren et al. 2017). It is typical to the region and could be due to due to the sensitivity to topography. The regional average annual precipitation standardized anomaly (PSA) and annual precipitation percent anomaly (PPA; Fig. 25.2; Zhan et al. 2017) show fluctuations from one year to another, but the fluctuation became relatively larger from 1930 to 1960 (Fig. 25.2). While the overall trend was negative for the HKH the trends were small and not significant at the 0.05 confidence level (Ren et al. 2017).

25.3.3   Past Changes in Extremes

25.3.3.1   Temperature

Changes in extremes are important as adapting to extremes, both by humans and ecosystem is more challenging. Our studies have suggested that most parts of the HKH underwent significant long-term changes in extreme temperature events over the past decades. Temperature extreme indices have shown changing trends, with extreme cold events significantly decreasing and extreme warm events significantly increasing over the HKH during the past six decades (Sun et al. 2017). The trends of the extreme events related to minimum temperature were greater in magnitude than those related to maximum temperature. Similar results were found in the Koshi Basin by Rajbhandari et al. (2016). Further, extreme values of the highest Tmax (TXx) and the lowest Tmin (TNn) showed increasing trends in the HKH, with the rising rate of TNn double that of TXx (Sun et al. 2017). In addition, summer day (SU) frequency also show increasing trend, while annual frost day (FD) frequency decreased (Sun et al. 2017). The minimum temperature is rising more rapidly than maximum temperature and as a result the daily temperature range (DTR) is decreasing. Like in the HKH almost all the extreme temperature indices in the Tibetan Plateau region showed statistically significant trends over the past half century (You et al. 2008; Zhou and Ren 2011; Sun et al. 2017).

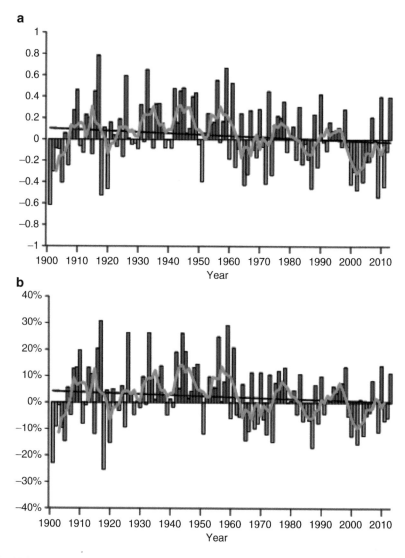

Fig. 25.2 The regional average annual PSA (**a**) and PPA (**b**) over 113 years (1901–2013) in the HKH. Time series, with the green line denoting the 5-year moving average, and the black line the linear trend

25.3.3.2 Precipitation

Like temperature the extremes in the precipitation are also changing in the HKH, although changes in temperature are more homogeneous at than the patterns in precipitation due to the sensitivity to topography, for example The regional average annual amount, and day and annual intensity anomalies for the percentile based light (below the 50th percentile), moderate (between the 50th and 90th percentiles), and intense (beyond the 90th percentile) precipitation over the period 1961–2013 in the HKH (Zhan et al. 2017). The intense precipitation show most significant increasing trend (p ≥ 0.95). The increase in in annual intense precipitation amount, days and intensity are 5.28 mm per decade, 0.14 day per decade and 0.39 mm/day per decade respectively. The light precipitation indices also show increasing trend but not of the similar significance as intense precipitation, while moderate precipitation do not show any trend.

25.3.4   Elevation Dependent Warming

The elevation-dependent warming (EDW) phenomenon is reported globally (Wang et al. 2014; Fan et al. 2015; Pepin et al. 2015). Similar phenomenon is also reported in the HKH as a whole and different parts of HKH (Shrestha et al. 1999; Liu and Chen 2000; Yan and Liu 2014; Wang et al. 2014, 2016; Duan and Xiao 2015; Guo et al. 2016; Yan et al. 2016; Sun et al. 2017). The exact physical mechanism driving EDW is not understood properly and needs further investigation. Study by Yan et al. (2016) on EDW over the Tibetan Plateau, and suggested that the increase in surface net radiation is driving the EDW phenomenon. One hypothesis suggests that the positive feedback associated with a diminishing cryosphere, particularly the snow cover is the cause of EDW (You et al. 2010). As the HKH has the largest extent of cryosphere (glaciers and ice caps, snow, river and lake ice, and frozen ground) outside the Polar Regions EDW is likely to have strong impact, which could further impact various ecosystem services of this region.

25.3.5   Future Climate Change

25.3.5.1   Temperature

The climate projections are unequivocal in suggesting continued warming in the future (Kumar et al. 2011; Kulkarni et al. 2013; Rajbhandari et al. 2015, 2016). A recent modelling work done under the Coordinated Regional Climate Downscaling Experiment (CORDEX) initiative suggested continued warming into the future (Sanjay et al. 2017). Sanjay et al. (2017) conducted analysis for the HKH domain based on CORDEX data. Present review is based on their results The magnitude of the projected seasonal warming is found to different for different part of the region, and dependent on season, averaging period, and scenario. Higher warming is projected during winter and the projected warming differs by more than 1 °C between the eastern and western HKH, with relatively higher values during winter (Sanjay et al. 2017). The highest warming is projected to be over the central Himalaya for the far-future period with the RCP8.5 scenario.

Table 25.1 Seasonal ensemble mean projected changes in near-surface air temperature (°C) relative to 1976–2005 in three HKH sub-regions defined by grid cells within each sub-region above 2500 m a.s.l.: northwest Himalayas and Karakoram (HKH1); central Himalayas (HKH2); southeast Himalayas and Tibetan Plateau (HKH3). The ranges for the 10 GCMs and 13 RCMs analysed are given in brackets

Scenario	Period	Multi-model ensemble mean	Summer monsoon season (June–September) (°C)			Summer monsoon season (June–September) (°C)		
			HKH1	HKH2	HKH3	HKH1	HKH2	HKH3
RCP4.5	2036–2065	Downscaled CORDEX RCMs	**2.0** (1.2, 3.3)	**1.7** (1.1, 2.4)	**1.7** (1.2, 2.2)	**2.3** (1.4, 3.2)	**2.4** (1.4, 3.4)	**2.4** (1.4, 3.1)
		Driving CMIP5 GCMs	**2.6** (1.7, 3.3)	**2.1** (1.6, 2.7)	**2.0** (1.6, 2.4)	**2.1** (1.2, 3.2)	**2.7** (1.6, 3.9)	**2.5** (1.4, 3.3)
	2066–2095	Downscaled CORDEX RCMs	**2.6** (1.4, 3.7)	**2.2** (1.4, 3.2)	**2.2** (1.7, 2.9)	3.1 (2.2, 4.1)	**3.3** (2.3, 4.5)	**3.1** (2.0, 4.8)
		Driving CMIP5 GCMs	**3.3** (2.5, 4.1)	**2.7** (2.2, 3.2)	**2.5** (1.9, 2.9)	**3.0** (2.1, 3.4)	**3.6** (2.4, 4.6)	**3.3** (2.1, 4.1)
RCP8.5	2036–2065	Downscaled CORDEX RCMs	**2.7** (1.7, 4.3)	**2.3** (1.4, 3.2)	**2.3** (1.5, 2.9)	**3.2** (1.8, 4.4)	**3.3** (2.1, 4.6)	**3.2** (2.0, 4.6)
		Driving CMIP5 GCMs	**3.3** (2.5, 4.3)	**2.7** (2.0, 3.4)	**2.5** (1.9, 3.0)	**3.0** (2.2, 3.9)	**3.4** (2.3, 4.7)	**3.2** (2.1, 4.2)
	2066–2095	Downscaled CORDEX RCMs	**4.9** (3.0, 7.7)	**4.3** (3.1, 6.1)	**4.2** (3.1, 5.4)	**5.4** (3.9, 8.2)	**6.0** (4.4, 9.0)	**5.6** (4.2, 8.4)
		Driving CMIP5 GCMs	**5.7** (4.0, 7.1)	**4.7** (3.9, 5.6)	**4.4** (3.5, 5.3)	**5.1** (3.8, 6.3)	**5.8** (4.2, 7.8)	**5.4** (3.8, 6.9)

The projections made by the study for the near-future and far-future periods for HKH are relatively higher than the seasonal global means based on the same subset of CMIP5 GCMs (Table 25.1). The summer season global mean projected change for the far-future period is 1.9 °C (RCP4.5) and 3.3 °C (RCP8.5), while for the winter season global mean projected changes are 2.0 °C (RCP4.5) and 3.5 °C (RCP8.5; Table 25.1 Sanjay et al. 2017). Analysis of an ensemble of five General Circulation Model runs projecting a global temperature increase of 1.5 °C by the end of the twenty-first century reveals would mean a temperature increase of 1.8 ± 0.4 °C averaged over the region (Fig. 25.3). Moreover this enhanced warming is even more pronounced for the mountain regions, for example for the Karakoram, Central Himalayas, and Southeast Himalayas, a 1.5 °C global temperature increase would imply regional temperature increases of 2.2 ± 0.4 °C, 2.0 ± 0.5 °C, and 2.0 ± 0.5 °C, respectively (Fig. 25.3).

Fig. 25.3 Comparison of results of models projecting 1.5 °C increase in near-surface air temperature (°C) globally and for the HKH and the three sub-domains: northwest Himalayas and Karakoram (HKH1); central Himalayas (HKH2); southeast Himalayas and Tibetan Plateau (HKH3). The temperature changes are for the end-of-century from the pre-industrial period (2071–2100 vs 1851–1880)

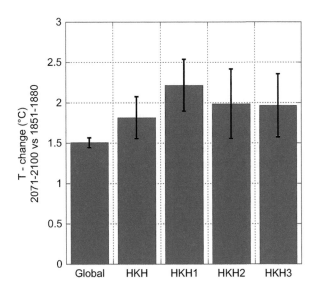

25.3.5.2 Precipitation

CORDEX projections for the HKH suggest summer season increase in precipitation over the hilly regions in the central Himalayas and southeast Himalayas and Tibetan Plateau for both RCP4.5 and RCP8.5 scenarios in the near-future and far-future periods. The largest projected seasonal ensemble mean total precipitation increase during the summer monsoon season is about 10% over HKH2 with RCP4.5 scenario and about 22% over HKH3 with RCP8.5 scenario. During winter season the largest projected increase in precipitation is over HKH1 of about 14% and 13% with RCP4.5 and RCP8.5 scenarios respectively. However, it has to be noted that the spread among the models is large for the high resolution CORDEX RCMs, as well as for their driving CMIP5 GCMs resulting in higher uncertainty in precipitation projection compared to temperature (Sanjay et al. 2017). The observed and projected future climate change can have profound impact in the ecosystem of this region directly and indirectly through changes in water availability, extreme events, cryospheric changes, etc. Adapting to long- and short-term climate-related problems requires a thorough understanding of climate changes in the past and possible changes in the future (You et al. 2017).

25.4   Implication of Climate Change in Ecosystem Resilience

Climate change affects human wellbeing in two ways, directly through altering local weather conditions inviting extreme weather events and hazards, and indirectly through its effects on ecosystems and ecosystem services that people need for their sustenance (Xu et al. 2009). The Earth's atmosphere has a natural greenhouse effect, without which the global mean surface temperature would be about 33 °C lower and life would not be possible. Human activities have increased atmospheric concentrations of carbon dioxide, methane, and other gases which has enhanced the greenhouse effect, resulting in surface warming (Serreze 2010). There are numerous implications from impact of climate change on ecosystems and biodiversity at different levels and scales. An anecdotal list of climate change impacts have been reviewed and included in Table 25.2. Here, we bring some examples where both social and ecological facets of resilience are being challenged with the past trends and projected future climate change:-.

25.4.1 Ecosystem Resilience

In the HKH, and elsewhere, ecosystems influence human societies, leading people to manage ecosystems for human benefit and poor environmental management can lead to reduced ecological resilience and social–ecological collapse (Cumming and Peterson 2017). One of the central questions in resilience science is how ecological function related to human wellbeing. This question has become increasingly relevant in the HKH as climatic and anthropogenic transformation of the region has intensified (Miehe et al. 2009). The distribution and abundance of species have been radically transformed due to climate change and massive land-use changes which have eliminated numerous endemic species (Chettri et al. 2010), and the expansion of developmental activities has added challenges in management of the natural ecosystem (Pandit et al. 2007). The variations on climatic variables such as temperature and precipitation have directly or indirectly affected the natural ecosystems (Table 25.2). This biotic reorganization is co-occurring with a variety of other changes, including climate change, alteration of nutrient cycles, and chemical contamination of the ecosystems (Xu et al. 2009; Chettri and Sharma 2016). Maintaining the ecosystem services that support humanity, and other lifeforms, during this extensive and rapid ecological reorganization requires understanding how ecology interacts with human society (Wangchuk 2007; Cumming and Peterson 2017). Here, we bring some empirical evidence of changes interpreted as a result of climate change.

25.4.1.1 Vegetation Shift

Mountains ecosystems are both fragile and sensitive. But is also provide opportunity for adaptation for many species sensitive to climate change with the provision to change the range shift to higher elevation. A number of studies have evolved indicating northward and high altitude movements of species due to climate change (Beckage et al. 2008; Pauchard et al. 2016; Lamsal et al. 2017b; Pecl et al. 2017). Evidences showed that a wide number of species have been reported to change their

Table 25.2 Impact of climate change on ecosystem and biodiversity at various systems, levels and range

Description	System		Level			Range		Impact mechanism/ hypothesis	Climate change driver
	Terrestrial	Freshwater	Ecosystem	Species	Genetic	Local	Widespread		
Loss and fragmentation of habitat	●		●				●	Ecological shift, land use change, exploitation	Temperature change
Vertical species migration and extinction	○			○		○		Ecological shifts	Temperature change
Decrease in aquatic biodiversity		●		●		●		Less oxygen, siltation	Temperature, extremes
Reduced forest biodiversity	○●		○●				○●	Ecological shift, habitat alteration, forest fire, phonological changes	Temperature change, precipitation change, land use change, overexploitation
Change in ecotone and micro-environmental endemism	○		○	○		○		Ecological shifts, microclimate	Temperature change
Peculiar tendencies in phenophases, in terms of synchronization and temporal variability	●			●		●		Phenological changes	Temperature change
Wetland degradation across the Himalayas		●	●		●			Siltation	Precipitation change
Degradation of riverine island ecosystems (Majuli) and associated aquatic biodiversity (refuted, but not overlooked)		●	●		●			Flooding	Extreme weather events

(continued)

Table 25.2 (continued)

Description	System		Level			Range		Impact mechanism/hypothesis	Climate change driver
	Terrestrial	Freshwater	Ecosystem	Species	Genetic	Local	Widespread		
Loss or degradation of natural scenic beauty	○	○	○				○	Drought, reduced snowfall	Less precipitation
Reduced agrobiodiversity	○●		○●				○●	Monoculture, inorganic chemicals, modern crop varieties, degeneration of crop wild relatives	Higher temperature and more precipitation
Change in utility values of alpine and sub-alpine meadows	○	○	○			○		Biomass productivity, species displacement, phonological changes	Higher temperature and more precipitation
Loss of species	○	○		○			○	Deforestation, land use change, land degradation	Higher temperature and more precipitation
Increase in exotic, invasive, noxious weeds	●			●		●		Species introduction and removal, land use change, tourism	Higher temperature and more precipitation
Decline in other resources (forage and fodder) leading to resource conflicts	●		●				●	Reduced net primary productivity	Higher temperature and less precipitation
Successional shift from wetlands to terrestrial ecosystems and shrinkage of wetlands at low altitudes (Loktak Lake, Deepor Beel, Koshi Tappu)	●○		●○			●○		Habitat alteration, drought, eutrophication	Higher temperature and less precipitation

Description	System		Level			Range		Impact mechanism/hypothesis	Climate change driver
	Terrestrial	Freshwater	Ecosystem	Species	Genetic	Local	Widespread		
Increase in forest fire	●		●			●		Forest fire, land degradation	Higher temperature and extreme weather events like long dry spells
Invasion by alien or introduced species with declining competency of extant and dominance by xeric species (Mikania, Eupatorium, Lantana, etc.)	●○			●○			●○	Species introduction, land use change	Higher temperature and less precipitation
Increased crop diversity and cropping pattern	○			○		○		Demographic and socio-economic change	Variable temperature and variable precipitation
Drying and desertification of alpine zones			○				○	Drought, overgrazing	Higher temperature and less precipitation
Change in land use patterns	●		●				●	Development policy, socioeconomic change	Variable temperature and precipitation
Soil fertility degradation	●		●				●	External inputs, land use intensification, desertification	Higher temperature and less precipitation
High species mortality	○	○		○		○		Range shift, pollution, deforestation	Higher temperature and less precipitation, less days/hours of sunshine

(continued)

Table 25.2 (continued)

Description	System		Level			Range		Impact mechanism/ hypothesis	Climate change driver
	Terrestrial	Freshwater	Ecosystem	Species	Genetic	Local	Widespread		
More growth/biomass production in forests, variable productivity in agriculture (orange)	○●		○●			○		Carbon enrichment, external input, reduced grazing	Increased CO_2 level, higher temperature
Net methane emission from wetlands		○●	○●			○●		Resource use, drainage, eutrophication, flow obstruction	Increased CO_2 level, higher temperature
Increased degradation and destruction of peat-land (bog, marshland, swamps, bayou)		○	○		○			Land conversion, drainage, removal of ground cover	Higher temperature and less precipitation
Land use change that increases soil degradation	●		●				●	Overpopulation, unsustainable agriculture	Variable temperature and precipitation
	● Observed response								
	○ Projected/Perceived								
	n/a or not yet identified								

Source: Sharma et al. (2009)

reported altitudinal range and moving northwards and higher altitude (Telwala et al. 2013). One of the recent study on sub-alpine species from Nepal, namely (i) Himalayan Fir (*Abies spectabilis*), (ii) Maple (*Acer campbellii*), (iii) Birch (*Betula utilis*), (iv) Black Juniper (*Juniperus indica*, (v) Brown oak (*Quercus semecarpifolia*), (vi) Himalayan hemlock (*Tsuga dumosa*), (vii) Bell rhododendron (*Rhododendron campanulatum*), (viii) Gerard jointfir (*Ephedra gerardiana*), and (ix) Himalayan heather (*Cassiope fastigiata*) have shown potential northward movements with projected climate change till 2100 (see Fig. 25.4; Lamsal et al. 2017b). Such upward or northward movement have major implications to alpine ecosystem, which accounts about 60% of the HKH (Chettri et al. 2008) and the cryosphere which are taken over by tree line advancement (Baker and Moseley 2007; Gaire et al. 2017). Alpine ecosystems are particularly vulnerable to warming, as species occurring near the mountain tops will have no space for their upward march and available areas are encroached from the south of low elevation (Pauchard et al. 2009; Singh et al. 2011). Such movements even change the land cover, bioclimatic zones and ecoregions (Sharma et al. 2009; Fig. 25.5) as also predicted by Zomer et al. (2014) and enabling invasive species to make inroad to such fragile areas (Lamsal et al. 2017b) making the ecosystems more vulnerable.

25.4.1.2 Degradation of Fragile Ecosystems

Approximately 39% of the HKH is comprised of grassland, 20% forest, 15% shrub land, and 5% agricultural land. The remaining 21% is other types of land cover such as barren land, rock outcrops, built-up areas, snow cover, and water bodies (Chettri et al. 2008). Among these, rangeland constitute to major part of the HKH with combined shared of grassland and shrub land. The Tibetan Plateau and Greater Himalayas is mostly dominated by rangeland where pastoralists' communities are dependent (Chettri et al. 2012). Intensification of water stress because of higher warming temperatures and changing precipitations are adversely affecting phenology, regeneration and productivity of many of these ecosystems, specially the high altitude rangeland (Singh et al. 2011; Yu et al. 2010; Wang et al. 2017). As a result, the entire mosaic of ecosystems found in the HKH have been witnessing changes manifested by climate change and other anthropogenic drivers.

Forest ecosystem, which covers about 20% of the total areas of HKH is one of the most important ecosystems for the region. It is sources of a wide range of provisioning ecosystem services and regulatory services including carbon sequestration (Lamsal et al. 2017c; Chaudhary et al. 2018). However, the forest degradation, fragmentation and loss have been widely reported with varied scenarios (Pandit et al. 2007; Hansen et al. 2013). Increasing fragmentation trend has been reported from the region such as India (Reddy et al. 2013), Nepal (Uddin et al. 2015; Reddy et al. 2018) and Bhutan (Reddy et al. 2016; Sharma et al. 2017). The rate of deforestation along the HKH has been reported to be 0.5% in Bhutan to 1.7% in Myanmar (Fig. 25.6).

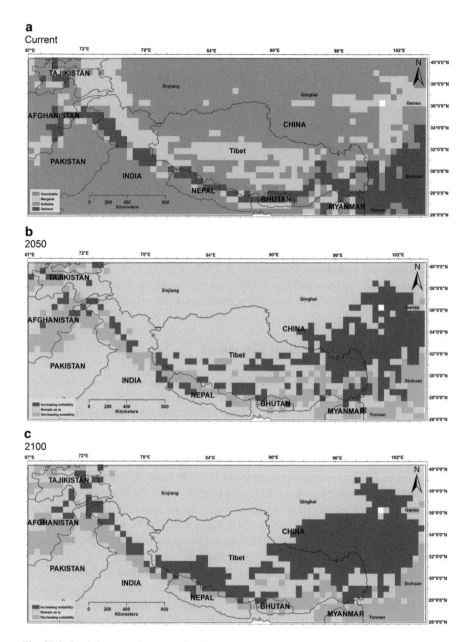

Fig. 25.4 Spatial extent of current climatic suitability (**a**) and changes for 2050 (**b**) and 2100 (**c**) for the nine species (Source Lamsal et al. 2017b)

Wetlands, an important ecosystem for the millions people living on the valley floors and plateau areas of the HKH, wetlands are central to their livelihoods (Trisal 2009). Lakes, floodplains, and peat lands support wide range of ecosystem services including tourism to the region's poorest communities (Sharma et al. 2015; Chaudhary et al. 2017). The wetlands maintain water quality, regulate water flow

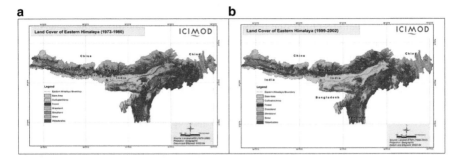

Fig. 25.5 Decadal land cover change in the Eastern Himalayas during 1998 (**a**) and 2002 (**b**). (Source Chettri et al. 2010)

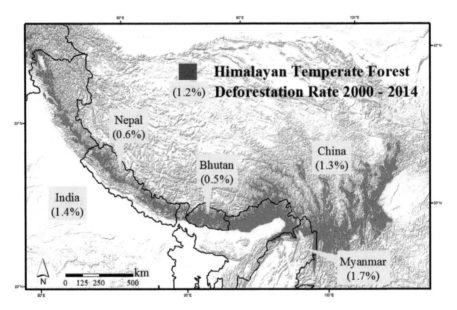

Fig. 25.6 The Himalayan temperate forest zone and deforestation rates (in parentheses) from 2000 to 2014 in forests that are not officially protected. (Source Brandt et al. 2017)

(floods and droughts), and in the case of high-altitude peat lands, regulate the global climate (storage of carbon in peat) and also support both regional and global biodiversity (Chettri et al. 2013).

However, the wetland ecosystem in the region is rapidly degrading: in some areas as many as 30% of the lakes and marshes have disappeared because of over-exploitation of wetland resources and climate change during the past few decades (Trisal 2009). Climate change, along with land use change, anthropogenic pressure for ecosystem services and natural hazards are collectively changing the wetland ecosystems in many parts of the HKH (Rashid and Romshoo 2013; Chettri et al. 2013; Lamsal et al. 2017c; Chaudhary et al. 2017).

Likewise, the agro-ecosystem, which is merely 5% of the total HKH area has also witnessed changes either due to agricultural intensification in some areas or by leaving it fallow due to social restructuring of the communities in the remote villages. However, the impact of climate change is yet to be studied and ascertain. At the global level it is estimated that the major crops such as rice, wheat and maize could see 26–36% decrease based on the trend of past climate change (Iizumi et al. 2013). The brunt of climate change has been witnessed in many parts of the HKH evident as reported on studies from people's perceptions (Chaudhary et al. 2011; Negi et al. 2017; Suberi et al. 2018), reported low productivity (Vedwan 2006) and loss of traditional varieties such as finger millets, buckwheat's etc. (Aryal et al. 2017). Owing to the higher than average warming, the HKH region is facing adverse challenges for vulnerability from climate change and other drivers of change impacting in a wide spectrum of fragile ecosystems.

25.4.2 Social-Ecological Interdependency

In the recent years, the significance of mountain ecosystems for its ecosystem services for human wellbeing has been realized (Grêt-Regamey et al. 2008, 2012; Palomo 2017). The inhabitant of the HKH is extensively dependent on the natural ecosystems for ecosystem services contributing to wellbeing. This is evident from a recent series of documentation on people's dependency on ecosystems across the HKH (Gosain et al. 2015; Paudyal et al. 2015; Chaudhary et al. 2016, 2017; Murali et al. 2017). Following standard frameworks (De Groot et al. 2010; Muller et al. 2010), some of the recent studies have shown strong interdependency between the local communities and surrounding ecosystems and also being impacted due to change of such dependent ecosystems (see Chaudhary et al. 2016, 2017; Kandel et al. 2018). It was observed that about 85% of the provision services are contributed by the Koshi Tappu wetland ecosystem where 7.5% is from cultural services alone (Sharma et al. 2015).

There was significant change observed in the surrounding ecosystems of Koshi Tappu which has threatened the flow of these ecosystems services. These is evident as the forest ecosystem has reduced by almost 85% over the past 30 years making people dependent on forest for services more vulnerable (ICIMOD and MoFSC 2014). Similar results were also documented from other wetlands (ICIMOD and RSPN 2014; ICIMOD and MoNREC 2017) and forest and agro-ecosystems (ICIMOD and BCN 2017; ICIMOD and RSPN 2017). These studies categorically indicates that there is a strong social-ecological interdependency and the strategies for addressing climate change including others drovers of change need new innovations.

25.5 Discussion and Conclusion

Climate change has been identified as one of the major drivers of global change and reported to be more pronounced in the HKH with higher warming rate (Krusic et al. 2015; Sanjay et al. 2017). These changes have already made cascading effects to the fragile ecosystems of the HKH with increasing temperature trend and projections, with variable precipitation patterns across the region (Srivastava et al. 2017; Zhan et al. 2017). It is also evident that the changing climate has influenced directly or indirectly the ecosystems health, functions and productivity which are directly linked to human wellbeing through its ecosystem services flow (Xu et al. 2009; Chettri and Sharma 2016). The diverse ecosystems of the HKH, the source of wide range of ecosystems services including water and habitat for numerous globally significant species, are witnessing increasing vulnerability (Chettri et al. 2010; Pandey and Bardsley 2015; Forrest et al. 2012). The study from Eastern Himalayas have already indicated that some ecosystems such as freshwater, forest, savannas grassland are categorized as vulnerable (see Chettri et al. 2010). The roots causes of driver of change – the anthropogenic activities have no sign of improvement, at least in immediate near future (Ceballos et al. 2017). However, there is little debate in scientific circles about the importance of human influence on ecosystems. To improve overall resilience, an understanding of contributing factors such as economic, social and environmental resource is necessary (Gardner and Dekens 2007; Norman et al. 2012).

High mountain areas are arguably the region most affected by climate change (Shrestha et al. 1999; Beniston 2003; Keller et al. 2005; Xu et al. 2009). Assessments of climate change impacts in these regions have been mostly single-disciplined. Biophysical studies have focused on temperature changes, glacier retreat, hazards, and biodiversity (Dullinger et al. 2012; Pepin et al. 2015). Social research has focused on impacts associated with water availability and livelihoods, and these impacts have been described more for downstream communities (Xu et al. 2009; Immerzeel et al. 2015) than upstream inhabitants (Beniston et al. 1997; Kohler et al. 2010). Integrative approaches that focus on climate change impacts on multiple ecosystem services in high mountain areas are still limited (Chettri and Sharma 2016; Chakraborty et al. 2018).

Climate change effects varied across the HKH but mostly they are devastating for mountain communities, accentuating their vulnerability to disasters, conflicts, migration and poverty (Gerlitz et al. 2015; Hoy et al. 2016; Sharma et al. 2017). As the HKH ecosystems are contiguous and the impacts are not limited to one national border, to address climate change, it is crucial to develop innovative joint research programmes at landscapes or river basins considering transboundary approach combined with efficient communication platforms to raise public awareness about its impacts in the mountains (Molden et al. 2017).

It is well known that the rate of temperature change with increased levels of greenhouse gases in the atmosphere is amplified at high latitudes, but there is growing evidence that the rate of warming is also amplified with elevation, such as in mountain environments (Shrestha et al. 1999; Pepin et al. 2015). However, because of sparse high-elevation observations, there is a danger that we may not be monitoring some regions of the globe that are warming the most (Wijngaard et al. 2017). In response to the prevailing threat from climate change, new regional and innovative efforts on climate change adaptation in terms of policy and practices have begun to emerge (Gerlitz et al. 2015; Hoy et al. 2016; Rasul and Sharma 2016). However, many of the current wave of contributions are observational and correlational, and few are experimental in nature, and too often at conceptual levels in which convincing results are lacking. We conclude with recommending a future strategy for multidisciplinary research at landscape or basin levels to reduce current uncertainties and to ensure that the changes taking place in remote high-elevation regions of the HKH are adequately observed and accounted for. The use of contemporary geospatial science and modelling tools may add value for informed decision making.

Acknowledgement The authors would like to express their gratitude to Dr. David Molden, Director General of ICIMOD, for his inspiration and support. The continuous support and commitment from ICIMOD's eight regional member countries is also acknowledged, as is the support of the Austrian Development Agency (ADA) and the German Federal Ministry for Economic Cooperation and Development through its German Agency for International Cooperation (GIZ), which made this publication possible.

References

Aryal K, Poudel S, Chaudhary RP et al (2017) Conservation and management practices of local crop genetic diversity by the farmers: a case from Kailash Sacred Landscape. Nepal J Agric Environ 18:15–28

Aukema JE, Pricope NG, Husak GJ et al (2017) Biodiversity areas under threat: overlap of climate change and population pressures on the world's biodiversity priorities. PLoS One 12(1):e0170615. https://doi.org/10.1371/journal.pone.0170615

Bajracharya AR, Bajracharya SR, Shrestha AB, Maharjan SB (2018) Climate change impact assessment on the hydrological regime of the Kaligandaki Basin, Nepal. Sci Total Environ 625:837–848. https://doi.org/10.1016/j.scitotenv.2017.12.332

Baker BB, Moseley RK (2007) Advancing treeline and retreating glaciers: implications for conservation in Yunnan, PR China. Arct Antarct Alp Res 39(2):200–209

Barnett TP, Adam JC, Lettenmaier DP (2005) Potential impacts of a warming climate on water availability in snow-dominated regions. Nature 438(7066):303–309

Beckage B, Osborne B, Gavin DG et al (2008) A rapid upward shift of a forest ecotone during 40 years of warming in the Green Mountains of Vermont. PNAS 105(11):4197–4202

Beniston M (2003) Climatic change in mountain regions: a review of possible impacts. In: Diaz et al (eds) Climate variability and change in high elevation regions: past, present and future. Springer, Dordrecht, pp 5–13

Beniston M, Diaz HF, Bradley RS (1997) Climatic change at high elevation sites: an overview. Clim Chang 36(3–4):233–251

Bhagawati R, Bhagawati K, Jini D et al (2017) Review on climate change and its impact on agriculture of Arunachal Pradesh in the Northeastern Himalayan region of India. Nat Environ Pollut Tech 16(2):535–539

Bharali S, Khan ML (2011) Climate change and its impact on biodiversity; some management options for mitigation in Arunachal Pradesh. Curr Sci 101(7):855–860

Bookhagen B, Burbank DW (2010) Toward a complete Himalayan hydrological budget: spatio-temporal distribution of snowmelt and rainfall and their impact on river discharge. J Geophys Res 115:F3. https://doi.org/10.1029/2009JF001426

Brandt JS, Haynes MA, Kuemmerle T et al (2013) Regime shift on the roof of the world: alpine meadows converting to shrublands in the southern Himalayas. Biol Conserv 158:116–127

Brandt JS, Allendorf A, Radeloff V, Brooks J (2017) Effects of national forest-management regimes on unprotected forests of the Himalaya. Conserv Biol 31(6):1271–1282

Brooks TM, Mittermeier RA, da Fonseca GA et al (2006) Global biodiversity conservation priorities. Science 313(5783):58–61

Burbank DW, Blythe AE, Putkonen J et al (2003) Decoupling of erosion and precipitation in the Himalayas. Nature 426(6967):652–655

Cánovas JB, Trappmann D, Shekhar M et al (2017) Regional flood-frequency reconstruction for Kullu district, Western Indian Himalayas. J Hydrol 546:140–149

Ceballos G, Ehrlich PR, Dirzo R (2017) Biological annihilation via the ongoing sixth mass extinction signaled by vertebrate population losses and declines. Proc Natl Acad Sci U S A 114(30):E6089–E6096

Chakraborty A, Saha S, Sachdeva K, Joshi PK (2018) Vulnerability of forests in the Himalayan region to climate change impacts and anthropogenic disturbances: a systematic review. Reg Environ Chang 18(6):1783–1799

Chaudhary P, Rai S, Wangdi S et al (2011) Consistency of local perceptions of climate change in the Kangchenjunga Himalaya landscape. Curr Sci 101(4):504–513

Chaudhary S, Chettri N, Uddin K et al (2016) Implications of land cover change on ecosystem services and people's dependency. A case study from the Koshi Tappu Wildlife Reserve, Nepal. Ecol Complex 28:200–211

Chaudhary S, Tshering D, Phuntsho T, Uddin K, Shakya B, Chettri N (2017) Implications of land cover change on mountain ecosystem services: a case study of the Phobjikha valley of Bhutan. Ecosyst Health Sustain 3:1–12

Chaudhary S, MacGregor K, Houston D et al (2018) Environmental justice and ecosystem services: a disaggregated analysis of community access to forest benefits in Nepal. Ecosyst Serv 20:99–115

Chettri N, Sharma E (2016) Reconciling the mountain biodiversity conservation and human well-being: drivers of biodiversity loss and new approaches in the Hindu-Kush Himalayas. Proc Indian Natl Sci Acad 82:53–73

Chettri N, Shakya B, Thapa R et al (2008) Status of protected area system in the Hindu Kush Himalaya: an analysis of PA coverage. Int J Biodivers Sci Manag 4(3):164–178

Chettri N, Sharma E, Shakya B et al (2010) Biodiversity in the Eastern Himalayas: status, trends and vulnerability to climate change, Climate change impact and vulnerability in the Eastern Himalayas – technical report 2. ICIMOD, Kathmandu

Chettri N et al (2012) Real world protection for the third pole and its people. In: Huettmann F (ed) Protection of polar regions. Springer, pp 113–133

Chettri N, Uddin K, Chaudhary S et al (2013) Linking spatio-temporal land cover change to biodiversity conservation in Koshi Tappu Wildlife Reserve, Nepal. Diversity 5:335–351

Cumming GS, Peterson GD (2017) Unifying research on social–ecological resilience and collapse. Trends Ecol Evol 32(9):695–713

De Groot RS, Alkemade R, Braat L et al (2010) Challenges in integrating the concept of ecosystem services and values in landscape planning, management and decision making. Ecol Complex 7(3):260–272

Duan AM, Xiao ZX (2015) Does the climate warming hiatus exist over the Tibetan Plateau? Sci Rep 5:13711

Dullinger S, Gattringer A, Thuiller W et al (2012) Extinction debt of high-mountain plants under twenty-first-century climate change. Nat Clim Chang 2(8):619–622

Eriksson M, Xu J, Shrestha AB et al (2009) The changing Himalayas: impact of climate change on water resources and livelihoods in the greater Himalayas. ICIMOD, Kathmandu

Fan X, Wang Q, Wang M et al (2015) Warming amplification of minimum and maximum temperatures over high-elevation regions across the globe. PLoS One 10(10). https://doi.org/10.1371/journal.pone.0140213

Forrest JL, Wikramanayake E, Shrestha R et al (2012) Conservation and climate change: assessing the vulnerability of snow leopard habitat to treeline shift in the Himalaya. Biol Conserv 150(1):129–135

Forsythe N, Fowler HJ, Li XF et al (2017) Karakoram temperature and glacial melt driven by regional atmospheric circulation variability. Nat Clim Chang 7:664–670

Gaire NP, Koirala M, Bhuju DR et al (2017) Site-and species-specific treeline responses to climatic variability in eastern Nepal Himalaya. Dendrochronologia 41:44–56

Gardner JS, Dekens J (2007) Mountain hazards and the resilience of social–ecological systems: lessons learned in India and Canada. Nat Hazards 41(2):317–336

Gerlitz JY, Banerjee S, Brooks N et al (2015) An approach to measure vulnerability and adaptation to climate change in the Hindu Kush Himalayas. In: Handbook of climate change adaptation. Springer, Berlin/Heidelberg, pp 151–176

Gosain BG, Negi GCS, Dhyani PP et al (2015) Ecosystem services of forests: Carbon Stock in vegetation and soil components in a watershed of Kumaun Himalaya, India. Int J Ecol Environ Sci 41(3–4):177–188

Grêt-Regamey A, Walz A, Bebi P (2008) Valuing ecosystem services for sustainable landscape planning in alpine regions. Mt Res Dev 28(2):156–165

Grêt-Regamey A, Brunner SH, Kienast F (2012) Mountain ecosystem services: who cares? Mt Res Dev 32(S1):S23–S34

Guo D, Yu E, Wang H (2016) Will the Tibetan Plateau warming depend on elevation in the future? J Geophys Res Atmos 121:3969–3978

Hansen MC, Potapov PV, Moore R et al (2013) High-resolution global maps of 21st-century forest cover change. Science 342(6160):850–853

Hoy A, Katel O, Thapa P et al (2016) Climatic changes and their impact on socio-economic sectors in the Bhutan Himalayas: an implementation strategy. Reg Environ Chang 16(5):1401–1415

Hudson AM, Olsen JW, Quade J et al (2016) A regional record of expanded Holocene wetlands and prehistoric human occupation from paleowetland deposits of the western Yarlung Tsangpo valley, southern Tibetan Plateau. Quat Res 86(1):13–33

ICIMOD, BCN (2017) A multi-dimensional assessment of ecosystems and ecosystem services at Udayapur, Nepal. ICIMOD Working Paper 2017/20. ICIMOD, Kathmandu

ICIMOD; MoFSC (2014) An integrated assessment of the effects of natural and human disturbances on a wetland ecosystem: a retrospective from the Koshi Tappu Wildlife Reserve, Nepal. ICIMOD, Kathmandu

ICIMOD, MoNREC (2017) A multi-dimensional assessment of ecosystems and ecosystem services at Inle Lake, Myanmar. ICIMOD Working Paper 2017/17. ICIMOD, Kathmandu

ICIMOD; RSPN (2014) An integrated assessment of the effects of natural and human disturbances on a wetland ecosystem: a retrospective from Phobjikha Conservation Area, Bhutan. ICIMOD, Kathmandu

ICIMOD, RSPN (2017) A multi-dimensional assessment of ecosystems and ecosystem services in Barshong, Bhutan. ICIMOD Working Paper 2017/6. ICIMOD, Kathmandu

Iizumi T, Sakuma H, Yokozawa M et al (2013) Prediction of seasonal climate-induced variations in global food production. Nat Clim Chang 3(10):904–908

Immerzeel WW, Wanders N, Lutz AF et al (2015) Reconciling high-altitude precipitation in the upper Indus basin with glacier mass balances and runoff. Hydrol Earth Syst Sci 19(11):4673–4687

IPCC (2007) Climate change. The physical sciences basis. Summary for policymakers (summary for policymakers). Intergovernmental Panel on Climate Change (IPCC), p 21

Joshi V, Kumar K (2006) Extreme rainfall events and associated natural hazards in Alaknanda valley, Indian Himalayan region. J Mt Sci 3(3):228–236

Joshi PK, Rawat A, Narula S, Sinha V (2012) Assessing impact of climate change on forest cover type shifts in Western Himalayan Eco-region. J For Res 23(1):75–80

Kandel P, Tshering D, Uddin K et al (2018) Understanding social–ecological interdependencies through ecosystem services value perspectives in Bhutan, Eastern Himalaya. Ecosphere 9(2). https://doi.org/10.1002/ecs2.2121

Keller F, Goyette S, Beniston M (2005) Sensitivity analysis of snow cover to climate change scenarios and their impact on plant habitats in alpine terrain. Clim Chang 72(3):299–319

Kohler T, Giger M, Hurni H et al (2010) Mountains and climate change: a global concern. Mt Res Dev 30(1):53–55

Krusic PJ, Cook ER, Dukpa D et al (2015) Six hundred thirty-eight years of summer temperature variability over the Bhutanese Himalaya. Geophys Res Lett 42(8):2988–2994

Kulkarni A, Patwardhan S, Kumar KK et al (2013) Projected climate change in the Hindu Kush–Himalayan region by using the high-resolution regional climate model PRECIS. Mt Res Dev 33(2):142–151

Kumar KK, Patwardhan SK, Kulkarni A et al (2011) Simulated projections for summer monsoon climate over India by a high-resolution regional climate model (PRECIS). Curr Sci 101:312–326

Lamsal P, Kumar L, Atreya K (2017a) Historical evidence of climatic variability and changes, and its effect on high-altitude regions: insights from Rara and Langtang, Nepal. Int J Sust Dev World Ecol 24(6):471–484

Lamsal P, Kumar L, Atreya K et al (2017b) The greening of the Himalayas and Tibetan Plateau under climate change. Glob Planet Chang 159:77–92

Lamsal P, Kumar L, Atreya K et al (2017c) Vulnerability and impacts of climate change on forest and freshwater wetland ecosystems in Nepal: a review. Ambio 46(8):915–930

Liu X, Chen B (2000) Climatic warming in the Tibetan Plateau during recent decades. Int J Climatol 20:1729–1742

Liu XD, Cheng ZG, Yan L et al (2009) Elevation dependency of recent and future minimum surface air temperature trends in the Tibetan Plateau and its surroundings. Glob Planet Chang 68(3):164–174

Lo V (2016) Synthesis report on experiences with ecosystem-based approaches to climate change adaptation and disaster risk reduction, Technical series no. 85. Secretariat of the Convention on Biological Diversity, Montreal, p 106

Messerli B, Ives JD (1997) Mountains of the world. Parthenon Publishing, New York

Miehe G, Miehe S, Schlütz F (2009) Early human impact in the forest ecotone of southern High Asia (Hindu Kush, Himalaya). Quat Res 71(3):255–265

Miehe G, Miehe S, Böhner J et al (2014) How old is the human footprint in the world's largest alpine ecosystem? A review of multiproxy records from the Tibetan Plateau from the ecologists' viewpoint. Quat Sci Rev 86:190–209

Mittermeier RA et al (2004) Hotspots revisited. Earth's biologically richest and most endangered terrestrial ecoregions. Cemex, Mexico City

Molden DJ, Vaidya RA, Shrestha AB et al (2014) Water infrastructure for the Hindu Kush Himalayas. Int J Water Resour Dev 30(1):60–77

Molden D, Sharma E, Shrestha AB et al (2017) Advancing regional and transboundary cooperation in the conflict-prone Hindu Kush–Himalaya. Mt Res Dev 37(4):502–508

Mukherji A, Molden D, Nepal S et al (2015) Himalayan waters at the crossroads: issues and challenges. Int J Water Resour Dev 31(2):151–160

Muller F, de Groot RS, Willemen L (2010) Ecosystem services at the landscape scale: the need for integrative approaches. Landsc Online 23:1–11. https://doi.org/10.3097/LO.201023

Murali R, Redpath S, Mishra C (2017) The value of ecosystem services in the high altitude Spiti Valley, Indian Trans-Himalaya. Ecosyst Serv 28:115–123

Negi VS, Maikhuri RK, Pharswan D et al (2017) Climate change impact in the Western Himalaya: people's perception and adaptive strategies. J Mt Sci 14(2):403–416

Nogués-Bravo D, Araújo MB, Romdal T et al (2008) Scale effects and human impact on the eleva-
tional species richness gradients. Nature 453(7192):216–219
Norman LM, Villarreal ML, Lara-Valencia F et al (2012) Mapping socio-environmentally vulner-
able populations access and exposure to ecosystem services at the US–Mexico borderlands.
Appl Geogr 34:413–424
Olson DM, Dinerstein E (2002) The global 200: priority ecoregions for global conservation. Ann
Mo Bot Gard 89(2):199–224
Palomo I (2017) Climate change impacts on ecosystem services in high mountain areas: a litera-
ture review. Mt Res Dev 37(2):179–187
Pandey R, Bardsley DK (2015) Social-ecological vulnerability to climate change in the Nepali
Himalaya. Appl Geogr 64:74–86
Pandey R, Jha S (2012) Climate vulnerability index-measure of climate change vulnerabil-
ity to communities: a case of rural Lower Himalaya, India. Mitig Adapt Strat Glob Chang
17(5):487–506
Pandit MK, Sodhi NS, Koh LP et al (2007) Unreported yet massive deforestation driving loss of
endemic biodiversity in Indian Himalaya. Biodivers Conserv 16(1):153–163
Parmesan C (2006) Ecological and evolutionary responses to recent climate change. Annu Rev
Ecol Evol Syst 37:637–669
Pauchard A, Kueffer C, Dietz H et al (2009) Ain't no mountain high enough: plant invasions reach-
ing new elevations. Front Ecol Environ 7(9):479–486
Pauchard A, Milbau A, Albihn A et al (2016) Non-native and native organisms moving into high
elevation and high latitude ecosystems in an era of climate change: new challenges for ecology
and conservation. Biol Invasions 18(2):345–353
Paudyal K, Baral H, Burkhard B et al (2015) Participatory assessment and mapping of ecosystem
services in a data-poor region: case study of community-managed forests in central Nepal.
Ecosyst Serv 13:81–92
Pecl GT, Araújo MB, Bell JD et al (2017) Biodiversity redistribution under climate change:
impacts on ecosystems and human well-being. Science 355(6332):1–9
Pepin N, Bradley RS, Diaz HF et al (2015) Elevation-dependent warming in mountain regions of
the world. Nat Clim Chang 5(5):424–430
Quyang H (2009) The Himalayas – water storage under threat. Sustain Mt Dev 56:3–5
Rajbhandari R, Shrestha AB, Kulkarni A et al (2015) Projected changes in climate over the Indus
river basin using a high resolution regional climate model (PRECIS). Clim Dyn 44(1):339–357
Rajbhandari R, Shrestha AB, Nepal S et al (2016) Projection of future Climate over the Koshi
River basin based on CMIP5 GCMs. Atmos Clim Sci 6:190–204
Rashid I, Romshoo SA (2013) Impact of anthropogenic activities on water quality of Lidder River
in Kashmir Himalayas. Environ Monit Assess 185(6):4705–4719
Rasul G, Sharma B (2016) The nexus approach to water–energy–food security: an option for adap-
tation to climate change. Clim Pol 16(6):682–702
Ravindranath NH, Chaturvedi RK, Joshi NV et al (2011) Implications of climate change on mitiga-
tion potential estimates for forest sector in India. Mitig Adapt Strat Glob Chang 16(2):211–227
Reddy CS, Sreelekshmi S, Jha CS et al (2013) National assessment of forest fragmentation in
India: landscape indices as measures of the effects of fragmentation and forest cover change.
Ecol Eng 60:453–464
Reddy CS, Satish KV, Jha CS et al (2016) Development of deforestation and land cover data-
base for Bhutan (1930–2014). Environ Monit Assess 188(12):658. https://doi.org/10.1007/
s10661-016-5676-6
Reddy CS, Pasha SV, Satish KV, Saranya KRL, Jha CS, Murthy YK (2018) Quantifying nation-
wide land cover and historical changes in forests of Nepal (1930–2014): implications on forest
fragmentation. Biodivers Conserv 27(1):91–107. https://doi.org/10.1007/s10531-017-1423-8
Ren GY, Shrestha AB (2017) Climate change in the Hindu Kush Himalaya. Adv Clim Chang Res
8:137–140

Ren YY, Ren GY, Sun XB et al (2017) Observed changes in surface air temperature and precipitation in the Hindu Kush Himalayan region over the last 100-plus years. Adv Clim Chang Res 8(3):148–156

Rodríguez-Rodríguez D, Bomhard B (2012) Mapping direct human influence on the world's mountain areas. Mt Res Dev 32(2):197–202

Royal Monetary Authority (RMA), Bhutan (2011) Annu Rep 2009/10

Sandhu H, Sandhu S (2014) Linking ecosystem services with the constituents of human Well-being for poverty alleviation in Eastern Himalayas. Ecol Econ 107:65–75

Sanjay J, Krishman R, Shrestha AB et al (2017) Downscaled climate change projections for the Hindu Kush Himalayan region using CORDEX South Asia regional climate models. Adv Clim Chang Res 8(3):185–198

Serreze MC (2010) Understanding recent climate change. Conserv Biol 24(1):10–17

Sharma E, Chettri N, Tse-ring K et al (2009) Climate change impacts and vulnerability in the Eastern Himalayas. ICIMOD, Kathmandu, p 27

Sharma E, Chettri N, Oli KP (2010) Mountain biodiversity conservation and management: a paradigm shift in policies and practices in the Hindu Kush-Himalayas. Ecol Res 25(5):909–923

Sharma B, Rasul G, Chettri N (2015) The economic value of wetland ecosystem services: evidence from Koshi Tappu Wildlife Reserve, Nepal. Ecosyst Serv 12:84–93

Sharma K, Robeson SM, Thapa P et al (2017) Land-use/land-cover change and forest fragmentation in the Jigme Dorji National Park, Bhutan. Phys Geogr 38(1):18–35

Shrestha AB, Aryal R (2011) Climate change in Nepal and its impact on Himalayan glaciers. Reg Environ Chang 11:S65–S77

Shrestha AB, Wake CP, Mayewski PA et al (1999) Maximum temperature trends in the Himalaya and its vicinity: an analysis based on temperature records from Nepal for the period 1971–94. J Clim 12:2775–2786

Shrestha UB, Gautam S, Bawa KS (2012) Widespread climate change in the Himalayas and associated changes in local ecosystems. PLoS One 7(5):e36741

Singh SP, Singh V, Skutsch M (2010) Rapid warming in the Himalayas: ecosystem responses and development options. Clim Dev 2(3):221–232

Singh SP et al (2011) Climate change in the Hindu Kush-Himalayas: the state of current knowledge. International Centre for Integrated Mountain Development (ICIMOD), Kathmandu

Srivastava P, Agnihotri R, Sharma D et al (2017) 8000-year monsoonal record from Himalaya revealing reinforcement of tropical and global climate systems since mid-Holocene. Sci Rep 7(1):14515

Suberi B, Tiwari KR, Gurung DB et al (2018) People's perception of climate change impacts and their adaptation practices in Khotokha valley, Wangdue, Bhutan. Indian J Tradit Knowl 17(1):97–105

Sun XB, Ren GY, Shrestha AB et al (2017) Changes in extreme temperature events over the Hindu Kush Himalaya during 1961–2015. Adv Clim Chang Res 8(3):157–165

Telwala Y, Brook BW, Manish K et al (2013) Climate-induced elevational range shifts and increase in plant species richness in a Himalayan biodiversity epicentre. PLoS One 8(2):e57103

Tewari VP, Verma RK, von Gadow K (2017) Climate change effects in the Western Himalayan ecosystems of India: evidence and strategies. For Ecosyst 4(1):13

Tiwari PC, Joshi B (2012) Environmental changes and sustainable development of water resources in the Himalayan headwaters of India. Water Resour Manag 26(4):883–907

Trisal C (2009) Wetland of the Hindu Kush Himalayas: ecosystem functions, services and implications of climate change. In: Sharma E (ed) Proceedings on international mountain biodiversity conference, 16–18 November 2008. ICIMOD, Kathmandu, pp 169–178

Uddin K, Chaudhary S, Chettri N et al (2015) The changing land cover and fragmenting forest on the roof of the world: a case study in Nepal's Kailash Sacred Landscape. Landsc Urban Plan 141:1–10

UN (1992) Report of the United Nations conference on environment and development. Rio de Janeiro 3–14 June 1992. A/CONF.151/26 (Vol. 1)

Vedwan N (2006) Culture, climate and the environment: local knowledge and perception of climate change among apple growers in northwestern India. J Ecol Anthropol 10(1):4–18

Viviroli D, Archer DR, Buytaert W et al (2011) Climate change and mountain water resources: overview and recommendations for research, management and policy. Hydrol Earth Syst Sci 15(2):471–504

Wang B, Bao Q, Hoskins B et al (2008) Tibetan plateau warming and precipitation changes in East Asia. Geophys Res Lett 35:L14702

Wang Q, Fan X, Wang M (2014) Recent warming amplification over high elevation regions across the globe. Clim Dyn 43(1–2):87–101

Wang Q, Fan X, Wang M (2016) Evidence of high-elevation amplification versus Arctic amplification. Sci Rep 19219. https://doi.org/10.1038/srep19219

Wang W, Jia M, Wang G et al (2017) Rapid warming forces contrasting growth trends of subalpine fir (Abies fabri) at higher-and lower-elevations in the eastern Tibetan Plateau. For Ecol Manag 402:135–144

Wangchuk S (2007) Maintaining ecological resilience by linking protected areas through biological corridors in Bhutan. Trop Ecol 48(2):176–187

Wijngaard RR, Lutz AF, Nepal S et al (2017) Future changes in hydro-climatic extremes in the Upper Indus, Ganges, and Brahmaputra River basins. PLoS One 12(12):e0190224. https://doi.org/10.1371/journal.pone.0190224

Xu J, Grumbine RE (2014) Building ecosystem resilience for climate change adaptation in the Asian highlands. Wiley Interdiscip Rev Clim Chang 5(6):709–718

Xu J, Grumbine RE, Shrestha A et al (2009) The melting Himalayas: cascading effects of climate change on water, biodiversity, and livelihoods. Conserv Biol 23(3):520–530

Yan LB, Liu XD (2014) Has climatic warming over the Tibetan Plateau paused or continued in recent years? J Earth Ocean Atmos Sci 1:13–28

Yan LB, Liu Z, Chen G et al (2016) Mechanisms of elevation-dependent warming over the Tibetan plateau in quadrupled CO2 experiments. Clim Chang 135:509–519

Yao T, Thompson GL, Mosbrugger V et al (2012) Third pole environment (TPE). Environ Dev 3:52e64

You QL, Kang SC, Aguilar E et al (2008) Changes in daily climate extremes in the eastern and central Tibetan Plateau during 1961–2005. J Geophys Res Atmos 113

You QL, Kang SC, Pepin N et al (2010) Relationship between temperature trend magnitude, elevation and mean temperature in the Tibetan Plateau from homogenized surface stations and reanalysis data. Glob Planet Chang 71:124–133

You QL, Ren GY, Zhang YQ et al (2017) An overview of studies of observed climate change in the Hindu Kush Himalayan (HKH) region. Adv Clim Chang Res 8(3):141–147

Yu H, Luedeling E, Xu J (2010) Winter and spring warming result in delayed spring phenology on the Tibetan Plateau. Proc Natl Acad Sci U S A 107(51):22151–22156

Zhan YJ, Ren GY, Shrestha AB et al (2017) Change in extreme precipitation events over the Hindu Kush Himalayan region during 1961–2012. Adv Clim Chang Res 8(3):166–175

Zhou YQ, Ren GY (2011) Change in extreme temperature events frequency over mainland China during 1961–2008. Clim Res 50:125–139

Zomer RJ, Trabucco A, Metzger MJ et al (2014) Projected climate change impacts on spatial distribution of bioclimatic zones and ecoregions within the Kailash Sacred Landscape of China, India, Nepal. Clim Chang 125(3–4):445–460

Chapter 26
An Integrative and Joint Approach to Climate Impacts, Hydrological Risks and Adaptation in the Indian Himalayan Region

Christian Huggel, Simon Allen, Susanne Wymann von Dach, A. P. Dimri, Suraj Mal, Andreas Linbauer, Nadine Salzmann, and Tobias Bolch

Abstract Climate change has enormous impacts on the cryosphere In the Indian Himalayan Region (IHR) which have been increasingly documented over the past years. The effects of cryosphere change on people, ecosystems and economic sectors is less clear but bears important risks. Adaptation to changing conditions and risks is a priority for the region. Here we draw on experiences of Indo-Swiss collaborations in the field of climate change, cryosphere, risks and adaptation in the IHR. First, we provide a synthesis of the climate and cryosphere in the IHR, and related impacts on downstream communities and systems. Second, we analyze the associated risks from a conceptual and adaptation perspective. We then introduce

C. Huggel (✉) · T. Bolch
Department of Geography, University of Zurich, Zurich, Switzerland
e-mail: christian.huggel@geo.uzh.ch

S. Allen
Department of Geography, University of Zurich, Zurich, Switzerland

Institute for Environmental Sciences, University of Geneva, Geneva, Switzerland

S. Wymann von Dach
Centre for Development and Environment (CDE), University of Bern, Bern, Switzerland

A. P. Dimri
School of Environmental Sciences, Jawaharlal Nehru University, New Delhi, India

S. Mal
Department of Geography, Shaheed Bhagat Singh College, University of Delhi, New Delhi, India

CEN Center for Earth System Research and Sustainability, Institute of Geography, University of Hamburg, Hamburg, Germany

A. Linbauer · N. Salzmann
Department of Geography, University of Zurich, Zurich, Switzerland

Department of Geoscience, University of Fribourg, Fribourg, Switzerland

© Springer Nature Switzerland AG 2020
A. P. Dimri et al. (eds.), *Himalayan Weather and Climate and their Impact on the Environment*, https://doi.org/10.1007/978-3-030-29684-1_26

concepts of co-production of knowledge as an approach to an inclusive and sustainable adaptation process which includes the development of future scenarios with a wide range of stakeholders. We visualize this approach using examples of the water resource sector.

26.1 Introduction

Weather and climate are a determinant element of the natural environment in the Indian Himalayan Region (IHR). Like in other high-mountain regions of the world weather and climate influence a broad range of physical processes and landscape elements over different temporal and spatial scales. Over long periods of time (millions of years) climate, tectonics and landscape evolution are interlinked while on much shorter time scales (days to century) weather and climate, their variability and changes, drive processes and changes of mass movements and mass flow processes. The Himalayan mountain cryosphere with its components glaciers, snow and permafrost is closely coupled to weather and climate. Changes in climate and weather patterns directly act on the cryosphere, but again over different periods of time. The widespread loss of glacier mass over the past years and decades is a function of increasing temperatures but different precipitation patterns from east to west also substantially influence the state and health of glaciers.

People in the IHR live downstream of the mountain cryosphere and are thus influenced by cryosphere changes and processes to a variable degree, most notably in terms of water resources and natural hazards such as from large avalanches and landslides or far-reaching floods from glacier lake outbursts. Risks are present where weather, climate and cryosphere related hazards meet exposed and vulnerable people and assets. Changes in risks develop as any of the aforementioned components change, i.e. hazard, exposure or vulnerability. Climatic and cryosphere changes can result in changing hazards while changes in exposure or vulnerability are typically a function of socio-economic developments. For instance, changes in economic demand for resources such as water or energy can significantly influence the risks. An integrative perspective on risks is therefore crucial. Climate change adaptation has the objective to reduce related risks, or identify and enhance options or opportunities.

The Government of India has addressed the climate change adaptation challenge by the establishment of the National Action Plan on Climate Change (NAPCC) which defined eight missions. One of these missions is the National Mission for Sustaining the Himalayan Ecosystem (NMSHE) whose implementation is coordinated by India's Department of Science and Technology (DST). NMSHE is a cross-cutting mission across different sectors such as water resources, agriculture, tourism, health, and others. The objective of the NMSHE is to scientifically assess the vulnerability of the IHR to climate change and to build corresponding capacities at the central and state levels, with the aim to develop response and adaptation strategies and measures.

Experiences in the IHR and worldwide, however, have shown that developing climate change adaptation strategies and especially implementing them is a complex and multi-faceted challenge. For instance, a simple transfer of scientific studies and results to decision and policy makers has not proven to be a successful approach. More frequent and more in-depth interactions between different actors, including science, policy and civil society, are necessary. However, experiences on how to design such a comprehensive process in the context of climate change adaptation and how to implement it are scarce (Huggel et al. 2015).

Here we reflect on experiences of Indo-Swiss collaborations in the field of climate change, cryosphere, risks and adaptation in the IHR. Some of the experiences are based on the Indian Himalayas Climate Adaptation Programme (IHCAP), an initiative of the Swiss Agency for Development and Cooperation (SDC) and the Indian Department of Science and Technology (DST). We first provide a synthesis of the climate and cryosphere in the IHR, and related impacts on downstream communities and systems. We then analyze the associated risks from a conceptual and adaptation perspective. In consideration of the aforementioned (i.e. the challenge of science-policy-civil society approaches) we revise concepts of co-production of knowledge and how they are valuable in our context. We conclude by presenting an integrative approach on risks and adaptation which is rooted in the co-production of knowledge and interdisciplinary and transdisciplinary perspectives. This approach has originally been developed within a larger consortium of Swiss and Indian universities and research institutions.

26.2 The IHR Climate and Cryosphere Related Impacts

Climate, cryosphere and water resources in the IHR have a tremendous importance for various economic sectors of the country and local livelihood, including water demand for irrigation in agriculture, drinking water for domestic use, or water for energy production. However, dynamic socio-economic and environmental changes, particularly climate change are expected to have significant adverse effects in the region. The scientific understanding of the relevant processes, spatio-temporal scales and regional diversity involved and the impacts is a key to design and implement adequate response and adaptation strategies and measures.

Much of the IHR receives precipitation during Indian summer whereas winter precipitation is driven by westerlies (Palazzi et al. 2013). During summer time most of the precipitation received is in liquid form whereas during winters it is mainly in solid-snow- form. This precipitation is important for downstream catchment regions for socio-economic growth, in particular agriculture, river ecosystems and for other water resource management perspectives. Previous research has shaped the understanding that during summer time the south-westerly flow of the Indian summer monsoon interacts with the Himalayan orography and turns to easterly or south-easterly flow along the Himalayan topography (Pant and Kumar 1997). In the case of Indian winter monsoon dominated by western disturbances embedded with large scale westerlies interact with the Himalayan topography. There are cases when these western disturbances co-exist with the Indian summer monsoon flow leading to heavy precipitation and extreme flood situations as seen during summer of 2013 (Uttarakhand flood) and 2014 (Jammu and Kashmir flood) and forming monsoon break-like situations. Topography and heterogeneous land-use modulate existing large-scale flow, and this interaction is detrimental for precipitation forming mechanisms and associated processes. Furthermore, the present context of warming temperatures also affects the hydrological balance at sub-catchment and regional catchment scale. Such impacts need to be evaluated as they affect the snow regime of the region. Therefore, in addition to the structured flow of summer and/or winter monsoon, it is imperative to understand such interactions at sub-regional scales.

The IHR has an important function for the storage and supply of water. Precipitation is stored in seasonal snow and perennial ice (glaciers). Release of water from snow and glacier melt is important during summer season to meet water demand for irrigation and drinking water in rural and urban agglomerations. Summer monsoon generally dominates the runoff across the central and eastern Himalaya region, and changes in runoff over the past decades have been variable, a decrease in runoff has for instance been observed in several catchments such in the Sutlej river basin (north-west Ganges) where a 30–40% runoff reduction from 1960s to 2000s was observed in relation with reduced summer precipitation (Collins et al. 2013). However, reliable long-term stream gauge records are scarce or not openly available, and climate-related influences on runoff are generally difficult to disentangle from other anthropogenic factors (e.g., exploitation for irrigation, hydropower, and land-use changes). Generally, there are relatively few measurements from glacier fed catchments in the IHR available but understanding the current water storage of the glaciers and its variation is of fundamental importance for defining urgently needed water and land management strategies.

Knowledge about the glaciers and their changes has significantly increased during the last few years revealing that glaciers in India have lost significant average mass during the last decade, but the rate of loses is spatially heterogeneous an cannot be generalized (Azam et al. 2018; Bolch et al. 2012; Kääb et al. 2012; Kulkarni and Karyakarte 2014). Although temperature and (monsoon) precipitation are the major drivers, the response of glaciers to climate and climate change depends on manifold factors such as the glacier size, the thermal regime, the topographic settings and debris-coverage. Debris-cover significantly alters the melt of the glaciers as thick debris cover insulates the ice. This results in more stable glacier terminus positions but thinning is also observed on debris-covered glaciers, e.g. with a mass loss of 0.32 m w.e. a-1 (1970–2007) for a central Himalayan region which comes close to rates found on debris-free glaciers (Bolch et al. 2011; Ragettli et al. 2016) (Kääb et al. 2012; Nuimura et al. 2011). The reasons are not yet fully understood, but are probably a combination of significant melt at ice cliffs and supraglacial lakes and changes in glacier flow. The limited knowledge of the key physical processes is reflected in a poor integration of debris-covered glaciers in glacio-hydrological models applied (Schauwecker et al. 2015).

Snow is a key component of the hydrological cycle in the IHR, and like glaciers is highly sensitive to climatic changes, but yet responding to climate change on different spatial and temporal scales than glaciers. Assessment of changes in snow melt related runoff across the IHR is complex and in general is hampered by limited available in-situ data (Rohrer et al. 2013). In the lowlands, the contribution of snow and glacier melt to the major Indian rivers has been estimated to 5–45% of average flows, with a larger significance during dry months or low monsoon years (Barnett et al. 2005; Immerzeel et al. 2010; Xu et al. 2009). Quantitative projections of downstream effects of changing water flow regimes in the headwaters are still rare but recent years have seen important research progress, indicating the importance of integrated cryo-hydrological models, and sustained uncertainties relating to future precipitation changes (Immerzeel et al. 2013; Moors et al. 2011). Some modeling studies found up to 20% decrease of melt from snow-dominated catchments but an initial increase of melt up to ca. 30% from strongly glacierized catchments for a scenarios of temperature increase of +1 to +3 °C (Singh and Bengtsson 2005). It is likely that ongoing climate change will significantly affect seasonal patterns of stream discharge.

The disappearance of mountain glaciers and expansion of large glacial lakes are amongst the most recognizable and dynamic impacts of climate warming in the alpine environment. In combination with altered stability of surrounding rock and ice walls, the potential threat from glacial lake outburst flooding (GLOF) is thus evolving over time (Haeberli et al. 2017). With residential, tourism, and particularly hydropower infrastructure expanding higher into alpine valleys, human demand on hydrological resources are intensifying, and conflicts over scarce, safe living space are increasing leading to higher exposure to risks (Zimmermann and Keiler 2015). There is a clear need for improved climate change adaptation planning by integrating GLOF risk reduction strategies (Haeberli et al. 2016; Khanal et al. 2015).

Glacial lake inventories with prioritization of detected lakes are important as they allow non-specialist local authorities to roughly identify lakes where more detailed and comprehensive studies should be directed (Ikeda et al. 2016; Worni et al. 2013). An accurate and objective classification of glacial lakes is challenging but essential to maintain credibility towards stakeholders and local populations. Glacial lake inventories provide information on the distribution of mapped lakes, their size, type, dam characteristics and other factors that may be relevant for a hazard assessment. Where possible, this knowledge can be supplemented with information on past lake outburst disasters in a given region, to identify, for example, common triggering and outburst mechanisms, event timing and seasonal components, and discharge characteristics (Vilímek et al. 2014).

Larger-scale coordinated national or trans-national efforts to map and assess the threat of Himalayan glacial lakes have primarily been led by the International Centre for Integrated Mountain Development (ICIMOD), see (Ives et al. 2010) for an overview. In total 8790 glacial lakes have been mapped across the Hindu-Kush Himalayan countries of Bhutan, China, Nepal, Pakistan, and India (excluding the states of Arunachal Pradesh and Jammu and Kashmir) (Ives et al. 2010). The number and average size of lakes typically decreases from east to west across the region, and in the IHR, the greatest density of potentially dangerous lakes are located in Sikkim.

In the recent Indo-Swiss climate adaptation programme (Indian Himalayas Climate Adaptation programme, IHCAP) glacial lakes across Himachal Pradesh have been remapped, providing an important baseline against which future changes in GLOF hazard can be assessed, with 120 lakes being identified for 2013/2014 (Allen et al. 2016b; IHCAP 2016) (Fig. 26.1). Research shows the high dynamic of lake formation and growth processes and the need for regular monitoring and prospective modelling to anticipate future potentially hazardous conditions. It also indicated that methodologies must be repeatable, objective, and transferable across large spatial scales.

In terms of hazard assessments, an important investigation is the assessment of future landscape evolution and lake formation in deglaciating mountain ranges (Frey et al. 2010; Linsbauer et al. 2012). As the erosive power of glaciers can form numerous and sometimes large closed topographic depressions, many overdeepenings are commonly found in formerly glaciated mountain ranges (Cook and Swift 2012). New lakes develop where such overdeepened areas become exposed and filled with water rather than sediments. The GIS-based model GlabTop was developed for estimating ice-thickness distribution and bed topography across the Swiss Alps (Linsbauer et al. 2012) and an improved version (Glabtop2) has subsequently been applied for large scale modelling in the Himalaya, which resulted in an an estimated total ice volume of ~2955 km^3 ± 30% and about 16,000 overdeepenings larger than 10^4m^2 (covering an area of ~2200 km^2 and having a total volume of ~120 km^3) detected in the modelled glacier-bed topographies of the entire Himalaya-Karakoram (Frey et al. 2014; Linsbauer et al. 2016). For Himachal Pradesh, the ~4000 modelled overdeepenings have been reduced down to 279 potential new lakes which are considered most likely to develop in the coming decades (approximately 10–50 years) (Allen et al. 2016b). Information on existing and potential future lakes forms the basis for assessing possible impacts, risks and opportunities related to such water bodies (e.g. hydropower, tourism, outburst hazards; cf. (Bajracharya and Mool 2009; Haeberli et al. 2016; Loriaux and Casassa 2013; Terrier et al. 2011).

Permafrost is recognised as an essential climate variable (ECV) by the World Meteorological Organisation, and is defined as any ground material that remains at or below 0 °C for at least two consecutive years. In response to global climate change, and the close coupling between atmospheric and ground temperatures, permafrost is warming and thawing in many regions (Vaughan et al. 2013). The thawing of permafrost can have widespread impacts relating to destabilization of steep slopes, changes in sub-surface hydrology, and increased sediment load in rivers (Harris et al. 2009). As a consequence, permafrost thawing can imply surface subsidence or differential creep affecting on-site infrastructure and thus tourism or avalanche protection installations (Bommer et al. 2010), slope failures from rock destabilization which can impact downstream areas and communities (Haeberli et al. 2017), or threatening residential areas, hydropower or transport infrastructure

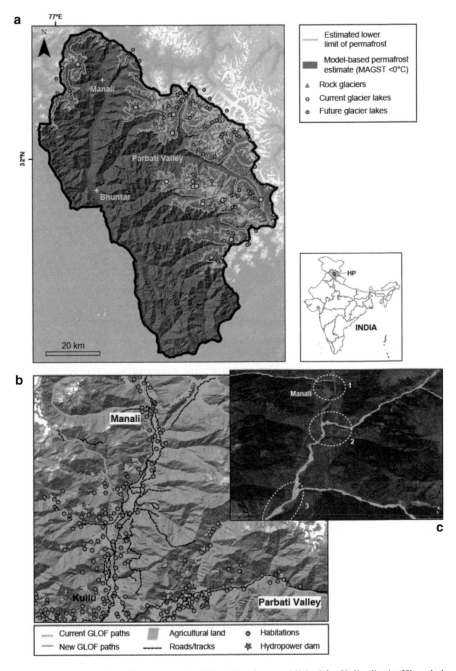

Fig. 26.1 (a) Overview of new cryospheric baseline data established for Kullu district Himachal Pradesh under phase I of IHCAP. (b) Zoom in on area downstream of Manali, demonstrating how this new data provided the basis for a first-order assessment of risk from glacial lake outburst floods (GLOFs) under both current and future (deglaciated) conditions. These studies identified three situations and thereby opportunities for implementation of adaptation strategies (c). Firstly, locations currently threatened by GLOFs but where no emerging new threats are anticipated (1), secondly, locations where entirely new threats are anticipated to emerge (2), and thirdly, locations currently threatened by GLOFs where the threat is anticipated to increase in response to continued upstream deglaciation (3)

by sediment flux (GAPHAZ 2017; Huggel et al. 2012). Ice-rich permafrost as a source of water is poorly studied but recently efforts have been reinforced, and for the Nepalese Himalaya it has been found that rock glaciers store a water volume of 16–25 billion m^3, a factor of 9 less than the water stored in glacier ice (Jones et al. 2018b). This has sparked a discussion whether water from rock glaciers could compensate for the reduction in runoff due to shrinking glaciers (Jones et al. 2018a), however, it needs to be considered that water release from rock glaciers is much slower than from glacier ice.

However, determining the likely spatial extent of permafrost should be considered a fundamental component of any risk assessment across mountain regions. Because sub-surface temperatures are difficult and costly to measure directly, permafrost distribution is commonly inferred from measured and modelled air and ground surface temperatures. In the Himalaya, however, little research has been conducted on permafrost to date.

ICIMOD initiated a first coordinated special project on permafrost to improve understanding of permafrost and related impacts in the region, initiated with a first scoping meeting with international experts in 2014, followed by a number of research and monitoring activities (icimod.org). The current best estimate of permafrost distribution across the HKH region is based on a coarse ~1 km resolution Global Permafrost Zonation Index (Gruber 2012) which generally agrees well with the distribution of mapped rock glaciers (providing a reasonable first order indicator of permafrost) (Jones et al. 2018b; Schmid et al. 2015). However, for impact-relevant studies related to permafrost, baseline information at a higher spatial resolution is required, ideally driven by local climate data, and validated against in-situ measurements.

In a pioneering Indo-Swiss collaboration implemented under phase I of IHCAP, (Allen et al. 2016a) produced a first local map of estimated permafrost distribution in Kullu district (Himachal Pradesh), combining simple topographic and climatic principles, more sophisticated physically based modelling, and rock glacier mapping. Overall, 9% (420 km^2) of the land area in Kullu district was classified as permafrost terrain, with permafrost likely extending down as low as ~4200 m a.s.l in isolated instances. The Kullu pilot study thereby provided some preliminary understanding of the possible extent and relevance of permafrost in this area of the Indian Himalayas, while generating fruitful new scientific partnerships and discussions on novel concepts and methodologies (Fig. 26.1).

The assessment of hydrological risks (related both to high-flow and low-flow conditions) and opportunities (for economic sector-wise water use) are typically supported by hydrological models. A large number of models representing different complexity, concepts and application potential exists. For hydrological models to be tools to eventually support planning of water resource management and adaptation they need to capture the relevant topical, spatial and temporal scales. Climatic and hydrological observations are indispensable elements of hydrological models and assessments. In high-mountain regions, cryosphere related data are additional requirements.

Several hydrological models have found widespread application in India, including the Soil and Water Assessment Tool (SWAT), and many large river catchments in India were modelled with the purpose of quantifying the climate change impact on hydrology (Masood et al. 2015; Pandey et al. 2017). (Siderius et al. 2013), for instance, used SWAT to quantify the snowmelt contributions to discharges of Ganges river, and other hydrological models were used to model impacts of climate change on Brahmaputra and Ganges river systems (Masood et al. 2015).

26.3   Risks and adaptation

Climate change impacts the cryosphere in many ways as outlined above, but the most relevant societal effects are a consequence of impacts propagating further downstream such as in the form of hydrological change or hazards. (Huss et al. 2017) distinguished several levels, from the cryosphere to hydrological, ecological and societal effects (Fig. 26.2). All three components of the cryosphere, i.e. glaciers, snow and permafrost, are affected but their response to climate change occurs on different spatial and temporal scales. Furthermore, effects on and from different

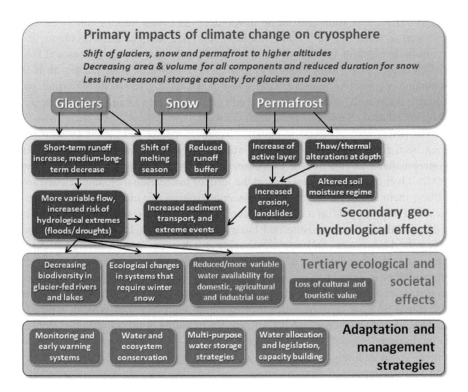

Fig. 26.2 Cascading impacts of climate change on the mountain cryosphere to societal effects and response. (Modified from Huss et al. 2017)

cryosphere components can accumulate, such as from glaciers and permafrost that can have cumulative destabilization effects on slopes. Hydrological changes are a predominantly observed impact from cryosphere changes and can propagate over large distances downstream, affecting both aquatic and terrestrial ecosystems and economic sectors and livelihoods (Milner et al. 2017).

Disaster risk, however, is essentially socially constructed (von Wymann et al. 2017). Reducing disaster risk requires therefore an in-depth understanding of risks and their root-causes, as well as collaboration between science, policy and civil society (UNISDR [The United Nations Office for Disaster Risk Reduction] 2015). However, comprehensive, forward-looking and widely accepted risk assessments are still rare and highly demanding in view of the multiple perspectives that need to be considered. It is particularly crucial to appropriately take into account the perspective of most disadvantaged stakeholders and considering non-economic or non-monetary aspects of risk. In high-mountain and downstream regions such as the IHR, hydrological risks and opportunities play a fundamental role for sustainable development in line with the 2030 Agenda (United Nations Member States of UN 2015). While water from snow, ice, lakes and precipitation provides an essential resource for sustaining lives and livelihoods and in mountain regions such as the Himalaya, hydrological risks, further influenced by climate and global changes, can also have a significant impact on mountain people and ecosystems (IPCC 2014). Large volumes of water cascading down steep mountain valleys can erode and transport sediment, cause or contribute to flooding and sedimentation problems, and resulting in negative impacts or devastation of downstream communities and land (Allen et al. 2016c; Ruiz-Villanueva et al. 2017). However, data scarcity in high mountain regions has tended to hinder any quantitative assessment of current and future risks as well as the development and implementation of adaptation strategies (McDowell et al. 2014; Salzmann et al. 2014).

The assessment of effects of climate change on natural and human systems has long been framed under a concept of vulnerability, which has traditionally been conceptualized as function of exposure, sensitivity and adaptive capacity (Parry 2007). Over recent years, however, at the level of international climate science and policy, there has been a shift in the conceptualization of vulnerability with emergence of 'climate risk' as a central concept, where risk is considered as the interaction of vulnerability, exposure and hazard (IPCC 2014). Climate change is now recognized as a major challenge in managing risk that requires the differentiation of physical and societal drivers of this risk. Within this new conceptualization of risk, vulnerability is linked to the context and inherent conditions to which a society or system is exposed, while the changes in the climate system influence hazards and the trends of their characteristics. For example, high levels of poverty typically increase vulnerability because poor people have limited or no capacities to recover from disasters, or to effectively protect themselves. On the other hand, strong institutions are more likely to develop the structures to appropriately cope with disasters and adapt to impacts of climate change, thus featuring lower levels of vulnerability. This differentiation between hazards, exposure and vulnerability provides a sound basis for the development of adaptation strategies that need to consider both the changes in frequency and magnitude of hazards due to climate change as well as societal dynamics that shape and determine the exposure and vulnerability of people and social-ecological systems.

However, the assessment of risks is a complex and challenging task, especially if pursuing inclusive and integrative stakeholder processes and the co-production of knowledge. Technically, recent studies have applied index approaches to assess risks related to water resources and scarcity, primarily in regions with adequate data availability (Martin-Carrasco et al. 2013). Nonetheless, in regions with limited data availability, in particular those including high-mountain areas, comprehensive water resources risk studies have barely been developed (Reynard et al. 2014), and the latest IPCC concept of climate risk is yet to be applied in a broad, holistic and integrated context. Risks can only be quantified (or at least reasonably categorized) if sensible risk metrics, categories or units are defined which can be estimated or calculated, given that the necessary data are available. Data required includes exposure of assets, e.g. number of people exposed to floods, or anything that is assigned value by a stakeholder or a community. How vulnerability is assessed is a wide field of research. Often an indicator based approach is taken (e.g. (Allen et al. 2016b) for current and future GLOF hazards and risks in Himachal Pradesh), where vulnerability is 'operationalized' along different dimensions, like poverty, social status, gender, age, health, education, etc. However, care has to be taken in applying technical risk assessment for identifying local adaptation and risk reduction measures. Local people often perceive climate-related risks in a broader perspective of (daily) risks they are exposed to. They value risks related to extreme events (e.g. floods) often as less important than more frequent risks (e.g. access to water). To achieve effective adaptation it is therefore important that different local stakeholders and scientists identify and assess risks in a joint learning process, as we sketch subsequently in this paper.

26.4   Co-Production of Knowledge

Undoubtedly, in the context of climate change adaptation, science has an important role in generating and improving the knowledge that is useful and necessary to design adaptation strategies and measures. Science is thus operating in an applied context and needs to interact not only across a range of disciplines but also with a range of actors and stakeholders concerned with adaptation to climate change. However, how exactly science should interact and be part of the adaptation process is still an ongoing discussion. Transdisciplinary research has generated evidence of knowledge production processes in environmental fields (Moser and Ekstrom 2010; Paavola 2008; Pohl et al. 2010). Essentially, the question is how a knowledge production process should be framed to facilitate and better inform adaptation processes. In contrast to earlier models or practice, it is commonly agreed that it is more effective if knowledge is produced in a joint process between different actors (including science), rather than knowledge being commissioned by policy (policy pull model), or pushed by science with limited applicability to the solution of problems (science push or loading dock model) (Cash et al. 2006; Hegger et al. 2012). In the IHR and India in general, previous experiences related to science-policy/practice processes are available and it is important to adequately consider them.

The IHCAP has generated and facilitated a substantial number of opportunities and points of interactions between science, policy, and society. A series of stakeholder consultations were undertaken under the Programme to effectively understand and record communities' voices, with regard to their perception of climate change and its impacts on water, food production and security, livelihood and health. Furthermore, IHCAP tried to analyze people's adaptation techniques and approaches, their risk coping mechanisms with risks, the challenges they face on a daily basis, their altering lifestyles, the impact on their future livelihoods, their uncertainties over being faced with situations beyond their coping capacities, their insecurities as a result of poverty and illiteracy, and their vulnerability to natural hazards and their fear of the unknown. The consultations were held across central and eastern Himalayas at three locations (Shimla, Almora and Gangtok) and were attended by professionals from different walks of life, all striving towards the same goal – sustainable mountain development. Besides, in order to identify the climate change adaptation priorities and knowledge need of different sectors, consultations were also held with government line departments at district and lower levels; and a range of programmes for their capacity building and knowledge transfer were organized in Himachal Pradesh, Uttarakhand and Sikkim.

Further experiences have previously been generated by other projects such as HighNoon, a collaboration between several scientific institutions in India and Europe, which enabled stakeholder processes for the purpose of developing adaptation strategies on integrated management of water resources in India (Moors et al. 2011) (Fig. 26.3).

In another fruitful example of international collaboration aiming to identify science and policy gaps in climate change adaptation, a range of stakeholders, particularly representatives of local grassroot institutions (Village Panchayats, Forest Panchayats, Mahila Mangal Dals and other local institutions), officials of government

Fig. 26.3 Stakeholder workshops held in Kullu, Himachal Pradesh, on jointly defining climate adaptation strategies and actions as promoted in the IIHCAP (ihcap.in)

departments, local political representatives, NGOs, CSOs, private sectors, and scientists have been brought around the table through several workshops organized in Nainital District of Uttarakhand by the Kumaun University in collaboration with the Australian National University, Australia. Similarly, local communities, grassroot institutions and policy planners were involved in the generation of climate data, their dissemination and monitoring at village level in Kumaon Himalaya under the UK India Education and Research Initiative (UKIERI) being implemented by Kumaun University, Nainital and Newcastle University, UK. Furthermore, the recently established Uttarakhand State Centre for Climate Change (SCCC) tries to involve scientists, sectoral agencies and local communities in evolving a framework for climate change adaptation mainstreaming in Uttarakhand State.

These initiatives also need to be attentive to the many local efforts of autonomous adaptation that can be found and occasionally are documented in the IHR. Autonomous adaptation occurs without formal adaptation plans, strategies or policies, typically at the local community level. A prominent example for this type of adaptation in the IHR are the artificial ice reservoirs constructed in the Ladakh region that provide meltwater during critical low-flow periods of the year (Nüsser et al. 2018).

26.5 Scenarios for Informing Adaptation Options

In recent years, the recognition has increased that forward-looking adaptation schemes leading towards sustainable development need to consider not only consequences of climate change but also, most importantly, socio-economic changes (IPCC 2014; Kriegler et al. 2012). This is especially crucial in a dynamic context such as the IHR where economic, socio-cultural, demographic, institutional, and environmental factors interact in a complex way and shape the regional development. Resulting changes for example in people's livelihoods, land-use patterns, economic structure, urban development are dynamic, context-specific and inherently uncertain, and eventually affect people's exposure to hazards and vulnerability depending on their assets and capabilities in an unequal way (Olsson et al. 2014). In the context of climate change research, integrative scenarios have recently been applied to better understand future trends (O'Neill et al. 2014). Scenarios describe how a system might unfold in future and are not an exact prediction of the future state. However, the development of plausible and consistent scenarios is not straightforward due to the inherent uncertainty and context-specific characteristics of development trends. Different scenario approaches and techniques have been developed that aim at different functions (Birkmann et al. 2015; Börjeson et al. 2006): analyzing how different factors influence future development (analytical), systematically reflecting and discussing trends (explorative), or defining desirable future states (normative). Their overall purpose typically is to support policy and decision-making anticipating future developments and thus allowing for effective adaptation measures, support to policy-making with a view to reduce risk and strengthen resilience. For elaboration of future scenarios a joint knowledge production process is particularly important, especially in data-scarce regions such as mountains, and can involve the identification of the main drivers of social and economic change, the joint development of storylines and eventually the analysis of quantitative scenarios (Valdivia et al. 2010). Such participatory processes in fact enhance the understanding of the complex dynamics influencing future socio-economic conditions and allow for policy-making towards effective adaptation and sustainable development (Malek and Boerboom 2015; Muccione et al. 2016; Reynard et al. 2014; Schneider and Rist 2014; Wiek et al. 2006). One of the most recent experiences in this respect comes from the Hindu Kush Himalayan Monitoring and Assessment Programme (HIMAP), an initiative, coordinated by ICIMOD, across the eight countries of the region (Afghanistan, Bangladesh, Bhutan, China, India, Myanmar, Nepal, Pakistan). Socio-economic scenarios have been designed and developed based on a participative stakeholder consultation process which increases the acceptance of the resulting scenarios and their possible up-take in policy processes. As a result of this process, three different scenarios were developed for the region, namely a 'Downhill' scenario which reflects conditions of strong climate change and a socially, economically and politically unstable region with strong ecosystem degradation; a 'Muddling through' scenario with medium instability conditions combined with strong climate change; and a 'Prosperous' scenario representing conditions of weak climate change with a region of social, economic and political stability (hi-map.org).

26.6   An Integrative Approach to Risks and Adaptation

Above we have outlined and reviewed the cryosphere change in the IHR, its impacts on natural and human systems and related risks. Impacts and risks are manifold and broadly scale with the magnitude and to some extent with rate of current and future climate change. The assessment of risks needs to be done in a comprehensive way, involving diverse perspectives of people, communities and institutions affected.

Adaptation is a response to reduce negative impacts of climate change but highly challenging as it involves multiple stakeholders and requires science to transcend across disciplines and multiple stakeholders. Over the past years, experience in climate change adaptation has been greatly enriched and understanding advanced but reconciling interests and expectations of multiple stakeholders remains a major challenge in practice. We have furthermore seen that a joint knowledge production process can be a feasible approach to address the adaptation challenge. However, comprehensive risk assessments, coupled with joint knowledge processes to design and implement adaptation have still rarely been achieved so far.

Here we sketch a research design that could be suitable to accommodate an integrative joint research process, centering on water related risks in the IHR (Fig. 26.4), specifically on those influenced by changing water supply and demand. Evidence from practical experience and theoretical consideration suggests that an initial phase of scoping and framing with stakeholders is a key element of a successful

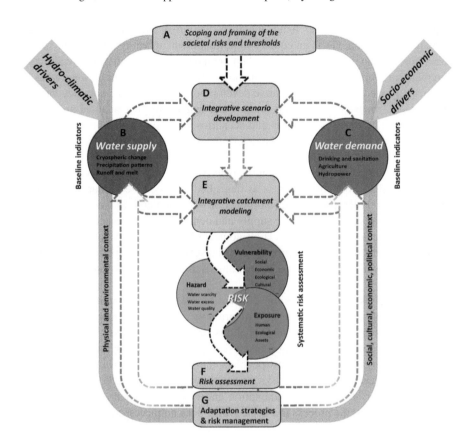

Fig. 26.4 Conceptual sketch of an integrated approach to hydrological risks and adaptation in the context of climatic and socio-economic changes (see text for explanation)

science-practice-policy process in climate impacts, risks and adaptation projects. Accordingly, component A in Fig. 26.4 ensures this necessary and critical interaction of scientists with policy makers and civil society groups, a knowledge co-production process defining the context, risks, metrics or categories coherent with perspectives of stakeholders from government, different social groups, and sectors. This phase and process may take a substantial amount of time, possibly more than a year, depending on how the social and political context in the respective region is. This long duration of the process may represent a major challenge, or even barrier, to the successful realization of adaptation because governments, donors or investors favor solutions that bring results over shorter periods of time. Experiences, however, show that if perspectives and interests cannot be sufficiently be brought in agreement, or reconciled, the success of any further action and implementation is strongly questioned. Component B of the process would represent a more typical scientific process involving analysis of (instrumental) observations and running a number of climate, cryosphere and hydrology models to determine the different aspects of water supply. On the other hand, the investigation of current and future water demand (component C), is an equally important element, yet typically done much less rigorous than the supply side although in many contexts, such as where agriculture plays an important role, water demand is a fundamental driver of water risks.

As mentioned above the integrative development of climate and socio-economic scenarios is crucial for assessing future risks and identifying adaptation options (component D). While projections and scenarios based on climate models are the realm of science it is crucial that the determinants of the scenarios are defined in a process involving stakeholders, e.g. to agree on emission scenarios to be considered, time horizons etc. Development of socio-economic scenarios is best done in a joint knowledge production process as described above; these scenarios typically determine largely the possible development of water demand. In a next step (component E), the results of the water supply and demand and the respective scenarios are integrated at a watershed level. Hydrological models that are capable to simulate water supply and demand at appropriate spatial scales are to be applied or adjusted, with the key hydrological dimensions and units defined in a joint knowledge sharing process.

The risk assessment follows the concept of risk being a function of climate-related hazards, and the exposure and vulnerability of people and assets (component F). Again it is crucial that the metrics, units or categories of risk are previously defined in a scoping and framing process (component A), taking into account that the risk is eventually subject of how people and institutions value things, rather than scientists technically assess the risks on their own. Finally, component G contributes to the joint elaboration of adaptation and risk management. Importantly, as Fig. 26.4 indicates, the process is a cycle which does not necessarily end with component G but would rather go in a loop up again through the different components. This is especially relevant a new knowledge and information becomes available through time which implies further refinement or changes to adaptation strategies and actions. The logic of the whole process is that a joint knowledge process with a sufficiently solid scoping and framing phase greatly facilitates designing, agreeing on and implementing adaptation and risk management.

Although presented here in the context of the IHR the proposed approach bares relevance beyond this regional context, and corresponding experiences were made in other mountain regions worldwide such in the Andes, or in the European Alps. However, while the overall approach may be robust against different backgrounds it is important to emphasize that the environmental, social, cultural, economic and political context is a decisive determinant in how the process and methodological steps are defined and carried out.

Acknowledgements Part of the experiences and outcomes described here were the results of the Indian Himalayan Climate Adaptation Program (IHCAP), an initiative of the Swiss Agency for Development and Cooperation (SDC) and the Indian Department of Science and Technology (DST), in collaboration with the University of Geneva, Meteodat Zurich and several Indian research institutions and universities. Some of the ideas and text in this chapter have emerged from fruitful collaboration with various Swiss and Indian colleagues during previous joint-activities. An earlier version of Fig. 26.4 has been drafted by Fabian Drenkhan (University of Zurich).

References

Allen SK, Fiddes J, Linsbauer A, Randhawa SS, Saklani B, Salzmann N (2016a) Permafrost studies in Kullu district, Himachal Pradesh. Curr Sci 11:257–260

Allen SK, Linsbauer A, Randhawa SS, Huggel C, Rana P, Kumari A (2016b) Glacial lake outburst flood risk in Himachal Pradesh, India: an integrative and anticipatory approach considering current and future threats. Nat Hazards 84:1741–1763. https://doi.org/10.1007/s11069-016-2511-x

Allen SK, Rastner P, Arora M, Huggel C, Stoffel M (2016c) Lake outburst and debris flow disaster at Kedarnath, June 2013: hydrometeorological triggering and topographic predisposition. Landslides 13:1479–1491. https://doi.org/10.1007/s10346-015-0584-3

Azam MF, Wagnon P, Berthier E, Vincent C, Fujita K, Kargel JS (2018) Review of the status and mass changes of Himalayan-Karakoram glaciers. J Glaciol 64:1–14. https://doi.org/10.1017/jog.2017.86

Bajracharya SR, Mool P (2009) Glaciers, glacial lakes and glacial lake outburst floods in the Mount Everest region, Nepal. Ann Glaciol 50:81–86

Barnett TP, Adam JC, Lettenmaier DP (2005) Potential impacts of a warming climate on water availability in snow-dominated regions. Nature 438:303–309

Birkmann J et al (2015) Scenarios for vulnerability: opportunities and constraints in the context of climate change and disaster risk. Clim Chang 133:53–68. https://doi.org/10.1007/s10584-013-0913-2

Bolch T, Pieczonka T, Benn DI (2011) Multi-decadal mass loss of glaciers in the Everest area (Nepal Himalaya) derived from stereo imagery. Cryosphere 5:349–358. https://doi.org/10.5194/tc-5-349-2011

Bolch T et al (2012) The state and fate of Himalayan Glaciers. Science 336:310–314. https://doi.org/10.1126/science.1215828

Bommer C, Phillips M, Arenson LU (2010) Practical recommendations for planning, constructing and maintaining infrastructure in mountain permafrost. Permafr Periglac Process 21:97–104. https://doi.org/10.1002/ppp.679

Börjeson L, Höjer M, Dreborg K-H, Ekvall T, Finnveden G (2006) Scenario types and techniques: towards a user's guide. Futures 38:723–739. https://doi.org/10.1016/j.futures.2005.12.002

Cash DW, Borck JC, Patt AG (2006) Countering the loading-dock approach to linking science and decision making comparative analysis of El Niño/Southern Oscillation (ENSO) forecasting systems. Sci Technol Hum Values 31:465–494. https://doi.org/10.1177/0162243906287547

Collins DN, Davenport JL, Stoffel M (2013) Climatic variation and runoff from partially-glacierised Himalayan tributary basins of the Ganges. Sci Total Environ 468–469:S48–S59. https://doi.org/10.1016/j.scitotenv.2013.10.126

Cook SJ, Swift DA (2012) Subglacial basins: their origin and importance in glacial systems and landscapes. Earth Sci Rev 115:332–372. https://doi.org/10.1016/j.earscirev.2012.09.009

Frey H, Haeberli W, Linsbauer A, Huggel C, Paul F (2010) A multi-level strategy for anticipating future glacier lake formation and associated hazard potentials. Nat Hazards Earth Syst Sci 10:339–352

Frey H, Machguth H, Huss M, Huggel C, Bajracharya S, Bolch T, Kulkarni A, Linsbauer A, Salzmann N, Stoffel M (2014) Estimating the volume of glaciers in the Himalayan–Karakoram region using different methods. Cryosphere 8:2313–2333. https://doi.org/10.5194/tc-8-2313-2014

GAPHAZ (2017) Assessment of glacier and permafrost hazards in mountain regions – technical guidance document. Prepared by Allen S, Frey H, Huggel C et al. Standing group on glacier and permafrost hazards in mountains (GAPHAZ) of the International Association of Cryospheric Sciences (IACS) and the International Permafrost Association (IPA). Zurich, Switzerland/Lima, Peru

Gruber S (2012) Derivation and analysis of a high-resolution estimate of global permafrost zonation. Cryosphere 6:221–233. https://doi.org/10.5194/tc-6-221-2012

Haeberli W, Buetler M, Huggel C, Friedli TL, Schaub Y, Schleiss AJ (2016) New lakes in deglaciating high-mountain regions – opportunities and risks. Clim Chang 139:201–214. https://doi.org/10.1007/s10584-016-1771-5

Haeberli W, Schaub Y, Huggel C (2017) Increasing risks related to landslides from degrading permafrost into new lakes in de-glaciating mountain ranges. Geomorphology 293:405–417. https://doi.org/10.1016/j.geomorph.2016.02.009

Harris C et al (2009) Permafrost and climate in Europe: monitoring and modelling thermal, geomorphological and geotechnical responses. Earth Sci Rev 92:117–171. https://doi.org/10.1016/j.earscirev.2008.12.002

Hegger D, Lamers M, Van Zeijl-Rozema A, Dieperink C (2012) Conceptualising joint knowledge production in regional climate change adaptation projects: success conditions and levers for action. Environ Sci Pol 18:52–65. https://doi.org/10.1016/j.envsci.2012.01.002

Huggel C, Clague JJ, Korup O (2012) Is climate change responsible for changing landslide activity in high mountains? Earth Surf Process Landf 37:77–91

Huggel C et al (2015) A framework for the science contribution in climate adaptation: experiences from science-policy processes in the Andes. Environ Sci Pol 47:80–94. https://doi.org/10.1016/j.envsci.2014.11.007

Huss M et al (2017) Toward mountains without permanent snow and ice. Earth's Future 5:418–435. https://doi.org/10.1002/2016EF000514

IHCAP (2016) Climate vulnerability, hazards and risk: An integrated pilot study in Kullu District, Himachal Pradesh, Synthesis report. Indian Himalayas Climate Adapation Programme (IHCAP), Delhi

Ikeda N, Narama C, Gyalson S (2016) Knowledge sharing for disaster risk reduction: insights from a Glacier Lake workshop in the Ladakh Region, Indian Himalayas. Mt Res Dev 36:31–40. https://doi.org/10.1659/MRD-JOURNAL-D-15-00035.1

Immerzeel WW, Van Beek LP, Bierkens MF (2010) Climate change will affect the Asian water towers. Science 328:1382–1385

Immerzeel WW, Pellicciotti F, Bierkens MFP (2013) Rising river flows throughout the twenty-first century in two Himalayan glacierized watersheds. Nat Geosci 6:742–745. https://doi.org/10.1038/ngeo1896

IPCC (2014) Climate change 2014: impacts, adaptation, and vulnerability. Part A: Global and sectoral aspects. In: Field CB (ed) Contribution of Working Group II to the fifth assessment report of the Intergovernmental Panel on Climate Change. Cambridge University Press: Cambridge/New York

Ives JD, Shresta RB, Mool PK (2010) Formation of glacial lakes in the Hindu Kush-Himalayas and GLOF risk assessment. International Centre for Integrated Mountain Development, Kathmandu

Jones DB, Harrison S, Anderson K, Betts RA (2018a) Mountain rock glaciers contain globally significant water stores. Sci Rep 8:2834. https://doi.org/10.1038/s41598-018-21244-w

Jones DB, Harrison S, Anderson K, Selley HL, Wood JL, Betts RA (2018b) The distribution and hydrological significance of rock glaciers in the Nepalese Himalaya. Glob Planet Chang 160:123–142. https://doi.org/10.1016/j.gloplacha.2017.11.005

Kääb A, Berthier E, Nuth C, Gardelle J, Arnaud Y (2012) Contrasting patterns of early twenty-first-century glacier mass change in the Himalayas. Nature 488:495–498. https://doi.org/10.1038/nature11324

Khanal NR, Mool PK, Shrestha AB, Rasul G, Ghimire PK, Shrestha RB, Joshi SP (2015) A comprehensive approach and methods for glacial lake outburst flood risk assessment, with examples from Nepal and the transboundary area. Int J Water Resour Dev 31:219–237. https://doi.org/10.1080/07900627.2014.994116

Kriegler E, O'Neill BC, Hallegatte S, Kram T, Lempert RJ, Moss RH, Wilbanks T (2012) The need for and use of socio-economic scenarios for climate change analysis: a new approach based on shared socio-economic pathways. Glob Environ Chang 22:807–822. https://doi.org/10.1016/j. gloenvcha.2012.05.005

Kulkarni AV, Karyakarte Y (2014) Observed changes in Himalayan glaciers. Curr Sci 106:237–244

Linsbauer A, Paul F, Haeberli W (2012) Modeling glacier thickness distribution and bed topography over entire mountain ranges with GlabTop: application of a fast and robust approach. J Geophys Res 117:F03007. https://doi.org/10.1029/2011JF002313

Linsbauer A, Frey H, Haeberli W, Machguth H, Azam MF, Allen S (2016) Modelling glacier-bed overdeepenings and possible future lakes for the glaciers in the Himalaya—Karakoram region. Ann Glaciol 57:119–130

Loriaux T, Casassa G (2013) Evolution of glacial lakes from the Northern Patagonia Icefield and terrestrial water storage in a Sea-level rise context. Glob Planet Chang 102:33–40. https://doi. org/10.1016/j.gloplacha.2012.12.012

Malek Ž, Boerboom L (2015) Participatory scenario development to address potential impacts of land use change: an example from the Italian Alps. Mt Res Dev 35:126–138. https://doi. org/10.1659/MRD-JOURNAL-D-14-00082.1

Martin-Carrasco F, Garrote L, Iglesias A, Mediero L (2013) Diagnosing causes of water scarcity in complex water resources systems and identifying risk management actions. Water Resour Manag 27:1693–1705. https://doi.org/10.1007/s11269-012-0081-6

Masood M, Yeh PJ-F, Hanasaki N, Takeuchi K (2015) Model study of the impacts of future climate change on the hydrology of Ganges–Brahmaputra–Meghna basin. Hydrol Earth Syst Sci 19:747–770. https://doi.org/10.5194/hess-19-747-2015

McDowell G, Stephenson E, Ford J (2014) Adaptation to climate change in glaciated mountain regions. Clim Chang 126:77–91. https://doi.org/10.1007/s10584-014-1215-z

Milner AM et al (2017) Glacier shrinkage driving global changes in downstream systems. Proc Natl Acad Sci 114:9770–9778. https://doi.org/10.1073/pnas.1619807114

Moors EJ et al (2011) Adaptation to changing water resources in the Ganges basin, Northern India. Environ Sci Pol 14:758–769. https://doi.org/10.1016/j.envsci.2011.03.005

Moser SC, Ekstrom JA (2010) A framework to diagnose barriers to climate change adaptation. Proc Natl Acad Sci 107:22026–22031. https://doi.org/10.1073/pnas.1007887107

Muccione V, Salzmann N, Huggel C (2016) Scientific knowledge and knowledge needs in climate adaptation policy: a case study of diverse Mountain Regions. Mountain Research and Development https://doi.org/10.1659/MRD-JOURNAL-D-15-00016.1 [online]. Available from: http://www.bioone.org/doi/abs/10.1659/MRD-JOURNAL-D-15-00016.1. Accessed 1 Dec 2016

Nuimura T, Fujita K, Fukui K, Asahi K, Aryal R, Ageta Y (2011) Temporal changes in elevation of the debris-covered ablation area of Khumbu Glacier in the Nepal Himalaya since 1978. Arct Antarct Alp Res 43:246–255. https://doi.org/10.1657/1938-4246-43.2.246

Nüsser M, Dame J, Kraus B, Baghel R, Schmidt S (2018) Socio-hydrology of "artificial glaciers" in Ladakh, India: assessing adaptive strategies in a changing cryosphere. Reg Environ Chang 19:1327–1337. https://doi.org/10.1007/s10113-018-1372-0. [online] Available from: https:// doi.org/10.1007/s10113-018-1372-0. Accessed 30 Aug 2018

O'Neill BC, Kriegler E, Riahi K, Ebi KL, Hallegatte S, Carter TR, Mathur R, van VDP (2014) A new scenario framework for climate change research: the concept of shared socioeconomic pathways. Clim Chang 122:387–400. https://doi.org/10.1007/s10584-013-0905-2

Olsson L, Opondo M, Tschakert P, Agrawal A, Eriksen SH, Ma S, Perch LN, Zakieldeen SA (2014) Livelihoods and poverty. In: Field CB et al (eds) Climate change 2014: impacts, adaptation, and vulnerability. Part A: Global and sectoral aspects. Contribution of Working Group II to the fifth assessment report of the Intergovernmental Panel of Climate Change. Cambridge University Press, Cambridge/New York, pp 793–832

Paavola J (2008) Science and social justice in the governance of adaptation to climate change. Environ Polit 17:644–659. https://doi.org/10.1080/09644010802193609

Palazzi E, von Hardenberg J, Provenzale A (2013) Precipitation in the Hindu-Kush Karakoram Himalaya: observations and future scenarios. J Geophys Res Atmos 118:85–100. https://doi.org/10.1029/2012JD018697

Pandey BK, Gosain AK, Paul G, Khare D (2017) Climate change impact assessment on hydrology of a small watershed using semi-distributed model. Appl Water Sci 7:2029–2041. https://doi.org/10.1007/s13201-016-0383-6

Pant GB, Kumar KR (1997) Climates of South Asia. Wiley, Chichester

Parry ML, Canziani OF, Palutikof JP, van der Linden PJ, Hanson CE (eds) (2007) Climate change 2007: impacts, adaptation and vulnerability. Contribution of Working Group II to the fourth assessment report of the Intergovernmental Panel on Climate Change. . Cambridge University Press: Cambridge

Pohl C et al (2010) Researchers' roles in knowledge co-production: experience from sustainability research in Kenya, Switzerland, Bolivia and Nepal. Sci Public Policy 37:267–281. https://doi.org/10.3152/030234210X496628

Ragettli S, Immerzeel WW, Pellicciotti F (2016) Contrasting climate change impact on river flows from high-altitude catchments in the Himalayan and Andes Mountains. Proc Natl Acad Sci 113:9222–9227. https://doi.org/10.1073/pnas.1606526113

Reynard E et al (2014) Interdisciplinary assessment of complex regional water systems and their future evolution: how socioeconomic drivers can matter more than climate. Wiley Interdiscip Rev Water 1:413–426. https://doi.org/10.1002/wat2.1032

Rohrer M, Salzmann N, Stoffel M, Kulkarni AV (2013) Missing (in-situ) snow cover data hampers climate change and runoff studies in the Greater Himalayas. Sci Total Environ 468–469., Supplement:S60–S70. https://doi.org/10.1016/j.scitotenv.2013.09.056

Ruiz-Villanueva V, Allen S, Arora M, Goel NK, Stoffel M (2017) Recent catastrophic landslide lake outburst floods in the Himalayan mountain range. Prog Phys Geogr Earth Environ 41:3–28. https://doi.org/10.1177/0309133316658614

Salzmann N, Huggel C, Rohrer M, Stoffel M (2014) Data and knowledge gaps in glacier, snow and related runoff research – a climate change adaptation perspective. J Hydrol 518,. Part B:225–234. https://doi.org/10.1016/j.jhydrol.2014.05.058

Schauwecker S, Rohrer M, Huggel C, Kulkarni A, Ramanathan A, Salzmann N, Stoffel M, Brock B (2015) Remotely sensed debris thickness mapping of Bara Shigri Glacier, Indian Himalaya. J Glaciol 61:675–688. https://doi.org/10.3189/2015JoG14J102

Schmid M-O, Baral P, Gruber S, Shahi S, Shrestha T, Stumm D, Wester P (2015) Assessment of permafrost distribution maps in the Hindu Kush Himalayan region using rock glaciers mapped in Google Earth. Cryosphere 9:2089–2099. https://doi.org/10.5194/tc-9-2089-2015

Schneider F, Rist S (2014) Envisioning sustainable water futures in a transdisciplinary learning process: combining normative, explorative, and participatory scenario approaches. Sustain Sci 9:463–481. https://doi.org/10.1007/s11625-013-0232-6

Siderius C, Biemans H, Wiltshire A, Rao S, Franssen WHP, Kumar P, Gosain AK, van Vliet MTH, Collins DN (2013) Snowmelt contributions to discharge of the Ganges. Sci Total Environ 468–469:S93–S101. https://doi.org/10.1016/j.scitotenv.2013.05.084

Singh P, Bengtsson L (2005) Impact of warmer climate on melt and evaporation for the rainfed, snowfed and glacierfed basins in the Himalayan region. J Hydrol 300:140–154

Terrier S, Jordan F, Schleiss A, Haeberli W, Huggel C, Künzler M (2011) Optimized and adaptated hydropower management considering glacier shrinkage in the Swiss Alps. In: Dams in Switzerland. Swiss committee on dams. Proceedings ICOLD conference, Luzern, Switzerland, pp 291–298

UNISDR [The United Nations Office for Disaster Risk Reduction] (2015) Sendai framework for disaster risk reduction 2015–2030. Geneva, Switzerland [online]. Available from: http://www.preventionweb.net/files/43291_sendaiframeworkfordrren.pdf

United Nations Member States of UN (2015) Transforming our world – The 2030 Agenda for sustainable development: Finalised text for adoption (1 August) [online]. Available from: https://

sustainabledevelopment.un.org/content/documents/7891TRANSFORMING%20OUR%20 WORLD.pdf

Valdivia C, Seth A, Gilles JL, García M, Jiménez E, Cusicanqui J, Navia F, Yucra E (2010) Adapting to climate change in Andean ecosystems: landscapes, capitals, and perceptions shaping rural livelihood strategies and linking knowledge systems. Ann Assoc Am Geogr 100:818–834. https://doi.org/10.1080/00045608.2010.500198

Vaughan DG, Comiso JC, Allison I, Carrasco J, Kaser G, Kwok R, Mote P, Murray T, Paul F, Ren J, Rignot E, Solomina O, Steffen K, Zhang T (2013) Observations: cryosphere. In: Qin STFD, Plattner GK, Tignor M, Allen SK, Boschung J, Nauels A, Xia Y, Bex V, Midgley PM (eds) Climate Change 2013: the physical science basis. Contribution of Working Group I to the fifth assessment report of the Intergovernmental Panel on Climate Change. Cambridge University Press, Cambridge/New York, pp 317–382

Vilímek V, Emmer A, Huggel C, Schaub Y, Würmli S (2014) Database of glacial lake outburst floods (GLOFs)–IPL project No. 179. Landslides 11:161–165. https://doi.org/10.1007/ s10346-013-0448-7

Wymann von Dach SW von Bachmann F, Alcántara-Ayala I, Fuchs S, Keiler M, Mishra A, Sötz E (2017) Safer lives and livelihoods in mountains: making the Sendai framework for disaster risk reduction work for sustainable mountain development. Centre for Development and Environment (CDE), University of Bern, Bern Open Publishing (BOP)

Wiek A, Binder C, Scholz RW (2006) Functions of scenarios in transition processes. Futures 38:740–766. https://doi.org/10.1016/j.futures.2005.12.003

Worni R, Huggel C, Stoffel M (2013) Glacial lakes in the Indian Himalayas — from an area-wide glacial lake inventory to on-site and modeling based risk assessment of critical glacial lakes. Sci Total Environ 468–469., Supplement:S71–S84. https://doi.org/10.1016/j.scitotenv.2012.11.043

Xu J, Grumbine R, Shrestha A, Eriksson M, Yang X, Wang Y, Wilkes A (2009) The melting Himalayas: cascading effects of climate change on water, biodiversity, and livelihoods. Conserv Biol 23:520–530

Zimmermann M, Keiler M (2015) International frameworks for disaster risk reduction: useful guidance for sustainable mountain development? Mt Res Dev 35:195–202. https://doi.org/10.1659/ MRD-JOURNAL-D-15-00006.1

Index